Industrial Automation from Scratch

A hands-on guide to using sensors, actuators, PLCs, HMIs, and SCADA to automate industrial processes

Olushola Akande

‹packt›

BIRMINGHAM—MUMBAI

Industrial Automation from Scratch

Group Product Manager: Preet Ahuja
Publishing Product Manager: Surbhi Suman
Senior Editor: Romy Dias
Technical Editor: Shruthi Shetty
Copy Editor: Safis Editing
Project Coordinator: Ashwin Kharwa
Proofreader: Safis Editing
Indexer: Rekha Nair
Production Designer: Arunkumar Govinda Bhat
Marketing Coordinator: Rohan Dhobal

First published: May 2023

Production reference: 1180523

Published by Packt Publishing Ltd.
Livery Place
35 Livery Street
Birmingham
B3 2PB, UK.

ISBN 978-1-80056-938-6

www.packtpub.com

This book is dedicated to industrial automation engineers, instrumentation and control engineers, programmable logic controller (PLC) specialists, production managers, chemical and process engineers, electrical engineers, electronics engineers, mechanical engineers, industrial maintenance engineers, installation and maintenance technicians, and all science-oriented people who have an interest in starting and building a career in industrial automation from scratch.

Contributors

About the author

Olushola Akande is an industrial automation and robotics expert with a wealth of experience in the design and development of automation and robotics-related systems. He is the CEO of Showlight Technologies Limited, an industrial automation, robotics, **artificial intelligence (AI)**, and **machine learning (ML)** training and servicing company in Lagos, Nigeria. He is a consultant to many industries in the area of automation and also an instructor in the field of industrial automation at various technical training centers. In 2021, he joined the Schneider Electric team to work on a large-scale project (Dangote Oil Refinery) for their **Integrated Control and Safety Systems (ICSS)** installation which includes high-integrity PLCs, HMIs, SCADA, and DCSs. He is currently working as an application engineer on the same project.

First, I would like to thank my wife, Fatimoh Akande, for her tremendous support and encouragement in completing this book.

I would also like to thank my parents, friends, and other members of my family for their support during those challenging years of my life and career, and I want to especially thank my friend, Stanley Ehidiamen, for his great contribution to my knowledge and skills in industrial automation. My special appreciation also goes to my lawyer, barrister Omotola Aderemi, and my brother, Olamide Akande, for their support on this book.

Thanks to the Packt Publishing team and all technical reviewers who contributed to this book.

I am grateful to God, the creator, who made this book a possibility.

About the reviewers

Sherif E. Nasr has worked in the field of industrial control and automation for more than 20 years and has achieved success and has been a pioneer in the curriculum, technical sales, after-market, project engineering, and design and commissioning stages. It has been his honor to do business in almost all market sectors; power, oil and gas, refineries, petrochemicals, fertilizers, water and wastewater, executing projects with remarkable **engineering, procurement, and construction (EPC)** contractors in such sectors. He has achieved projects either with locally known contractors in the region or in the **Middle East and North Africa (MENA)** area. He has experience working with many brands such as Metso Automation, Siemens, A+B, ABB, Endress+Hauser, SMAR, Magnetrol, UE, Ebro, Bray, ARCA, Dresser, and so on. He hopes to continue his achievements in the upcoming years.

I would like to thank my family; my father, mother, wife, and children, who understand the time requirements and create an environment for working hard even over home hours. I would also like to thank my professor, Dr. Magdy El Tawil, and other staff for their contribution and valuable advice in all stages. I am also thankful to my whole family for providing support and tolerating my busy schedule, and still standing by my side. God bless all of us.

Javeria Parwani is a skilled control systems specialist with extensive experience of more than eight years in the industrial automation domain. She is an electronics engineer by qualification, and through working with different industries, she has gained knowledge of the leading industrial automation trends and technologies.

I'd like to thank God for all the opportunities in life.

Table of Contents

3

Actuators and Their Applications in Industrial Automation   47

4

Overview of AC and DC Motors   69

5

Introduction to Variable Frequency Drive (VFD)   89

6

Drawing Schematic/Wiring Diagrams Using CAD Software   107

Part 2: Understanding PLC, HMI, and SCADA

7

8

9

Deep Dive into PLC Programming with TIA Portal — 207

10

Understanding Human Machine Interfaces (HMIs) — 249

Part 3: Process Control, Industrial Network, and Smart Factory

13

Industrial Network and Communication Protocols Fundamentals   415

14

Exploring Smart Factory (Industry 4.0) with 5G   441

Preface

This book is designed to provide what you need to know as an industrial automation engineer, from basic to advanced – beneficial for both beginners and professionals in the field of industrial automation. It is an indispensable book for industrial automation engineers or anyone that wants to start or build a career in industrial automation. This book is a valuable resource for both undergraduates and graduates in the engineering, physics, and computer departments of tertiary institutions as well as staff in electrical, mechanical, production, process control, and instrumentation engineering departments in the manufacturing, food processing, chemical, oil and gas, and other industries where automated machines are used. You will learn about and be able to use sensors, actuators, and **Programmable Logic Controllers** (**PLCs**) for industrial automation. The book is loaded with hands-on step-by-step guides on PLC wiring and programming, HMI design and development, SCADA system setup, and process control.

The book is organized into three parts. Each part contains several chapters on important topics.

Part 1 is devoted to the concepts and skills required to get started in industrial automation. This is an introductory part where you will learn what industrial automation is all about and how to identify electrical control components and wire them for basic industrial control operations. You will also learn how to draw a schematic/wiring diagram using CAD software.

Part 2 will take you through the major tools in industrial automation (PLC, HMI, and SCADA). You will learn by doing by following the step-by-step guide provided. Simulation is also possible in some of the steps, which makes it easy for those without the actual device to have an understanding of how it works. What seems complex is made simple using a hands-on approach.

Part 3 is devoted to process control, industrial networks, and smart factories, which you will gain fundamental knowledge about.

Who this book is for

I strongly believe that this book will be of great benefit to readers who are new to the field of industrial automation, as well as those who are experienced and practicing in the field as it covers topics from a beginner level to an advanced level in a simplified and practical way. You can further advance your skills in a specific area through articles, books, or resources that specifically relate to the topic you would like to go into in more depth after reading this book.

What this book covers

Chapter 1, Introduction to Industrial Automation, provides an overview of industrial automation.

Chapter 2, Switches and Sensors – Working Principles, Applications, and Wiring, deals with switches and sensors. Knowledge of switches and sensors is important in industrial automation. Switches and sensors provide information about the environment for a controller to process. In this chapter, you will learn what switches and sensors really are and how they work, as well as learning about their wiring.

Chapter 3, Actuators and Their Applications in Industrial Automation, explains various actuators used in industrial automation. You will learn about their basic principles of operation and their applications in industries.

Chapter 4, Overview of AC and DC Motors, deals with electric motors, such as AC motors, DC motors, servo motors, and stepper motors, used in industrial automation. Electric motors play a big role in industries. Therefore, it is necessary for automation engineers to have good knowledge of them.

Chapter 5, Introduction to Variable Frequency Drive (VFD), focuses on the basic practical use of VFD to control the speed and direction of an induction motor. Induction motors are the most widely used electric motors in manufacturing industries and there is always a need to vary their speed and direction to suit various applications.

Chapter 6, Drawing Schematic/Wiring Diagrams Using CAD Software, shows how to use CAD software (PCSCHEMATIC automation software) to draw schematic/wiring diagrams relating to industrial automation. You will learn how to create a control system drawing using the software.

Chapter 7, Understanding PLC Hardware and Wiring, focuses on the hardware aspect of PLCs. You will gain fundamental knowledge of PLCs and learn how to wire sensors and actuators.

Chapter 8, Understanding PLC Software and Programming with TIA Portal, focuses on the software aspect of PLCs. You will learn the basic things you need to know to start programming PLCs with the most common PLC programming language (ladder diagram).

Chapter 9, Deep Dive into PLC Programming with TIA Portal, delves deeper into the software aspect of PLCs. You will learn more about PLC programming using the step-by-step guide provided.

Chapter 10, Understanding Human Machine Interfaces (HMIs), deals with HMIs and how to interface with PLCs to give commands to machines and also get feedback from the machines.

Chapter 11, Exploring Supervisory Control And Data Acquisition (SCADA), focuses on SCADA systems. SCADA is an interesting and advanced aspect of industrial automation that's widely used in oil and gas, manufacturing, and other industries for monitoring, control, and other functions in large plants where machines to be controlled or monitored are in remote locations.

Chapter 12, Process Control – Essentials, deals with the control of industrial processes. Process control is also an important area in industrial automation. You cannot have a quality product without control. This chapter explains process control and the various measurements and devices applicable to industrial process control. You will learn how to control temperature using a temperature controller. You will understand the wiring and programming of analog input of Siemens S7 1200 PLCs through a step by step guide.

Chapter 13, Industrial Network and Communication Protocols Fundamentals, explores the fundamentals of industrial networks. Industrial networks connect industrial automation and control devices. SCADA and other advanced controls rely on industrial networks. It is therefore necessary to learn what an industrial network is and the common protocols available. The chapter also explains the basics of wireless networks and what 5G really is.

Chapter 14, Exploring Smart Factory (Industry 4.0) with 5G, gives an overview of smart factories, which are the current trend in manufacturing and allow automated machines or equipment in the factory to be controlled and monitored from anywhere in the world via the internet. The chapter explains smart factories and the technologies that power them (IoT, AI, robotics, cloud computing, and so on) in a comprehensive way.

To get the most out of this book

This book is written to accommodate beginners or those who are new to the field of industrial automation, just as the book title *Industrial Automation from Scratch* implies. However, it will be advantageous for readers to have basic electrical knowledge and computer skills to be able to follow simple steps on a computer with the Windows operating system. Also, the book requires you to have a personal computer, which can either be a desktop or a laptop with the Windows operating system installed.

The following hardware is required to get your hands dirty with some of the practical exercises but is not a must:

- SIMATIC S7-1200 PLC (CPU 1211C AC/DC/RELAY)
- SIMATIC HMI (KTP 400)
- Power supply unit (220V AC - 24V DC)
- RTD temperature transmitter for Pt100
- RTD temperature sensor (Pt100)
- 500 Ohm resistor
- Heater
- Circuit breaker
- Ethernet cable to connect between the PLC or HMI and PC
- **Normally open (NO)** and **normally closed (NC)** push buttons (two or more each)
- Pilot lamps, contactors, wires, and so on

Simulation is available to enable you to get a feel of the hands-on practical exercises even if the above-stated devices and materials are not available. The software required for practicing and also for simulation can be downloaded for free (trial version) or purchase the full version with no restrictions from the manufacturer.. Links to download the trial version of the software are provided in the chapters.

Download the color images

We also provide a PDF file that has color images of the screenshots and diagrams used in this book. You can download it here: `https://packt.link/WUaf3`.

Conventions used

There are a number of text conventions used throughout this book.

Bold: Indicates a new term, an important word, or words that you see onscreen. For instance, words in menus or dialog boxes appear in **bold**. Here is an example: "Select **System info** from the **Administration** panel."

> **Tips or Important Notes**
> Appear like this.

Get in touch

Feedback from our readers is always welcome.

General feedback: If you have questions about any aspect of this book, email us at `customercare@packtpub.com` and mention the book title in the subject of your message.

Errata: Although we have taken every care to ensure the accuracy of our content, mistakes do happen. If you have found a mistake in this book, we would be grateful if you would report this to us. Please visit `www.packtpub.com/support/errata` and fill in the form.

Piracy: If you come across any illegal copies of our works in any form on the internet, we would be grateful if you would provide us with the location address or website name. Please contact us at `copyright@packt.com` with a link to the material.

If you are interested in becoming an author: If there is a topic that you have expertise in and you are interested in either writing or contributing to a book, please visit `authors.packtpub.com`.

Share your thoughts

Once you've read *Industrial Automation from Scratch*, we'd love to hear your thoughts! Scan the QR code below to go straight to the Amazon review page for this book and share your feedback.

`https://packt.link/r/1800569386`

Your review is important to us and the tech community and will help us make sure we're delivering excellent quality content.

Download a free PDF copy of this book

Thanks for purchasing this book!

Do you like to read on the go but are unable to carry your print books everywhere? Is your eBook purchase not compatible with the device of your choice?

Don't worry, now with every Packt book you get a DRM-free PDF version of that book at no cost.

Read anywhere, any place, on any device. Search, copy, and paste code from your favorite technical books directly into your application.

The perks don't stop there, you can get exclusive access to discounts, newsletters, and great free content in your inbox daily.

Follow these simple steps to get the benefits:

1. Scan the QR code or visit the link below

https://packt.link/free-ebook/9781800569386

2. Submit your proof of purchase
3. That's it! We'll send your free PDF and other benefits to your email directly

Part 1:
Learning the Concepts and Skills Required to Get Started

This is an introductory part, in which you will learn what industrial automation is all about. You will learn about sensors and actuators and be able to identify electrical control components and learn how to wire them for basic industrial control operations. You will also be able to draw a schematic/wiring diagram using CAD software.

This part has the following chapters:

- *Chapter 1, Introduction to Industrial Automation*
- *Chapter 2, Switches and Sensors – Working Principles, Applications, and Wiring*
- *Chapter 3, Actuators and Their Applications in Industrial Automation*
- *Chapter 4, Overview of AC and DC Motors*
- *Chapter 5, Introduction to Variable Frequency Drive (VFD)*
- *Chapter 6, Drawing Schematic/Wiring Diagrams Using CAD Software*

1
Introduction to Industrial Automation

Manufacturing involves using raw materials or parts to create goods or products that will be sold to customers. Creating products from raw materials can be done by using hand tools that are operated by humans or by machines such as motors, pumps, and drills, which are also operated by humans. Before machines, items/products were made by hand using tools, and lots of time and effort was spent producing a single item. The development of manufacturing resulted in hand tools being replaced by machines operated by humans, which made production faster with minimum effort. Nowadays, human involvement in terms of operating machines in manufacturing is being reduced or replaced via **industrial automation**.

By completing this chapter, you should be able to understand what industrial automation is and identify the various types of industrial automation that exist. You will also be able to describe the basic levels of industrial automation and identify the benefits of industrial automation in industries. Finally, you will be able to identify the cons of industrial automation in society.

In this chapter, we will cover the following topics:

- Introducing industrial automation
- Exploring the types of industrial automation
- Understanding the levels of industrial automation
- Discovering the advantages and disadvantages of industrial automation

Introducing industrial automation

Industrial automation is where computers, robots, and other control devices are used to operate machines and equipment that's used in manufacturing or processing plant with little human intervention.

A set of sequential operations are established in a factory where components are assembled to make a product or where materials are put through a refining process to produce an end product that is suitable for consumption (that is, a **production line** or **assembly line**). This usually consists of machines and tools that need to be operated to get the *finished product*. Before industrial automation, items (such as mugs, spoons, pots, and so on) were hand-made by individual craftsmen and women and it could take hours or days to manufacture a single item.

There was a development in the 18th century where, instead of items being produced by hand, processes were invented that allowed items to be produced by machines. These machines/tools were operated by humans. Hence, we can refer to the items that are produced through this process as man-machine made. This process dramatically reduces production time and costs compared to the hand-made method. A further shift in manufacturing was the introduction of assembly lines/production lines in the 19th century, which further reduced production time and costs as different workers could operate different machines in different sections of the production or assembly lines, leading to mass production. Nowadays, many factories have machines/tools in the production line being operated automatically by devices and control systems. Repetitive tasks such as filling, capping, and stamping are now automated.

A worker in a factory assembly line uses their eyes to see, their ears to hear, their brain to think, and their hands to move things or perform the required actions on the assembly line. With industrial automation, the activities/operations that are typically performed by the worker can be replaced by a control system where sensors do the job of the eyes and ears, a controller does the job of the brain, and an actuator does the job of the hands. Hence, **sensors**, **controllers**, and **actuators**, which we will look at in more detail in the subsequent chapters of this book, can replace a worker in a factory assembly line. In some manufacturing processes, an assembly line that requires one hundred staff members can eventually require just one or two people doing that through industrial automation. With the introduction of **artificial intelligence** (**AI**), machines that can even think like humans are being developed to handle human operations that we could not previously even imagine being handled by a machine.

Industrial automation can also be defined as being able to control machinery and equipment in an industry by using **personal computers** (**PC**), **programmable logic controllers** (**PLCs**), sensors, actuators, and other control devices to minimize the human involvement in the manufacturing or production processes. PCs, PLCs, and other forms of controllers replace human decision-making; the sensors replace the eyes and ears, while the actuators replace the hands, as explained previously. PCs, PLCs and other forms of controllers have programs (a set of instructions) written into them that enable them to make decisions based on the state of the sensors. The results of these decisions are used by the actuator to perform the action that the worker would have done.

The following figure shows an example of industrial automation using a KUKA industrial robot. Jobs, which are typically done by humans, are being carried out by robots (programmed machines):

Figure 1 1 – Industrial automation in a vehicle manufacturing assembly line using a KUKA industrial robot

This image is licensed under Wikimedia Commons: `https://creativecommons.org/licenses/by-sa/4.0/deed.en`.

Besides manufacturing, the transportation industry is another area where industrial automation can be applied, where self-operating vehicles are used for both private and commercial purposes – autopilot control in commercial jets enables jets to travel without human pilots. Warehouses, buildings, and other industries also benefit from industrial automation.

With that, you have learned that industrial automation involves using devices that operate machines automatically with little human involvement, which has significantly improved manufacturing and other industries. The knowledge you have acquired in this section will help you understand the various forms of automation we will look at in the next section.

Exploring the types of industrial automation

Various types of industrial automation are applied in the world today. Industrial automation solutions can fall into any of the following four categories:

- **Fixed automation system**
- **Programmable automation system**
- **Flexible automation system**
- **Integrated automation system**

In the next few sections, you will discover what these categories are in detail. Let's have a look.

Fixed automation system

In a fixed automation system, the series of operations that must be performed on the raw material are fixed. This type of automation is used to perform fixed and repetitive operations to achieve high production rates. The machines involved are programmed or configured to produce a specific design for a product. Once adopted, it is relatively difficult to change the design or style of the product. Fixed automation systems are used when the same style of product is to be manufactured or assembled. The series of operations that are involved in the production or manufacturing process are designed or programmed based on the product. This type of automation system is characterized by a *high production rate*, *high efficiency*, and *high initial cost* and is suitable for manufacturing a large volume of products that will have a low cost per product.

Examples of fixed automation systems are as follows:

- Paint and coating automation process
- Automatic assembly machines
- Bread production lines
- Steel rolling mills
- Papermills
- Metal pressing/stamping machines in a vehicle assembly line in the automobile industry

Let's have a look at programmable automation systems.

Programmable automation system

In a programmable automation system, the machines involved can be used to manufacture different styles of products. However, this requires reprogramming and changeover for each new style of product, which takes time to accomplish and creates downtime in production.

This type of automation is suitable for manufacturing where identical or similar styles of products are produced within a certain time frame. This is usually referred to as **batch production**. A long setup time is required to modify the program or reconfigure the sequence of operations for the new product design or batch to be manufactured.

Examples of programmable automation systems are as follows:

- **Industrial robot**: This is a type of robot (programmed machine) that's used in the manufacturing industry. It consists of a power supply, a controller, and a mechanical arm that can move in three or more axes. It can be programmed and configured for different kinds of tasks.

    The following figure shows a robot designed to be able to do any work that can be programmed (within its limits and work envelope) simply by changing the program. Here, the robot was programmed for writing:

Figure 1.2 – KUKA industrial robot

This image is licensed under Wikimedia Commons: `https://creativecommons.org/licenses/by-sa/4.0/deed.en`.

- **Computer Numerical Control (CNC) machine**: This is a type of automated machine in which geometric code (G-code) and miscellaneous code (M-code), which are alphanumeric, form the basic program instructions for different kinds of tasks. This type of machine can automate drilling, milling, or 3D printing using a computer or controller.

The following figure shows a CNC machine (KVR-4020A). This can be used for practically any type of application, including manufacturing or production, die-molding, heavy and hard milling, and more:

Figure 1.3 – CNC vertical machining center (Kent CNC KVR 4020A)

This image is licensed under Wikimedia Commons:

https://creativecommons.org/licenses/by-sa/4.0/deed.en.

Let's proceed to learn about flexible automation systems.

Flexible automation system

Flexible automation systems are an advanced form of programmable automation systems. They also require reprogramming and changeover for each new style of product, but this does not take time to accomplish as it is done in a programmable automation system. Hence, a flexible automation system reduces the downtime that's experienced in a programmable automation system. The product styles that a flexible automation system can produce are sufficiently limited so that the changeover can be accomplished very quickly and automatically. A flexible automation system allows a different range of products to flow through the line with little downtime.

Examples of flexible automation systems are as follows:

- **Automated guided vehicles (AGVs)**: AGVs are vehicles without an onboard driver that are used mostly to transport materials/goods in a factory or warehouse. They can navigate along a pre-defined path using several guidance technologies, which makes it easy for them to change routes and expand the operation of the system in response to changes to a scalable and flexible material handling solution. AGVs can be used alongside an industrial robot to provide a very cost-effective material handling solution in a factory or warehouse. The AGV can transport pallets, cartons, and products that have been loaded onto it using an industrial robot, and then send them to various areas of the manufacturing facility or warehouse.

  The following figure shows an example of an AGV:

Figure 1.4 – Automated guided vehicle by AIUT

This image is licensed under Wikimedia Commons: `https://creativecommons.org/licenses/by-sa/4.0/deed.en`.

- **Flexible Assembly System (FAS)**: A FAS is an assembly system that can produce a variety of products in small to medium batches with rapid changeover and reprogramming for a new set of product designs or styles.

Let's now have a look at integrated automation systems.

Integrated automation system

As the name implies, an integrated automation system integrates various machines and tools such as CAD, robots, cranes, conveyors and other automated machineries to work under a single control system to execute an automation system of a production process. In this, separate machines, data, and processes are made to work together and controlled by a single control system. It allows the entire manufacturing plant to be automated and controlled by computers with less human intervention.

Integrated automation system is used in **computer integrated manufacturing (CIM)** in which computers control the entire production process with little human intervention. It is also used in various kinds of advanced process automation systems. Integrated automation system can give room for the implementation of various advanced technologies such as automated material handling systems, **Radio Frequency Identification (RFID)**, barcode tracking systems, **Manufacturing Execution System (MES)**, **Computer Aided Process Planning (CAPP)**, automated conveyors and cranes, and many others.

In this section, you learned about the various types of industrial automation available while looking at relevant examples. This knowledge will help you have a better understanding of the next section and other topics that will be covered in this book.

Understanding the levels of industrial automation

Industrial automation is a complex system with several devices communicating and working with each other to provide the desired result. The simplest way to describe the levels/hierarchy of industrial automation is by using a *three-level* representation, as shown here:

- **Field level**
- **Control level**
- **Supervisory and production level**

We will describe each of these levels in the following section and provide the relevant examples for ease of understanding.

Field level

This is the lowest level in the hierarchy of industrial automation. It consists of field devices such as sensors and actuators, which are used in industrial automation. Sensors convert physical characteristics into electrical signals (digital or analog). They are input devices and can be referred to as the eyes and ears of automation. Examples of sensors include **proximity sensors**, **temperature sensors**, **pressure sensors**, **level sensors**, **flow sensors**, and **limit switches**. Actuators, on the other hand, convert electrical signals (digital or analog output signals) into physical characteristics, which can be in the form of motion. Examples include **AC/DC motors**, **servo motors**, **stepper motors**, **pumps**, **control valves**, **solenoids**, **contactors**, and **relays**. The job of the field devices (sensors and actuators) is to transfer machine and process data to the next level (control level) for *monitoring* and *analysis*.

This level can be referred to as the eyes and arms (hands) of an industrial automation system. The sensor is acting as the eyes, while the actuator is acting as the arms. Real-time process parameters such as temperature, pressure, level, and flow are converted into electrical signals by the sensors. Data that's collected from the sensors is transferred to the controller for further monitoring and analysis. The actuators control the process parameter through a signal from the controller.

Control level

This level consists of **programmable logic controllers** (**PLCs**) or other forms of controllers. PLCs are the brains behind modern industrial automation. They are used to carry out control functions in industries. They take data from different kinds of sensors, make decisions using the program written into it, and output a control signal that the actuator will use to carry out the required task. They can be programmed to deliver automatic control functions based on the signals they receive from sensors. More details on PLCs will be provided in *Chapter 7, Understanding PLC Hardware and Wiring*, and *Chapter 8, Understanding PLC Software and Programming with TIA Portal*.

Supervisory and production level

This level consists of **Supervisory Control And Data Acquisition** (**SCADA**) and **Human Machine Interface** (**HMI**), among others, for monitoring and controlling various parameters and setting production targets. *Chapter 10, Understanding Human Machine Interfaces (HMIs)*, explains HMI while *Chapter 11, Exploring Supervisory Control And Data Acquisition (SCADA)*, explains SCADA.

To help you gain a better understanding of HMI and SCADA, this book has been arranged to cover some basic knowledge of industrial automation before venturing into HMI and SCADA.

The following diagram represents the basic levels of industrial automation:

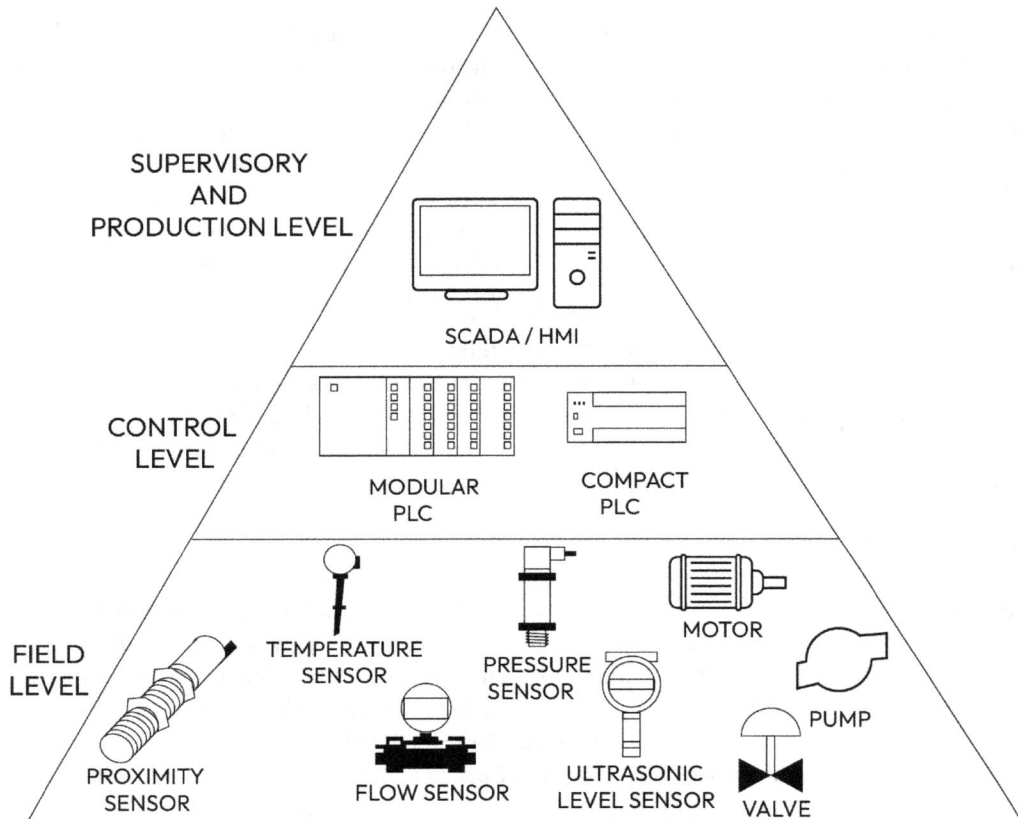

Figure 1.5 – Levels/hierarchy of industrial automation

In this section, you learned about the basic levels of industrial automation, which include the field level, the control level, and the supervisory and production level. *Chapter 2, Switches and Sensors – Working Principles, Applications, and Wiring,* and *Chapter 3, Actuators and Their Applications in Industrial Automation,* will give more detailed explanation on field level. *Chapter 7, Understanding PLC Hardware and Wiring, Chapter 8, Understanding PLC Software and Programming Using the TIA Portal,* and *Chapter 9, Deep Dive into PLC Programming with TIA Portal,* will give detailed explanation on the control level using hands-on approach while *Chapter 11, Exploring Supervisory Control And Data Acquisition (SCADA),* will further explain the supervisory and production level and include a hands-on project/practical project (interfacing SCADA with S7-1200 PLC using mySCADA software) that will give you a practical experience and understanding of the concept.

Discovering the advantages and disadvantages of industrial automation

There's no doubt that industrial automation has transformed manufacturing and other industries by allowing things to be done faster and better. Industrial automation is required for any industry to remain competitive. Although automation has led to great improvements in manufacturing and other industries, there are some negative effects of industrial automation that need to be looked into so that you are prepared and equipped to overcome them.

In the following sections, we will discover what these advantages and disadvantages are.

Advantages of industrial automation

Let's look at how people can (or do) benefit from industrial automation:

- It reduces the time taken between starting and completing a process.

- It improves workers' safety. Automation saves workers from being exposed to hazardous environments in the factory.

- The use of robots in industrial automation increases production output. Robots can work 24/7 at a constant speed.

- Industrial automation reduces operating costs. Adding automated machines to an operation means less human effort (that is, fewer employees are needed to get the job done). Also, there will be less material waste due to the accuracy of the robots in performing a task.

- Industrial automation provides a faster **return on investment** (**ROI**) due to reduced processing time, reduced operating cost, increased production output, and so on.

- The use of **automated guided vehicles** (**AGVs**) and other automated machinery in the manufacturing industries reduces accidents that would have been caused by human error.

- Industrial automation allows managers to focus more on other aspects of their jobs since robots/automated systems require less supervision.

- Automated systems/robots don't get distracted, exhausted, or bored while working, so delays are eliminated and consistent performance can be expected.

Disadvantages of industrial automation

It is a good thing to relieve humans from certain activities through industrial automation; however, this threatens human workforces.

The following are some of the disadvantages of industrial automation:

- Industrial automation requires **high initial capital** (that is, the cost of designing, fabricating, installing, and commissioning automated systems is high).

- Industrial automation can get rid of jobs. Jobs done by humans can be taken over by an automated machine or robot. However, this might not necessarily be a disadvantage as the staff can be trained to service the machinery/ automated system or work in other areas of the business.

- Skilled personnel are required for *maintaining* and *troubleshooting* automated systems.

In this section, we learned how people and industries can benefit from industrial automation, as well as the various side effects of industrial automation. It is important to take these advantages and disadvantages into consideration when you are thinking about integrating automation technology in an industry.

Summary

Now that you have completed this chapter, which provided you with a basic understanding of industrial automation, you have achieved the first leg of this book's journey. Well done! You should now be able to explain industrial automation, describe the different types of industrial automation, describe the levels of industrial automation, and describe the advantages and disadvantages of industrial automation.

The topics that were covered in this chapter are relevant for automation engineers and will help you understand the next chapter, which will give you an overview of switches and sensors.

Questions

The following questions will help you test your understanding of this chapter. Ensure that you have read and understood the topics in this chapter before attempting these questions:

1. _____ is used to control machinery and equipment in an industry by using personal computers (PC), programmable logic controllers (PLCs), sensors, actuators, and other control devices to minimize the human involvement in the manufacturing or production processes.

2. _____ automation systems reduce the downtime that's experienced in programmable automation systems.

3. In the hierarchy of industrial automation, the _____ level consists of PLCs or other form of controllers.

4. _____ uses various forms of devices and control systems to operate machines and equipment that's used in manufacturing or processing plant with little human intervention.

5. _____ is a set of sequential operations that's established in a factory where components are assembled to make a finished product or where materials are put through a refining process to produce an end product that is suitable for consumption.

6. _____, _____, and _____ are the three types of industrial automation.

7. A method of manufacturing where identical or similar style of products are produced together within a time frame is referred to as _____.

8. Vehicles without an on-board driver, which are used mostly to transport materials/goods in a factory or warehouse, are called _____.

9. In the industrial automation hierarchy, the _____ level consists of sensors and actuators.

10. _____ convert electrical signals (digital or analog signals) into physical characteristics that can be in the form of motion or heat.

11. _____ convert physical characteristics into electrical signals (digital or analog).

12. CNC is an acronym for _____.

13. PLC is an acronym for _____.

14. SCADA is an acronym for _____.

15. HMI is an acronym for _____.

Switches and Sensors – Working Principles, Applications, and Wiring

Understanding how human being functions will help you understand automation. Having a brain is not enough for a human being to operate. Sight, hearing, touch, and smell are required to provide information about the environment for the brain to process. Similarly, **switches and sensors** provide information about the environment for a controller to process. The result of processing determines the actions that will be carried out by an *actuator*, which will be discussed in the next chapter. This chapter is specifically reserved for switches and sensors to enable us to discuss actuators in depth. Just like sight, hearing, touch, and smell are important for humans to function, switches and sensors are important components that are required to automate an industrial process.

In this chapter, we are going to cover the following main topics:

- Introducing switches and sensors
- Describing manually operated switches
- Describing mechanically operated switches
- Understanding proximity sensors (capacitive, inductive, and photoelectric)

Introducing switches and sensors

A **switch** is an electrical component that connects or disconnects a signal path in an electrical circuit. It is used to make or break an electrical circuit. A switch has two states: an ON state and an OFF state. In an electrical circuit, the current will flow when the switch is in the ON state. When in the OFF state, the current will not flow. Hence, the ON state will connect a signal path, while the OFF state will disconnect a signal path. A switch can be referred to as a device that's used to turn equipment ON or OFF.

Switches can be categorized as follows:

- **Manually operated switch**
- **Mechanically operated switch**

Sensors are devices that detect or sense the presence or absence of objects. A sensor can be thought of as an automatic switch. They gather information from their environment and convert it into a form of signal that can be read and/or seen by an observer or equipment. In a human being, their eyes, nose, ears, tongue, and skin can be referred to as sensors. Their eyes detect light energy, their nose can detect chemicals or various kinds of smell, their ears can detect sound, and their skin can detect temperature or pressure. Likewise, in automation, various types of sensors detect various physical quantities such as light, sound, temperature, and so on. These physical quantities will be converted into a digital or analog signal that can be read and/or seen by an observer or piece of equipment (controller). There are various types of sensors, but we will cover proximity sensors later in this chapter.

In this section, we learned about switches and can summarize them as devices that are used to turn equipment either ON or OFF. We also learned that sensors can be thought of as automatic switches. They are not operated by humans. Having an idea of switches and sensors will give you a better understanding of the types of switches that will be explained in the next section. It will also help you understand proximity sensors, which will be explained later in this chapter.

Describing manually operated switches

In manufacturing, there is always the need for an operator to start or stop a machine and there could be a need to manually perform some other form of control aside from starting and stopping. These usually require switches that can be operated by hand. Switches that are operated by hand are referred to as **manually operated switches**. Some examples are as follows:

- **Push button**
- **Rocker switch**
- **Toggle switch**
- **Selector switch**
- **Knife switch**

We will explore each of these in the following sections.

Push button

A **push button** is a type of switch that closes or opens a circuit when a button is pressed or depressed. The button is designed to operate in two states (that is, ON or OFF). When it's in the ON position, an internal metal spring contacts the two terminals of the switch, allowing the current to flow. When it's in the CFF position, the internal switch retracts and the current will stop flowing.

Push buttons exist as normally *open* or normally *closed*.

Normally open or normally closed buttons can be used to control a single electrical circuit.

In a normally open push button, when the push button is not pressed, the contacts are open, and the connection is broken (that is, the switch is OFF). When the push button is pressed, the contacts are closed, and the connection is made (that is, the switch is ON).

In a normally closed push button, when the push button is not pressed, the contacts are closed, and a connection is made (that is, the switch is ON). When the push button is pressed, the contacts are open, and the connection is broken (that is, the switch is OFF).

The following diagram shows the symbols for a **normally open (NO)** and **normally closed (NC)** push button:

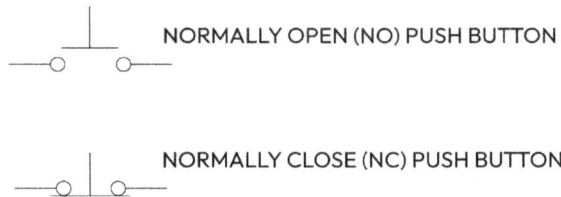

Figure 2.1 – Symbols for normally open (NO) and normally closed (NC) push buttons

The following figure shows the front and rear of a NO push button. Terminal 3 and terminal 4 are used to connect the switch to a circuit:

Figure 2.2 – The front and rear of a NO push button

The credit for this image goes to *Showlight Technologies LTD* www.showlight.com.ng.

The following wiring diagram shows the use of a NO push button to light a lamp. The lamp will come on when the push button is pressed and go off when it is released or depressed:

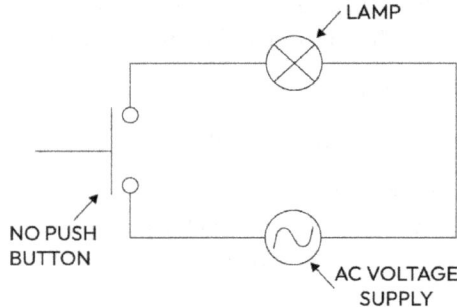

Figure 2.3 – Using a NO push button to light a lamp

The following figure shows the front and rear of a NC push button. Terminal 1 and terminal 2 are used to connect the switch to a circuit:

Figure 2.4 – The front and rear of an NC push button

This credit for this image goes to *Showlight Technologies LTD* www.showlight.com.ng.

Rocker switch

A **rocker switch** is an on and off switch in which one side is raised and the other side is depressed when pressed. It rocks like a rocking horse rocks back and forth. The following figure shows what it looks like from the front and rear view:

Figure 2.5 – The front and rear of a DPST rocker switch

This credit for this image goes to *Showlight Technologies LTD* www.showlight.com.ng.

Rocker switches are available in various forms:

- SPST
- SPDT
- DPST
- DPDT

Let's explore what they are, one by one.

Single Pole Single Throw (SPST) can switch (turn on or off) only one circuit. It is the simplest form of a switch in that it can only have one input and can connect to only one output.

The following diagram shows the symbol of an SPST switch:

SINGLE POLE SINGLE THROW (SPST)

Figure 2.6 – Symbol of an SPST

The following wiring diagram shows the simple wiring for an SPST switch to light a lamp:

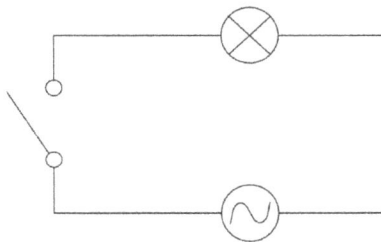

Figure 2.7 – Simple wiring diagram showing the use of an SPST switch to light a lamp

Single Pole Double Throw (SPDT) is a type of switch that has only one input (usually referred to as *common*) and can connect to and switch between two different outputs. A contact is usually connected to the input by default, and this is called a *normally closed* contact. The other contact, which is connected to the input during operation, is referred to as a *normally open* contact.

The following diagram shows the symbol of an SPDT switch:

SINGLE POLE DOUBLE THROW (SPDT)

Figure 2.8 – Symbol of an SPDT switch

The following is a simple wiring diagram of an SPDT switch being used to light two lamps (one at a time):

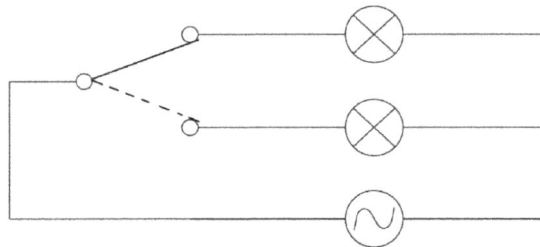

Fig. 2.9 – Using an SPDT switch to light two lamps

Double Pole, Single Throw (DPST) is a type of switch that has two inputs (usually referred to as *common*) and each input can connect and switch between one corresponding output. It consists of two SPST switches that operate simultaneously in one package.

The following diagram shows the symbol for a DPST switch:

DOUBLE POLE SINGLE THROW (DPST)

Figure 2.10 – Symbol of a DPST switch

The following is a simple wiring diagram of a DPST switch being used to switch a lamp and an induction motor on or off simultaneously:

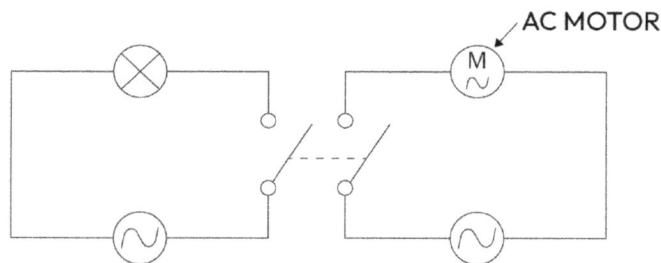

AC MOTOR

Figure 2.11 – Simple wiring of a DPST switch being used to switch a
lamp and an induction motor on or off simultaneously

Double Pole Double Throw (DPDT) is a type of switch that has two inputs (usually referred to as *common)* and each input can connect and switch between two corresponding outputs. It consists of two SPDP switches that operate simultaneously on one package.

The following diagram shows the symbol of a DPDT switch:

DOUBLE POLE DOUBLE THROW (DPDT)

Figure 2.12 – Symbol of a DPDT switch

The following wiring diagram shows the use of a DPDT switch to run a **DC motor** in both forward and reverse directions:

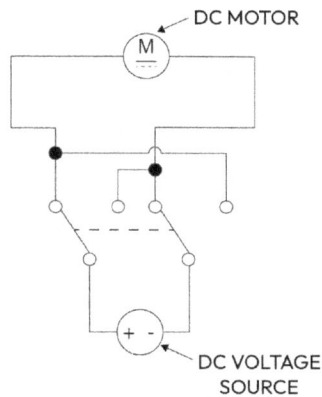

Figure 2.13 – Using a DPDT switch to run a DC motor

Next, let's look at the toggle switch.

Toggle switch

A **toggle switch** is a type of switch consisting of a lever that is moved back and forth to open or close an electrical circuit. They can have more than one lever position.

Toggle switches also exist as SPST, SPDP, DPST, and DPDT switches. The following figure shows what a toggle switch looks like:

Figure 2.14 – A toggle switch

The credit for this image goes to *Showlight Technologies LTD* www.showlight.com.ng.

Slide switch

A **slide switch** uses a slide action to produce the same connections as a toggle switch. A slide switch can also exist as an SPST, SPDP, DPST, or DPDT switch. The following figure shows what a slide switch looks like:

Figure 2.15 – Mountable slide switch

This credit for this image goes to *Wikimedia Commons* and can be found at https://creativecommons.org/licenses/by-sa/4.0/deed.en.

Selector switch

A **selector switch** may have two or more selector positions. It can control the ON or OFF state of a different circuit by rotating the handle. It is used where more than one control option is needed.

The following figure shows a selector switch's rotating handle and its connection terminals:

Figure 2.16 – Selector switch

The credit for this image goes to *Showlight Technologies LTD* www.showlight.com.ng.

Knife switch

A **knife switch** is a switching device that makes, breaks, or changes the course of electric current. It consists of one or more movable copper blades that are hinged and make contact with stationary forked contact jaws by being forced between them.

The following figure shows what a knife switch looks like:

Figure 2.17 – Open SPST knife switch

The credit for this image goes to *Wikimedia Commons* and can be found at https://creativecommons.org/licenses/by-sa/4.0/deed.en.

In this section, you learned about push buttons, rocker switches, toggle switches, selector switches, and knife switches. You must familiarize yourself with them and understand how they can be used in a circuit. This knowledge will help you gain a better understanding of mechanically operated switches, which we will look at in the next section.

Describing mechanically operated switches

A **mechanically operated switch** is controlled automatically by factors such as pressure, position, or temperature. This differs from a manually operated switch, which is operated by hand or by a human operator. They are used in situations where automatic control is required. The switch operates automatically by pressure, temperature, or position via a mechanism in the switch. They can also be referred to as sensors.

A mechanically operated switch takes the place of a *human operator*. Some examples of this type of switch are as follows:

- **Limit switch**
- **Level switch**
- **Pressure switch**
- **Temperature switch**

Let's have a look at each.

Limit switches

This is a switch that is operated by the motion of a machine part. It operates when a predetermined limit is reached. The standard limit switch is a mechanical device that uses physical contact to detect the presence or absence of an object. It consists of a switch body, an actuator, and an operating head.

The following diagram shows a limit switch and its parts:

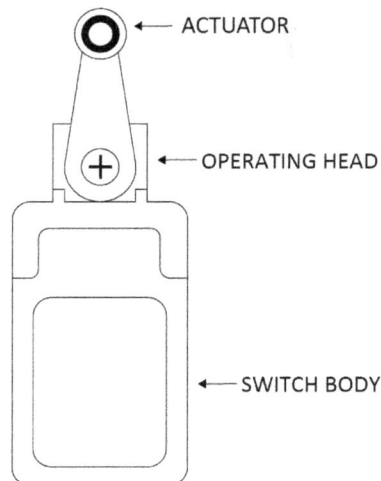

Figure 2.18 – A limit switch and its parts

The **switch body** contains the electrical contacts. The **actuator** is the part of the limit switch that contacts the target and moves to cause connection or disconnection in the electrical contacts. The **operating head** is where the rotation occurs, and it consists of a mechanism that connects or disconnects the electrical contacts.

Limit switches have two basic electrical contact configurations – **Single Pole, Double Throw (SPDT)** and **Double Pole, Double Throw (DPDT)**. SPDT consists of one normally open and one normally closed contact, while DPDT consists of two normally open and two normally closed contacts.

The following diagram illustrates the basic contact configuration of a limit switch:

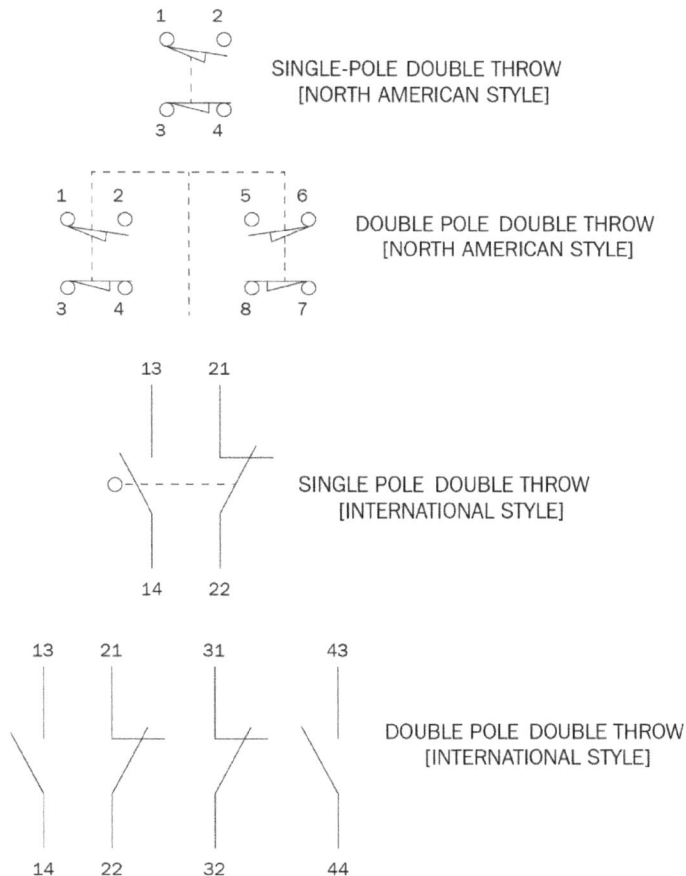

Figure 2.19 – Illustration of a limit switch

One type of actuator that a limit switch can use is **momentary operation**. With this type of actuator, when no force is applied, the device is in its free position and the contacts are in their normal state.

When force is applied, the device will be in its operating position – in this case, the electrical contacts will change from their normal state to their operating state. When force is removed, the device will return to the free position and the electrical contacts will return to their normal state.

The following diagram shows the operation of a momentary operation type of limit switch:

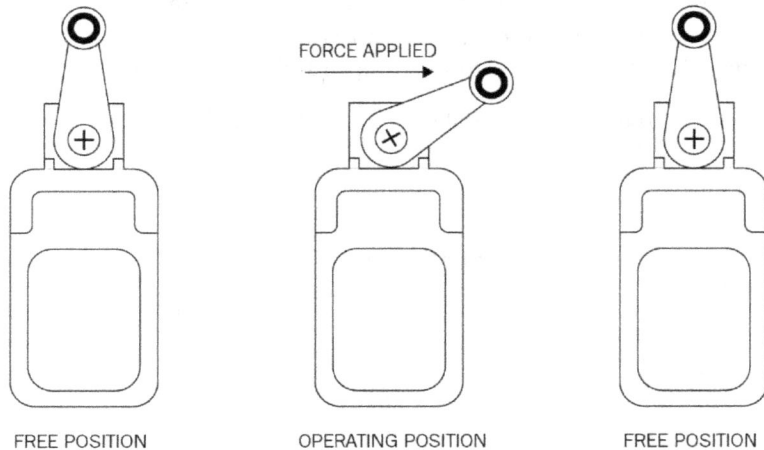

Figure 2.20 – The operation of a momentary operation type of limit switch

Another type of actuator is **maintained operation**. With this type of actuator, the actuator lever and the electrical contacts remain in their operating state after the actuator is no longer in contact with the target. When force is applied in the opposite direction, the actuator lever and electrical contacts return to their free position.

The following diagram shows the wiring of a limit switch with terminal 1 and terminal 2 being used to control the start and stop of a single-phase induction motor:

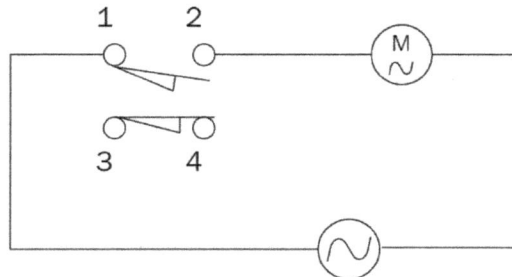

Figure 2.21 – Wiring of a limit switch with terminal 1 and terminal 2 being used
to control the start and stop of a single-phase induction motor

> **Important Note**
>
> In the preceding diagram, terminal 1 and terminal 2 of the limit switch is being used in the circuit. The motor will not run when the limit is not reached because terminals 1 and terminal 2 are open. When the limit is reached, the opened terminals (1 and 2) will close, and the motor will start running.

The following diagram shows the wiring of a limit switch with terminal 3 and terminal 4 being used to control the start and stop of a single-phase induction motor:

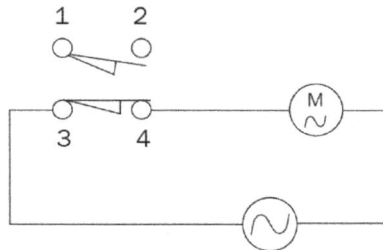

Fig. 2.22 – Wiring of a limit switch with terminal 3 and terminal 4 being used to control the start and stop of a single-phase induction motor

> **Important Note**
>
> In the preceding diagram, terminal 3 and terminal 4 of the limit switch is being used in the circuit. The motor will run when the limit is not reached because terminals 3 and terminal 4 are closed (that is, connected). When the limit is reached, the closed terminals (3 and 4) will open, and the motor will stop running.

The following figure shows a limit switch with the front cover screwed to the case:

Figure 2.23 – A limit switch

The credit for this image goes to *Showlight Technologies LTD* (www.showlight.com.ng).

The following figure shows a limit switch with the front cover removed:

Figure 2.24 – A limit switch with the front cover removed

The credit for this image goes to *Showlight Technologies LTD* www.showlight.com.ng.

Level switch

A **level switch** can also be referred to as a level sensor. It can detect the presence of liquids, powder, or granulated materials in a vessel. It can provide automatic control of motors or pumps when the level is high or low. Various level switches are available, including *float switch*, *capacitive*, *ultrasonic*, and *vibrating fork*. A float switch, as shown in the following diagram, is a common type of level switch that consists of a hollow floating body and an internal mechanical switch, also known as a sensor, and an external counterweight. The fixed point or external counterweight will be placed at a fixed position in the vessel or tank so that the internal mechanical switch in the floating body goes between the open and close as the water level rises or falls. The opening and closing of the switch can be made to stop or start a pump automatically to control the level of water in a tank:

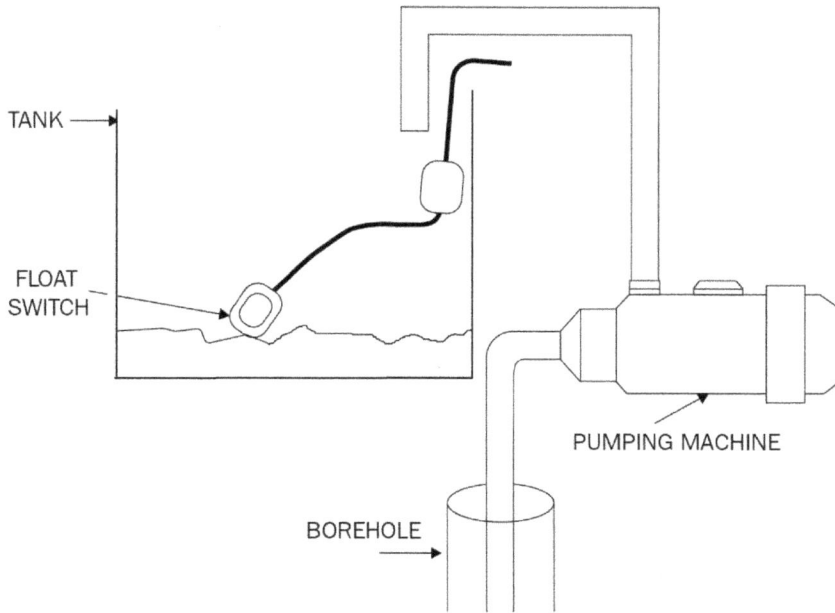

Figure 2.25 – The float switch and a pump

The following figure shows what a float switch looks like:

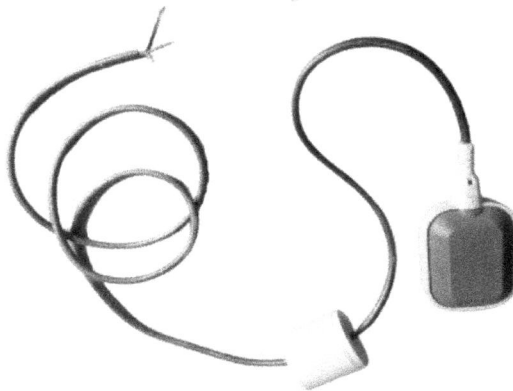

Fig. 2.26 – A float switch

The credit for this image goes to *Showlight Technologies LTD* www.showlight.com.ng.

The most basic float switch is a **two-wire, single-pole, single-throw (SPST)** float switch, which can exist as normally closed or normally open. An SPDP float switch, which has three wires, also exists. The following diagram shows the symbols for various forms of float switches that are available:

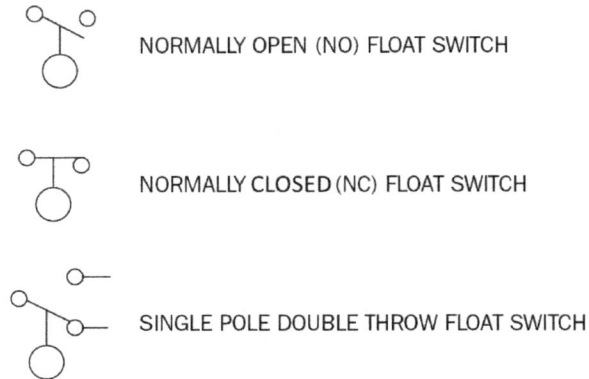

NORMALLY OPEN (NO) FLOAT SWITCH

NORMALLY CLOSED (NC) FLOAT SWITCH

SINGLE POLE DOUBLE THROW FLOAT SWITCH

Figure 2.27 – Symbols for various forms of float switches

Next, we will look at each float switch mentioned in this section.

Normally closed float switch

In a **normally closed float switch**, the electrical circuit is closed in a downward position and opened in an upward position. So, with gravity pulling it down, it is closed; as the liquid level rises to the preset level, it will open, as shown in the following diagram:

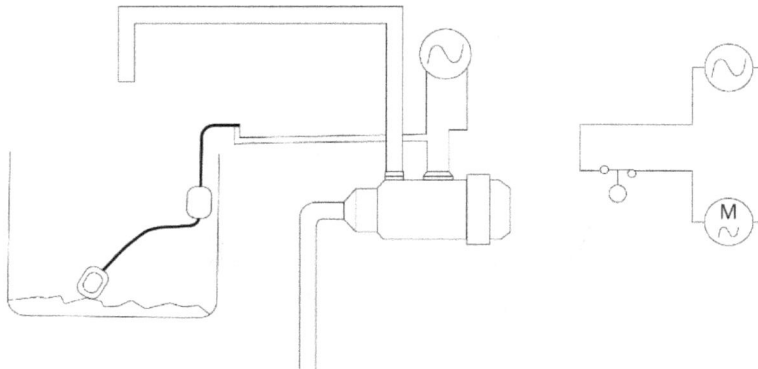

Figure 2.28 – A normally closed float switch being used to turn off a pump that fills a tank

In the preceding diagram, when the water level is low, the float switch will close, and the power will flow to the pump and the pump will begin to fill the tank. As the water level rises, the float switch will become open at a preset point, and the pump will stop.

Normally open float switch

In a **normally open float switch**, the electrical circuit is open in the downward position and closed in the upward position. So, with gravity pulling it down, it is open, and as the liquid level rises to the preset level, it will close:

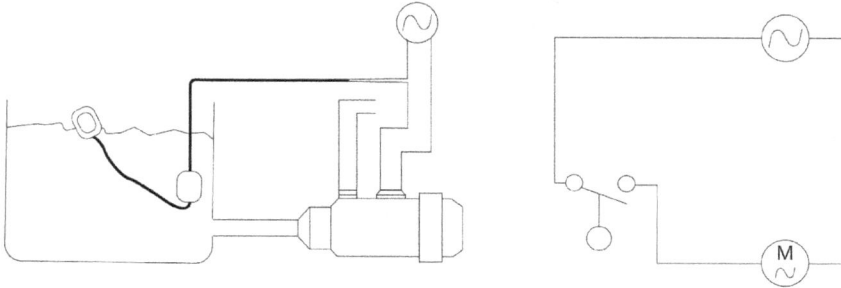

Figure 2.29 – A normally open float switch that automatically empties a tank when the water level is full

In the preceding diagram, when the water level is full, the float switch will close, power will flow to the pump and the pump will begin to empty the tank (that is, it will pump water out of the tank). As the water level reduces, the float switch will become open at a preset point, which will cause the pump to stop operating.

SPDT float switch

A **SPDT float switch** has three wires – usually, they are *black*, *brown*, and *blue*. When the float is low, the black wire disconnects from the red wire and gets connected to the blue wire. When the float is high, the black wire gets disconnected from the blue wire and gets connected to the red wire.

The following diagram shows the symbols of an SPDT float switch:

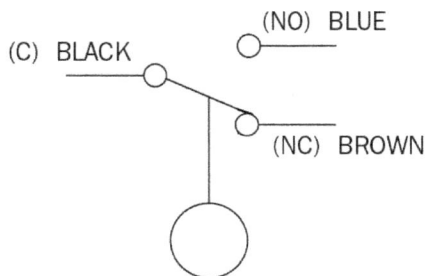

Figure 2.30 – SPDT float switch

This switch can be used to either fill a tank automatically when empty or empty a filled tank automatically.

We'll look at these two scenarios in the following sections.

Filling a tank

Here, the black and blue wires are used to turn the pump on or off. When the water level is low, the black wire gets connected to the blue wire and the pump starts to operate. When the pump is full, the black wire gets disconnected from the blue wire and the pump stops.

The following diagram shows an SPDT float switch when the water level is low:

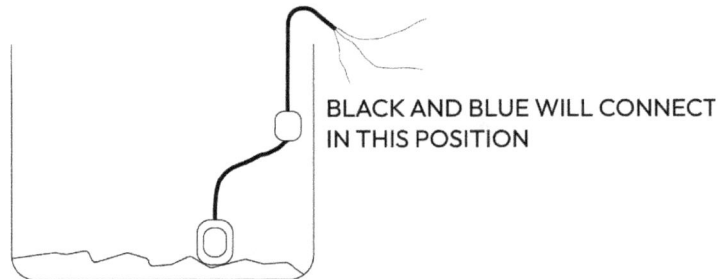

BLACK AND BLUE WILL CONNECT IN THIS POSITION

Figure 2.31 – SPDT float switch when the water level is low

Emptying a tank

Here, the black and brown wires of the float switch are used to turn the pump on or off. When the water level is high, the black wire gets connected to the brown wire and the pump starts to operate. When the water level is low, the black wire disconnects from the brown wire and the pump stops.

The following diagram shows an SPDT float switch when the pump is full:

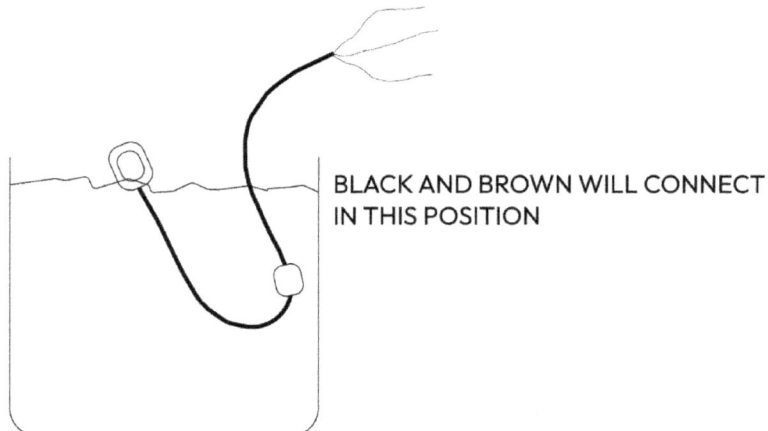

BLACK AND BROWN WILL CONNECT IN THIS POSITION

Figure 2.32 – SPDT float switch when the pump is full

In the next section, we will explore another mechanically operated switch.

Pressure switch

A **pressure switch** is a form of sensor that closes or opens an electrical contact when a certain pressure is obtained through an increase or decrease in pressure.

A mechanical pressure switch consists of a **process connection, measuring element, switch contacts, electrical connection**, and a **setpoint adjustment screw**.

The core part of a mechanical pressure switch is its measuring element. Elastic diaphragms or stainless-steel pistons are usually used for this purpose. The diaphragm is designed for vacuum or low pressure ranges up to 16 bars, while the piston is suitable for higher pressure ranges up to 350 bars. The process connection and case are usually made of galvanized steel or brass. The process connection is the part that connects to the air supply or for the process to control. The switch contacts are the internal contacts that open or close as pressure increases or decreases. The electrical connection connects to the internal electrical contact, which opens or closes as pressure rises or falls. The setpoint adjustment screw can be used to calibrate or adjust the switch (that is, set the pressure at which the switch opens or close).

The following diagram shows the cross-section of a pressure switch:

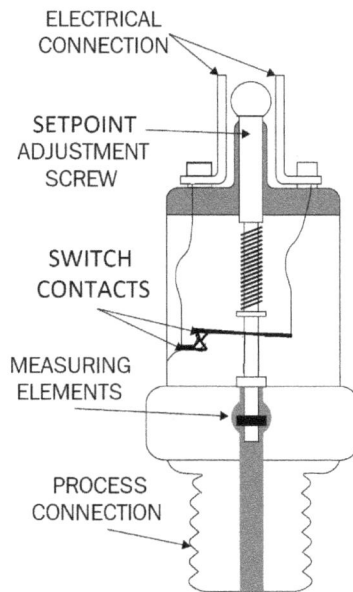

Figure 2.33 – Cross-section of a pressure switch

The *single pole single throw* form of a pressure switch can exist as NO and NC.

In an *NO* type, the switch contacts are open (not connected) when no pressure is applied, and they close when the applied pressure is equal to the setpoint.

In an NC type, the switch contacts are closed (connected) when no pressure is applied, and they open when the applied pressure is equal to the setpoint.

Your choice of switch depends on the type of circuit you intend to drive with the switch.

The following diagram shows the symbols of a pressure switch:

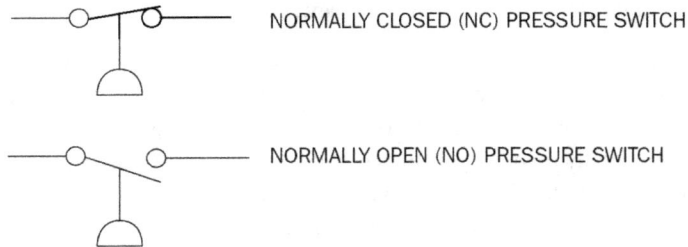

NORMALLY CLOSED (NC) PRESSURE SWITCH

NORMALLY OPEN (NO) PRESSURE SWITCH

Figure 2.34 – Symbols of a pressure switch

The following diagram shows the wiring of a pressure switch being used to switch off a lamp when a setpoint has been reached:

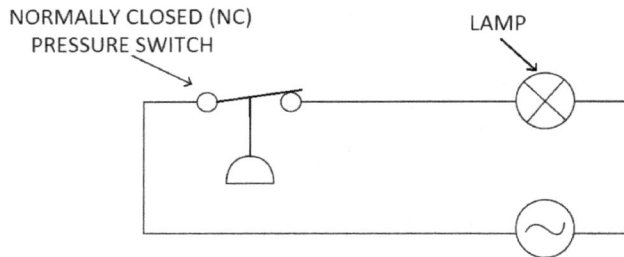

NORMALLY CLOSED (NC)
PRESSURE SWITCH

LAMP

Figure 2.35 – A pressure switch (NC) being used to switch off a lamp when a setpoint has been reached

The following diagram shows the wiring of a pressure switch being used to switch on a buzzer when a setpoint has been reached:

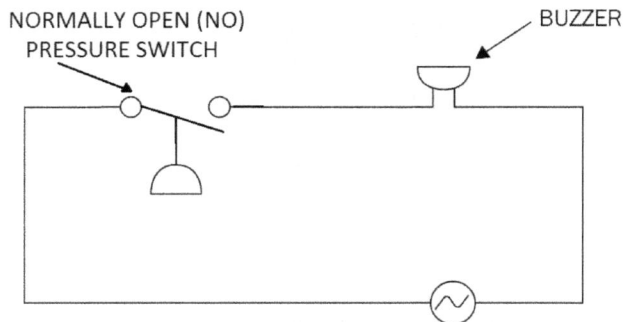

NORMALLY OPEN (NO)
PRESSURE SWITCH

BUZZER

Figure 2.36 – A pressure switch (NO) being used to switch on a buzzer when a setpoint has been reached

Lastly, we will look at what a *temperature switch* is.

Temperature switch

A **temperature switch** is a switch that opens or closes when the temperature reaches a setpoint or falls below a setpoint. The design of a temperature switch consists of two parts:

- **Sensor**: This can be a sensing bulb filled with a fluid – liquid, gas, or a bimetallic strip. It's usually immersed in the process whose temperature is to be controlled.

- **Snap-action contacts**: These switch the electrical power of the device controlling the temperature of the process on or off.

There are various types of temperature switches, including bimetallic strip temperature switches and liquid-filled temperature switches.

A **bimetallic strip temperature switch** is one of the most common temperature switches. It consists of a thin rectangular strip formed of two dissimilar metals that are bonded back to back. The two dissimilar metals experience dissimilar thermal expansion rates. Hence, when the temperature increases, one metal will expand faster than the other. This causes the strip to bend. The bending strip is used to activate a snap action switch that opens or closes a circuit.

The following diagram shows a simple bimetallic strip temperature switch:

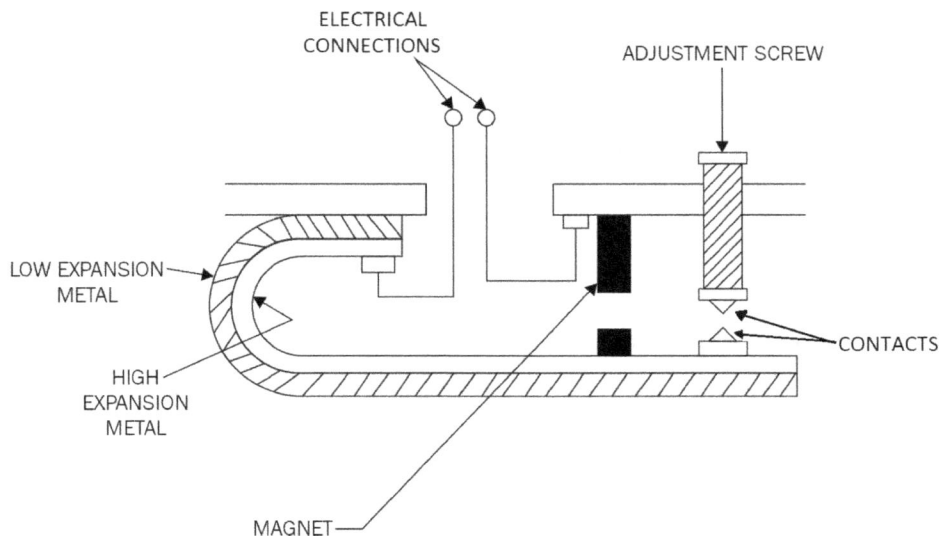

Figure 2.37 – A bimetallic strip temperature switch

A **liquid-filled temperature switch** (**capillary thermostat**) consists of a fluid encapsulated in a metal tube (bulb and capillary). The fluid expands or contracts when the temperature changes. The change in fluid moves the head, which triggers a snap action switch, either opening or closing the circuit.

The following diagram shows a capillary thermostat (a liquid-filled temperature switch):

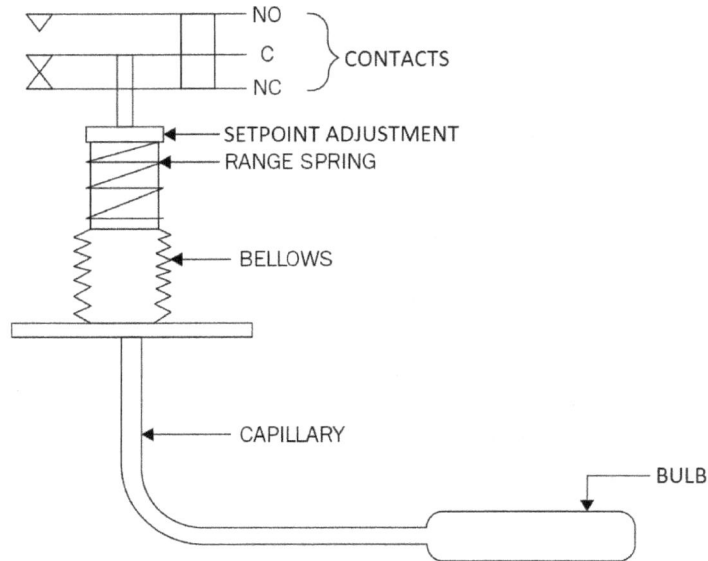

Figure 2.38 – A capillary thermostat temperature switch

The following diagram shows the symbols of a temperature switch:

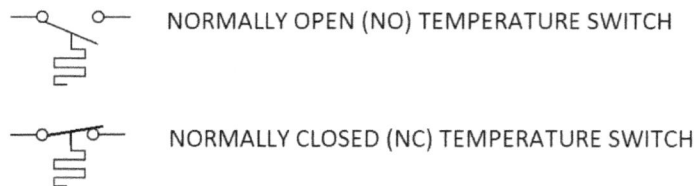

NORMALLY OPEN (NO) TEMPERATURE SWITCH

NORMALLY CLOSED (NC) TEMPERATURE SWITCH

Figure 2.39 – Symbols of a temperature switch

In this section, you learned about some of the various mechanically operated switches that can be used in industries to automate certain processes. You also saw the various symbols that are used for different switches. These symbols are important for reading schematic diagrams. Knowing how to wire these switches is also beneficial. The knowledge you've gained from this section will help you have a better understanding of the next section.

Understanding proximity sensors (capacitive, inductive, and photoelectric)

Proximity sensors are commonly used in industrial applications. Knowledge of proximity sensors can help you understand other types of sensors.

Proximity sensors detect the presence or absence of objects using electromagnetic fields, light, or sound. There are various types, each suited to a specific environment or application.

The available types of proximity sensors are as follows:

- **Inductive proximity sensors**
- **Capacitive proximity sensors**
- **Photoelectric proximity sensors**

In the next few sections, we are going to discover each of these proximity sensors.

Inductive proximity sensor

This is a *non-contact sensor* that detects ferrous metals, such as *carbon steel, stainless steel*, and *cast iron*.

An inductive proximity sensor consists of a **coil, oscillator, detection circuit**, and **output**. The oscillator provides an *oscillating magnetic field* that radiates around the winding of the coil, which is located at the sensor's surface. When a ferrous metal gets near the magnetic field, a small current (eddy current) is induced on the metal's surface. This small current changes the natural frequency of the magnetic circuit, which, in turn, reduces the oscillating amplitude. The detection circuit monitors the amplitude of the oscillation and triggers an output from the *output circuitry* when the oscillation becomes reduced to a sufficient level.

If the sensor has a *NO* configuration, it will produce an *ON* signal when a metal enters the sensing region. With an *NC* configuration, its output is an *OFF* signal when the metal enters the sensing region. This output can be read by a programmable logic controller (PLC) or other form of controller that will convert the *ON* or *OFF* state into useable information.

The following diagram shows the main circuits in an inductive proximity sensor:

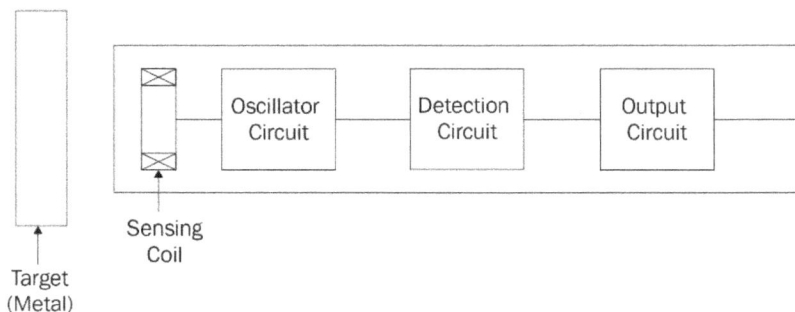

Figure 2.40 – Inductive proximity sensor

Now that we have explored what an inductive proximity sensor is, we will investigate the definition of a capacitive proximity sensor, which you should understand with ease due to the knowledge you have taken from this section. Let's dive right in.

Capacitive proximity sensor

This is a *non-contact sensor* that detects both metallic and non-metallic objects in the form of powder, granulate, liquids, and solids.

Capacitive sensors act like capacitors. They make use of the electrical properties of capacitance and the change of capacitance based on a change in the electrical field around the active face of the sensor.

A metal plate in the sensing face of the sensor acts as the first plate of a capacitor, and it's electrically connected to an internal oscillator circuit. The object to be sensed acts as the second plate of the capacitor.

The external capacitance between the target and the internal sensor plate forms a part of the feedback capacitance in the oscillator's circuit. As the target approaches the sensor's face, the oscillations increase until they reach a threshold level and activate the output.

The following diagram shows the main circuits in a capacitive proximity sensor:

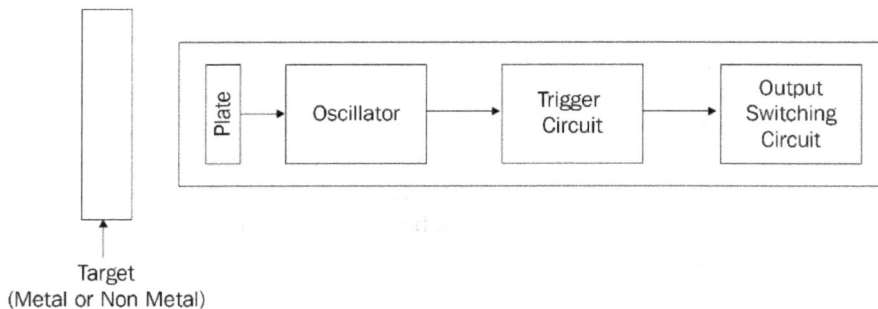

Figure 2.41 – Capacitive proximity sensor

The following figure shows a three-wire capacitive sensor. These three wires are as follows:

- **BN** (brown)
- **BK** (black)
- **BU** (blue)

Figure 2.42 – A capacitive proximity sensor (Omron-E2K-X8ME1)

The credit for this image goes to *Showlight Technologies LTD* www.showlight.com.ng.

The following wiring diagram is usually found on the body of a three-wire proximity sensor, which can exist as *NPN* or *PNP*. For NPN, the power supply will be connected to **BN** and **BU**, while the load will be connected across **BN** and **BK**. For PNP, the power supply will be connected to **BN** and **BU**, while the load will be connected across BN and BK:

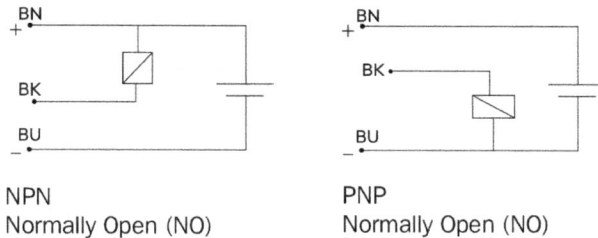

Figure 2.43 – Wiring diagram of a proximity sensor (NPN and PNP)

A simple wiring diagram of how to use a three-wire proximity sensor (NPN) to turn *on* or *off* a DC lamp is shown in the following diagram:

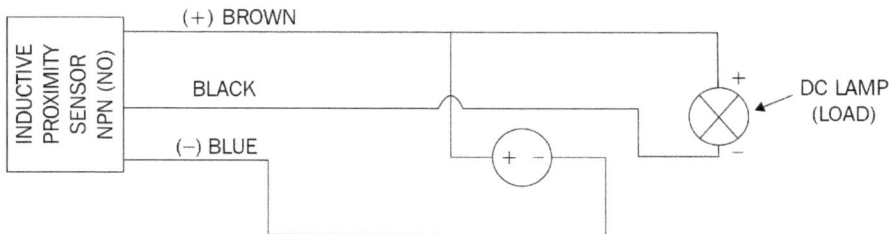

Figure 2.44 – Using a three-wire proximity sensor (NPN) to turn a DC lamp on or off

On the other hand, a simple wiring diagram of how to use a three-wire proximity sensor (PNP) to turn on or off a DC lamp is shown here:

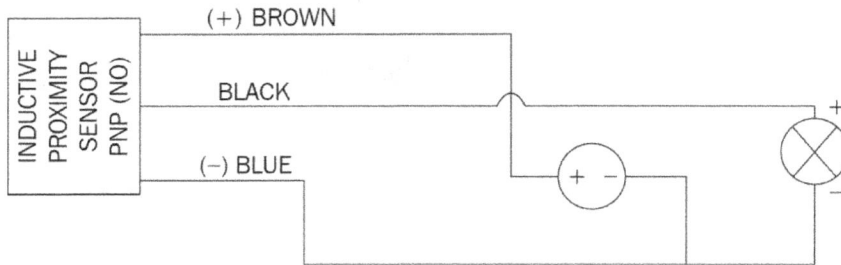

Figure 2.45 – Using a three-wire proximity sensor (PNP) to turn a DC lamp on or off

Now, let's look at the next proximity sensor.

Photoelectric

A **photoelectric** sensor consists of a light emitter (*sender*), photoreceiver, and supporting circuitry. They are used to detect the presence or absence of objects by using a light emitter and a receiver.

The emitter transmits a beam of either visible or invisible light to the detecting receiver.

There are three types of photoelectric sensors, as follows:

1. **Through beam**: With this sensor, the emitter (sender) and the receiver are in a separate housing and positioned to face each other. The sender provides a constant beam of light; detection occurs when an object passing in-between the sender and receiver breaks the beam. The following diagram illustrates this:

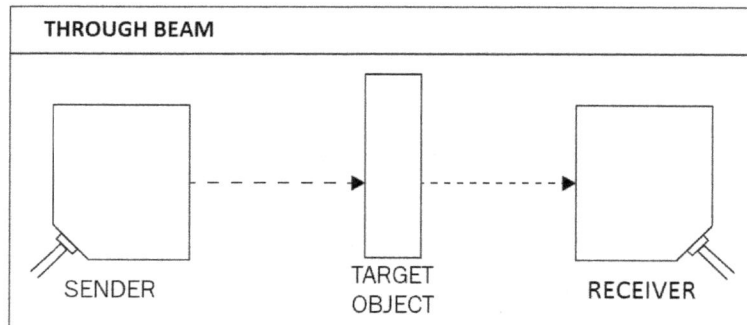

Figure 2.46 – Through beam photoelectric sensor

2. **Retro-reflective**: The operation of a retro-reflective sensor is like that of a *through beam*. Here, the sender and receiver are in the same housing facing the same direction, unlike the *through beam*. The emitter produces a light beam and projects it toward a reflector, which then deflects the beam back to the receiver. Detection occurs when the light path is disturbed or broken. Here's what this looks like:

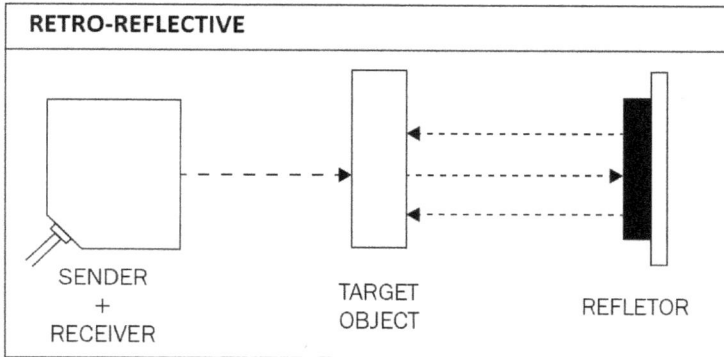

Figure 2.47 – Retro-reflective photoelectric sensor

3. **Diffused reflective**: With this sensor, the emitter and the receiver are in the same housing and facing the same direction. The emitter sends out a constant beam of light that diffuses in all directions, filling a detection area. When a target object enters the area, it deflects part of the beam back to the receiver. The target acts as the reflector. Detection occurs when light is reflected off the disturbance (or target) object. Let's have a look at this:

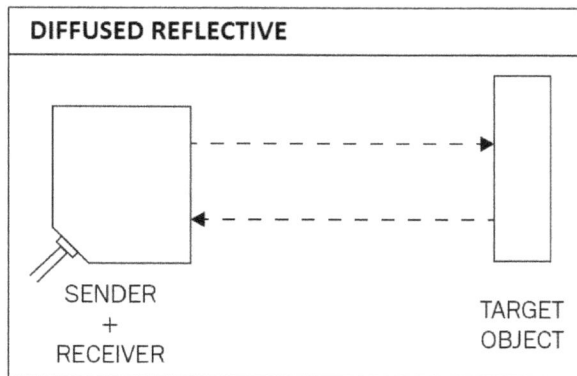

Figure 2.48 – Diffused photoelectric sensor

The following diagram shows the wiring of a retro-reflective photoelectric sensor. Retro-reflective proximity sensors are common in the industry today. They are good for long-distance detection:

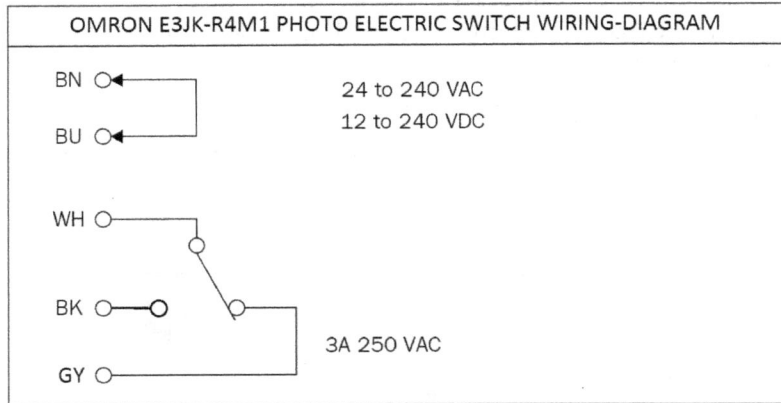

Figure 2.49 – Wiring diagram of a retro-reflective photoelectric sensor

In this wiring diagram, the sensor has five wires, as follows:

- **BN** (brown)
- **BU** (blue)
- **WH** (white)
- **BK** (black)
- **GY** (gray)

The BN and BU wires will be connected to the power. WH is common and it will be connected to either the positive or the negative, depending on what you require at the output terminal of the sensor (BK and GY).

When WH is connected to the positive, you will have the positive already at GY and when there is an obstruction, you will have the positive at BK.

When WH is connected to the negative, you will have the negative already at GY and when there is an obstruction, you will have the negative at BK.

The following diagram shows how to use the sensor to turn a DC lamp on or off automatically by connecting the common (WH) to the positive:

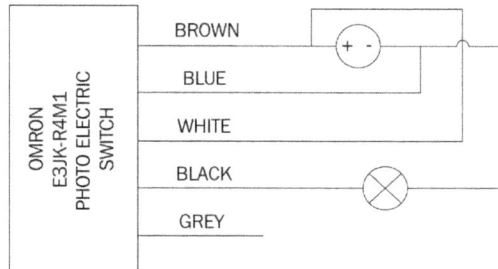

Figure 2.50 – A retro-reflective photoelectric sensor being used
to turn a DC lamp on or off (WH on positive)

In this section, you learned about some of the various proximity sensors that are used in the industry. The operation principles, symbols, and simple wirings that you learned about in this section will help you have a better understanding of industrial automation.

Summary

Congratulations on completing this chapter! You should now be able to explain switches and sensors, describe manually operated switches and their various types, describe mechanically operated switches and their various types, and describe proximity sensors and their various types. You should also be able to identify the symbols for various switches and understand how they are wired to perform control functions.

The topics that were covered in this chapter are relevant for automation engineers and will help you understand the next chapter.

Questions

The following are questions to test your understanding of chapter two. Ensure you have read and understood the topics in this chapter before attempting the questions.

1. _____ provide information about the environment for a controller to process.

2. Devices that detect the presence or absence of object are referred to as _____

3. In human being, eyes, nose, ear, tongue, and skin can be referred to as _____

4. Switches that are operated by hand are referred to as _____

5. A _____ switch is controlled automatically by factors such as pressure, position, or temperature.

6. _____is a type of switch that has only one input (usually referred to as common) and can connect to and switch between two different outputs.

7. _____ can detect the presence of liquids, powder, or granulated materials in a vessel.

8. A non-contact sensor that detects ferrous metals, such as carbon steel, stainless steel, and cast iron, is called _____

9. A _____ is a switch that opens or closes when temperature reaches a set-point or falls below a set point.

10. A switch that is operated by the motion of a machine part is called _____

11. The symbol in *Figure 2.51* represents _____

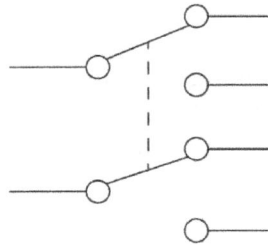

Figure 2.51

12. The symbol in *Figure 2.52* represents _____

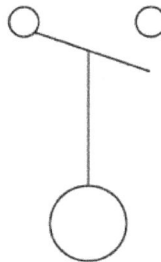

Figure 2.52

3

Actuators and Their Applications in Industrial Automation

In the previous chapter, we related the way humans function to automation. The eyes, touch, and smell were likened to switches and sensors while the controller was likened to the brain. In this chapter, we will be looking at actuators, which can be likened to the hands. In our bodily system, the hands do what the brain tells them to do. Similarly, in automation, the actuator does what the controller tells it to do. An actuator is simply a mover. It can move, carry, or perform other similar functions just like the hand. It requires a control signal from a controller to carry out its actions. The control signal is the result of the processing done by the controller through the program (a set of instructions) written into it. Industrial automation is not complete without an actuator.

This chapter will explain various actuators used in industrial automation. You will learn their basic principle of operation and about their application in industry. Actuators are one of the key components in industrial automation that need to be given enough attention. This chapter will provide relevant information that an automation engineer needs to know about actuators.

In this chapter, we are going to cover these main topics:

- Introducing actuators
- Learning about electrical actuators
- Exploring pneumatic actuators
- Getting to know about hydraulic actuators

Introducing actuators

An **actuator** is a device or component that produces an action or motion. It can be referred to as a **mover** because it moves to control a system. It usually requires a control signal from a controller and an energy source, which can be **electric**, **pneumatic** (air), or **hydraulic** (liquid). When it receives a control signal, it converts its own energy (electric, pneumatic, or hydraulic) to a form of energy or motion that can perform the required action.

One major application of an actuator in industry is the opening and closing of **valves**.

Manually operated valves use hand-wheels, manual gearboxes, knobs, and so on as their actuators for operation. The hand-wheel or knob can be turned clockwise or counterclockwise by humans to open or close the valve.

In situations where manual actuation of a valve to open or close it is not realistic, such as remote locations or on very large valves, a pneumatic, electric, or hydraulic actuator is required.

The actuator (pneumatic, electric, or hydraulic) provides the power or motion required to open or close the valve by converting its own energy to a mechanical force (or motion) when it receives a control signal from a controller. *Figure 3.16* shows a valve with a hand-wheel on the left side of the figure that can be operated manually and a valve on the right side of the image with a pneumatic actuator that can be operated automatically via a control signal.

Figure 3.1 shows the functional block diagram of an actuator:

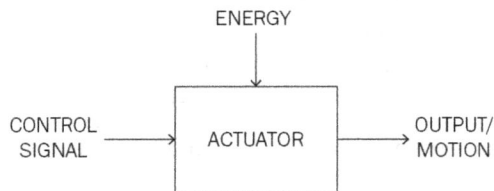

Figure 3.1 – Functional block diagram of an actuator

The motion produced by an actuator can either be linear or rotational. There are various types of actuators whose produced motion can either be linear or rotational.

There are basically three types of actuators:

- Electric actuators

- Pneumatic actuators

- Hydraulic actuators

We will look at what these three types of actuators are in the next topics of the chapter. We will learn about their basic construction as well as their application in industry. Having this knowledge will help in choosing the right actuator for your industrial automation needs.

Learning about electric actuators

Electric actuators are used in industry to convert electrical energy to kinetic energy to cause the movement of loads or perform actions that required motion or force. The operation of most electric actuators is based on the interaction between a magnetic field and a current carrying a conductor to create a *turning force*. Industrial fans, blowers, pumps, dampers, and so on are common applications of such an electric actuator. They utilize electric motors to perform the necessary action. Electric actuators are used in robotics and other assembly applications due to their accuracy, flexibility, and low operating cost. They are used in industrial machines that require circular motion. They are also used for clamping, pressing, cutting, stamping, and other forms of applications that require linear motion. An electric actuator can be referred to as a device that converts electrical energy into a linear or circular motion. Hence, electric actuators are of two types:

- **Electric linear actuator:** This is an actuator that will move in a linear direction. In this, rotational motion is first generated by an electric motor at high speed. It is then reduced by a gearbox to increase the torque, which will turn a leadscrew in a rotational motion, which in turn makes a drive nut move linearly. Limit switches are used to ensure the drive nut does not move beyond a certain point in the forward or reverse direction.

  Figure 3.2 shows the parts of a linear actuator:

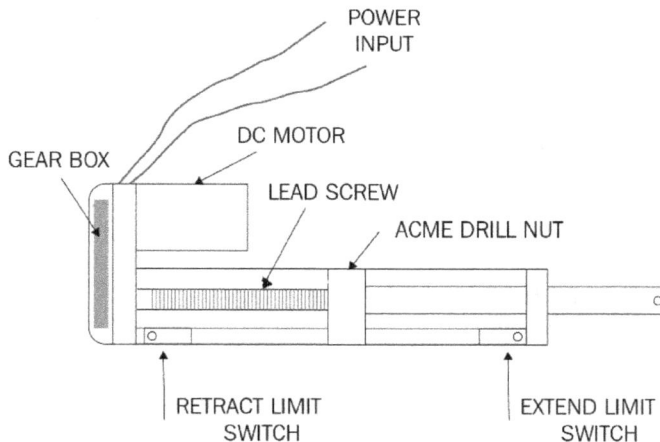

Figure 3.2 – Parts of an electric linear actuator

- **Electric rotary actuator:** Electric rotary actuators are commonly used in automation systems. Examples include **direct current (DC)** motors, AC motors, stepper motors, and servo motors.

We will look at examples of electric rotary actuators in the next sections.

Direct current motors

Direct current (DC) motors convert electrical energy (through DC) to mechanical energy (motion). A DC motor consists of a stator, an armature, split ring commutators, and brushes. The stator is the stationary part, which usually consists of permanent magnets, and the armature is the rotating part, consisting of a coil of wire. The windings of the coil are connected to split ring commutators. The split-ring commutators ensure the direction of the current in the coil is reversed each half turn.

When a current is passed to the coil through the brushes and split ring commutators, a rotating magnetic field is created that interacts with the differing fields of the magnets in the stator to create a turning force (*torque*), which causes the armature (coil) to rotate.

Figure 3.3 shows the internal structure of a simple DC motor.

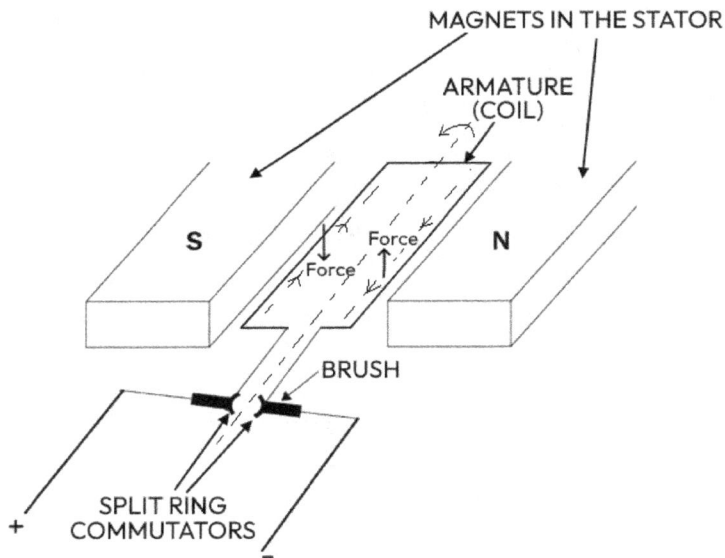

Figure 3.3 – Internal structure of a simple DC motor

We will go into further details of a DC motor in *Chapter 4, Overview Of AC and DC Motors*.

Alternating current motors

Alternating current (AC) motors convert electrical energy (through AC) to mechanical energy (motion). Unlike a DC motor, which makes use of DC as its energy source, an AC motor uses AC as its energy source. AC motors generally consist of two main parts, which are the stator and rotor.

The stator is a stationary part with coils supplied with AC. The flow of the AC in the stator coil creates a rotating magnetic field. The rotor is the rotating part and is attached to an output shaft. The rotor also produces its own rotating magnetic field through the induced current from the stator or other means. The interaction between the rotating magnetic field of the stator and the field of the rotor creates a turning force that causes rotation. An AC motor can be single-phase or three-phase. A single-phase AC motor uses a single-phase supply (live and neutral) while a three-phase AC motor uses a three-phase supply (L1, L2, and L3). *Chapter 4, Overview of AC and DC Motors*, contains more information on AC motors.

The following figure shows an **AC induction motor**:

Figure 3.4 – AC induction motor

The credit for this image goes to *Wikimedia Commons* and can be found at the following link: https://creativecommons.org/licenses/by-sa/4.0/deed.en

Stepper motor

A **stepper motor** is a type of DC motor that rotates in steps. It does not turn arbitrarily when its terminal is connected to a source, rather it turns in steps. It also consists of a stator and a rotor. The stator consists of windings (coils) while the rotor is usually made of a permanent magnet. Activating the windings of the stator step by step in a particular order will create a turning force as the electromagnetic poles of the stator winding react with the poles of the permanent magnet of the rotor.

The following figure shows a stepper motor:

Figure 3.5 – Stepper motor

The credit for this image goes to *Wikimedia Commons* and can be found at the following link: `https://creativecommons.org/licenses/by-sa/4.0/deed.en`

Servo motor

A **servo motor** is a rotary actuator that can rotate with high efficiency and great precision. It is good for rotating objects at some specific angle or distance. They are commonly used in robotics and automation technology. They consist of a control circuit that provides feedback on the current position of the shaft. The feedback signal from the control circuit enables the motor to control the rotational or linear speed and position. Servo motor types include DC servo motors and AC servo motors, amongst other types. A **DC servo motor** is powered by a DC supply. Common DC servo motors used in toys and other robotics-related products consist of a DC motor, gears, a potentiometer, and a control circuit that ensures it turns at a specific angle via a control signal.

The following diagram shows the main parts of a common DC servo motor:

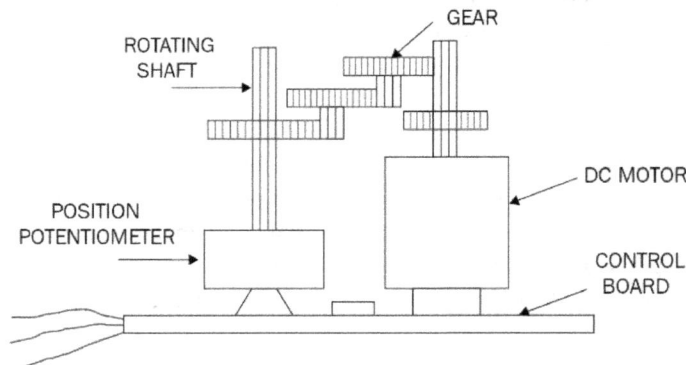

Figure 3.6 – Internal parts of a DC servo motor

An **AC servo motor** is powered by an AC supply. It has an encoder and is used with controllers to provide feedback and close-loop control. The following diagram shows some of the parts of an AC servo motor:

Figure 3.7 – AC Servo motor parts

Other forms of electrical actuators include solenoid, electro-mechanical relay, and contactor, among others. Let's have a look at what each of them is in the next sections.

A **solenoid** can be referred to as a device that converts electrical energy into mechanical energy (motion). It consists of a coil of wire, plunger, and spring. The coil is formed in such a way that the plunger can move in and out of the coil's body. When a current passes through the coil, a magnetic field is created and is used to create a linear or angular motion depending on the type. The spring returns the plunger back to its original position when power is removed. In *Figure 3.8*, when current flows through the coil, an electromagnetic field is generated, which pulls the plunger inward. The spring pushes it to its original position when power is removed.

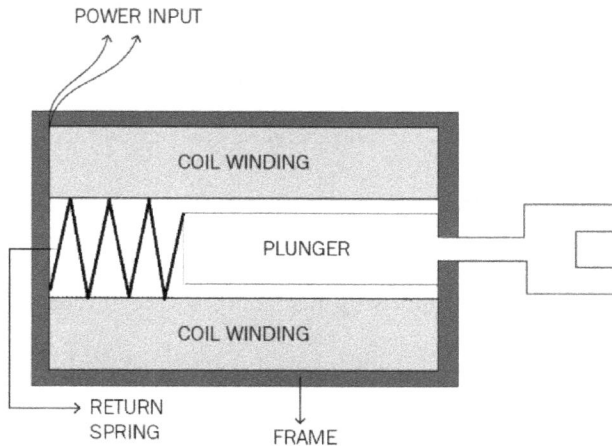

Figure 3.8 – Parts of solenoid

A solenoid is used in an automated system to create *motion*. The system can be applied to open or close a valve as shown in the solenoid valve in the following diagram:

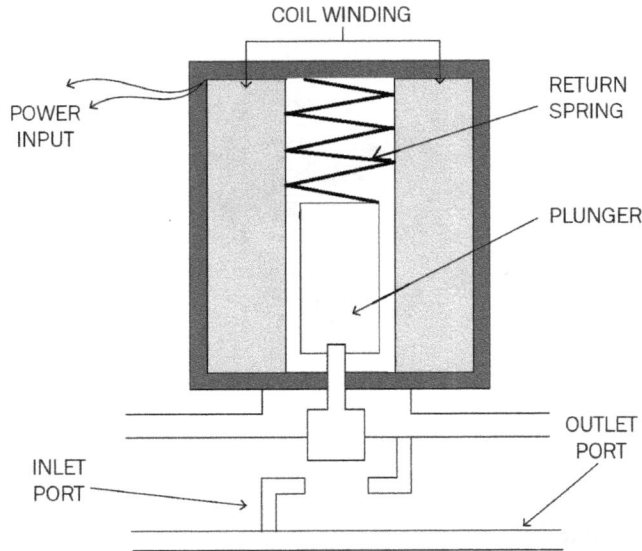

Figure 3.9 – Solenoid valve

In *Figure 3.9*, when current flows through the coil, the magnetic field of the energized coil pulls the plunger to open the valve. When the power is removed, the coil becomes de-energized and the spring forces the plunger to close the valve. Solenoid valves are useful for controlling the flow of compressed air into a *pneumatic cylinder* or the flow of hydraulic fluid into a *hydraulic cylinder*. Other applications of solenoid valves include a door locking system that offers secure door closure and a pusher on a conveyor belt diverting an object or package to a specific location.

Electro-mechanical relay

An **electro-mechanical relay** is an electrically operated switch that can open or close a circuit with the physical movement of a contact. It can be referred to as a switch that is operated electrically, unlike the manual or mechanically operated switch explained in *Chapter 2, Switches and Sensors – Working Principles, Applications, and Wiring*. The following diagram shows the internal structure of a relay:

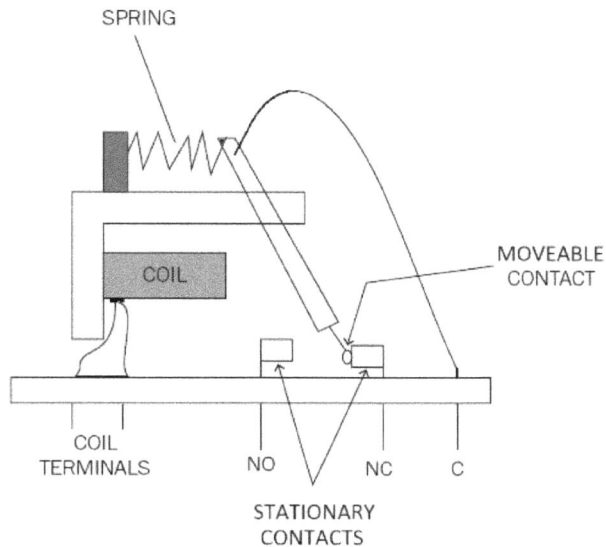

Figure 3.10 – Internal structure of a relay

It consists of a **coil** (electromagnet), a **moveable contact**, **stationary contacts**, and a **spring**.

When the relay coil is energized by allowing a current to flow through the terminal, the moveable **contact** (**C**) will be attracted towards the **normally open** (**NO**) of the stationary contact, making them contact each other. When the power is removed, the moveable contact returns to the **normally closed** (**NC**) position. The moveable contact and the fixed contact can act as a switch for equipment that needs power, as shown in the following diagram:

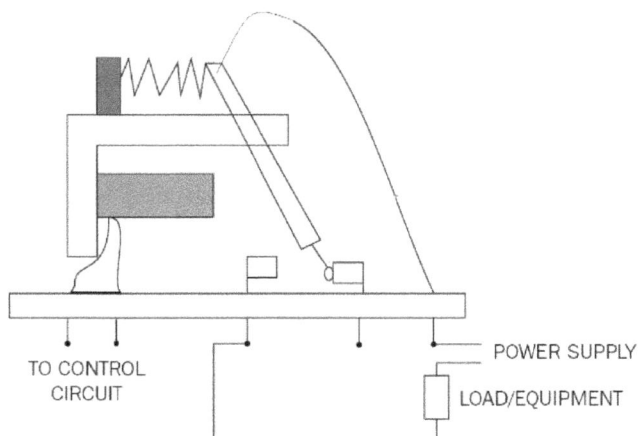

Figure 3.11 – Relay acting as a switch

In *Figure 3.11*, when there is a supply to the coil from a control circuit, the coil becomes energized, the moveable contact gets connected to the **NO** of the stationary contact, and the load turns on. When there is no more supply from the control circuit, the coil becomes de-energized, the moveable contact returns to the **NC** contact of the stationary contacts, and the load turns off.

Relays are of various configurations, just like a switch. The various types include SPST, SPDT, DPST, and DPDT. The relay that I explained is a **single pole double throw** relay. Refer to *Chapter 2, Switches and Sensors – Working Principles, Applications, and Wiring*, to have a good understanding of *SPST, SPDT, DPST*, and *DPDT*.

The following diagram shows the symbol of various relay configuration:

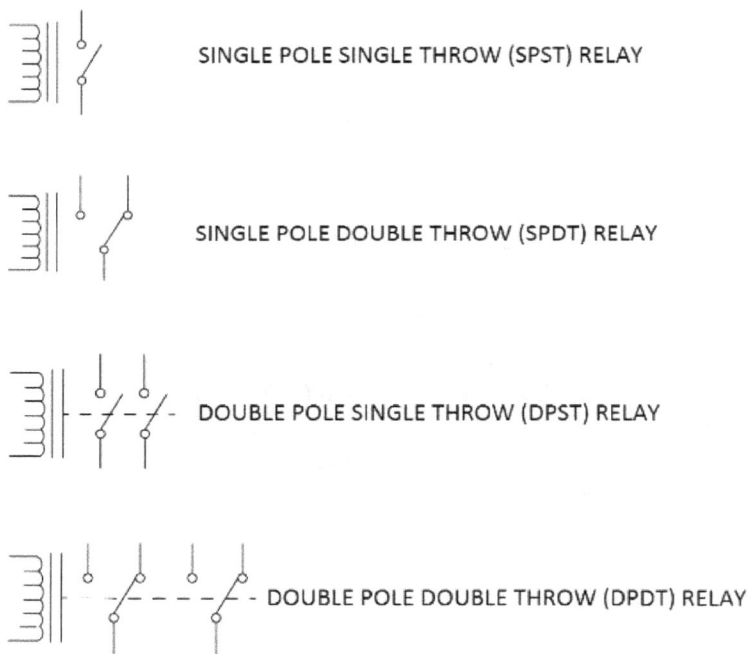

SINGLE POLE SINGLE THROW (SPST) RELAY

SINGLE POLE DOUBLE THROW (SPDT) RELAY

DOUBLE POLE SINGLE THROW (DPST) RELAY

DOUBLE POLE DOUBLE THROW (DPDT) RELAY

Figure 3.12 – Relay configuration

The following figure shows a single pole double throw relay with the cover removed:

Figure 3.13 – Single-pole-double-throw relay with the cover removed

The credit for this image goes to *Wikimedia Commons* and can be found at the following link:

`https://creativecommons.org/licenses/by-sa/4.0/deed.en`

The **C** can move between **NO** and **NC**. When the coil is not energized, the **C** is connected to the **NC** contact and when the coil is energized, the **C** returns to the **NC** contact.

Electromechanical relays are used where high power and high current are required in an automation system.

Contactor

A **contactor** is an electrically operated switch just like a relay. It is a switch that can be operated electrically unlike manual and mechanically operated switches. It usually consists of at least one coil, and up to three NO contacts. In some instances, an auxiliary contact that operates with the main contact is included. The terminals of the coil are usually labeled A1 and A2. When a current is supplied to the coil through terminals A1 and A2, it becomes energized and the magnetic field forces the contacts to close simultaneously. When the power is removed, the contacts become open.

Relays and contactors have similar basic operations and functions. The main difference is their current handling capacity. Relays can switch smaller current loads of up to about 20 A or more while a contactor can handle larger current loads of up to 12,500 A.

The following diagram shows the symbol of a contactor:

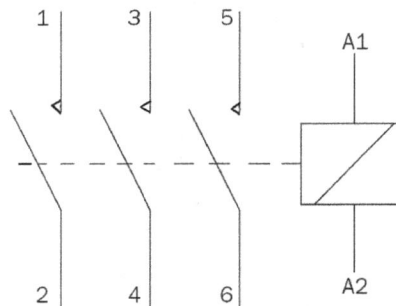

Figure 3.14 – Symbol of a contactor

And if you're curious, the following figure shows what a contactor looks like:

Figure 3.15 – Contactor

The credit for this image goes to *Wikimedia Commons* and can be found at the following link: `https://creativecommons.org/licenses/by-sa/4.0/deed.en`

In this section, you have learned what actuators are and the various categories of actuators in industrial automation, which include electrical, pneumatic, and hydraulic actuators. This section has covered an electrical actuator and you will have gained a well-rounded understanding of the topic. The next section will focus on what a *pneumatic actuator* is.

Exploring pneumatic actuators

Pneumatic actuators use pressurized air or gas as an energy or power source to produce rotary or linear motion. They are very reliable, efficient, and safe for use in an environment where electricity may cause a fire hazard. They have various industrial applications, which include the regular opening and closing of valves, pick and place handlers, and so on. A pneumatic actuator can also be referred to as a **pneumatic cylinder** or **air cylinder**.

In a pneumatic actuator, compressed air or pressurized gas enters a chamber, and the gas builds up pressure in contrast to the outside atmospheric pressure, which results in the motion of a device, which could be a piston or gear. The motion created by a pneumatic actuator can be *linear* or *circular*. Hence, a pneumatic actuator simply converts the energy in compressed air into a linear or circular motion.

Figure 3.16 shows a valve on the left side of the figure that can be operated manually and a valve on the right side of the figure with a pneumatic actuator that can be operated automatically via a *control signal*. The pneumatic actuator converts its own energy into a motion that opens or closes the valve automatically via a control signal.

Figure 3.16 – A gate valve with a hand-wheel (left) and a linear pneumatic actuator (right)

The credit for this image goes to *Wikimedia Commons* and can be found at the following link: `https://creativecommons.org/licenses/by-sa/4.0/deed.en`

There are basically two types of pneumatic actuators:

- Pneumatic linear actuator
- Pneumatic rotary actuator

In the next sections, we will discover what they are.

Pneumatic linear actuator

In a **pneumatic linear actuator**, compressed air entering a chamber or cylinder builds up pressure and can cause a piston to move forward or backward in a linear motion when the pressure built up in the chamber contrasts with the outside atmospheric pressure.

Pneumatic linear actuators come in two types:

- Single-acting cylinders
- Double-acting cylinders

In a **single-acting cylinder**, there is only one port for air to flow into the cylinder. When air enters through this port, the pressure increases and pushes a piston forward for the push type or backward for the pull type. A spring that is positioned in or outside the cylinder returns the piston to its original position to make it ready for another burst of pressure.

The following diagram shows the symbol of a single-acting cylinder:

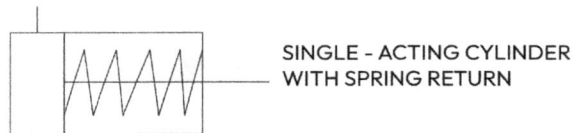

SINGLE - ACTING CYLINDER
WITH SPRING RETURN

Figure 3.17 – Symbol of a single-acting cylinder with spring return

Figure 3.18 shows a practical application of a single-acting cylinder (also known as a **pneumatic hammer** or **jack drill**). It's a pneumatic tool that combines a hammer and a chisel and is used to break apart rocks, asphalt, and concrete.

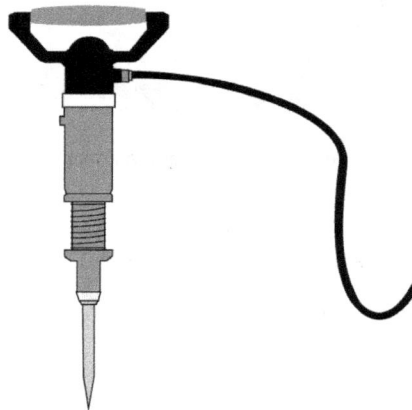

Figure 3.18 – Pneumatic drill or jack hammer

In a **double-action cylinder**, there are two ports, one at each end of the cylinder. The first application of air through one of the ports pushes the piston forward while the second application of air through the other port pushes the piston back into its original position. This does not require a spring since air can be applied at each end to make the piston move forward or backward.

The following diagram shows the symbol of a double-acting cylinder:

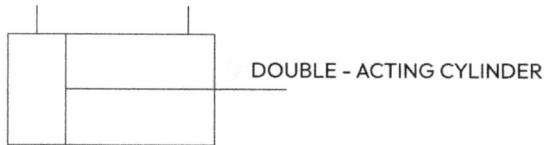

DOUBLE - ACTING CYLINDER

Figure 3.19 – Symbol of a double-acting cylinder

The following figure shows a double-acting cylinder:

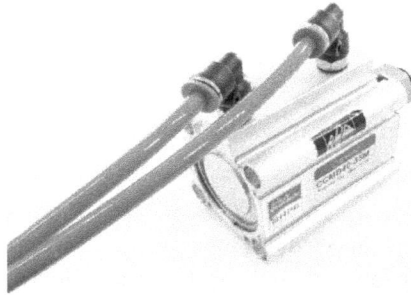

Figure 3.20 – Double-acting cylinder

The credit for this image goes to *Wikimedia Commons* and can be found at the following link: https://creativecommons.org/licenses/by-sa/4.0/deed.en

Now let's move on to the next type of pneumatic actuator.

Pneumatic rotary actuator

A **pneumatic rotary actuator** rotates rather than moving in a straight line. Hence the name *rotary* is given to it. They can be designed to mount onto a valve that requires the rotation of their stem to open or close the valve remotely or automatically via a *control signal*.

It consists of an air chamber on one or both sides of a piston or *diaphragm*. An increase in the air pressure pushes the piston to create a linear or *circular* motion. If the motion created is linear, gears or a cam (a machine component used to convert rotary motion into linear motion or vice versa) is used to convert the linear motion to circular motion internally.

The following figure gives you an idea of what a pneumatic rotary actuator looks like if you are curious:

Figure 3.21 – Pneumatic rotary actuator

There are three types of pneumatic rotary actuators. Let's explore them in the next sections.

Scotch yoke

In a **scotch yoke** actuator, a piston is connected to a rotary shaft. An increase in the air pressure moves the piston linearly, which rotates the rotary shaft through a cam. A double-acting type will require air pressure on the other side to cause rotation in the opposite direction.

The following diagram shows the parts of a scotch yoke pneumatic rotary actuator:

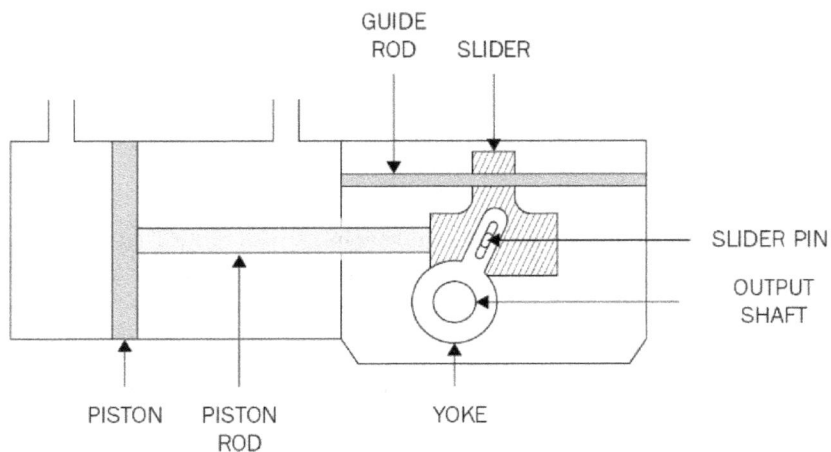

Figure 3.22 – Scotch yoke pneumatic rotary actuator

Vane

A **vane** is a type of rotary actuator that doesn't use a piston that will move linearly before converting to circular motion internally. It creates circular motion directly from compressed air through a movable vane in the air chamber.

The following diagram shows the parts of a pneumatic vane actuator:

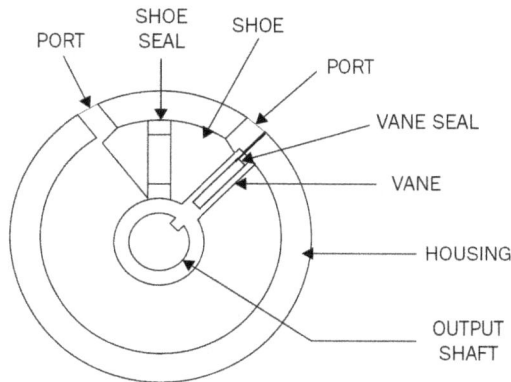

Figure 3.23 – Vane pneumatic rotary actuator

Rack and pinion

In a **rack-and-pinion** actuator, the piston moves a rack past a gear pinion. This action rotates the pinion to produce a circular motion at the output shaft. The following diagram shows the parts of a rack and pinion pneumatic rotary actuator:

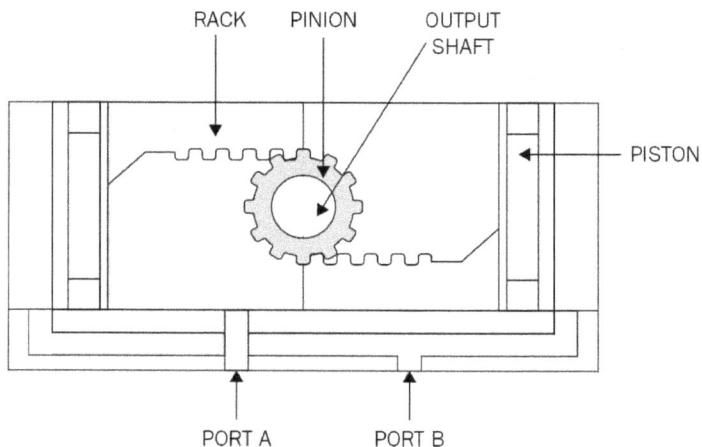

Figure 3.24 – Rack-and-pinion pneumatic rotary actuator

In *Figure 3.24*, when the pressurized air enters the middle chamber through **Port A**, the two pistons are separated and will move to the two ends of the cylinder. The air in the air chambers at each end is discharged via **Port B**. The rack of the two pistons drives the shaft simultaneously and rotates it counterclockwise. When pressurized air enters the air chamber at both ends of the piston through **Port B**, the two pistons are shifted inward. The rack of the two pistons drives the shaft simultaneously and rotates it clockwise.

In this section, you have discovered how compressed air can be used to cause motion using pneumatic actuators. You have also learned about the types of pneumatic actuators and their application in industry. Knowledge of pneumatic actuators will help you have a quick understanding of *hydraulic actuators*, which is covered in the next section.

Getting to know about hydraulic actuators

Hydraulic actuators use pressurized hydraulic fluid (that is, oil) as an energy source to produce rotary or linear motion for an application consisting of high force and ruggedness. A hydraulic actuator can also be referred to as a *hydraulic cylinder*. It converts hydraulic energy (that is, the energy of fluid) into motion. It also consists of a cylinder in which a piston connected to a piston rod moves back and forth by pumping hydraulic fluid into a port at one end of the piston or at the other end.

Similar to a pneumatic actuator, a hydraulic actuator has two types:

- Hydraulic linear actuator
- Hydraulic rotary actuator

The principle of operation of each type is similar to the ones explained about pneumatic actuators in the previous section of this chapter, only that hydraulic actuators utilize hydraulic fluid (that is, oil) unlike compressed air or gas used in pneumatic systems.

The following diagram shows a single-acting hydraulic linear actuator:

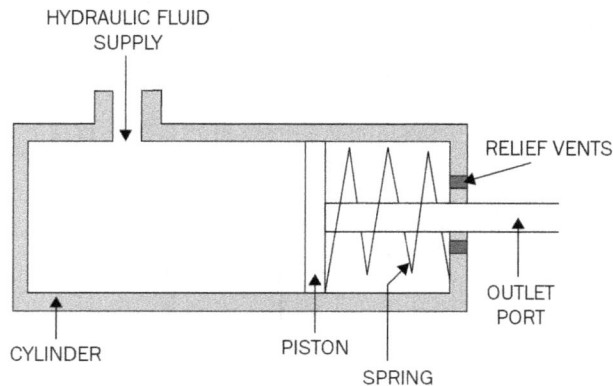

Figure 3.25 – Hydraulic linear actuator

Figure 3.25 demonstrates the various parts of a single-acting hydraulic actuator. It consists of a **cylinder**, **piston**, a piston rod, a spring, and so on. It has only one port for fluid (**hydraulic fluid supply**) to flow into the cylinder. When the hydraulic fluid enters through this port, the pressure increases and pushes a piston forward for a **push type** or backward for a **pull type**. A spring that is positioned in or outside the cylinder returns the piston to its original position to make it ready for another burst of pressure.

A hydraulic actuator can also be employed to open or close a valve as shown here:

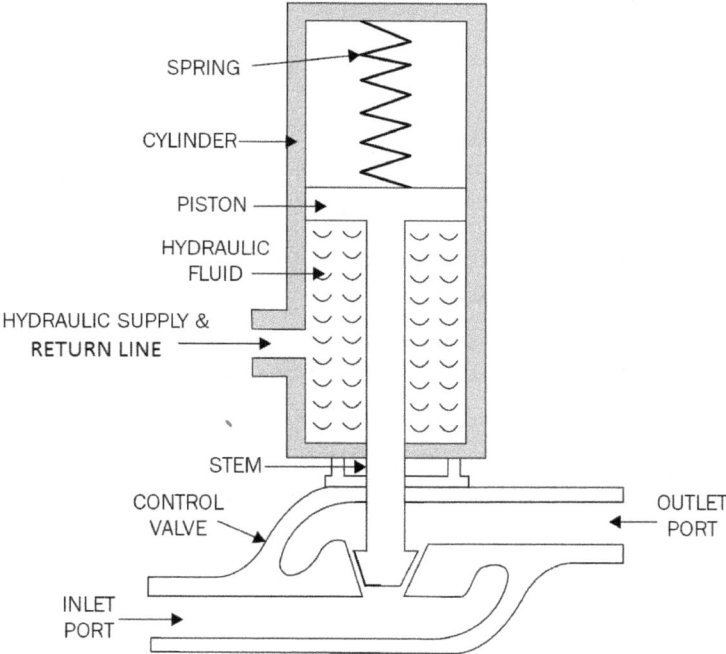

Figure 3.26 – Hydraulic valve

In this section, you have learned about the concept of using pressurized fluid to create motion. You have also discovered that the operation of a hydraulic actuator is like a pneumatic actuator. The major difference is the kind of fluid used. Pneumatic actuators use compressed air or gas while hydraulic actuators use hydraulic fluid and are used for high-force and rugged applications.

Summary

You have successfully completed this chapter of the book. Well done! With having reached the end of this chapter, you should now be able to explain actuators and the basic types available (electric actuators, pneumatic actuators, and hydraulic actuators).

Electric actuators use electricity, pneumatic actuators use compressed air, while hydraulic actuators use pressurized fluid (oil). They all perform actions through motion, which can either be linear or rotational. Some actuators are better suited for a particular application than others. For instance, a hydraulic actuator is useful for tasks requiring high force. It is good to have knowledge of all three types mentioned in this book. You might find any of the types in the industry you happen to find yourself working in as an automation engineer.

The topics touched on in this chapter are relevant for automation engineers and will help you to better understand the next chapter in this book.

Questions

The following are questions to test your understanding of this chapter. Ensure you have read and understood the topics in this chapter before attempting the questions.

1. A device that converts electrical energy into linear or circular motion is called_____.

2. A DC motor that rotates in steps is called _____.

3. _____ utilize pressurized air or gas as an energy or power source to produce rotary or linear motion.

4. The two main parts of an AC motor are _____, and _____.

5. _____ use pressurized hydraulic fluid (oil) as an energy source to produce rotary or linear motion for high-force and rugged applications.

6. The symbol in *Figure 3.27* represents _____.

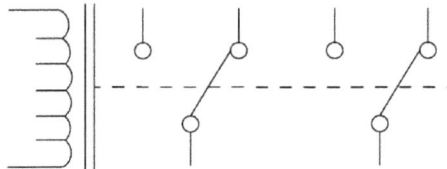

Figure 3.27

7. The symbol in *Figure 3.28* represents _____.

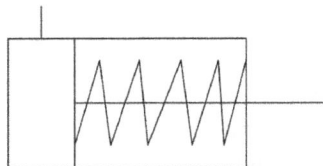

Figure 3.28

8.   The symbol in *Figure 3.29* represents _____.

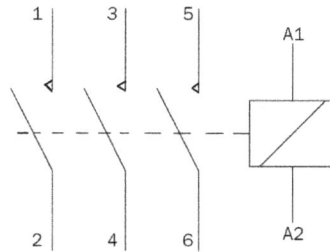

Figure 3.29

4

Overview of AC and DC Motors

Electric motors are generally used to convert electrical energy into mechanical energy. Mechanical energy can be used to move an electric car, rotate a fan, and so on. An electric motor can be referred to as an actuator that converts electricity into motion. Hence, it is an electric actuator. Electric motors are very common in our homes. You will find them in most appliances, such as ceiling fans, standing fans, washing machines, microwaves, electric blenders, electric can openers, toys, and so on.

In industry, they are widely used for pumps, blowers, mixers, agitators, conveyors, and so on. The use of electric motors in various equipment and machines in homes and industry has proven them to be among the most important electrical inventions of all time. Electric motors may be classified by their source of power, construction method, application, and the types of motion they produce. The following block diagram shows a simple classification of electric motors:

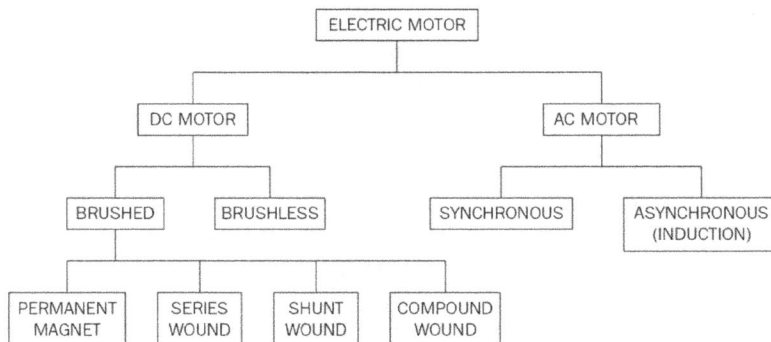

Figure 4.1 – Block diagram showing basic types of electric motors

This chapter will explain various electric motors, such as AC motors, DC motors, servo motors, and stepper motors, used in industrial automation. You will learn about the basic types of electric motors, the differences between synchronous and asynchronous motors, and the wiring of asynchronous (induction) motors in a star or delta connection. The chapter also covers the differences between star and delta connections and direct online starters, including components such as a circuit breaker, contactor, and overload relay. Electric motors play a big role in industry. It is therefore necessary for automation engineers to have a good knowledge of them.

In this chapter, we are going to cover the following main topics:

- Understanding AC motors
- Understanding DC motors
- Understanding stepper motors and servo motors

Understanding AC motors

An AC motor is a type of electric motor that is powered by **alternating current** (**AC**) supplied from a power grid or other AC electricity source. Check out *Chapter 3, Actuators and Their Applications in Industrial Automation*, to learn about the basic construction of an AC motor. Here, we will be looking at the various types of AC motors, which include the following:

- Synchronous AC motors
- Asynchronous AC motors (induction motors)

Let's start by understanding the synchronous AC motor.

Synchronous AC motors

In a synchronous motor, the stator winding (winding fitted in the stator) is usually connected to a three-phase AC supply. The AC supplied to the stator establishes a rotating magnetic field in the stator and it rotates at synchronous speed. The rotor is supplied with **direct current** (**DC**) to make it act as a permanent magnet. Sometimes, the rotor can be made of permanent magnets. Supplying DC to the rotor or making it a permanent magnet makes its magnetic field stationary (that is, the north and south poles are fixed).

As the magnetic field of the stator rotates, its poles attract the opposite poles of the rotor and this creates a turning force that causes the rotor to rotate at the same speed as that of the rotating magnetic field (synchronous speed). The term synchronous motor means that the speed of the rotor of the motor is the same as the speed of the rotating magnetic field of the stator.

The synchronous speed of an AC motor is the speed of the rotating magnetic field. It is determined by the frequency of the voltage source and the number of poles, as shown in the following formula:

Synchronous speed = 120f/P

Where *f* = frequency in Hz and *P* = number of poles.

Synchronous motors can be used for applying constant speed from no load to full load. They are good for use in high-precision applications in industry and are also well suited for robotic actuators.

The following diagram shows a synchronous motor:

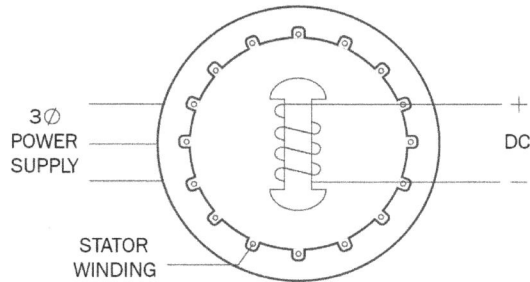

Figure 4.2 – Synchronous motor

Next, let's check how an asynchronous motor works.

Asynchronous AC motors (induction motors)

In an asynchronous motor, the stator winding is usually connected to a three-phase AC supply. The AC supplied to the stator establishes a rotating magnetic field in the stator and it rotates at synchronous speed. The rotor is not directly connected to a supply (that is, there is no electrical connection to the rotor). The rotor current required to produce a turning force is obtained by the electromagnetic induction from the magnetic field of the stator according to Faraday's law of electromagnetic induction, which states that a current will be induced in a conductor placed in a changing magnetic field.

An induction motor can simply be referred to as a motor that works on the principle of electromagnetic induction. **Electromagnetic induction** is the phenomenon in which current is induced in a conductor placed in a changing magnetic field according to Faraday. The following diagram shows a current induced in coil 2 due to a changing magnetic field caused by the AC supply in coil 1. Coil 2 is placed in the changing magnetic field. Hence, current will be induced in the coil (coil 2) according to Faraday:

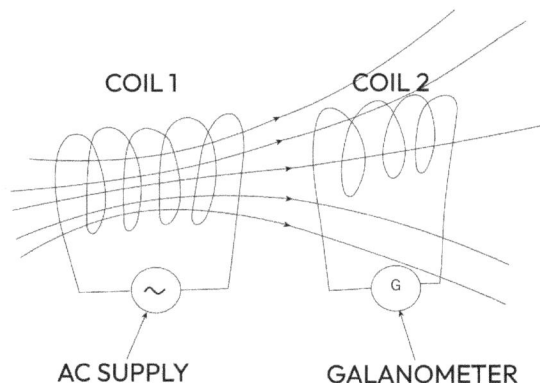

Figure 4.3 – Current induced in coil 2 due to a changing magnetic field caused by the AC supply in coil 1

The interaction between the rotating magnetic field of the stator and the magnetic field of the rotor creates a turning force that causes the rotor to rotate, but the rotor never runs at the same speed as that of the rotating magnetic field of the stator. An induction motor usually operates more slowly than synchronous speed (the speed of the rotating magnetic field of the stator) since rotation at synchronous speed would result in no induced current in the rotor.

The difference between the rotor's speed and the speed of the rotating magnetic field of the stator is referred to as slip. One major fact about induction motors is that the rotor current is created by induction instead of being separately excited as in synchronous motors or being self-magnetized using a permanent magnet.

The following diagram shows a three-phase asynchronous (induction) motor:

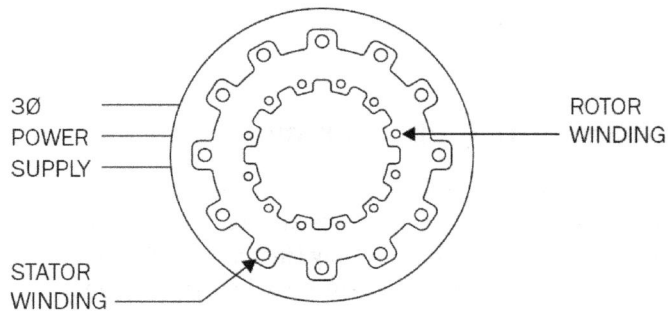

Figure 4.4 – Three-phase asynchronous (induction) motor

Depending upon the power supply input, there are basically two types of induction motor, which include single-phase and three-phase. Three-phase induction motors are more commonly used in manufacturing and other industries. They have a high starting torque, low costs, low maintenance costs, and are durable. They can be used in conveyors, pumps, blowers, mixers, lifts, crushers, and so on.

We will now look at the stator windings of a three-phase motor and its terminals.

A three-phase motor has six terminals and one earth and can be connected either in star or delta. The six terminals are as follows:

- U1 (start 1)
- U2 (end 1)
- V1 (start 2)
- V2 (end 2)
- W1 (start 3)
- W2 (end 3)

The following diagram shows the internal wiring of the stator winding and the six terminals:

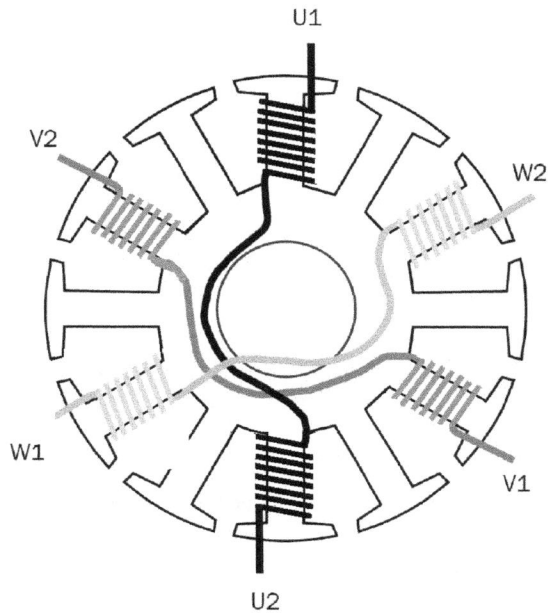

Figure 4.5 – Internal wiring of the stator winding and the six terminals

A motor terminal block with six terminals is shown in the following diagram:

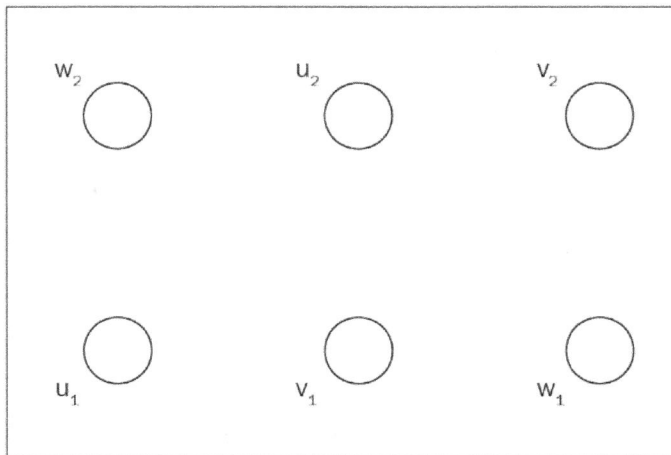

Figure 4.6 – Three-phase motor terminal block showing the six terminals

The following diagram shows a three-phase motor wiring in star connection (left) and delta connection (right):

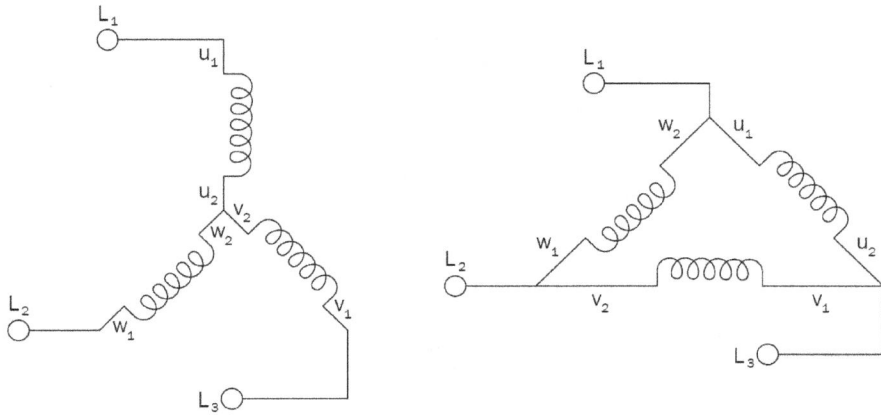

Figure 4.7 – Three-phase motor wiring in star connection (left) and delta connection (right)

The following diagram shows a three phase motor wiring showing terminal block and jumper bars in star connection (left) and delta connection (right):

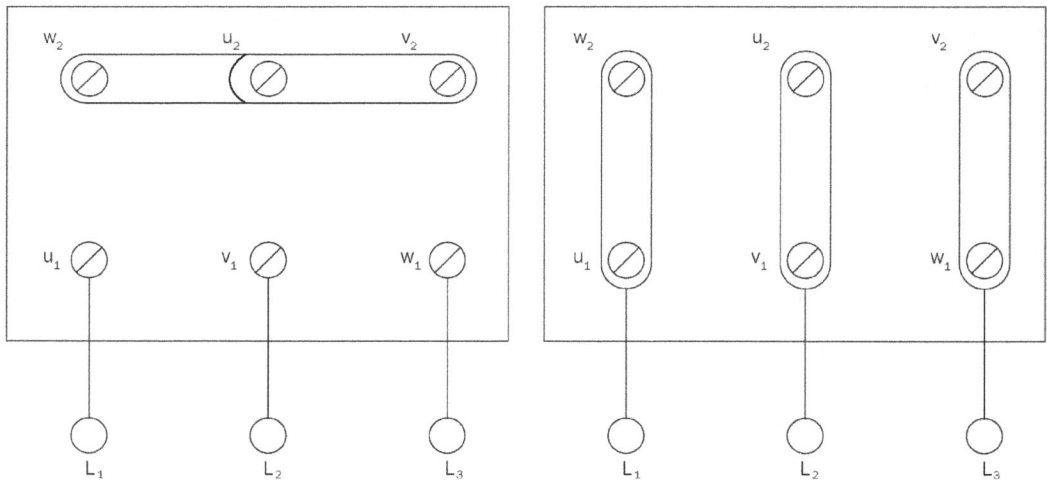

Figure 4.8 – Three-phase motor wiring showing terminal block and jumper bars in star connection (left) and delta connection (right)

The following figure shows a three-phase motor in star connection:

Figure 4.9 – A three-phase motor connection in star

The credit for this photo goes to *Showlight Technologies Ltd:* www.showlight.com.ng.

In the preceding figure, it can be seen that W2, U2, and V2 are connected together via jumper bars and the three phase lines (L1, L2, and L3) are connected to U1, V1, and W1 respectively, making it a star connection.

The jumper bars can be rearranged to connect W2 and U1, U2 and V1, and V2 and W1 while still connecting the three phase lines (L1, L2, and L3) to U1, V1, and W1 respectively, making it a delta connection.

Differences between star and delta connections

In this section, let's look at the differences between star and delta connections:

- In a star connection, similar points (start or end points) of three coils (that is, U1, V1, and W1 or U2, V2, and W2) are connected together to form a neutral point and three wires run from the remaining three terminals. In delta, the end point of each coil is connected to the starting point of another coil to form a closed loop and three wires are taken out from the coil joints.

- A star connection has a neutral point while a delta connection has no neutral point.

- In a star connection, the phase voltage is $1/\sqrt{3}$ line voltage. In delta, phase voltage is equal to line voltage.

- The speed of a star-connected motor is slow because it receives $1/\sqrt{3}$ of line voltage while the speed of a delta-connected motor is fast because each phase gets the total line voltage.

- A star connection is commonly used in power transmission while a delta connection is common in power distribution and industries.

Let's have a look at **direct on line (DOL)** starters.

Direct on line (DOL) starters

A DOL starter is a method for starting a three-phase induction motor. It can also be referred to as an across the line starter. It is the simplest type of starter, which connects the three-phase motor directly to the power supply through a contactor. A DOL starter consists of a circuit breaker, contactor, overload relay, and push buttons (normally open (NO) and normally closed (NC) push buttons). Let's take a look at some of these in detail in the following sections.

Circuit breaker

This is a switching device that can automatically interrupt current flow whenever a fault occurs. It protects a circuit from damage caused by overcurrent/overload or a short circuit. It can also be operated manually to supply or disconnect power from a circuit. It connects the power from a source to the next device (contactor) in a direct online starter.

The following figure shows a photo of a two-pole circuit breaker mounted on a **Deutsches Institut für Normung** (**DIN**) rail, which is the German institute for standardization. A DOL starter for a three-phase motor will use a three-pole circuit breaker:

Figure 4.10 – Image of a two-pole circuit breaker mounted on a DIN rail

The credit for this photo goes to *Wikimedia Commons* and can be found at the following link: https://creativecommons.org/licenses/by-sa/4.0/deed.en.

Contactor

This is an electrical switching device consisting of main contacts, auxiliary contacts, and a coil (A1 and A2). It provides operating power to a load through its main contacts when the coil terminals (A1 and A2) are energized. Contactors are mostly used for controlling electric motors. The following figure shows a contactor:

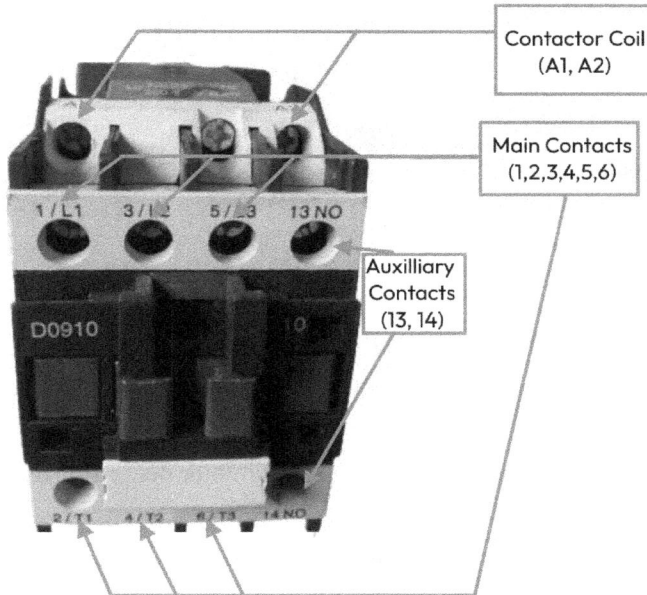

Figure 4.11 – A contactor

The credit for this photo goes to *Showlight Technologies Ltd:* www.showlight.com.ng.

Overload relay

This is a type of relay that opens a circuit when the load draws a current that exceeds a preset value. There is a provision for setting the preset value on the overload relay. The overload relay does not open instantaneously since the motor can draw up to 800% of its full load when starting. Overload relays are set to trip when the current drawn by the motor exceeds 115% to 125% of the full load current of the motor. The setting of the overload depends on the properties of the motor that is to be protected.

The following figure shows an overload relay:

Figure 4.12 – An overload relay

The credit for this photo goes to *Showlight Technologies Ltd:* www.showlight.com.ng.

Trip Control (95-96) is an NC contact. It opens when the motor draws a current that exceeds the preset value.

Trip Indicator (97-98) is an NO contact. It closes when the motor draws a current that exceeds the preset value.

Current Limit Setting can be used to set the current. If the current rises above the set value (limit) over a certain period of time, the overload relay will trip, and an auxiliary contact (95-96) will become open. DOL is wired such that the NC auxiliary contact interrupts the motor control circuit, de-energizing the contactor coil.

Overload relay is usually connected to a contactor as shown in the following figure:

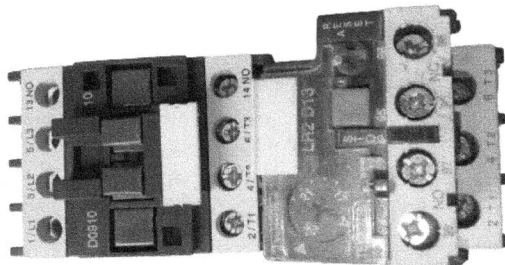

Figure 4.13 – Overload relay and a contactor

The credit for this photo goes to *Showlight Technologies Ltd:* www.showlight.com.ng.

The following figure shows the wiring diagram of a DOL starter:

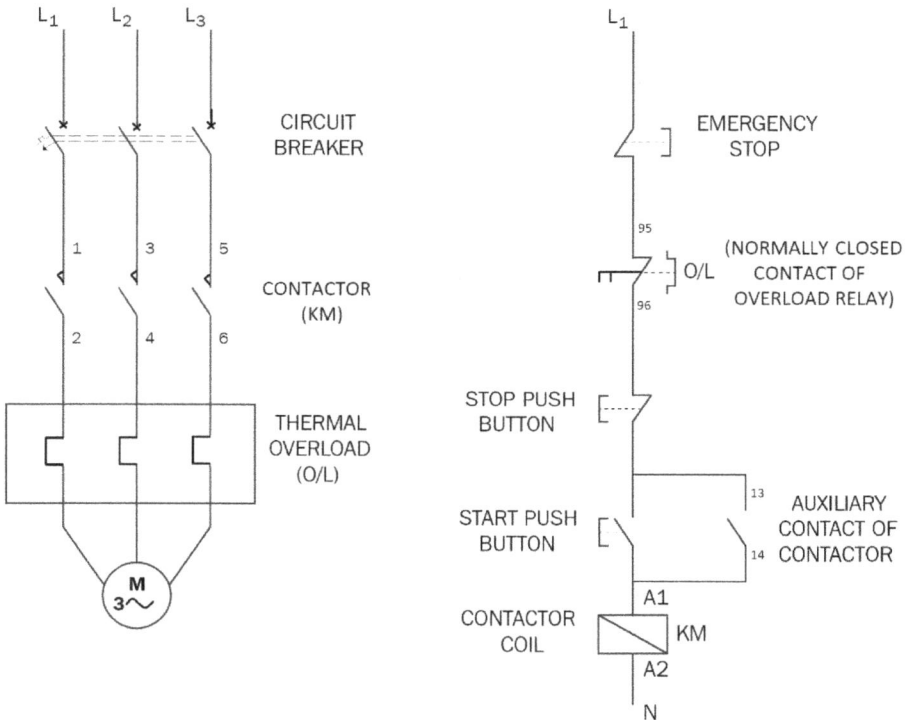

Figure 4.14 – Wiring diagram of a DOL starter

The DOL wiring diagram can be divided into two parts: the power circuit (left) and control circuit (right).

In the power circuit (left), the circuit breaker connects to a three-phase power supply (L1, L2, and L3). The other end of the circuit breaker connects to a contactor at terminals 1, 3, and 5. Terminals 2, 4, and 6 of the contactor then connect to an overload relay while the other end of the overload relay (T1, T2, and T3) connects to a three-phase motor as shown in the figure.

In the control circuit (right), L1 from the supply connects to one end of the emergency stop push button. The other end connects to 95 of the auxiliary contact of the overload relay while 96 of the auxiliary contact of the overload relay connects to one end of an NC push button. The other end of the NC push button connects to one end of an NO push button while the other end of the NO push button connects to A1 of the contactor coil. A2 of the contactor coil connects to **neutral (N)**. The auxiliary contacts of the contactor (13 and 14) are connected in parallel to the NO push button. A single phase (L1 and N) is used in the control circuit.

When the start button (NO push button) is pressed, current flows to the coil and it becomes energized. The main and auxiliary contacts of the contactor become closed, and the motor starts running. When the start button is released, current flows through an alternative path created by the auxiliary contact of the contactor (13 and 14), keeping the coil energized and the motor running. The motor will stop running when the stop button or emergency stop is pressed and also when there is an overload, and the NC auxiliary contact (95-96) becomes open.

Chapter 5, Introduction to Variable Frequency Drive (VFD), of this book will cover the use of AC frequency drive to start, stop, reverse, or vary the speed of an induction motor.

In this section, you have learned about AC motors, which can be classified as synchronous and asynchronous motors. Faraday's law of electromagnetic induction, which states that current will be induced in a conductor placed in a changing magnetic field, was used to explain the operation of an asynchronous (induction) motor. This section also explained the two methods of connecting a three-phase induction motor, that is, star and delta connection. The section concluded with DOL starter wiring, which is crucial for all automation engineers.

Let's now learn about DC motors.

Understanding DC motors

A DC motor is a type of electric motor that is powered by DC supplied by a battery or a DC source. Read *Chapter 3, Actuators and Their Applications in Industrial Automation*, to learn about the basic construction of a DC motor. Here, we will be looking at the two types of DC motor:

- Brushed DC motor
- Brushless DC motor

Let's start by looking at brushed DC motors.

Brushed DC motors

Brushed DC motors are used for simple control systems. They consist of a stator, rotor, commutator, brushes, and so on, and the brushes are usually made of carbon. A magnetic field is produced by passing a current through the commutator and the brush connected to the rotor. One of the drawbacks of a brush motor is the wear and tear of brushes. Types of brushed DC motor include the following:

- Permanent magnet
- Series wound
- Shunt wound
- Compound wound

Let's understand each of these in detail in the following sections.

Permanent magnet DC motors

This type of motor does not have a field winding on the stator, rather, it uses a permanent magnet to create the magnetic field of the stator, which the rotor field reacts with to produce the turning force that causes rotation.

The following figure shows a simplified circuit diagram of a permanent magnet DC motor:

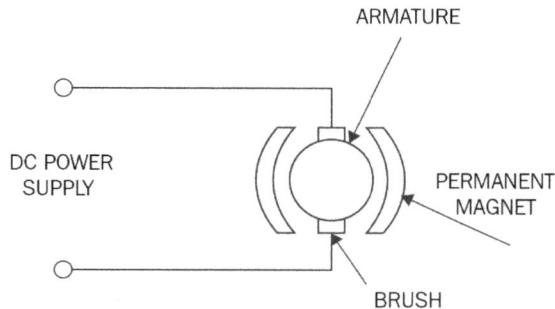

Figure 4.15 – Simplified circuit diagram of a permanent magnet DC motor

Let's now look at a series wound DC motor.

Series wound DC motors

This type of motor has the field winding (housed in the stator) in series with the armature winding. The field winding is powered by a DC supply as shown in the following diagram. The same amount of current flows through the field winding and the armature winding since they are in series. As current flows through the field winding and armature winding, strong magnetic fields are produced. The interaction between the magnetic field in, the field winding and the armature winding produce the turning force (torque) that creates rotation.

The following diagram shows a simplified circuit diagram of a series wound DC motor:

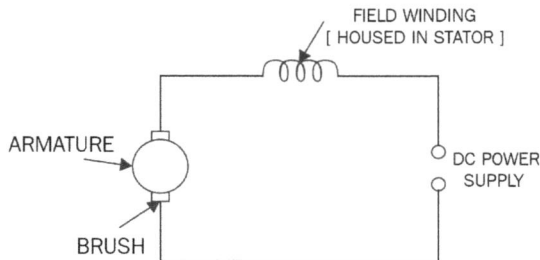

Figure 4.16 – Simplified circuit diagram of a series wound DC motor

Let's now have a look at a shunt DC motor.

Shunt DC motors

This type of motor has the field winding (housed in the stator) in parallel with the armature winding. It is powered by DC and the total current splits into two parallel paths to supply the field winding and the armature winding. As the current flows through the stator and armature winding, a magnetic field is produced. The interaction between the magnetic fields of the stator and rotor produce the turning force that creates rotation.

The following diagram shows a simplified circuit diagram of a shunt DC motor:

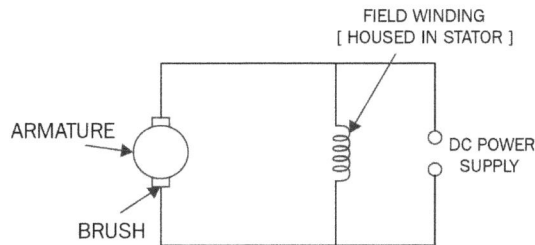

Figure 4.17 – Simplified circuit diagram of a shunt DC motor

Let's have a look at a compound wound DC motor.

Compound wound DC motors

This type of motor has the field winding (housed in the stator) in series with the armature winding and a shunt field that is excited separately. The shunt field winding is connected across the field supply and the series field winding is connected in series with the armature. When DC is supplied, both the shunt field winding and the series field winding produce the required magnetic field that interacts with the armature to produce the necessary turning force that creates rotation. Compound wound DC motor has the characteristics of both the series and shunt types of DC motor.

The following diagram shows a simplified circuit diagram of a compound wound DC motor:

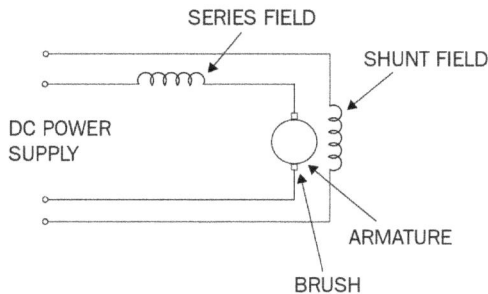

Figure 4.18 – Simplified circuit diagram of a compound wound DC motor

Now that we have an understanding of various brushed DC motors, let's get an understanding of brushless DC motors.

Brushless DC motors

This design does not contain brushes. Hence, there is no wear and tear of brushes as experienced in a brushed type of DC motor. The stator (armature) consists of windings of coil through which current is passed to produce a magnetic field. The rotor is made of a permanent magnet unlike the brushed type.

When a DC power is applied to the stator, it becomes an electromagnet. The force and reaction between the permanent magnet in the rotor and the electromagnet created in the stator produces the required rotation. A common example of a brushless DC motor is the type used in the fan of a computer power supply, as shown in the following figure:

Figure 4.19 – Brushless DC motor. Left – rotor with a permanent
magnet and blade. Right – stator showing coil windings

The credit for this photo goes to *Wikimedia Commons* and can be found at the following link: `https://creativecommons.org/licenses/by-sa/4.0/deed.en`.

In this section, we learned about DC motors. We discussed the two common types of DC motors, that is, the brush and brushless types. The brushed type of DC motor was further classified as a permanent magnet, series wound, shunt wound, and compound wound. DC motors are also used for industrial automation. Hence, it will be wise for any automation engineer to have knowledge of their basic structure.

Let's now learn about stepper motors and servo motors.

Understanding stepper motors and servo motors

A stepper motor can be referred to as a type of brushless DC motor that divides its full rotation into a number of equal steps and rotates in steps. They have multiple coils organized in phases. The motor will rotate a step at a time as each phase is energized in sequence.

Basically, stepper motors are of two types:

- Unipolar
- Bipolar

Let's look at them in detail in the following sections.

Unipolar stepper motors

This usually has six wires, but sometimes it has five, as shown in the following diagram. They have one winding with a center tap per phase:

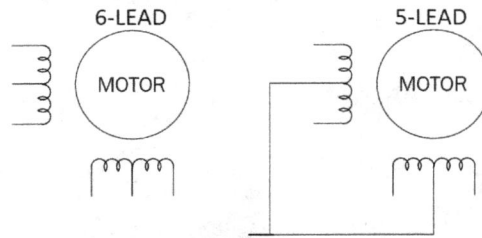

Figure 4.20 – Unipolar stepper motor

Let's have a look at a bipolar stepper motor.

Bipolar stepper motors

These usually have four wires. They also have one winding per phase but do not have a center tap. They have more torque and are more efficient than unipolar stepper motors:

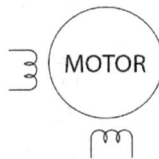

Figure 4.21 – Bipolar stepper motor

Let's now learn how to drive a stepper motor.

Driving a stepper motor

Unlike a regular brushed DC motor that can be connected directly to a DC source, driving a stepper motor is more complicated. It usually requires a stepper motor driver or stepper motor controller to energize the phases in a timely sequence to get it turned.

The following diagram shows a simple stepper motor driver that can be used for a bipolar stepper motor:

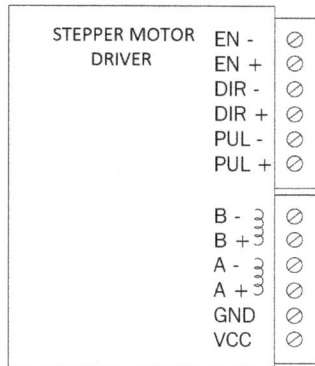

Figure 4.22 – Stepper motor driver

The four leads or wires of the stepper motor are connected to terminals A+, A-, B+, and B- and the power supply is connected to VCC and GND.

DIR +, DIR -, PUL +, and PUL - can be connected to a PLC to send the pulse and direction signal as shown in *Figure 4.23*:

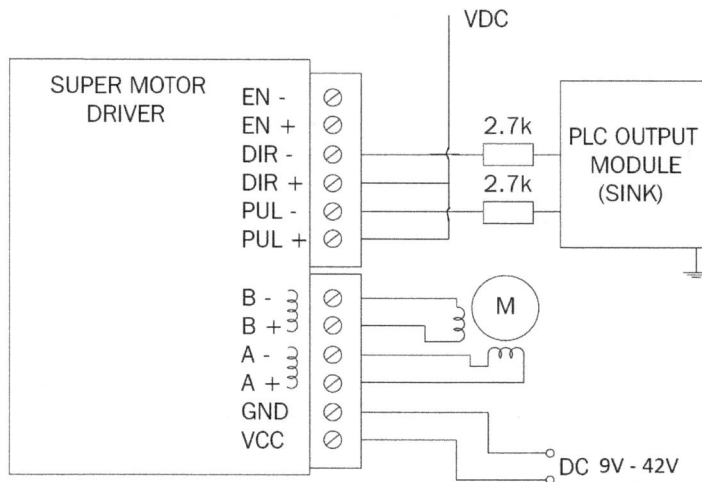

Figure 4.23 – Connection of the stepper motor driver, stepper motor, and PLC

Now that we have an understanding of stepper motors, including their types and how to drive them, let's have a look at servo motors.

Servo motors

Servo motors are specially designed for use in a feedback control system. They allow for a precise angular position, velocity, and acceleration. They contain a regular motor in addition to a feedback mechanism or encoder. The feedback allows the control system to establish whether the object being monitored is correctly positioned and allows actions to be taken based on the movement and position of the object. Both the AC and DC types of servo motors are used for industrial control.

A servo motor is also not simple to control like a DC motor that can just be connected to a DC source. A basic DC servo motor can be controlled with a microcontroller. AC servo motors used in industry usually require a PLC and a servo amplifier (driver) as shown in the following diagram:

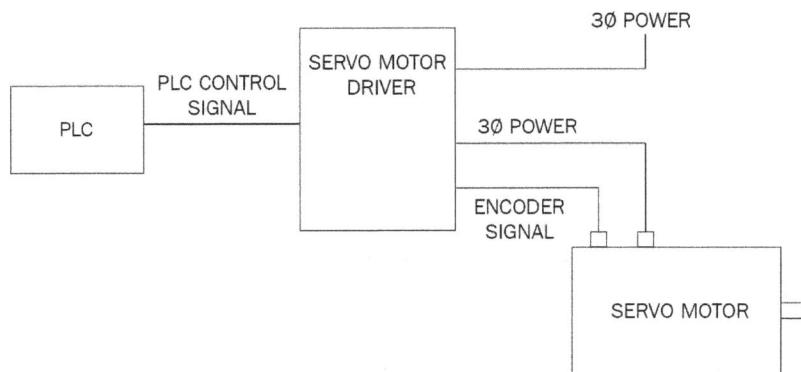

Figure 4.24 – Simplified wiring of PLC, servo motor driver, and servo motor

In this section, we have learned about stepper motors and servo motors. They are special types of motors used in industrial automation. Unlike the basic DC motor that only needs to be connected to a DC source to get it running, stepper motors and servo motors require a driver to get them running. They are very useful in the field of robotics and industrial automation. Hence, the importance of knowledge of stepper motors and servo motors cannot be over-emphasized for an industrial automation engineer.

Summary

Congratulations! You have successfully completed another chapter of this book. Well done. The chapter explained electric motors as machines used to convert electrical energy into mechanical energy. Classifying electric motors based on the type of power supply they use, we have AC motors and DC motors. AC motors can further be classified into synchronous and asynchronous while DC motors can be further classified as brushed and brushless.

You should now understand asynchronous motors and be able to connect in a star or delta connection. Also, you should be able to wire an asynchronous motor to start and stop it using the DOL starter method

The chapter also discussed stepper motors and servo motors. You should now understand stepper motors and how to drive them. And finally, you should now have an idea of how a servo motor can be connected to a PLC.

The knowledge acquired in this chapter will help in understanding **Variable Frequency Drive (VFD)** in the next chapter.

Questions

The following are questions to test your understanding of this chapter. Ensure you have read and understood the topics in this chapter before attempting the questions:

1. The two types of electric motor when classified by their source of power are _____ and _____.

2. A motor whose rotor speed is the same as the speed of the rotating magnetic field of the stator is called _____.

3. _____ is the phenomenon in which current is induced in a conductor placed in a changing magnetic field.

4. In an AC induction motor, the difference between the speed of the rotating magnetic field of the stator and the speed of the rotor is called _____.

5. A type of DC motor whose field winding (housed in the stator) is in series with the armature winding is called _____.

6. A type of brushless DC motor that divides its full rotation into a number of equal steps and rotates in steps is called _____.

7. The _____ motor allows for a precise angular position, velocity, and acceleration.

8. _____ is the simplest type of starter that connects the three-phase motor directly to the power supply through a contactor.

9. In a _____ connection, similar points (start or end points) of three coils (that is, U1, V1, and W1 or U2, V2, and W2) are connected together to form a neutral point and three wires run from the remaining three terminals.

10. In a _____ connection, phase voltage is equal to line voltage.

Introduction to Variable Frequency Drive (VFD)

In the previous chapter, we learned about the various types of electric motors. Among the various types, the asynchronous (induction) motor is the most commonly used motor in the industry. However, machines in the industry, such as conveyors, mixers, agitators, pumps, compressors, and others, that utilize an induction motor require a variable speed, which an induction motor alone cannot provide because induction motors are designed to run at a constant speed (that is, their speed remains almost the same for various load range).

A variable speed drive is an electronic device used to achieve the varying speed needed by either an asynchronous or synchronous motor for various applications.

This chapter looks at a **Variable Frequency Drive (VFD)**, its working principle, benefit, wiring, and programming. We will also look at some practical applications, including forward, reverse, and speed control of an induction motor.

In this chapter, we are going to cover the following main topics:

- An overview of a VFD

- Wiring a VFD

- Understanding a basic VFD parameter setting

- Wiring and programming a VFD to start/stop a three-phase motor and setting the frequency (speed) using a keypad

- Wiring and programming a VFD (VAT20) to use a push button or switch to run a three-phase motor in a forward/reverse direction and to use a keypad to set the frequency (speed)

- Wiring and programming a VFD (VAT20) to use a push button or switch to run a three-phase motor in a forward/reverse direction and to use a potentiometer to set the frequency (speed)

An overview of a VFD

A VFD is a device used to vary the speed of **Asynchronous or Synchronous (AC)** electric motors by varying the frequency and voltage supplied. A VFD can also be referred to as an **Adjustable Speed Drive (ASD)** or inverter. It is the most effective way of controlling the speed of an induction motor. The principle of operation of a VFD is based on the fact that the speed of an induction motor is directly proportional to the frequency of the voltage supplied to the motor. When frequency increases, speed also increases, and when frequency reduces, the speed will also reduce. A synchronous motor runs at synchronous speed (the speed of the rotating magnetic field of the stator) while an asynchronous (induction) motor's speed is slightly less than synchronous speed. In the previous chapter, it was mentioned that synchronous speed is determined by the frequency of the voltage source and the number of poles, as shown by the following formula:

Synchronous speed = 120f/P

Here, f = frequency in Hz and P = the number of poles.

From the preceding equation, it can be seen that a change in frequency will cause a change in speed. If an application requires a varying speed, a VFD can be used to increase or reduce the speed of the motor, and it does so by varying the frequency and voltage. Some of the companies in the VFD industry include Siemens, Rockwell Automation, Yaskawa Electric Corporation, and General Electric. This book focuses on the General Electric VFD.

The following image shows a VFD manufactured by General Electric (VAT 20-U20N0K4S):

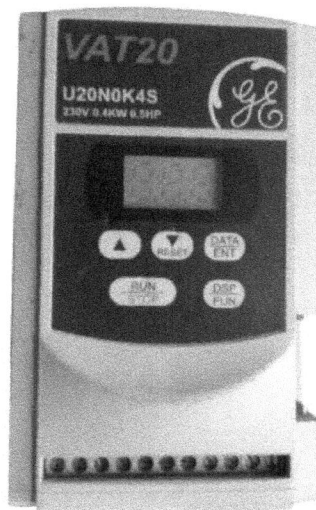

Figure 5.1 – A VFD from General Electric (VAT 20-U20N0K4S)

The credit for this photo goes to *Showlight Technologies Limited at* www.showlight.com.ng.

You should now have an understanding of what a VFD does, the relationship between the speed of an induction motor and the frequency, and the various companies in the VFD industry. Your knowledge here will help in understanding other sections in this chapter. Let's learn how a VFD works in the next section.

Understanding how a VFD works

The principle of operation of a VFD can be explained using the three basic stages that make it up, as shown in the following block diagram:

Figure 5.2 – A block diagram of a VFD

Let's have a look at the three stages to understand how a VFD works:

- **Converter**: This is the first stage of a VFD. It converts the AC input power to DC power. This stage can also be referred to as the rectifier stage. It can consist of diodes or an array of fast switches.

  The following circuit diagram shows the converter/rectifier stage of a VFD that is made up of diodes:

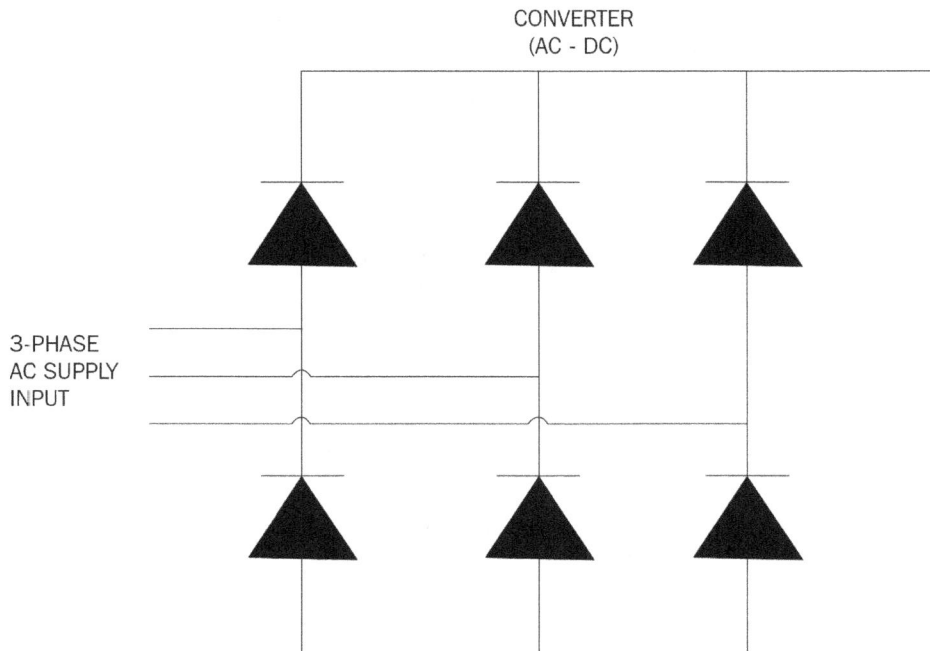

Figure 5.3 – The converter/rectifier stage

- **DC link**: Power from the rectifier is fed to the DC link stage. The DC link stage smoothens the DC from the converter/rectifier stage. It consists of a capacitor and may also include an inductor that remove ripples from the rectified signal to provide clean DC power for the next stage.

The following circuit diagram shows the DC link stage of a VFD:

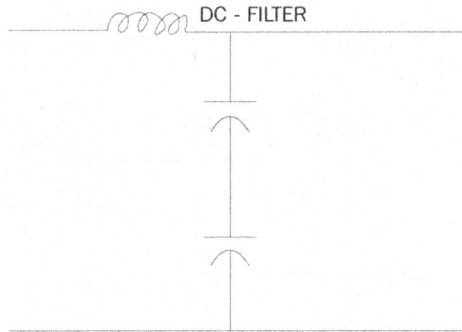

Figure 5.4 – The DC link stage of a VFD

- **Inverter**: This is the third stage, which converts the DC power from the DC link into a three-phase AC output. This stage comprises of a power-switching device such as a **Metal Oxide Semiconductor Field Effect Transistor (MOSFET)**, or an **Insulated Gate Bipolar Transistor (IGBT)**, or a thyristor. The output of an inverter is not a pure sine wave but an approximation, based on the principle of **Pulse Width Modulation (PWM)**, which is the major inverter technology.

The following circuit diagram shows the inverter stage of a VFD:

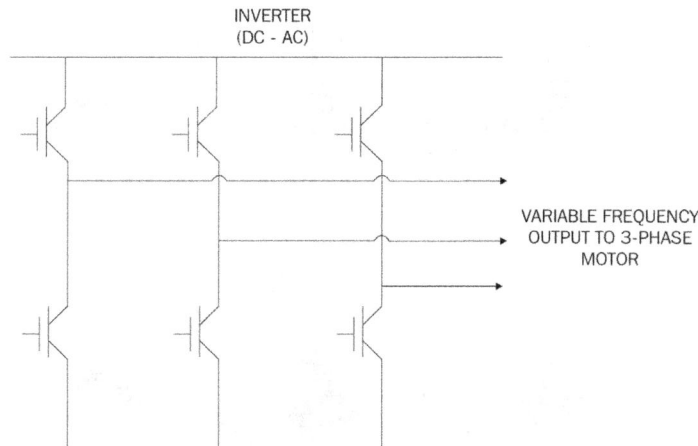

Figure 5.5 – The inverter stage of a VFD

Having understood how a VFD works, let's have a look at the advantages in the next section.

Advantages of a VFD

A VFD offers so many advantages, which account for its use in many industries. Some of the advantages of VFD include the following:

- A VFD extends motor life and reduces maintenance costs because it starts the motor by ramping it up to the operating speed, rather than starting it abruptly as a direct online starter would do. This starting method reduces mechanical and electrical stress that would have shortened the life of the motor or increased the maintenance costs of both the motor and the driven equipment.

- A VFD reduces energy consumption. Running a motor at full speed consumes a lot of energy, which increases energy costs. Some applications do not require a motor to run at full speed. A VFD can be used in such cases to match the speed of the motor to the process requirement. This will greatly reduce electricity costs.

- A VFD has built-in safety features that provide protection for both the motor and the VFD itself. The VFD can automatically stop operation when a fault occurs. The built-in safety features eliminate the cost of additional safety components.

- A VFD offers flexible control of AC motors. With VFD, various options are available for controlling a motor. It can use push buttons, selector switches, and a potentiometer, and can even be operated through a **Programmable Logic Controller** (**PLC**) or other form of industrial controller.

- A VFD reduces disturbances in the power line. When AC motors are started across the line, they can demand up to 600% of the full load current, depending on the capacity of the motor. This can cause a significant disturbance or a short-duration reduction in voltage (voltage sag) in the power system connected to it. Sensitive equipment connected to the same power distribution system can be interrupted due to the low voltage that occurs as a result of the disturbances. A VFD eliminates voltage sag or disturbance in a power line since it starts at 0V and ramps up to the required speed.

Let's discuss some disadvantages of a VFD as well.

Disadvantages of a VFD

Despite the many advantages that make VFDs useful for many applications in the industry, they also have some disadvantages that need to be known, understood, and mitigated. Some of the disadvantages include the following:

- **Harmonics**: Harmonics are the most common issue that has to be dealt with when using a VFD. They are high-frequency signals that are superimposed on the fundamental waveform, creating a distorted waveform. They can cause overheating of motors and transformers, premature failure of circuit breakers, incorrect operation of electronic instruments, and other issues. A remedy for harmonics caused by a VFD is the use of an AC line reactor (choke) or filter.

- **High initial cost**: Depending on how large a system is, the initial cost of purchasing a VFD could be a disadvantage. However, the cost issue can be neglected by the energy saving and protection offered by the built-in safety features.

This section explains some of the disadvantages of a VFD along with a way to mitigate each challenge. Let's proceed to the next section, where we will learn how to select a VFD.

Selecting a VFD

Before going too far into the selection of VFD, it is important to first know and understand basic motor information, which can be found on a motor nameplate. Here are some motor information that can be found on a nameplate:

- **Horsepower**: This is a measure of the output power of a motor. The output power of a motor can sometimes be given in watts, where 1HP = 746 watts. A 10HP motor is 7.46KW.

- **Full Load Amps (FLA)**: This is the maximum current expected to be drawn by a motor when operating at maximum torque and HP.

- **Voltage**: This is the voltage required to operate a motor for optimal performance. Operating the motor below or above the rated voltage will affect its performance.

- **Revolutions per Minute (RPM)**: This refers to the speed of the shaft at the rated frequency, the rated load, and the rated voltage. It is the number of complete rotations the rotor's shaft will make in 1 minute.

- **Frequency**: This is the frequency of the supply that the motor is designed to operate.

The following diagram gives a better understanding of a motor nameplate:

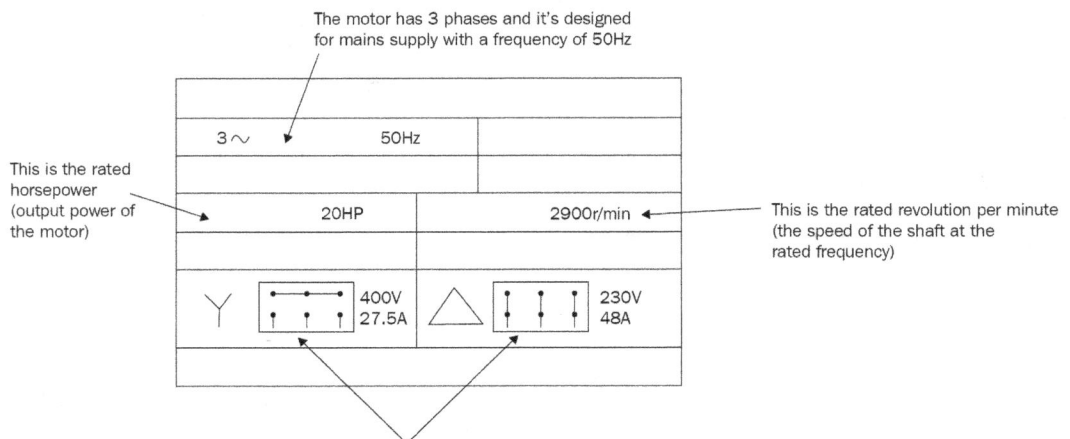

Figure 5.6 – An electric motor nameplate showing basic information

Now that we understand the basic information on a motor nameplate, let's look at some necessary things to be considered when selecting a VFD for a motor:

- **Input voltage**: This is the voltage required to power up a VFD. Some VFDs require a three-phase 330–400V input while others require a single-phase 220–230V input. The type you will choose depends on the source you have available at the site (whether it's a single-phase or three-phase).

- **Output voltage**: The output voltage is the voltage that a VFD can provide to operate a motor. The output is usually three-phase with a voltage of 200V, 220V, 380V, 400V, and more. Check the required voltage to power up the motor on the nameplate and select a VFD that can produce the voltage.

- **Load capacity**: This refers to the amount of load the VFD can bear without damage. The load capacity of the VFD you will select depends on the motor size. A small motor will require a VFD with a small load capacity, while a large motor will require a VFD with a large load capacity. It is usually given in HP or Amps. Knowing the HP of your motor will help in searching for a suitable VFD, which will be further narrowed down by other parameters such as FLA and voltage. It is very important to ensure that the VFD can handle the current demand of the motor. The proper way to select a VFD for a motor is to use the FLA of the motor rather than HP alone. Check the FLA of the motor on its nameplate and ensure the VFD is rated for at least that much current.

  If the drive will be fed with single-phase power, ensure you use a drive built for single-phase. VFDs are significantly derated for single-phase operation. A single-phase input/three-phase output VFD can be up to 3 or 5HP; if a three-phase motor is larger than 5HP, for example, and a single-phase is to be used to power up the VFD, you'll have to derate the three-phase VFD for single-phase power. Derating a three-phase VFD for single-phase input can simply be achieved by multiplying the FLA of the motor by 2 and then selecting a three-phase VFD with corresponding amps. For example, if a 10HP motor with FLA of 26.6A is to be used with a VFD and the site only has a single-phase power line voltage of 230V, we will have to derate the VFD to use the single-phase power; this can be done by multiplying 26.6 by 2, which will be 53.2A. Hence, a three-phase 230V VFD of 53.2A or higher can be used, which will be about 20HP and will run a 10HP/230V motor on a single-phase.

We have just completed an overview of VFD, which includes understanding the working of a VFD, the advantages of a VFD, the disadvantages of a VFD, and the selection of a VFD. This knowledge is important before doing any practical work on a VFD. The next sections cover practical work on VFD, and the theories in the previous section (an overview of a VFD) will be of great benefit.

Wiring a VFD

Before wiring a VFD, ensure you have properly sized the VFD for your motor by using the information in the previous section (*Selecting a VFD*).

Next, it is a good practice to look through the manual of the VFD to have an understanding of the VFD itself and its wiring.

The following diagram shows the wiring of VAT 20 in the manual provided by the device manufacturer:

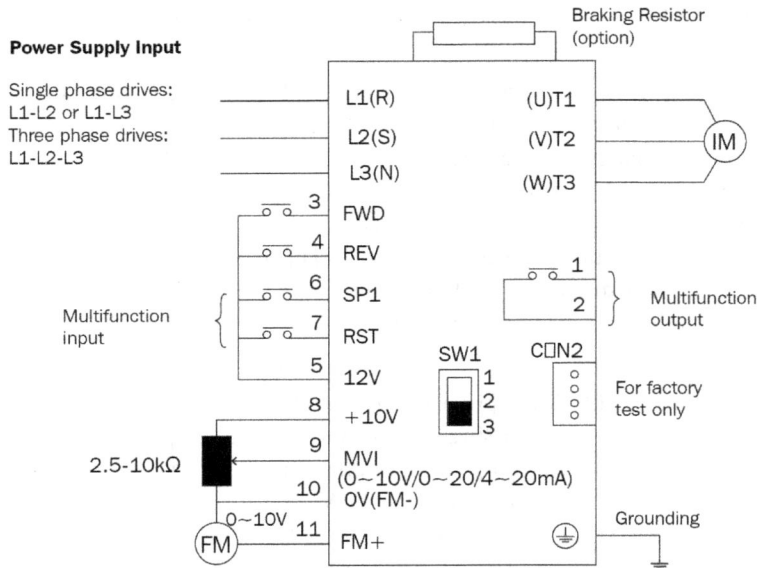

Figure 5.7 – A wiring diagram of a GE VFD (VAT 20)

Let's now have a look at some of the terminals available in a VFD and how to connect them:

- **Power input (L1, L2, and L3)**: This is the input for the AC that will power the drive. The AC power from the source should connect to the power input terminals L1 and L2 for the single-phase input type or L1, L2, and L3 for the three-phase input type. It is advisable to use a properly sized circuit breaker and an AC line reactor between the power source and the power input of the VFD. The breaker can be used to switch ON or OFF the VFD and also act as protection while the line reactor improves the power factor. An AC line reactor is necessary for this VFD if the capacity of the power source is above 600KVA according to its manual. Always check the manual of the drive you want to use for necessary additional devices that may be required.

- **Three-phase AC output terminal (U, V, and W)**: This is the AC power output terminal of the VFD, which should be connected to the three-phase wire (U, V, and W) of the three-phase motor.

- **Terminals 1 and 2 (trip relay)**: This is a relay that can be connected to an indicator to indicate fault.

- **Terminals 3 and 4 (forward and reverse):** These terminals are used to control the operation of a VFD. Push buttons, selector switches, or signals from a PLC can be connected to control the operation of the VFD. It can be made to make the motor run in a forward or reverse direction.

- **Terminal 5 (+12):** This is the common point of terminals 3, 4, 6, and 7. The VFD supplies +12V at this terminal. Hence, push buttons or selector switches can be connected between terminals 5 and 3, 4, 6, and 7, as shown in the preceding wiring diagram to send a control signal to the VFD to control the operation of the VFD for various operations.

- **Terminals 6 and 7:** This can as well be used to control the operation of the VFD. The parameter setting will give further details on their use. The wiring is similar to the wiring of terminals 3 and 4.

- **Terminal 8:** This is the power terminal of a potentiometer. This should connect to the third pin of a potentiometer.

- **Terminal 9:** This is the analog input point. This should connect to the second (middle) pin of a potentiometer.

- **Terminal 10:** This is the analog common point (0V). This should connect to the first pin of a potentiometer.

- **Terminal 11:** This is the analog output positive connection point. A voltmeter or ammeter can be connected between terminals 11 and 10 to read analog output from VFD.

The following photo shows the VFD showing the terminals for a better understanding:

Figure 5.8 – The button view of GE VFD (VAT 20)

The credit for this photo goes to *Showlight Technologies Limited* at www.showlight.com.ng.

Let's now do some basic wiring for this VFD using some of the knowledge we have acquired so far:

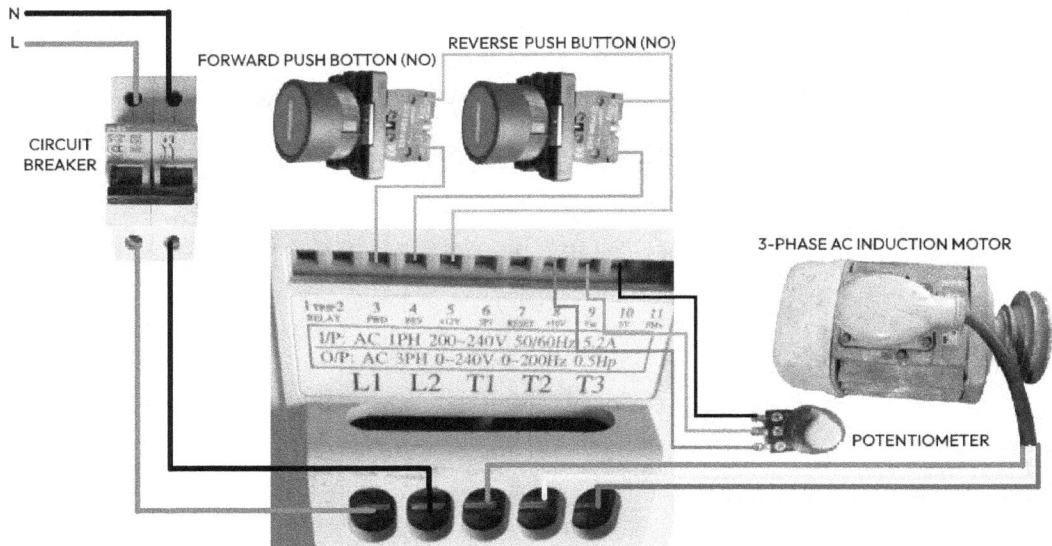

Figure 5.9 – The wiring of a VFD

The credit for this figure goes to *Showlight Technologies Limited at* www.showlight.com.ng.

We just completed learning how to wire a VFD. The wiring (pictorial diagram) in *Figure 5.9* is common and basic. You can run the AC induction motor in a forward or reverse direction using the push buttons, and you can use the potentiometer to vary the speed. Let's now proceed to the next section, where we will learn how to change parameter settings to make a VFD operate as desired.

Understanding a basic VFD parameter setting

It is not enough to know how to wire a VFD without knowing how to change a parameter setting. A properly wired VFD, as shown in *Figure 5.9*, will not function as desired if the parameter setting for the required function is not used.

A VFD parameter is a variable associated with the operation of a VFD that can be adjusted or altered to make the VFD function in a particular way. Adjusting or altering the values of parameters of a VFD to make it operate in a particular way can be referred to as **VFD programming**. Each VFD manufacturers and models have parameter settings that are peculiar to each model. It is important to consult the manual for the VFD manufacturer and model you want to use to be able to alter the parameter setting. We will be using the parameter setting of General Electric VFD (VAT 20) to explain a VFD parameter setting in this book.

The following table shows a few of the parameters in a GE VFD (VAT 20) manual (`https://inverterdrive.com/file/GE-VAT20-Manual-200401`):

List of parameters					
Function	FN_	Function description	Unit	Range	Factory setting
	0	Factory adjustment			0
Acceleration/ deceleration time	1	Acceleration time	0.1s	0.1–999s	5s
	2	Deceleration time	0.1s	0.1–999s	5s
Operation mode	3	0: Forward/Stop, Reverse/Stop 1: Run/Stop, Forward/Reverse	1	0 -- 1	0
Motor direction	4	0: Forward 1: Reverse	1	0 -- 1	0
Jog frequency	9	Jog frequency	0.1Hz	1.0–120Hz	6Hz
Operation control	10	0: Keypad 1: External terminal	1	0 -- 1	0
Frequency control	11	0: Keypad 1: External terminal (0–10V)/(0–20mA) 2: External terminal (4–20mA)	1	0 --2	0
Stop method	14	0: Deceleration stop 1: Free run stop	1	0 -- 1	0

Table 5.1 – The parameters in a GE VFD (VAT 20) manual

We will now discuss some of the parameters.

- **Operation mode (F03)**: This parameter allows you to specify the way you want the two push buttons connected to terminal 3 and terminal 4 to function. The two options available are as follows:

  - **Forward/stop and reverse/stop**: Value = 0
  - **Run/stop and forward/reverse**: Value = 1

  Specifying 0 as the value for parameter F03 will make the push button or switch connected to terminal 3 function as forward/stop, while the push button or switch connected to terminal 4 will function as reverse/stop. Specifying 1 as the value for the parameter will make the push button or switch connected to terminal 3 function as run/stop, while the push button connected to terminal 4 will function as forward/reverse.

- **Operation control (F10)**: This parameter allows you to specify your choice of operating the VFD. The two available options are as follows:

  - **Keypad**: Value = 0
  - **External terminal**: Value = 1

  Specifying 0 as the value for parameter F10 allows you to operate the VFD (run/stop) using the keypad or operator panel on the VFD. Specifying 1 as the value for the parameter allows you to operate the VFD using push buttons, switches, or via a signal from an external device (a PLC or controller).

- **Frequency control (F11)**: This parameter allows you to specify a frequency source. The three options available are as follows:

  - **Keypad**: Value = 0

  - **External terminal** (0–10V or 0–20mA): Value = 1

  - **External terminal** (4–20mA): Value = 2

  Specifying 0 as the value for parameter F11 allows you to set the frequency of the VFD using the keypad (operator panel). Specifying 1 as the value for the parameter allows you to set the frequency using a potentiometer connected to terminals 8, 9, and 10), as shown in the wiring (pictorial diagram) of *Figure 5.9*, or send a signal of 0–10V or 0–20mA from an external device. Specifying 2 as the value for the parameter allows you to set the frequency by sending a 4–20mA signal from an external device.

It is advisable to study the manual of any VFD you want to work with. We have learned about some VFD parameter settings of General Electric VFD (VAT 20). You can learn more from the manual of the VFD. Let's get our hands dirty with a practical example. We will be learning how to wire and program the General Electric VFD (VAT 20) to start/stop a three-phase motor with a keypad and also use a keypad to set the frequency (speed).

Wiring and programming a VFD to start/stop a three-phase motor and setting the frequency (speed) using a keypad

Before we start the programming, let's wire the system, as shown in the following connection diagram:

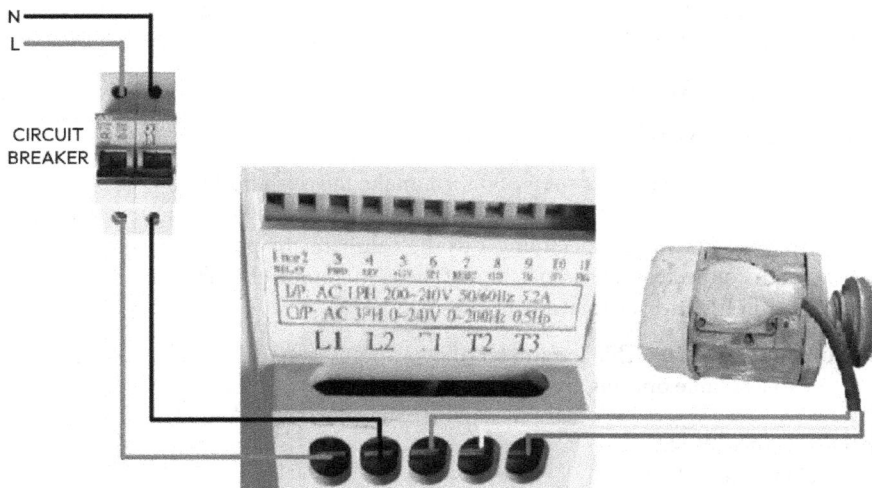

Figure 5.10 – The wiring required to start/stop a three-phase induction motor with a keypad and also using the keypad to set the frequency (speed)

The credit for this figure goes to *Showlight Technologies Limited at* www.showlight.com.ng.

The following photo shows the keypad and display of General Electric VFD (VAT20) that we are using in this book:

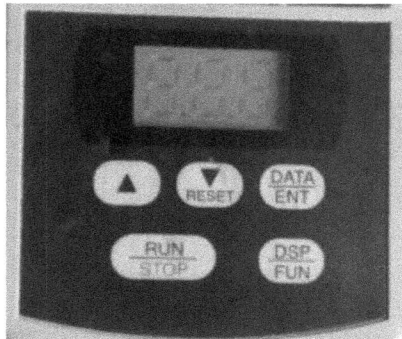

Figure 5.11 – The keypad and display of a General Electric VFD (VAT20)

The credit for this photo goes to *Showlight Technologies Limited* www.showlight.com.ng.

Switch on the circuit breaker to supply power to the drive and start the programming by following the steps:

1. Press the **DSP/FUN** key to enter the parameter setting and use the up/down arrow to scroll to the **F10** parameter (operation control).
2. Press **DATA/ENT** for the parameter value and use the up/down arrow to scroll to **0** (keypad).
3. Press **DATA/ENT** to accept the value. The display will show **END** and return to **F10**.
4. Use the up/down arrow to scroll to the **F11** parameter (frequency control).
5. Press **DATA/ENT** for the parameter value and use the up/down arrow to scroll to 0 (keypad).
6. Press **DATA/ENT** to accept the value. The display will show **END** and return to **F11**.
7. Press **DSP/FUN** key to exit the parameter setting.

You have just completed the programming and will see the frequency blinking. Use the up or down arrow key to specify a frequency and press **RUN/STOP** to start or stop the motor.

We have now learned how to use a keypad to start or stop a three-phase motor. Let's look at another practical example to further increase our knowledge.

Wiring and programming a VFD (VAT20) to use a push button or switch to run a three-phase motor in a forward/reverse direction and use a keypad to set the frequency (speed)

Before we start the programming, let's wire the system as shown in the following diagram:

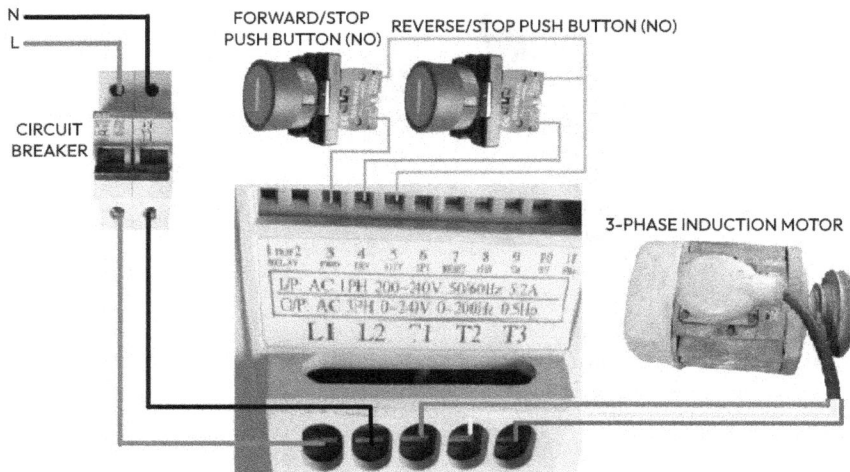

Figure 5.12 – The wiring required to use a push button or switch to run a three-phase motor in a forward/reverse direction and use a keypad to set the frequency (speed)

The credit for this figure goes to *Showlight Technologies Limited* at www.showlight.com.ng.

Switch on the circuit breaker to supply power to the drive and start the programming by following the steps:

1. Press the **DSP/FUN** key to enter the parameter setting and use the up/down arrow to scroll to the **F03** parameter (operation mode).

2. Press **DATA/ENT** for the parameter value and use the up/down arrow to scroll to **0** (forward/ stop and reverse/stop).

3. Press **DATA/ENT** to accept the value. The display will show **END** and return to **F03**).

4. Use the up/down arrow to scroll to the **F10** parameter (operation control).

5. Press **DATA/ENT** for the parameter value and use the up/down arrow to scroll to **1** (external terminal).

6. Press **DATA/ENT** to accept the value. The display will show **END** and return to **F10**.

7. Use the up/down arrow to scroll to the **F11** parameter (frequency control).

8. Press **DATA/ENT** for the parameter value and use the up/down arrow to scroll to **0** (keypad).

9. Press **DATA/ENT** to accept the value. The display will show **END** and return to **F11**.

10. Press **DSP/FUN** key to exit the parameter setting.

You have just completed the programming and will see the frequency blinking. Use the up or down arrow key to specify a frequency, and use the push button or switch connected to terminal 3 to run the motor in a forward direction or stop the motor. Use the push button or switch connected to terminal 4 to run the motor in a reverse direction or stop the motor.

We have now learned how to use a push button or switch to run a three-phase motor in a forward/reverse direction and use a keypad to set the frequency (speed). Let's further increase our knowledge by adding a potentiometer to set the frequency (speed).

Wiring and programming a VFD (VAT20) to use a push button or switch to run a three-phase motor in a forward/reverse direction and to use a potentiometer to set the frequency (speed)

Before we start the programming, let's wire the system as shown in the following diagram:

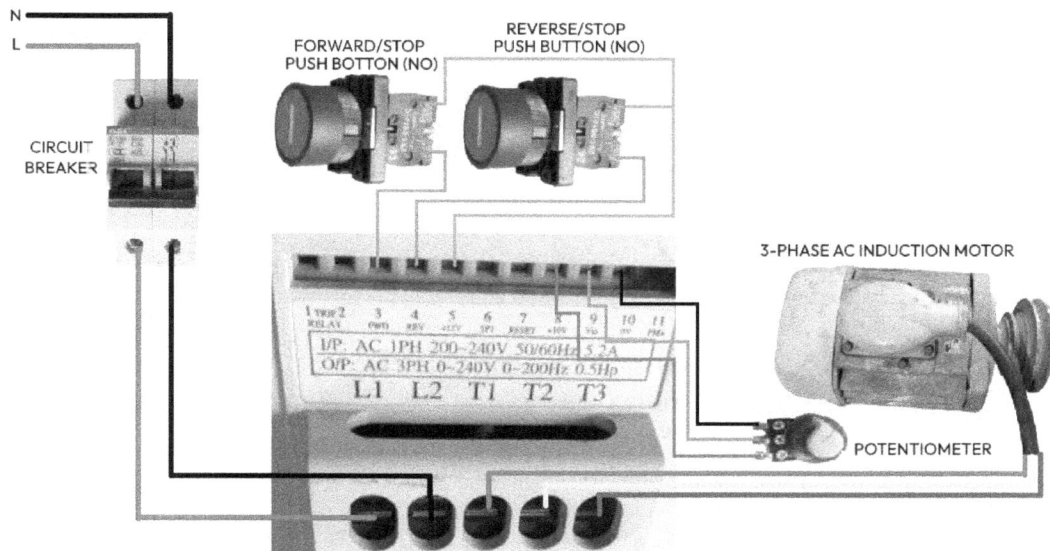

Figure 5.13 – The wiring required to use a push button or switch to run a three-phase motor in a forward/reverse direction and to use a potentiometer to set the frequency (speed)

The credit for this figure goes to *Showlight Technologies Limited at* www.showlight.com.ng.

Switch on the circuit breaker to supply power to the drive and start the programming by following the steps:

1. Press the **DSP/FUN** key to enter the parameter setting and use the up/down arrow to scroll to the **F03** parameter (operation mode).

2. Press **DATA/ENT** for the parameter value and use the up/down arrow to scroll to **0** (forward/stop and reverse/stop).

3. Press **DATA/ENT** to accept the value. The display will show **END** and return to **F03**).

4. Use the up/down arrow to scroll to the **F10** parameter (operation control).

5. Press **DATA/ENT** for the parameter value and use the up/down arrow to scroll to **1** (external terminal).

6. Press **DATA/ENT** to accept the value. The display will show **END** and return to **F10**.

7. Use the up/down arrow to scroll to the **F11** parameter (frequency control).

8. Press **DATA/ENT** for the parameter value and use the up/down arrow to scroll to **1** (external terminal).

9. Press **DATA/ENT** to accept the value. The display will show **END** and return to **F11**.

10. Press **DSP/FUN** key to exit the parameter setting.

You have just completed the programming and will see the frequency blinking. Use the potentiometer to specify a frequency, and use the push button or switch connected to terminal 3 to run the motor in a forward direction or stop the motor. Use the push button or switch connected to terminal 4 to run the motor in a reverse direction or stop the motor.

Summary

The chapter explained the VFD as a device that can be used to change the speed of an AC motor by varying its frequency and voltage. We have also learned about the advantages that made them popular in various industries. Also, we looked at the disadvantages that need to be known, understood, and mitigated. We also learned how to select a VFD, and finally, we learned how to wire and program a VFD (VAT 20) for some common applications.

In the next chapter, we will learn how to use **Computer-Aided Design (CAD)** software to draw electrical diagrams, which is very important for industrial automation engineers to communicate with each other.

Questions

The following are questions to test your understanding of this chapter. Ensure you have read and understood the topics in this chapter before attempting the questions:

1. Synchronous speed is determined by _____and _____.

2. _____ is a device used to vary the speed of AC electric motors (asynchronous or synchronous motors) by varying the frequency and voltage supplied.

3. The three basic stages that make up a VFD are _____, _____, and _____.

4. _____ smoothens the DC from the converter/rectifier stage.

5. High-frequency signals that are superimposed on a fundamental waveform, creating a distorted waveform, can be referred to as _____.

6. The number of complete rotations a rotor's shaft will make in 1 minute is called _____.

7. _____ is a variable associated with the operation of a VFD that can be adjusted or altered to make the VFD function in a particular way.

Drawing Schematic/Wiring Diagrams Using CAD Software

Computer-Aided Design (CAD) software is computer applications that help in creating, modifying, analyzing, and optimizing a design. CAD software is commonly used by engineers (mechanical, electrical, and computer engineers, among others) and in other fields that involve drawing. The need for CAD in industrial automation cannot be over-emphasized as designs or control systems need to be created, modified, analyzed, and optimized on a computer to achieve the desired result. Examples of CAD software include **AutoCAD**, **ProfiCAD**, **PCSCHEMATIC Automation**, **Automation Studio**, **SmartDraw**, and so on. In engineering or industrial automation-related drawing, CAD software saves time and drawings are more accurate than when using manual methods.

This chapter explains the use of the PCSCHEMATIC Automation software to draw schematic/wiring diagrams relating to industrial automation. We will learn how to create a control system drawing using the software.

In this chapter, we are going to cover the following main topics:

- Understanding electrical diagrams/drawings
- Overview of the PCSCHEMATIC Automation software
- Downloading and installing PCSCHEMATIC Automation
- Launching PCSCHEMATIC Automation and exploring the screen elements
- Creating and saving a project in PCSCHEMATIC Automation
- Using the symbol, line, text, arc/circle, and area tools
- Drawing power and control schematic of a DOL starter using PCSCHEMATIC

Technical requirements

Our focus in this chapter is the use of CAD software (PCSCHEMATIC Automation) to draw circuit/schematic diagrams.

While efforts were made to make the chapter comprehensive for readers without the knowledge of the previous chapters, knowledge of *Chapter 4, Overview Of AC and DC Motors*, will be of help.

Steps to install the CAD software (PCSCHEMATIC Automation) will be provided in this chapter. You will need a computer system or laptop with the following minimum system requirements:

- **Operating system**: Windows XP/Vista/7/8.1/10
- **Memory (RAM)**: 1 GB of RAM
- **Hard disk space**: 400 MB of free space
- **Processor**: Intel Pentium 4 or later

Understanding electrical diagrams/drawings

Before we start to learn how to draw schematics with CAD, let's get an understanding of the various types of electrical drawings/diagrams you are likely to come across in industrial automation and control.

Common electrical drawings/diagrams in industrial automation and control include the following:

- **Schematic diagrams** can be referred to as electrical diagrams that show the components in a circuit using standard symbols and also how they connect with each other but do not include where the components and wires are located in the real system or device:

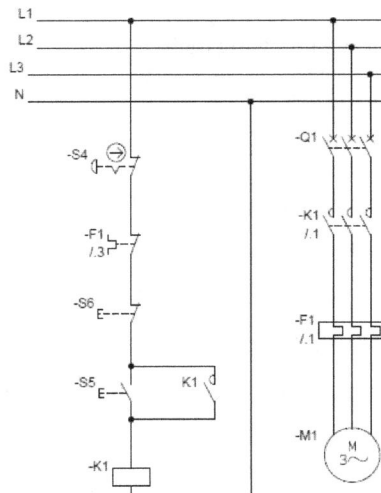

Figure 6.1 – Schematic diagram of a Direct On-Line (DOL) starter

- **Wiring diagrams** are electrical diagrams that show the components involved in a circuit, the physical connections between the components, and where the components and wires are located in the real system or device, unlike schematic diagrams, which do not show the real location of the components and wires involved:

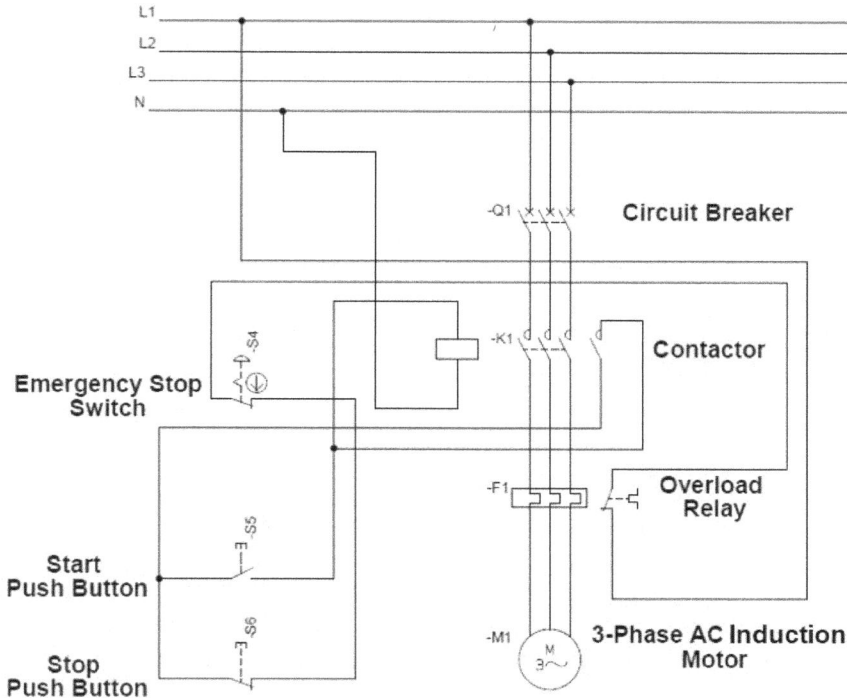

Figure 6.2 – Wiring diagram of a DOL starter

- **Pictorial diagrams** are electrical diagrams that show a picture or detailed drawing of the physical components and the wiring between components. They are used by people that do not have sound electrical drawing skills. A pictorial diagram is not suitable for troubleshooting circuits. *Figures 5.9, 5.10, 5.12, and 5.13* in *Chapter 5, Introduction to Variable Frequency Drive (VFD)*, shows the pictorial diagrams of a basic **Variable Frequency Drive's** (**VFD's**) wiring.

- **Ladder diagrams** consist of two vertical lines (power supply) and horizontal lines that represent the control circuit. They are called ladders because they look like a ladder. This diagram also does not include where the components and wires are located in the real system or device, but they show how the components interconnect. Ladder diagrams make complex circuits easy to read and understand. They are suitable for troubleshooting circuits:

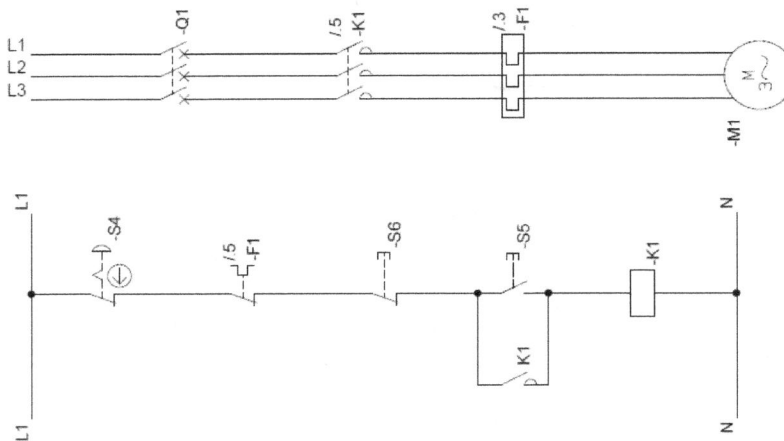

Figure 6.3 – Ladder diagram of a DOL starter

This section explains electrical diagrams and the various types available. You should now be able to differentiate between schematic diagrams, pictorial diagrams, ladder diagrams, and wiring diagrams, which most people find difficult to differentiate between. Many people think they all mean the same thing, but they actually have different meanings.

In the next section, we will discuss a software (PCSCHEMATIC Automation) that can be used to create electrical diagrams/drawings.

Overview of PCSCHEMATIC Automation software

PCSCHEMATIC Automation is CAD software used for creating professional and complex drawings for automation, electrical, hydraulic, and pneumatic related installations. PCSCHEMATIC Automation contains tools that simplify the creation of complex electrical drawings. It is a powerful and efficient application for creating electrical drawings with ease.

Some of the features of PCSCHEMATIC Automation include the following:

- A clean and neat circuit development environment

- A well-equipped database of electrical and mechanical components

- Complete solutions for all electrical circuit drawings

- Support for pneumatic and hydraulic circuit drawing
- Report generation
- Complex design support

You have now learned about PCSCHEMATIC Automation and its features. You should now have a good understanding of the capability of the software.

In the next section, we will be learning how to download and install the PCSCHEMATIC Automation software. Having the software downloaded and installed on your PC will enable you to have a hands-on experience with the software while reading the book. You will be able to practice the steps on your own PC.

Downloading and installing PCSCHEMATIC Automation

PCSCHEMATIC Automation has a free version of its software that can be used for your CAD needs. The free version comes with 25 electrical symbols and your drawings can have up to 10 pages and 200 connection points. If you need more than the available symbols, or if you need more pages or connections points for your drawing, you will need to check their paid version for the version that meets your needs.

Follow the step-by-step guide shown next to download the free version of the software:

1. Visit https://www.pcschematic.com/en/.

   The home page of PCSCHEMATIC Automation, as shown in the following screenshot, will be seen in your browser:

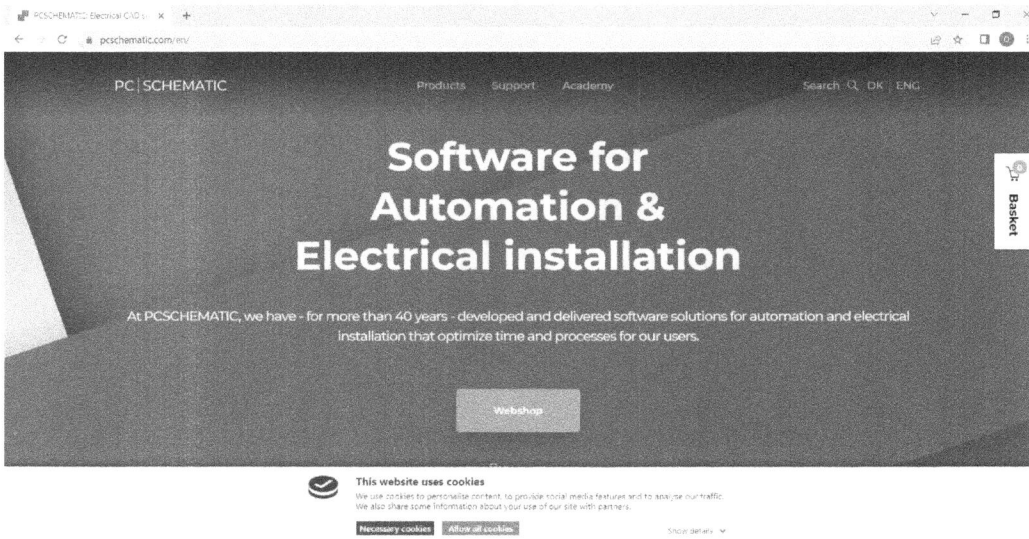

Figure 6.4 – Home page of PCSCHEMATIC Automation

2. Click **Support** and then **PC | Schematic** in the **Downloads** section. Next, click on **PCSCHEMATIC Automation Demo**.

3. Fill out the necessary details in the following form and click **Submit**. An email containing the download link for the **PCSCHEMATIC Automation Demo** software will be sent to you:

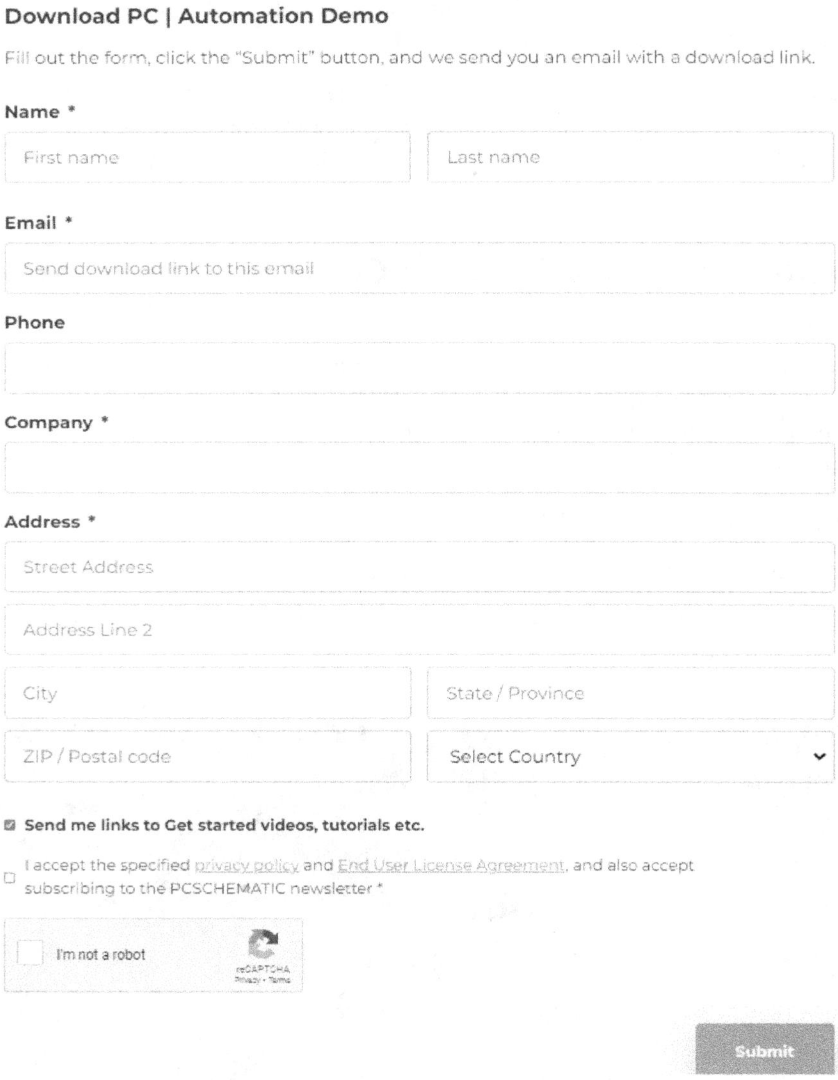

Download PC | Automation Demo

Fill out the form, click the "Submit" button, and we send you an email with a download link.

Name *

First name   Last name

Email *

Send download link to this email

Phone

Company *

Address *

Street Address

Address Line 2

City   State / Province

ZIP / Postal code   Select Country

☑ Send me links to Get started videos, tutorials etc.

☐ I accept the specified privacy policy and End User License Agreement, and also accept subscribing to the PCSCHEMATIC newsletter *

☐ I'm not a robot   reCAPTCHA
                     Privacy · Terms

Submit

Figure 6.5 – Form required to download PCSCHEMATIC Automation

4. Check your email and open the message from PCSCHEMATIC that contains the download link, as shown in the following screenshot:

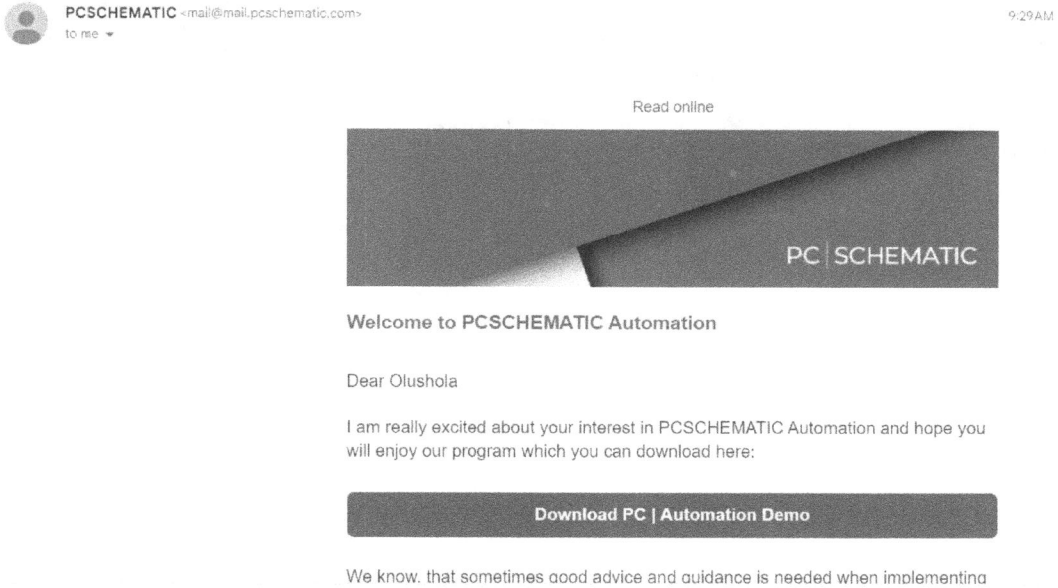

Figure 6.6 – Message from PCSCHEMATIC

5. Click **Download PC | Automation Demo**.

You will have PCSCHEMATIC Automation in the Downloads folder of your PC when the download is complete. We will learn how to install the application you have just downloaded in the next section.

Installing PCSCHEMATIC Automation

Follow these steps to install the PCSCHEMATIC Automation software on your PC:

1. Double-click the application file of the PCSCHEMATIC Automation software you downloaded (PCSCHEMATIC_Automation40_UK_V23) and click **Run**.

A setup screen, shown as follows, will appear:

Figure 6.7 – PCSCHEMATIC Automation Demo Ver 23.0 Installation screen

2. Click **Next** to continue the setup. Follow the instructions on the screen to complete the setup:

    A screen containing the license agreement will appear, as shown in the following screenshot:

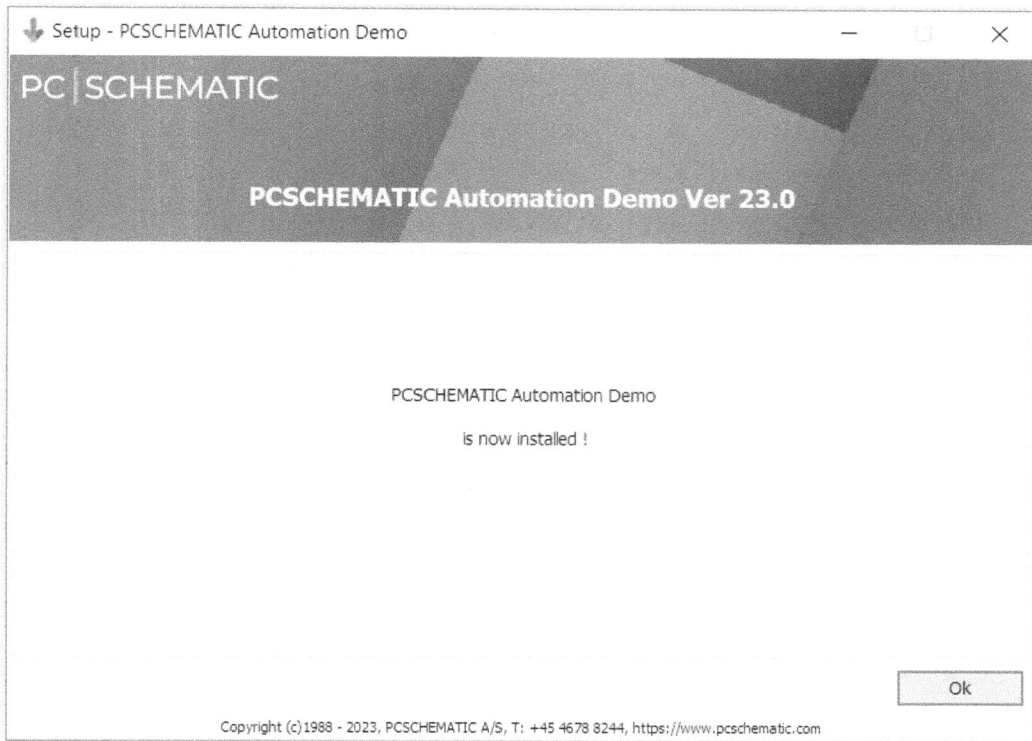

Figure 6.8 – Final installation screen

3.   Click **Ok** on the final screen and start enjoying the free CAD software.

This section explained how you can download and install the free version of the PCSCHEMATIC Automation software. This is very important to understand other parts of this book. You will be able to follow the step-by-step guide on your own when you have the software. The version used in the remaining sections of this chapter is lower than the version 23.0 we downloaded and installed in this section but the procedures explained in the older version of the remaining sections can still be used on the newer version (23.0). In the next section, we will learn how to launch the PCSCHEMATIC Automation and also explore the screen elements.

Launching PCSCHEMATIC Automation and exploring the screen elements

In this section, we will learn how to launch PCSCHEMATIC Automation and also explore the screen elements. Let's start by learning how to launch the application.

Launching the application

Follow these steps to launch PCSCHEMATIC Automation:

1. From the **Start** menu, locate **PCSCHEMATIC Automation.**
2. Click **PCSCHEMATIC Automation** to start the application.

Exploring the screen elements

The following screenshot shows the PCSCHEMATIC Automation interface, showing some of the important elements:

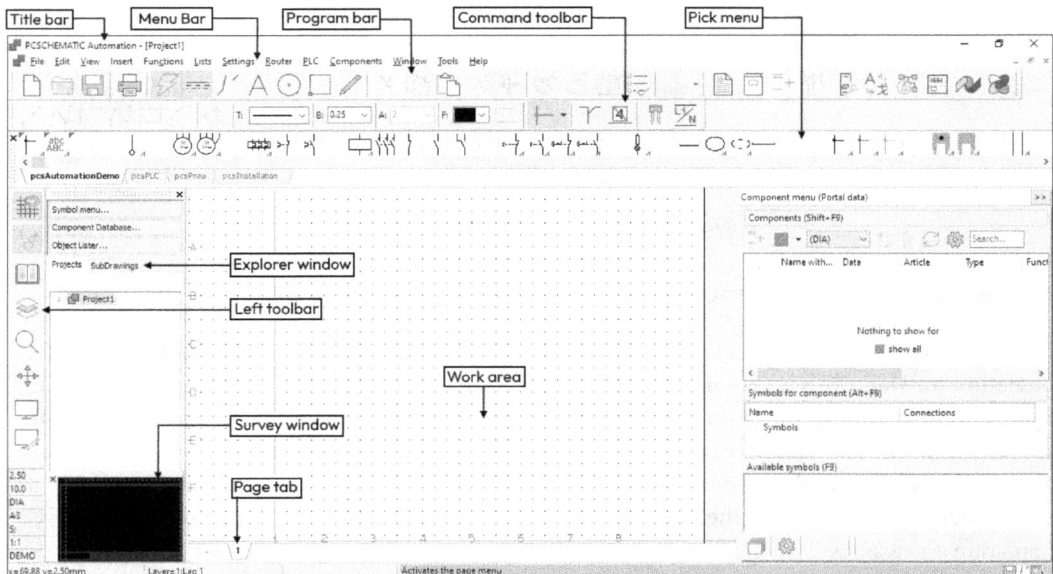

Figure 6.9 – Screen elements of PCSCHEMATIC Automation

Let's have a look at each screen element.

- **Title bar**: This bar displays the name of the program and the filename of the current project if saved.

- **Menu bar**: This contains menus (such as **File**, **Edit**, **View**, **Insert**, and others) that contain all the functions in the program.

- **Program toolbar**: This contains commonly used program functions such as save, print, and so on, as well as common drawing and editing tools.

- **Command toolbar**: This bar changes appearance depending on the tool or function selected on the program bar. It provides useful functions and editing tools for each tool or function selected in the program toolbar.

- **Pick menu**: You can place symbol, line, arc, and other tools that you will need frequently in the pick menu so you can easily pick them from there to use.

- **Explorer window**: The explorer window gives direct access to the Symbol Menu where you can select symbols you need in your drawing. You also have direct access to the component database where you can access all the components in the database. The explorer window also contains the **Project** tab where you can get information about all the opened projects.

- **Work area**: This can be seen as the paper for your drawing. Drawings are done in the work area. You can specify the size in **Settings | Page setup**.

- **Left toolbar**: The left toolbar contains the **Zoom** tool to magnify the work area, and the **Pan** tool to move the work area left, right, up, or down. It also includes other tools such as **Snap**, **Zoom to page**, and others.

- **Survey window**: This is the black box that shows the part of the page you are currently viewing on the screen.

- **Page tab**: This shows the name of a page and can be used to switch between pages.

In this section, you have learned how to launch PCSCHEMATIC Automation and have also explored the screen elements, which include the title bar, menu bar, program bar, command bar, pick menu, explorer window, work area, left toolbar, survey window, and page tab. You need to be familiar with the screen elements to be able to understand other sections of this chapter. In the next section, we will learn how to create and save a project in PCSCHEMATIC Automation.

Creating and saving a project in PCSCHEMATIC Automation

Follow these steps to create a new project:

1. Click on **File | New**:

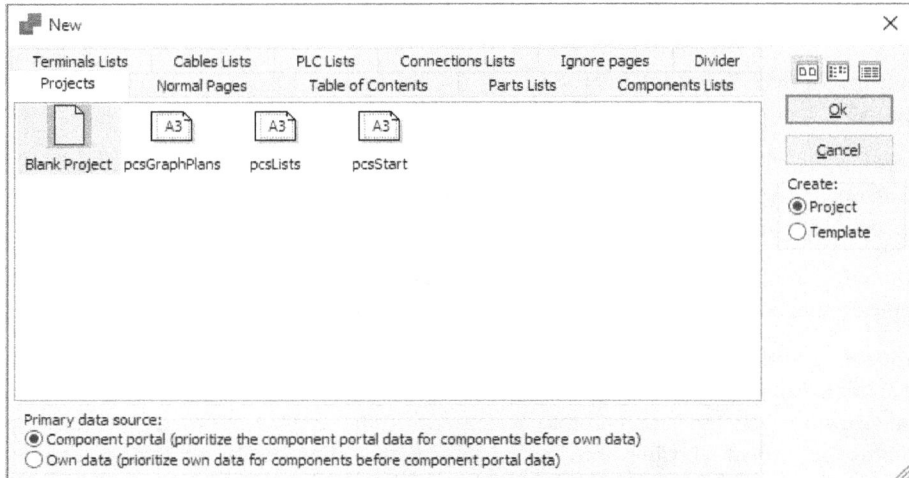

Figure 6.10– New dialogue box

2. Select **Blank Project** and click **Ok** to create a new project from scratch. The **Settings** dialogue box appears:

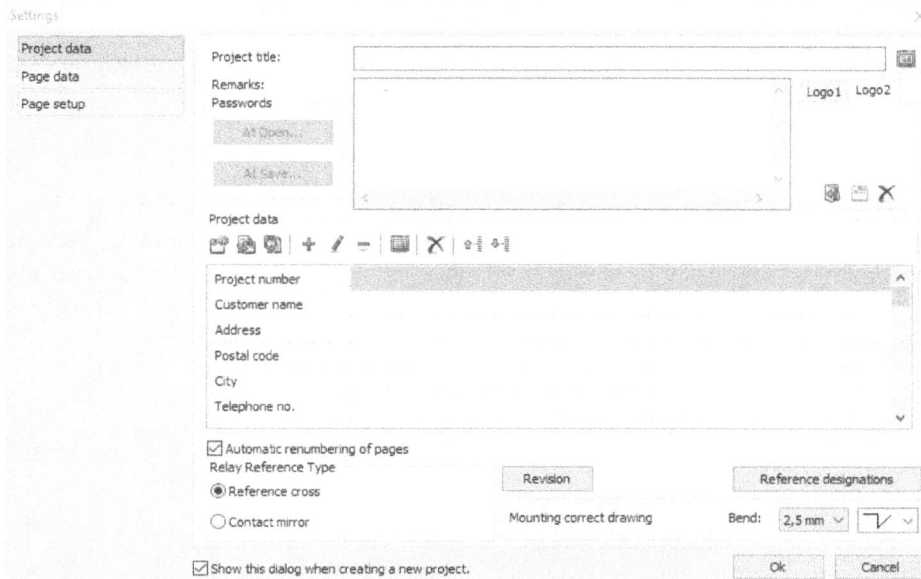

Figure 6.11 – Settings dialogue box – Project data

3. Click the **Project data** tab and fill in the **Project title**, **Customer name**, and other fields.

4. Click the **Page data** tab and check **With drawing header** on the **Primary header** tab to have a title block on your page. You can also make further changes and fill in the necessary details:

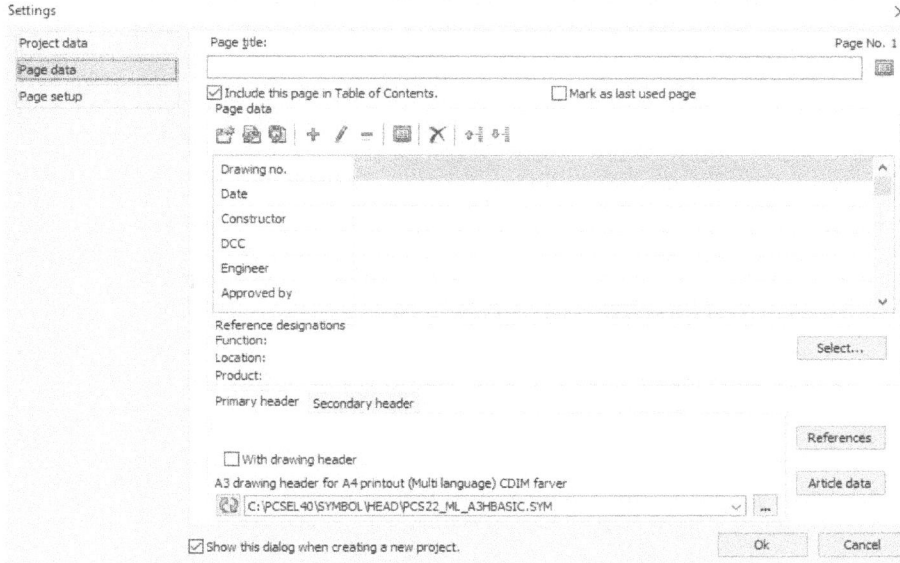

Figure 6.12 – Settings dialogue box – Page data

5. Click on **Ok**. If you followed the steps correctly, you should have something like what's shown in the following screenshot:

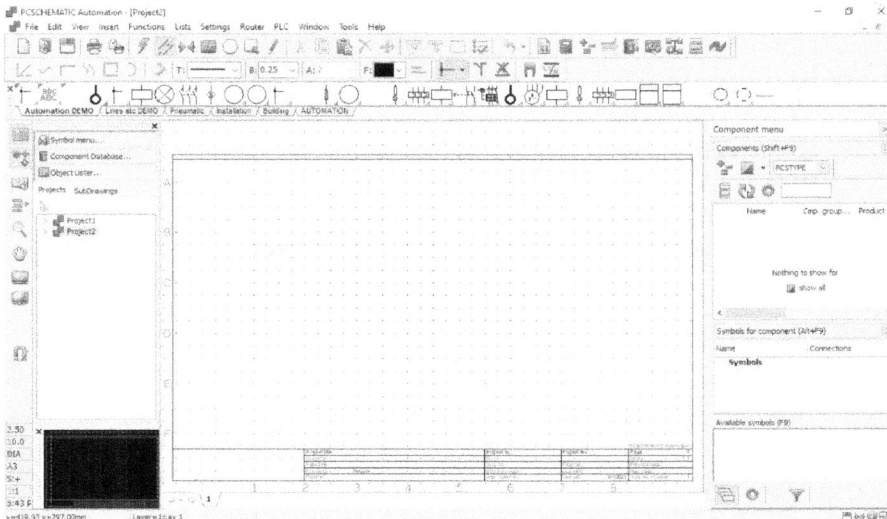

Figure 6.13 – New project in PCSCHEMATIC Automation

The preceding figure shows a new project in PCSCHEMATIC Automation.

Saving a project

Follow these steps to get your project saved:

1. Click the **File** menu and **Save As**. The **Save As** dialogue box will appear:

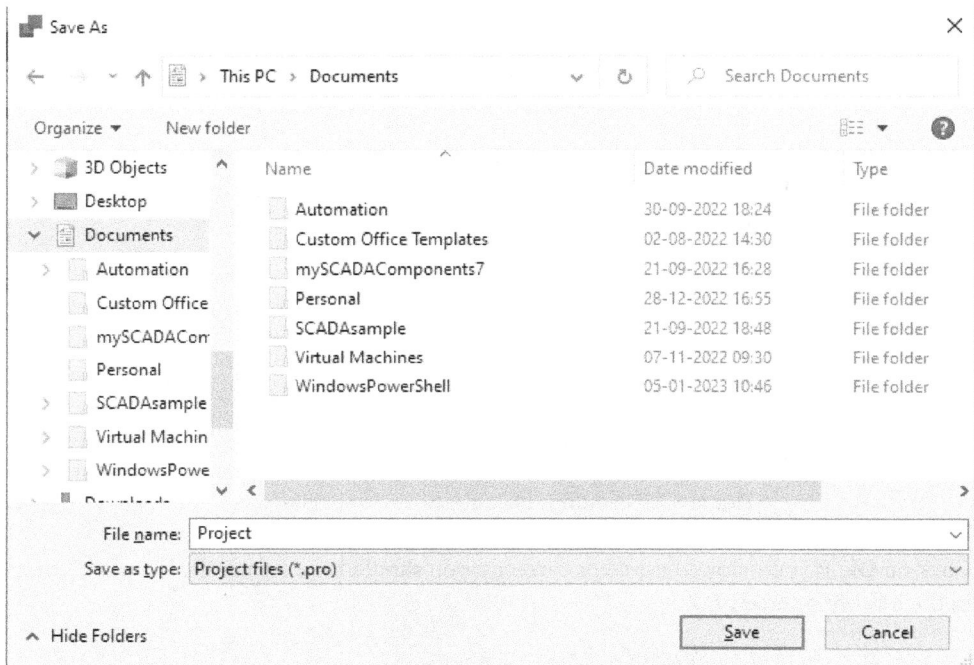

Figure 6.14 – Save As dialogue box

2. Select a location, for example, Documents.

3. Type a project name, for example, Water treatment plant.

4. Select the **Save As type,** for example, project file (.pro), Acad file (.dwr), PDF file (.pdf), or DXF file (.dxf).

5. Click on **Save**.

This section explained how to create a project from scratch and how to save it. It is important to know how to create the environment to draw your schematic and how to save it for future retrieval. In the next section, we will learn how to use the symbol, line, text, arc/circle, and area tools. Knowledge of these tools is required to create electrical diagrams/drawings in PCSCHEMATIC Automation.

Using the symbol, line, text, arcs/circles, and area tools

Symbol, line, text, arc/circle, and area buttons are available on the program bar, as shown in the following screenshot. They are very useful when working with objects and creating schematics:

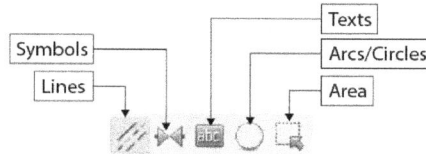

Figure 6.15 – Lines, symbols, texts, arcs/circles, and area buttons

Let's learn how to use each of these tools in the next section.

Using symbols

Symbols are the key parts of schematics. Electrical drawings/diagrams consist of symbols of components connected together with lines.

Follow these steps to place a symbol on the work area:

1. Click symbols on the program bar.

    The **Symbol Menu** dialog box will appear on the screen:

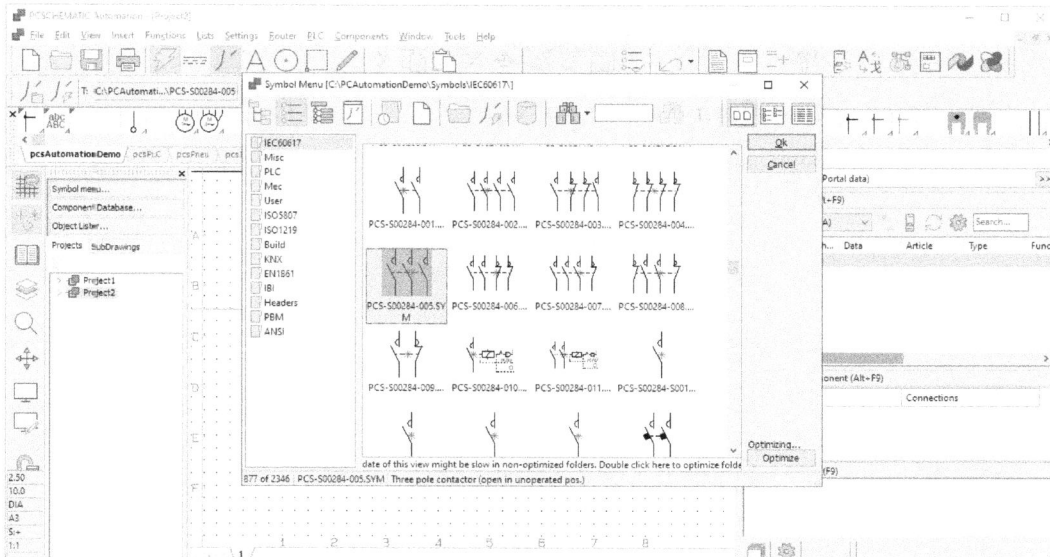

Figure 6.16 – Symbol Menu dialog box

2.   Locate and select the symbol you need from the **Symbol Menu** dialogue box and click **Ok**.

The symbol will appear with the crosshair in the work area, as shown in the following screenshot:

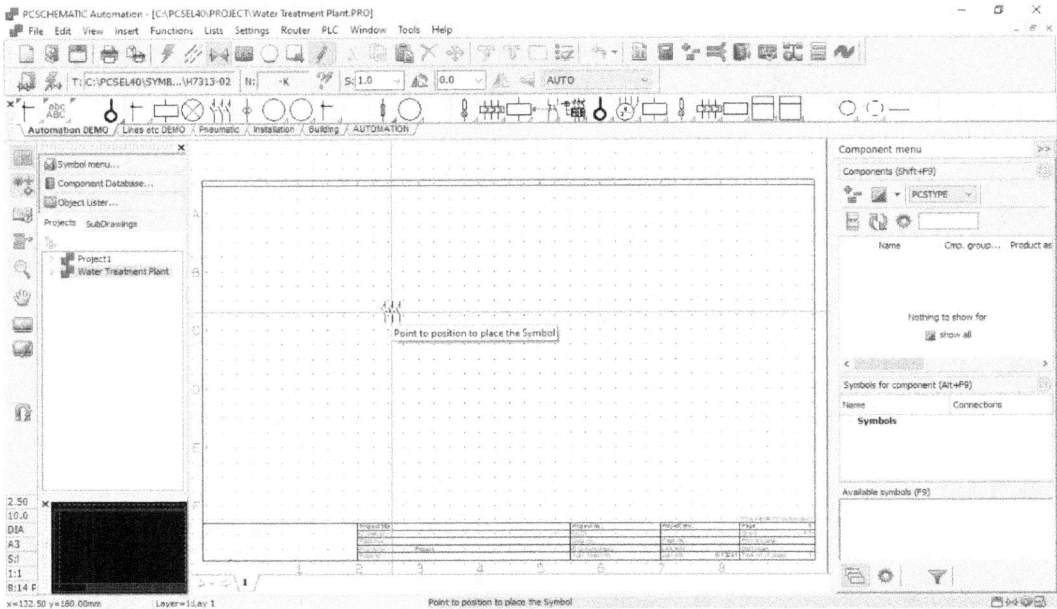

Figure 6.17 – Symbol with crosshair

3.   Click on the point you want to place the symbol.

The **Component data** dialog box will appear:

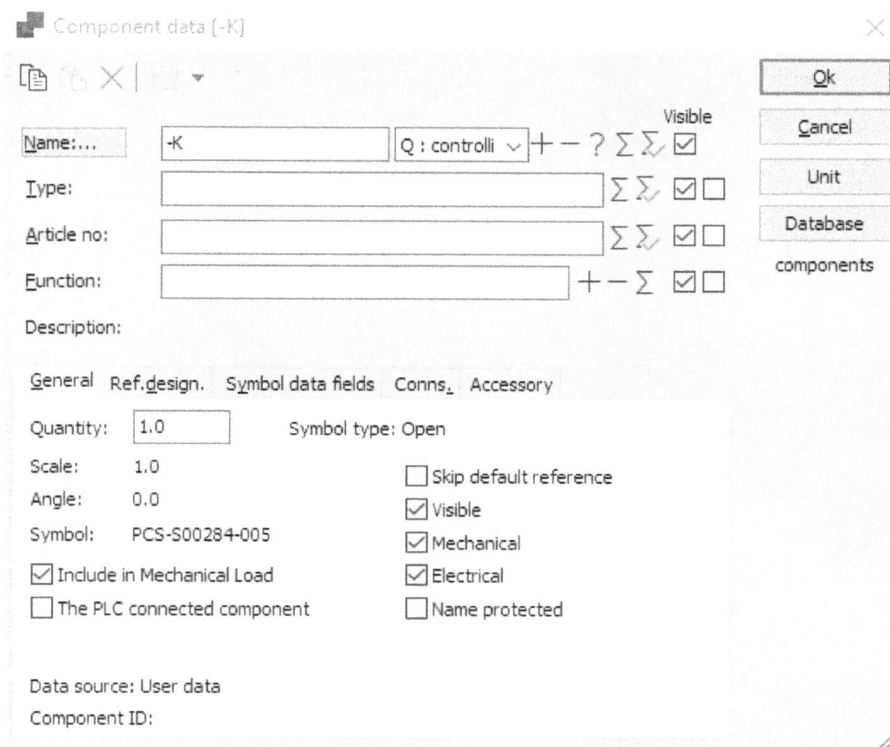

Figure 6.18 – Component data dialogue box

4. Type the component's name in the **Name** box. You can just click the question mark to add the next available number to the letter in the **Name** box.

5. Click on **Ok**.

    The symbol will appear in the work area. You can continue to add more of the same symbols by repeating *steps 3 to 5*. The following screenshot shows two of the same symbols placed in the work area:

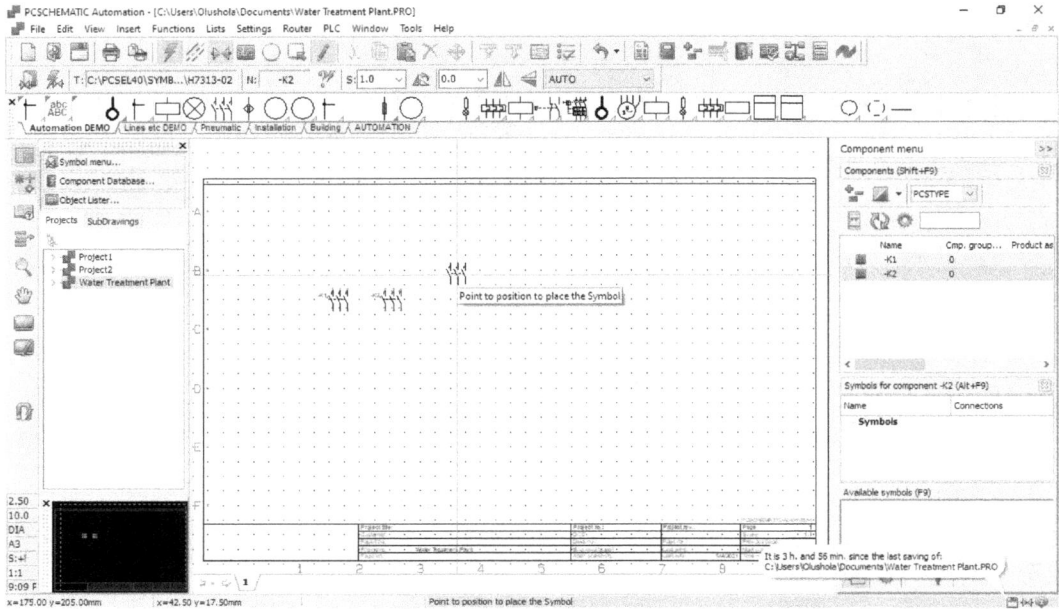

Figure 6.19 – Placing more symbols in the work area

6. Press *Esc* to end the command.

Let's now learn how to use the lines tool.

Using lines

The lines button can be used to draw a line (straight line, slanted line, angled line, and others).

Follow these steps to draw a line:

1. Click on the lines button on the program bar.

2. Click on the draw icon on the program bar.

3. On the **command bar**, select either a straight line, slanted line, angled line, curved line, and so on. Also select a line type, line width, and choose from the other options:

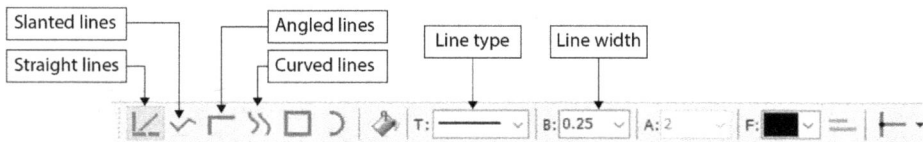

Figure 6.20 – Command bar showing lines related to tools

4. Click the start corner of the line.

    A dialogue box may appear depending on what you have selected on the command bar:

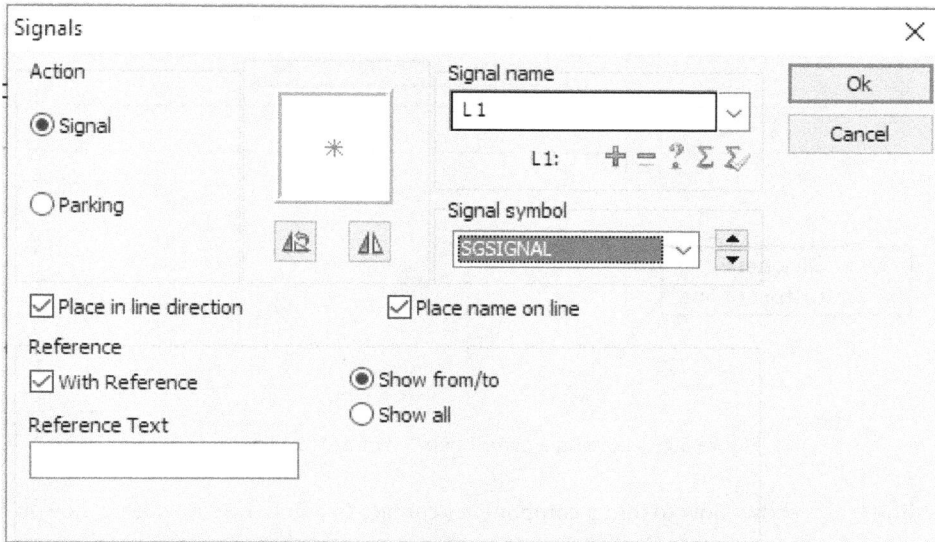

Figure 6.21 – Signals dialogue box

5. Type a signal name, for example, L1. You can type a letter (for example, L) and click the question mark to add the next free number to the letter and click **Ok**.

6. Click other points on the line.

7. Type a signal name, for example, L1, and click **Ok**.

8. Press *Esc* to end the line command.

    You can start another line with the signal name L2, L3, and so on, as shown in the following screenshot:

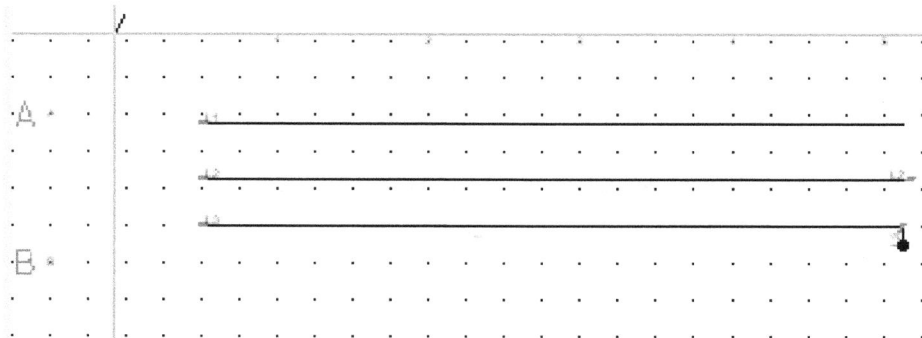

Figure 6.22 – Three lines with signal names L1, L2, and L3

9.   You can also use the same line command to join a component's contact to a line, as shown in the following screenshot, or join a component's contact to another component's contact:

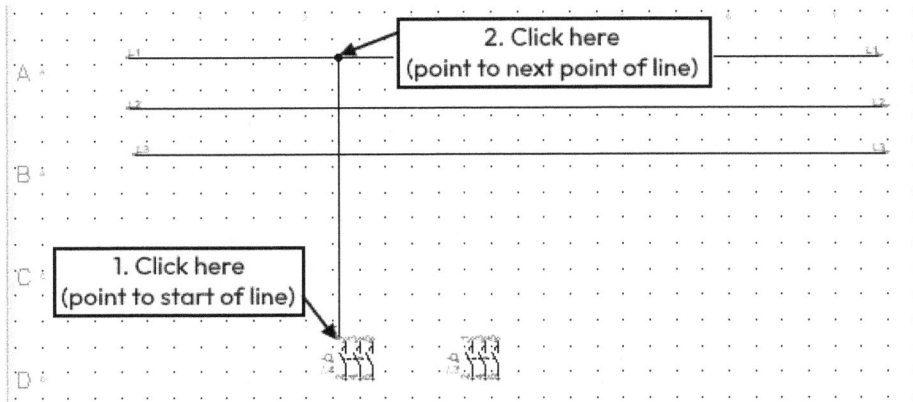

Figure 6.23 – Joining a component's contact to a line

The preceding figure shows how to join a component's contact to a line. Let's now learn how to use the texts tool in the next section.

Using texts

The **texts** button is used to activate the **texts** command for typing text. There is always the need to have text on a schematic diagram to describe things. We are going to learn how to add text (**contactor**) beside the contactor symbol.

Follow these steps to add text to your work area:

1.   Click on the **texts** tool on the program bar.

2.   On the command bar (*Figure 6.24*), click the **Text properties** button:

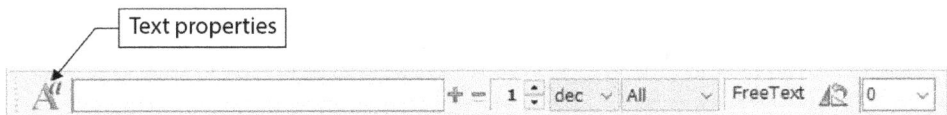

Figure 6.24 – Command bar with text-related tools

3. Make the necessary changes in the **Text properties** dialogue box and click **Ok**:

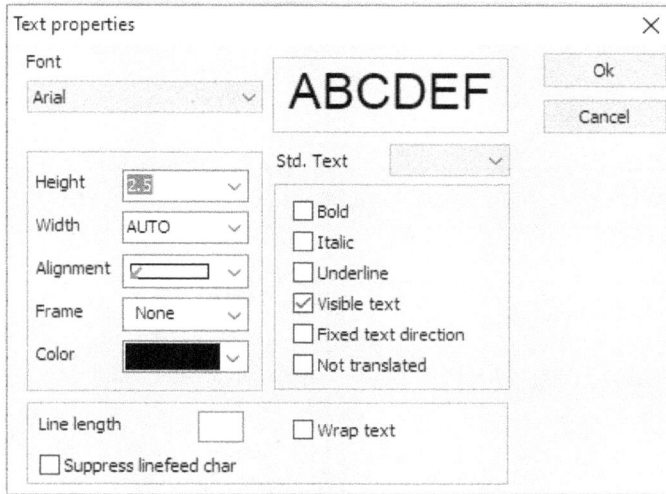

Figure 6.25 – Text properties dialogue box

4. Click inside the textbox on the command bar, type some text (for example, `Contactor`), and press *Enter*.

5. The crosshair appears with the text attached to it, as shown in the following screenshot:

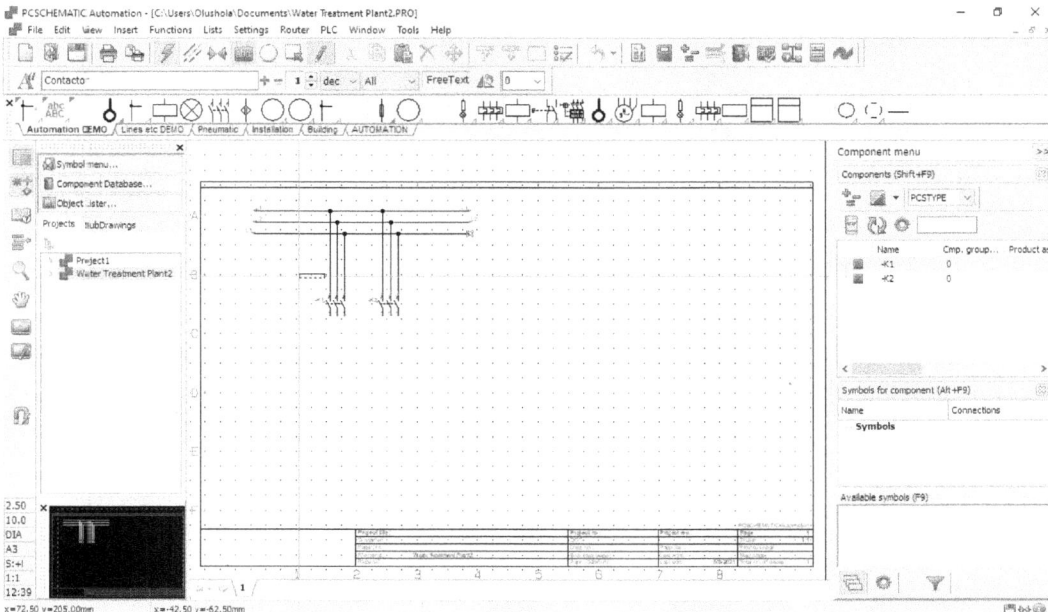

Figure 6.26 – Crosshair with the text attached to it

6. Move to the point you want to place the text and click on that point.

7. The text will appear in the position, as shown in the following screenshot. You can press *Esc* to end the command:

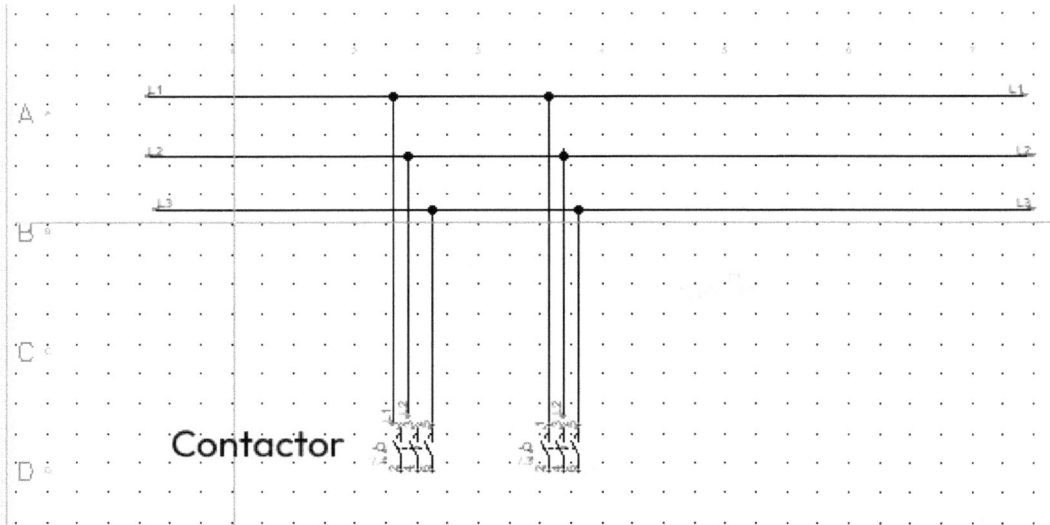

Figure 6.27 – Text placed in the work area

The preceding figure shows text (**Contactor**) placed in a work area in PCSCHEMATIC Automation. We will now learn how to use the arcs/circles tool in the next section.

Using arcs/circles

The arcs/circles tool on the program bar can be used to draw an arc or circle, as the name implies. There could be a need to draw an arc or circle in your schematic diagram.

Follow these steps to draw an arc or circle:

1. Click on the arcs/circles button on the program bar.

2. On the command bar, specify a radius (**R**), arc start angle (**V1**), arc end angle (**V2**), and choose from the other options.

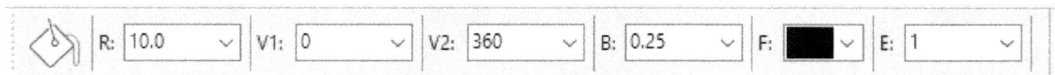

Figure 6.28 – Command bar with arcs/circles-related options

3. Click on the arcs/circles button on the program bar.

    The crosshair will show a circle attached to it as, shown in the following screenshot:

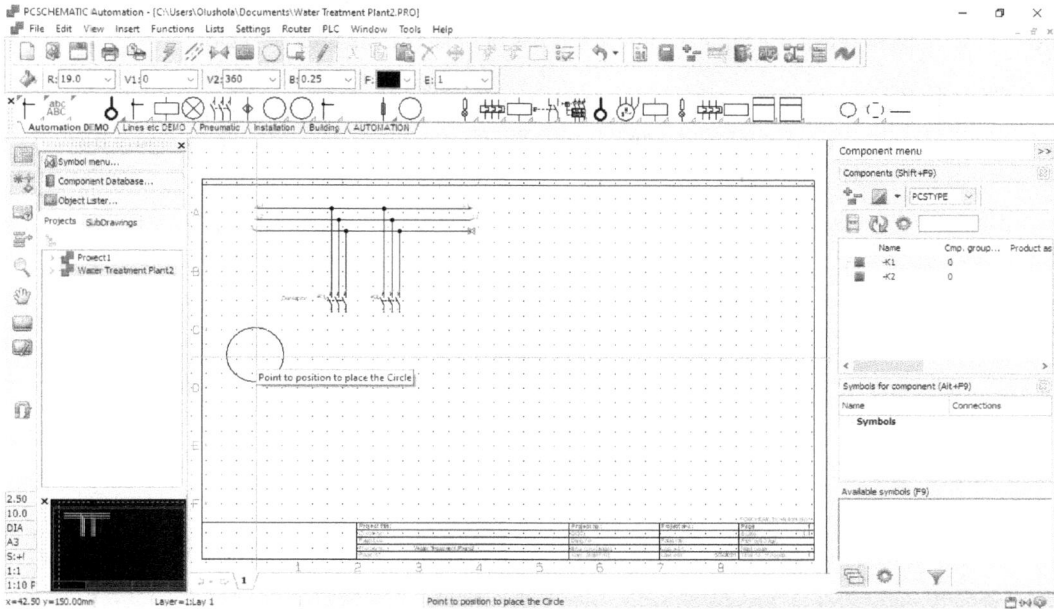

Figure 6.29 – Crosshair with a circle attached to it

4. Move to the position you want to place the circle and click on that point.
5. Press *Esc* to end the command.

> **Note**
> You can make an arc by setting **V1** (arc start angle) to 0 and **V2** (arc end angle) to a value below 360. For example, when **V1** is 0 and **V2** is 180, you will have a semi-circle.

In the next section, we will learn how to use the area tool.

Using area

The area button can be used to select the same or different types of objects in a region.

Follow these steps to select the same or different types of objects in an area or region:

1. Click on the area button on the program bar.

2. Drag a marquee rectangle around the objects to select the objects, as shown in *Figure 6.30*:

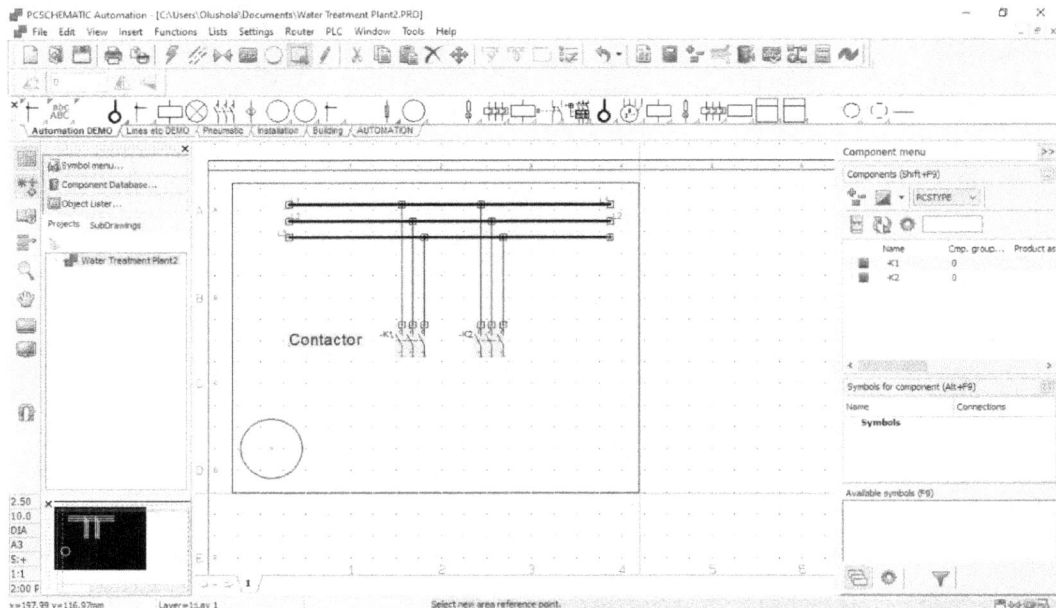

Figure 6.30 – Different objects selected using the area tool

The preceding figure shows that the objects are selected. You can click **Delete**, **Copy**, or any other button on the program toolbar depending on the action you want to carry out on the selected objects.

You have now learned how to use the line, symbol, text, arc/circle, and area tools. They are useful when drawing schematics. The knowledge in this section will help in understanding a simple schematic drawing of a DOL starter power and control circuit, explained in the next section.

Drawing a power and control schematic of a DOL starter using PCSCHEMATIC Automation

In this section, we will apply the knowledge we have so far gathered to draw a simple power and control schematic of a DOL starter using PCSCHEMATIC Automation. Let's get started:

1. Create a new project and save it with the filename DOL Starter using the method explained in the previous section.

    You should have what is shown in the following screenshot:

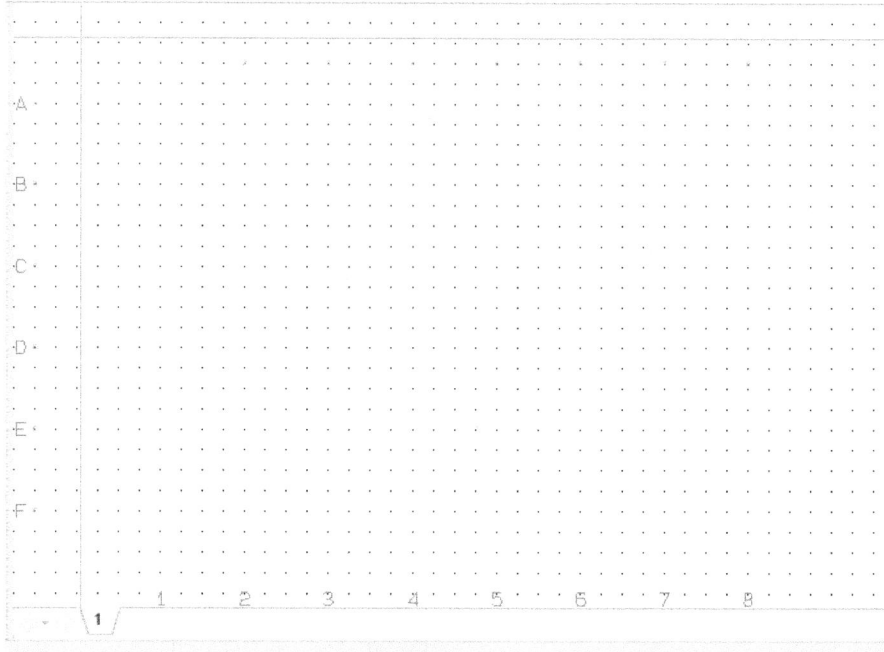

Figure 6.31 – New project page in PCSCHEMATIC Automation

2. Draw three phase lines (L1, L2, and L3) and a neutral line (N) using the line tool explained in the previous section. You should have something like the following:

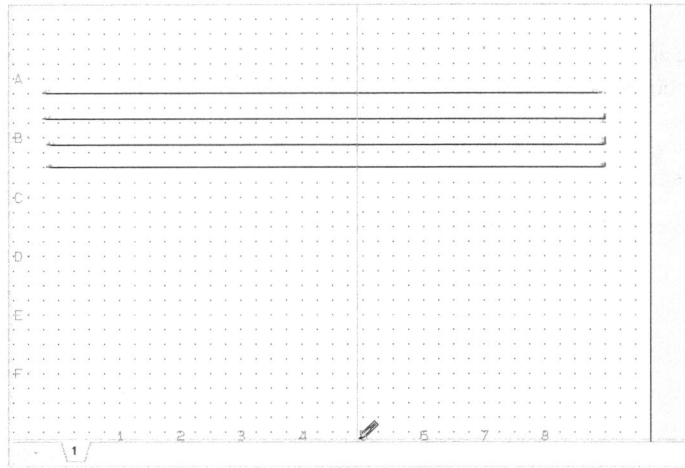

Figure 6.32 – Three phase lines (L1, L2, and L3) and a neutral line (N) drawn in the work area using the line tool

3. Insert the symbols for the breaker, contactor, overload relay, and three-phase motor using the steps in the previous section.

    Figure 6.33 shows the **Symbol Menu** dialogue box with a circuit breaker and contactor symbol.

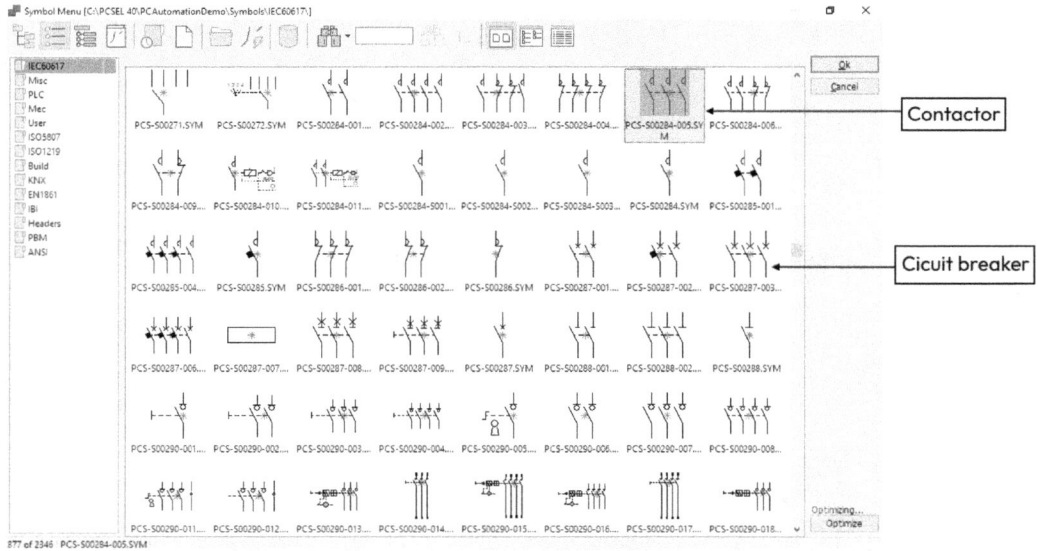

Figure 6.33 – Symbol Menu with circuit breaker and contactor symbol

Figure 6.34 shows the **Symbol Menu** dialogue box with the overload relay symbol.

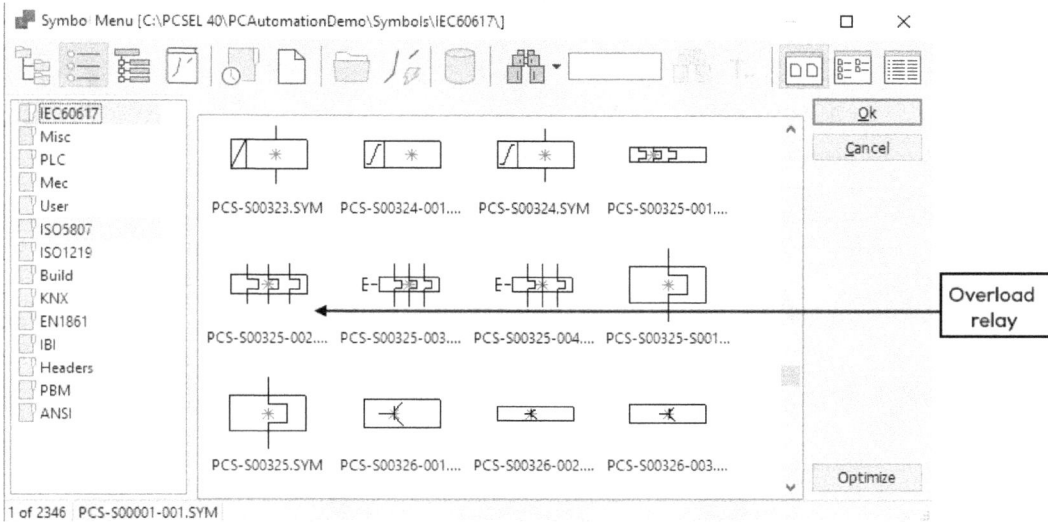

Figure 6.34 – Symbol Menu with the overload relay symbol

Figure 6.35 shows the **Symbol Menu** dialogue box with a 3-phase AC induction motor symbol:

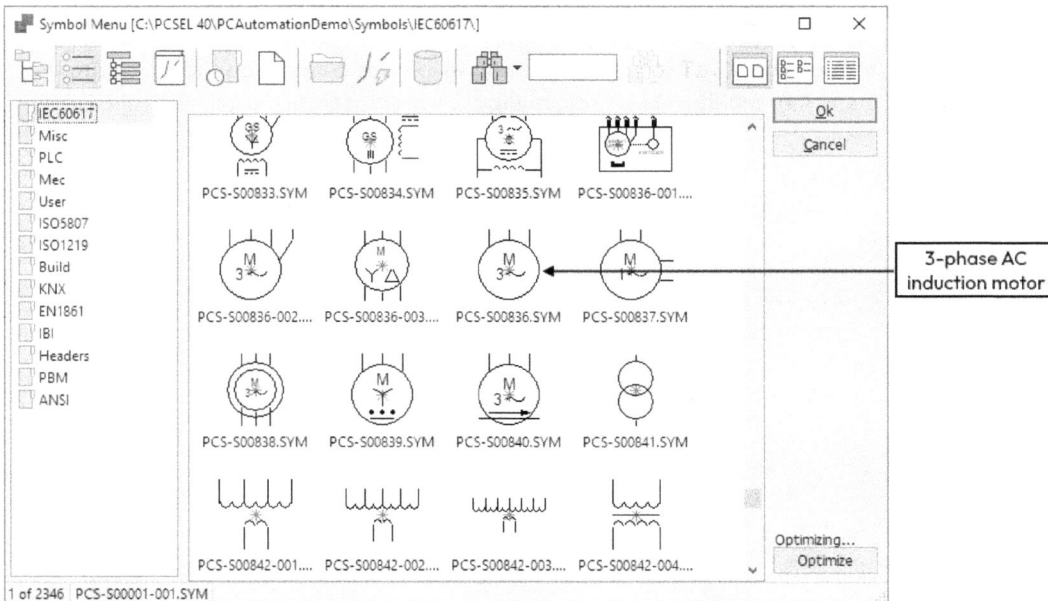

Figure 6.35 – Symbol Menu with a 3-phase AC induction motor symbol

4. After placing all the required components in the work area, you should have something like what's shown in the following screenshot:

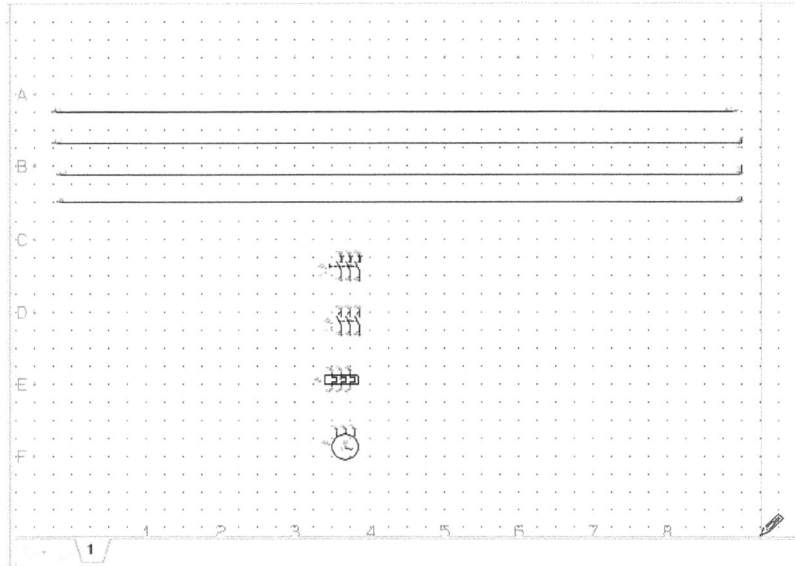

Figure 6.36 – Required symbols placed in the work area

5. Use the line tool explained in the previous section to join the contacts on the breaker to the phases (L1, L2, and L3). You should have something like what's shown in the following screenshot:

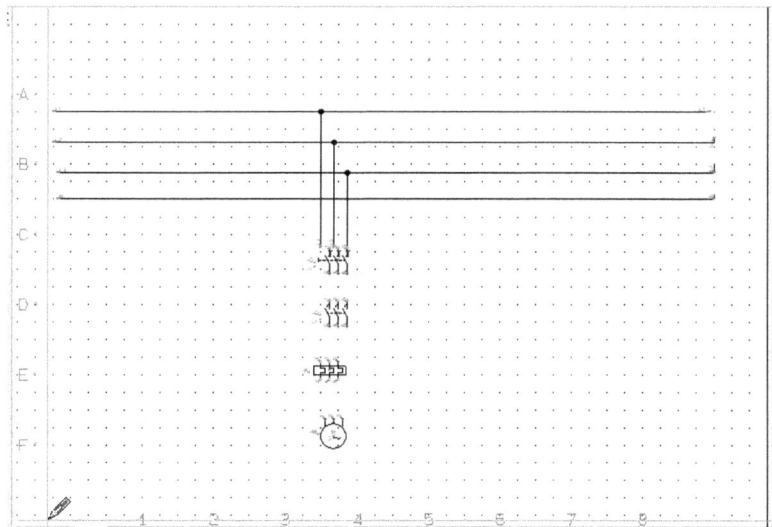

Figure 6.37 – Contactor's contacts joined to L1, L2, and L3

6. Use the same line tool to join each component's contact to the others. You should have something like what's shown in the following screenshot if you get it right:

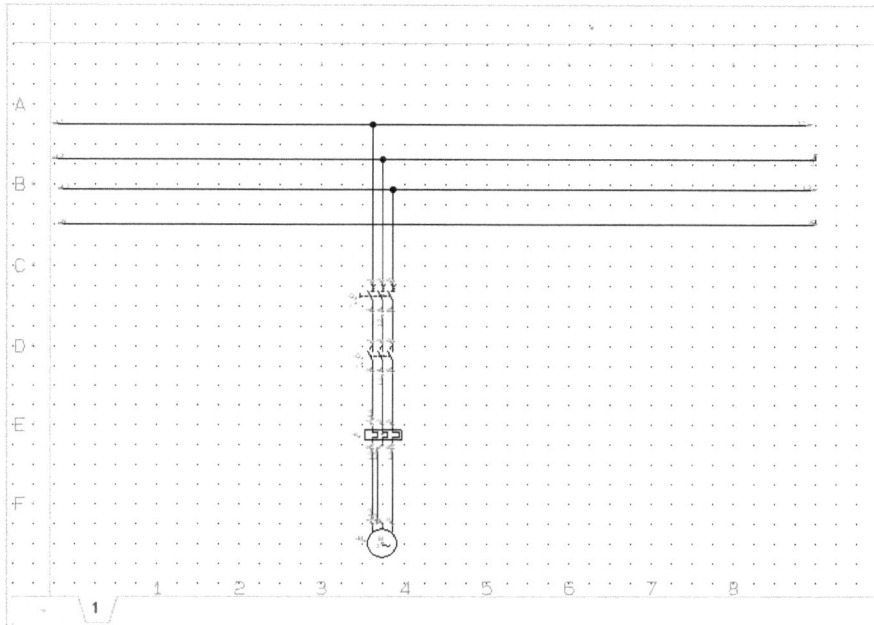

Figure 6.38 – Component contacts joined to each other

The preceding screenshot shows the power circuit of a DOL starter. Let's try to draw the control circuit on page two.

7. Click on **Insert** and click **Insert Page** as shown in *Figure 6.39*:

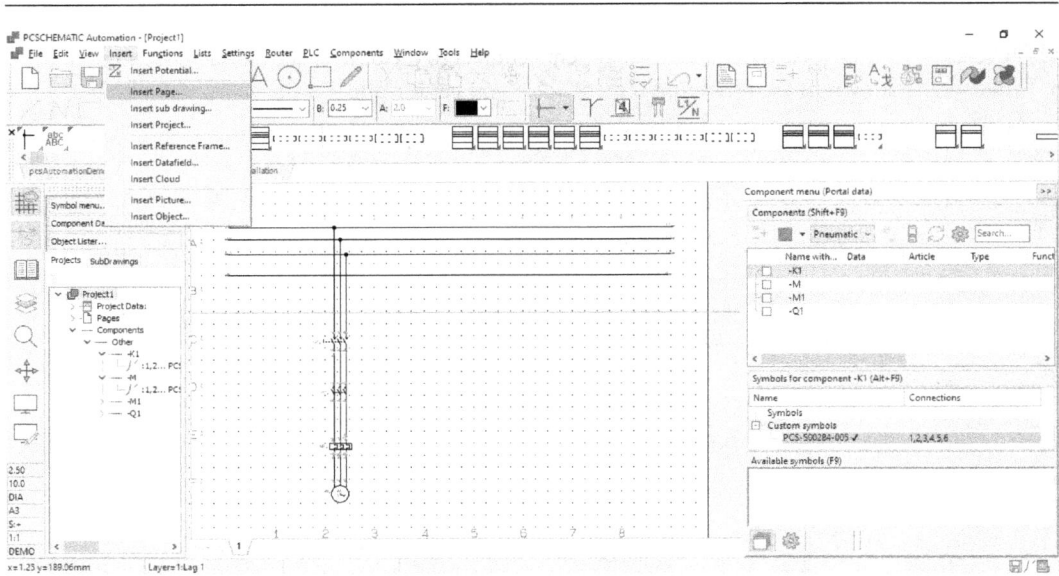

Figure 6.39 – Inserting a page

8.   Select **Normal** and click **Ok** in the **Page function** dialogue box:

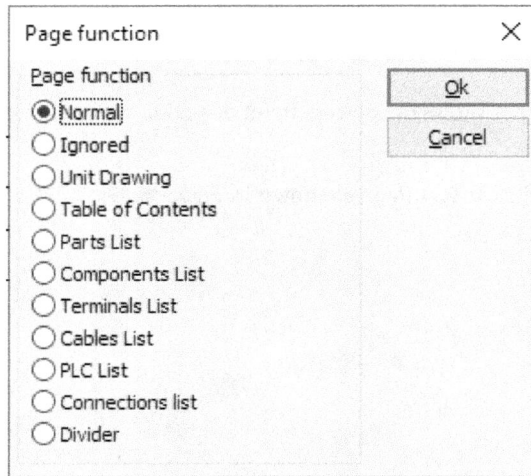

Figure 6.40 – Page function dialogue box

9. Select **Blank Page** and click **Ok** in the **New** dialogue box (*Figure 6.41*):

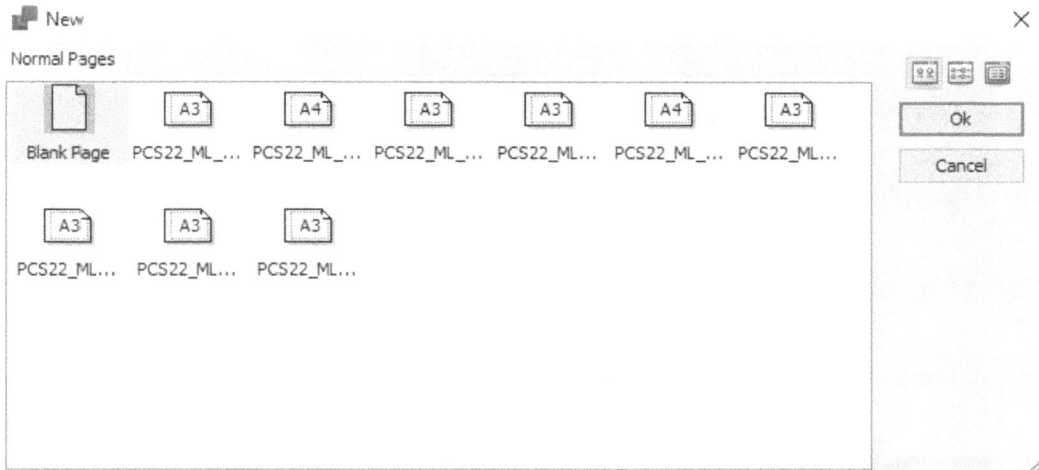

Figure 6.41 – New dialogue box

You should have a new page (page 2) as shown in the following screenshot:

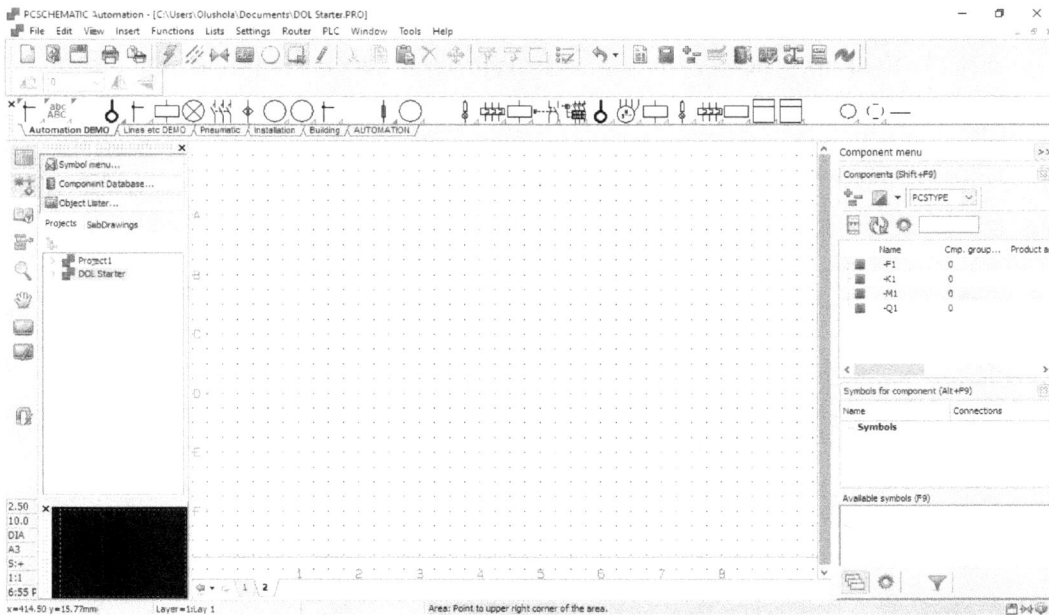

Figure 6.42 – New page (page 2) inserted

10. Right-click on the work area and click **Page data:**

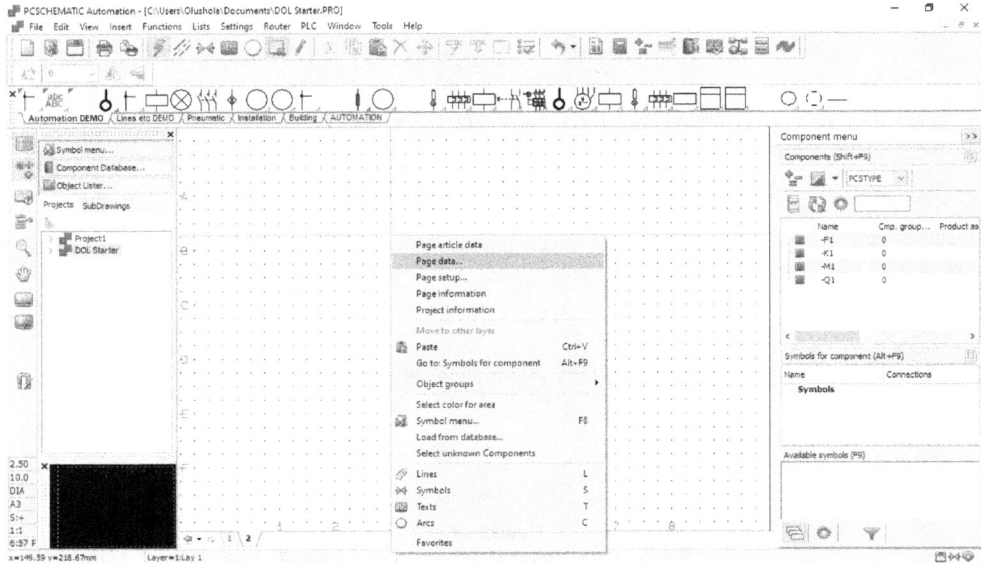

Figure 6.43 – Selecting the Page data option

11. Check **With drawing header** which is under **Primary header** tab in the **Settings** dialogue box and click **Ok:**

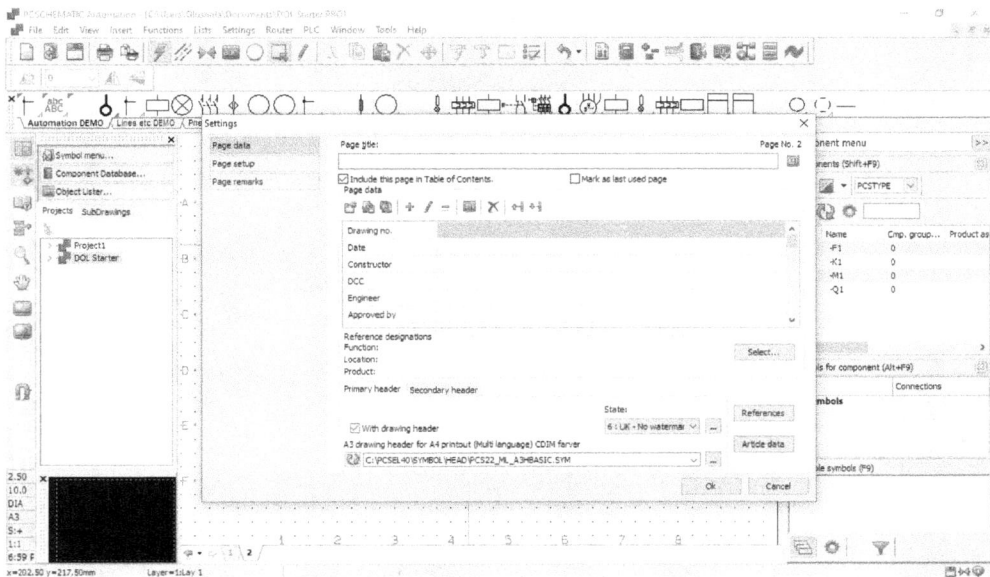

Figure 6.44 – Settings dialogue box

You should have something like what is shown in the following screenshot if everything is right:

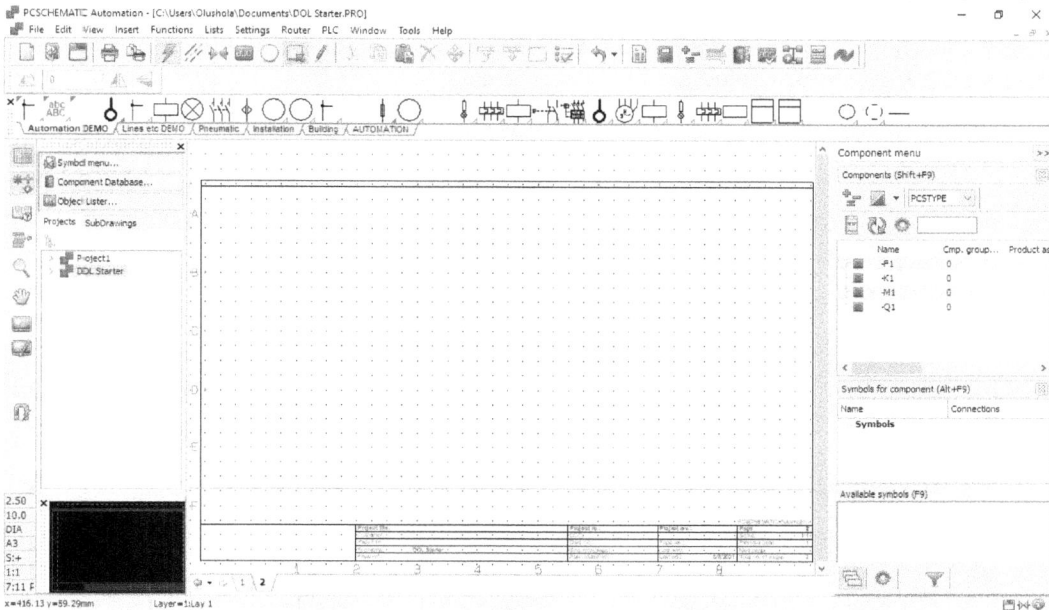

Figure 6.45 – Primary header included on page 2

12. Start by drawing line 1 (L1) and a neutral line (N) using the line tool. You should have what is shown in *Figure 6.46* if you get it right:

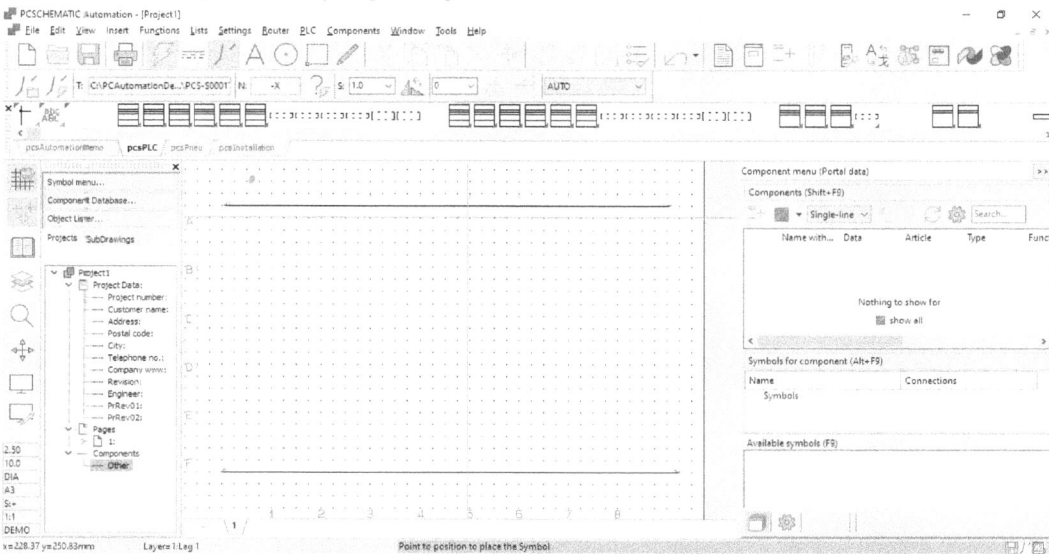

Figure 6.46 – Line 1 (L1) and a neutral line (N) drawn in the work area using the line tool

13. Place the necessary symbols, such as emergency stop, OL contact (NC), stop push button (NC), start push button (NO), auxiliary contact of contactor, and contactor coil in the work area and arrange them as shown in the following screenshot. The symbols are all available in the `Symbol Menu`. You just need to locate them:

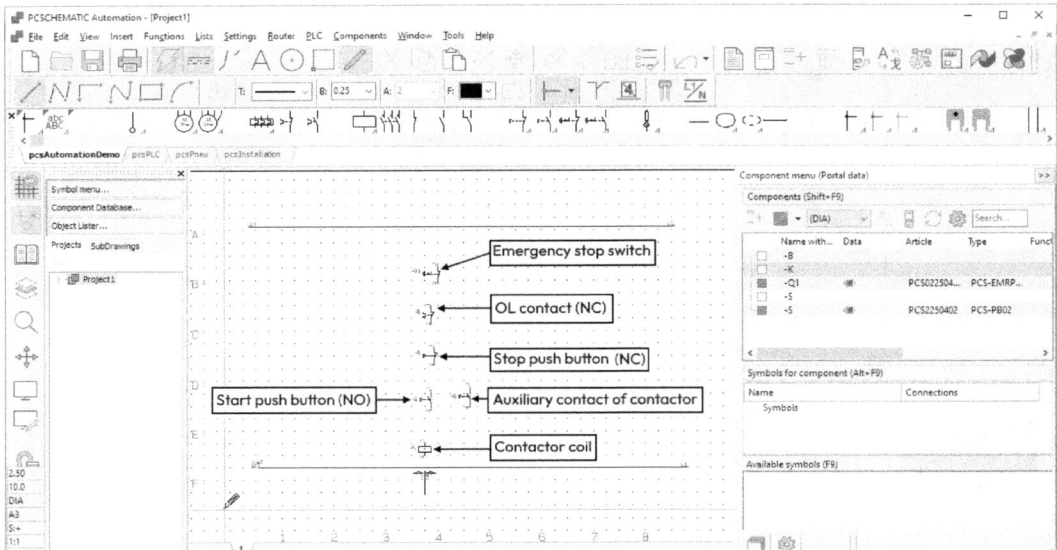

Figure 6.47 – Required symbols placed in the work area

14. Use the line tool to join one end of the emergency stop switch to line L1 and join one end of the contactor coil to N, as shown in *Figure 6.48*:

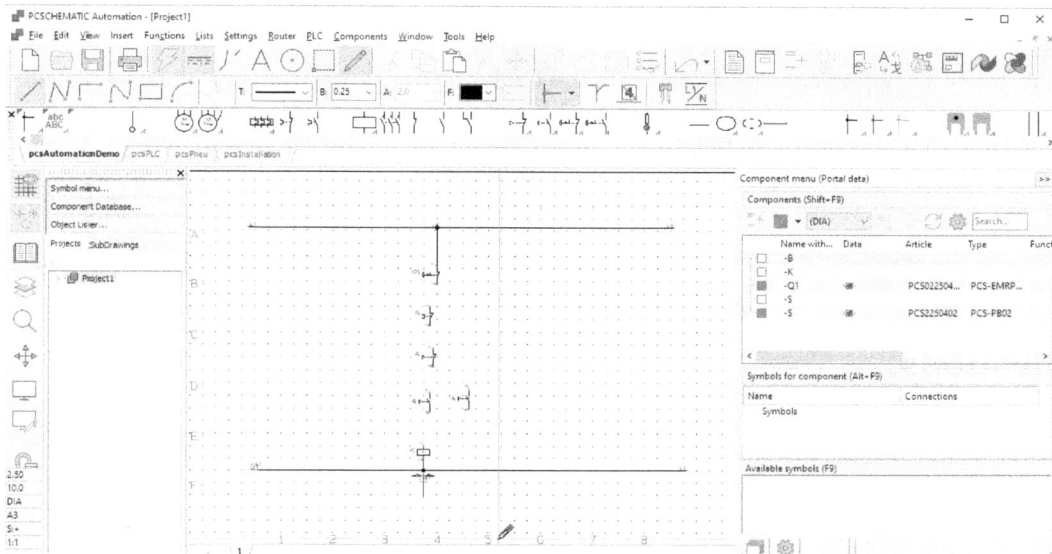

Figure 6.48 – One end of the emergency stop switch joined to L1 and one end of the contactor coil joined to N

15. Finally, use the same **Line** tool to join other contacts as shown in *Figure 6.49*.

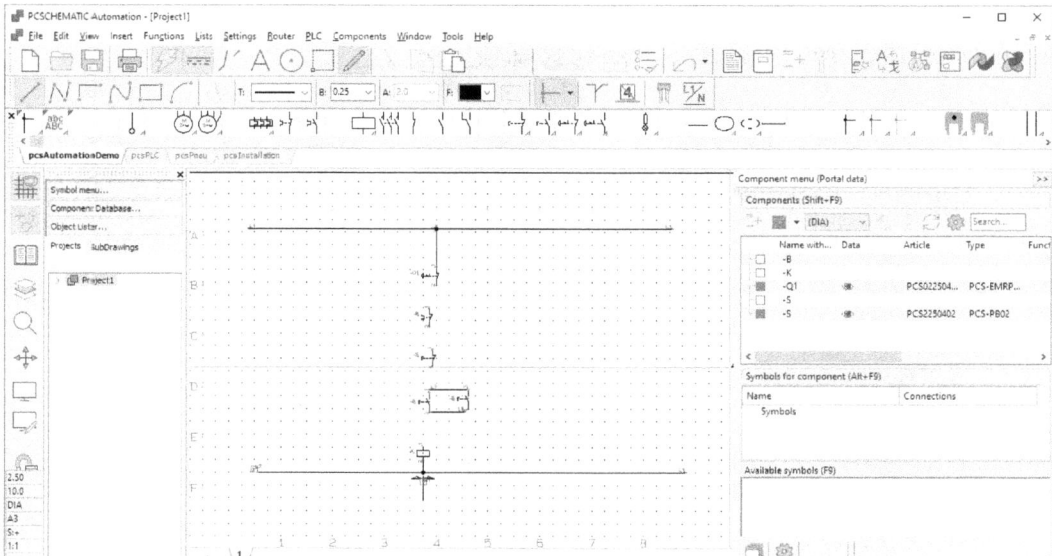

Figure 6.49 – Control circuit of a DOL starter

We have just completed learning how to draw the schematic of a DOL starter. You can use the techniques here to draw other schematic diagrams.

Summary

Great! You just completed the sixth chapter of this book. You have learned about **Computer Aided Design (CAD)** software, including the practical use of powerful and professional CAD software (PCSCHEMATIC Automation). The chapter covered the use of symbols, lines, arcs/circles, text, and area tools, and a step-by-step guide to creating a common industrial control schematic (DOL starter). The steps included screenshots to further aid understanding.

In the next chapter, we will be getting to know about **Programmable Logic Controllers (PLCs)**, the brains behind machines in various industries.

Questions

The following are questions to test your understanding of this chapter. Ensure you have read and understood the topics in this chapter before attempting the questions:

1. CAD is the acronym for _____.
2. _____ is a type of electrical diagram that makes complex circuits easy to troubleshoot.
3. An electrical diagram that shows a picture or detailed drawing of physical components and the wiring between them is known as _____.
4. In PCSCHEMATIC Automation, the black box that shows the part of the page you are currently viewing on the screen is called _____.
5. In PCSCHEMATIC Automation, *Figure 6.50* is _____.

Figure 6.50 – Screenshot showing a part of the PCSCHEMATIC Automation screen

Part 2: Understanding PLC, HMI, and SCADA

This part will take us through the major tools in industrial automation (PLC, HMI, and SCADA). You will learn by following the step-by-step instructions provided. Simulation is also available in some of the steps, which makes it easy for those without the actual device to have an understanding of how it works and how to use it. What seems complex has been made simple by using a hands-on approach.

This part has the following chapters:

- *Chapter 7, Understanding PLC Hardware and Wiring*
- *Chapter 8, Understanding PLC Software and Programming with TIA Portal*
- *Chapter 9, Deep Dive into PLC Programming with TIA Portal*
- *Chapter 10, Understanding Human Machine Interfaces (HMIs)*
- *Chapter 11, Exploring Supervisory Control And Data Acquisition (SCADA)*

Understanding PLC Hardware and Wiring

Industrial automation has been in existence since before the **programmable logic controller** (PLC) was developed in the late 1960s. Relays and timers were used for automating manufacturing processes prior to the introduction of PLCs but they had so many challenges. Some of these challenges included the large space occupied by relays and timers, the amount of time required to troubleshoot the system, and making changes to the system was very difficult due to the fact that so many relays were hardwired together in a specific order for the machine to operate. A PLC provides a solution to all these challenges by needing less space, being easy to troubleshoot, and changes can easily be made to a control system that uses a PLC. The following photograph shows a relay room with so many relays that take up much space, are not easy to troubleshoot, and changes are not easy to make on the control system:

Figure 7.1 – Relays in a relay room (Sergeev Pavel, CC BY-SA 3.0 `https://creativecommons.org/licenses/by-sa/3.0`, via Wikimedia Commons)

This chapter focuses on the hardware aspect of PLCs. We will understand what a PLC really is, the various parts that make up a PLC, the operation of a PLC, types of PLCs, PLC manufacturers/vendors, and how to wire sensors and actuators learned in the previous chapter to a PLC.

We are going to cover the following main topics in this chapter:

- Introducing a PLC
- Exploring PLC modules
- Exploring PLC types
- Understanding the PLC scan cycle
- Knowing about PLC vendors
- PLC wiring (1) – Wiring of switches, lights, and actuators to a PLC
- PLC wiring (2) – Wiring of proximity sensors (3-wire)
- PLC wiring (3) – Wiring of photoelectric sensors (retro-reflective)
- Wiring of the Siemens S7-1200 PLC (CPU 1211C AC/DC/Relay)

Technical requirements

While every part of this book is valuable, *Chapter 2, Switches and Sensors – Working Principles, Applications, and Wiring*, and *Chapter 3, Actuators and Their Applications in Industrial Automation*, are very important to have a better understanding of this chapter.

Introducing a PLC

A PLC is an industrial computer that consists of both hardware and software used for automating industrial processes. They are mostly used in manufacturing industries and can be found in other industries including transportation, manufacturing, warehousing, oil and gas, building, and so on. A PLC can also be referred to as an electronic device that accepts inputs from sensors and switches, processes them, and gives out signals to control actuators. PLCs take data from the plant floor through switches and sensors, they then execute the program logic written into them and give an output signal based on the result of the program logic to control actuators or machines connected to them. PLCs are the brains behind almost all modern automated machines in industries.

Let's now have a look at the parts that make up a PLC.

Exploring PLC modules

The hardware part of a PLC consists of various modules. The basic modules that make up a PLC include the following:

- The power supply
- The CPU

- Input module
- Output module

The following figure shows the block diagram of a PLC:

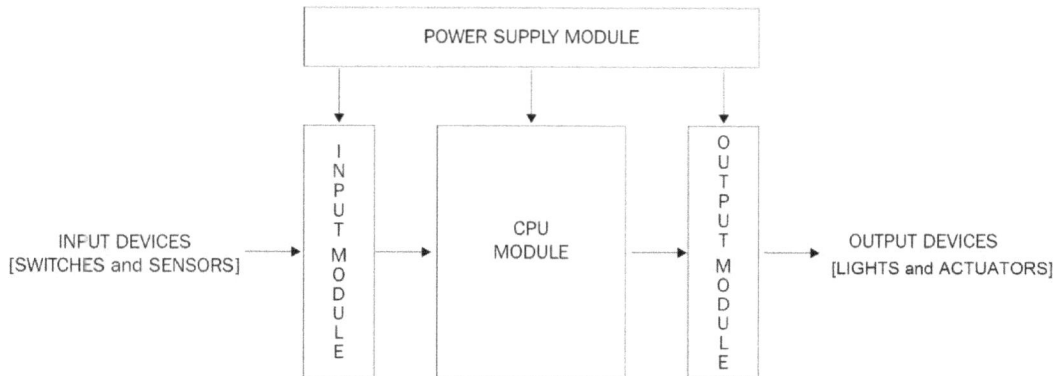

Figure 7.2 – Block diagram of a PLC

In the next section, we will be looking at each module.

The power supply

The power supply part of a PLC provides the appropriate voltage and current to other parts/modules of the PLC. Most PLCs use 24V DC and the supply from the grid is usually 110V or 220V AC, depending on the country. A power supply takes the 220V or 110V AC as input and provides the required voltage (usually 24V DC) and current as output to other parts of a PLC. The current rating of power supplies can range from 2 to about 50 amps, depending on the PLC size. The following figure shows a Siemens power supply module:

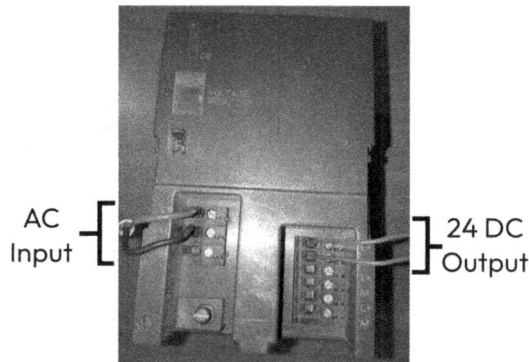

Figure 7.3 – Siemens power supply

The credit for this photo goes to *Showlight Technologies Ltd.* (www.showlight.com.ng).

The power supply module of a PLC provides the energy required for the PLC to function. A PLC without a power supply is just like a vehicle without fuel. The power supply works by stepping down the line voltage of usually 110V or 220V AC to a lower AC voltage with the aid of a transformer. A rectifier then converts the lower voltage AC to DC. A capacitor smooths or filters the DC, and a regulator circuit within the unit provides a regulated DC output of usually 24V DC.

Let's now have a look at the next module in our list, the CPU module.

The CPU module

The **central processing unit (CPU)** can be regarded as the brain of the PLC system. It consists of a processor, memory, and other integrated circuits that handle program execution, storage, and communication to other external devices including a programming device or **personal computer (PC)**, a **Human-Machine Interface (HMI)**, a **variable frequency drive (VFD)**, and so on. The CPU controls and coordinates the entire operation of the PLC. The following figure shows the CPU of an Siemens PLC (S7-300 CPU 313):

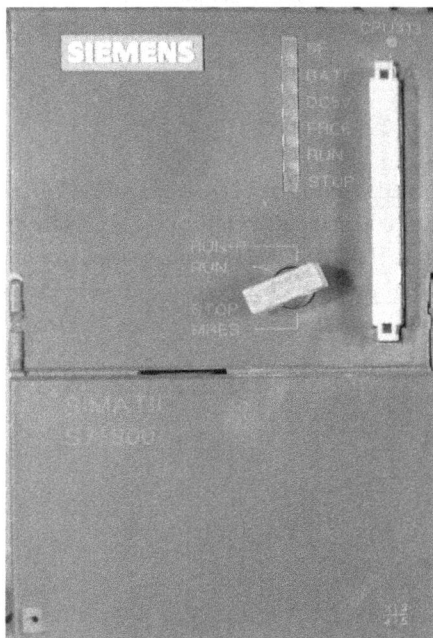

Figure 7.4 – CPU of Siemens PLC (S7-300 CPU 313)

The credit for this figure goes to *Showlight Technologies Ltd.* (www.showlight.com).

The CPU of a PLC functions like the brain of a human. The entire range of human activities (including walking, sitting, standing, and others) is controlled by the brain. Similarly, the entire operation of the PLC is controlled by the CPU. The CPU of a PLC can also be linked to the CPU of a PC, which controls all the activities of the computer. A computer without a CPU cannot function and likewise, a PLC without the CPU cannot function. The CPU executes the program written into the PLC and makes the decisions needed by the PLC to automate or operate machinery and communicate with other devices.

Let's have a look at the next module in our list, the input module.

The input module

The input module is the part that connects all the input devices (switches and sensors) to the PLC. It allows the PLC to monitor the current state of switches and sensors that are required to perform the necessary control functions when executing a program. Input modules can basically be classified as digital and analog. Digital input modules are used when the input devices to be connected produce a discrete signal (ON or OFF), while analog input modules are used when the input device to be connected produces an analog signal (that is 0-5V, 0-10V, 0-20mA, or 4-20mA).

The following figure shows a digital input module for a Siemens PLC (SM321, DI 16X DC24V):

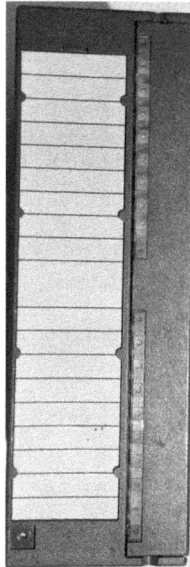

Figure 7.5 – Siemens digital input module (SM321, DI 16X DC24V)

The credit for this figure goes to *Showlight Technologies Ltd.* (www.showlight.com.ng).

The input module receives signals from switches and sensors and sends them to the CPU for processing.

Let's have a look at the output module.

The output module

This is the part that connects the PLC to the output devices or actuators. An output device can be a light, relay, contactor, solenoid valve, control valve, and so on. The PLC output module operates or controls the output devices based on the state of the inputs and the result of the written program in the CPU. Output modules can also be classified into two, namely digital output modules and analog output modules. The digital output module can control output devices or loads that are either ON or OFF. Examples of such loads or devices include lights, relays, contactors, solenoid valves, and others. The digital output module produces binary outputs (1 or 0), which means ON or OFF for loads that are either ON or OFF. Analog output modules, on the other hand, produce variable or changing signals that can range from 0-5V, 0-10V, 0-20mA, or 4-20mA. They are used for controlling output devices that require a control signal between full ON and full OFF. Examples of such loads or devices include analog voltmeters, analog ammeters, control valves, VFDs, and so on. For example, an output module can be connected to a control valve to vary the opening or closing of the control valve slowly between full ON and full OFF. You can decide to open a control valve to half or one-third of its full capacity depending on the need.

The output module of a PLC can be a transistor, a TRIAC, or a relay type, depending on the type of component used in the output circuit to switch the load ON or OFF.

In a transistor output type, a transistor is in the output circuitry for switching loads ON or OFF. A transistor type of output can only switch a DC load. A transistor-type output is fast, has a longer lifetime, and switches small currents.

In a TRIAC output type, TRIAC is used in the output circuitry for switching loads ON and OFF. It can only switch AC loads (that is, it cannot be used for DC loads). It also has a longer lifetime and is faster when compared to the relay output type.

In a relay type, relays are used in the output circuitry for switching loads. It can be used for both AC and DC loads. It is the most common type of PLC output.

The following figure shows the Siemens digital output module:

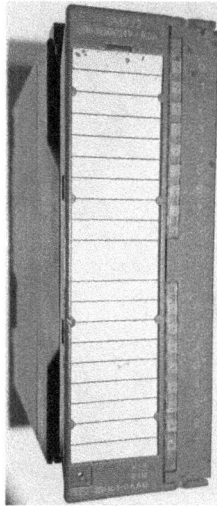

Figure 7.6 – Siemens digital output module (SM322, DO 16X DC24V/0.5A)

The credit for this figure goes to *Showlight Technologies Ltd. (*www.showlight.com*)*

The output module takes the processed signal from the CPU to control output devices such as lights, relays, contactors, and so on.

Let's have a look at other PLC modules.

Other PLC modules

Other than the above-mentioned PLC modules, there are other modules you can find in a PLC. These include the following:

- **Communication module**: This allows the PLC to communicate with other PLCs, PCs, and other devices within the factory facility or at a faraway distance. Communication module of a PLC usually have an interface or communication port that supports a particular communication or protocol (for example, PROFIBUS, PROFINET, DeviceNet, Modbus, AS-i or Ethernet). You can learn about network protocol in *Chapter 13, Industrial Network and Communication Protocols Fundamentals*. Communication module for S71200 PLC system include CM 1241 RS232, CM 1241 RS485, CM 1243-2 AS-i Master, CM 1242-5 PROFIBUS DP-Slave, CM 1243-5 PROFIBUS DP-Master, CP 1242-7 GPRS depending on the kind of communication or protocol you intend to use.

- **Positioning module**: This module allows various position control, speed control, torque control, and so on to be carried out in a motion control system. Examples of positioning modules for the Q-series Mitsubishi PLC are QD77GF4, QD77GF8, QD77MS16, and so on.

> **Note**
> Inductive loads can deliver back currents that can damage output relays. A diode, varistor, or another snubber circuit should be used to protect the PLC output from being damaged by the back current.

In this section, you have learned about the power supply module, CPU module, input module, and output module. They are the main parts of a PLC system. Having a good understanding of the parts explained in this section will help to better understand subsequent sections.

Exploring PLC types

Based on the hardware specifications, there are two basic types of PLC:

- Compact
- Modular

Let's have a look at each of the types.

Compact PLC

In this, the CPU, input module, output module, and sometimes the power supply are integrated into a single module.

It has a fixed/limited number of I/O by default, but some are designed to support expansion modules, which can be added to increase the number of inputs and outputs. A compact PLC can also be referred to as a fixed or integrated PLC. S7-200 and S7-1200 are examples of compact PLCs by Siemens, while FX 1N, FX2N, and others are examples of compact PLCs from Mitsubishi. Other PLC manufacturers also have their compact PLC series/models.

The following figure shows a compact PLC from Siemens (S7-1200, CPU 1211C, AC/DC/Rly):

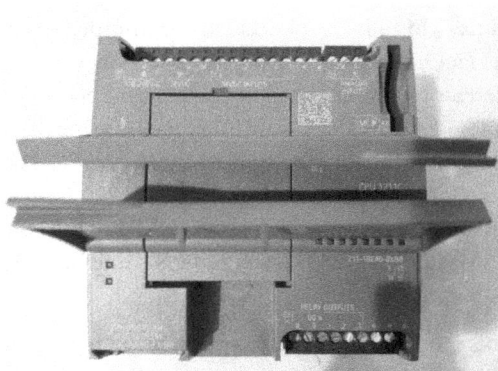

Figure 7.7 – Compact PLC from Siemens

The credit for this figure goes to *Showlight Technologies Ltd.* (www.showlight.com.ng).

Modular PLC

In this, the CPU, input module, output module, and power supply are separate, that is, they are not integrated together. Each module is plugged into a common rack/chassis or bus. The number of inputs and outputs can be increased by plugging in additional input and output modules depending on the system requirement. The rack/chassis serves as the backbone of the PLC system that plugs every module (power supply, CPU, input, output, and other modules) together. The following figure shows a modular PLC from Siemens (S7-300):

Figure 7.8 – Modular PLC from Siemens

The credit for this figure goes to *Showlight Technologies Ltd.* (http://www.showlight.com.ng).

In this section, you have learned about the two basic types of PLC: compact PLCs and modular PLCs. Let's proceed by learning about the PLC scan cycle in the next section.

Understanding the PLC scan cycle

The PLC scan cycle is a good way to explain how the PLC works. Let's have a look at it to have a better understanding of the operation of a PLC.

The PLC scan cycle is the cycle in which the PLC reads the inputs, runs the PLC programs, performs diagnostic and communication tasks, and updates the output. It's a repetitive process. The time taken by the PLC to complete one scan cycle is referred to as the scan time and is measured in milliseconds.

The following is a PLC scan cycle:

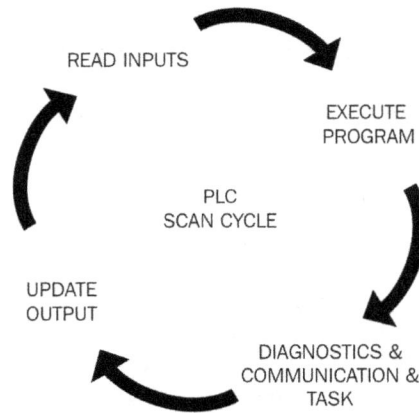

Figure 7.9 – PLC scan cycle

PLC scan cycle may vary slightly depending on the PLC in use. Basically, the PLC starts by reading the inputs, that is, checking the state of the connected switches and sensors to know which one is ON or OFF or the current value if it's analog. It then executes the user program based on the state of the inputs. Next, the PLC performs communication tasks, which include communicating necessary information to the control network such as Profibus, Modbus, Ethernet IP, and so on. It also performs diagnostic tasks to ensure the entire PLC system is functioning properly. It then updates the output before it starts a new scan.

It is important to know about the PLC scan cycle, which this section has explained. Let's now proceed to the next section to learn about some PLC manufacturers.

Knowing about PLC manufacturers

It is important to know the major manufacturers of PLC. We will discuss some of the major PLC manufacturers here:

- **Siemens**: Siemens is the biggest brand in the industrial automation industry. Their PLC series and automation system is called **SIMATIC**, which stands for **Siemens Automatic**. Hence, the brand name of Siemens PLC is SIMATIC. Siemens offers various ranges of products, which include the SIMATIC LOGO! PLC, SIMATIC S7-200 series PLC, SIMATIC S7-300 series PLC, SIMATIC S7-400 series PLC, SIMATIC S7-1200 series, and SIMATIC S7-1500 series PLC.

- **Rockwell Automation**: Rockwell Automation is an American company with headquarters in Milwaukee in the United States. The brand name of Rockwell Automation PLC's product is Allen Bradley. Their PLC product range includes the Allen Bradley Micrologix PLC system, Allen Bradley Compactlogix small PLC system, Allen Bradley Controllogix large PLC system, and so on.

- **Mitsubishi Electric**: Mitsubishi Electric is a Japanese company with headquarters located in Tokyo, Japan. They are also one of the top PLC manufacturers. The brand name of Mitsubishi PLC products is MELSEC, and their product ranges are the MELSEC-Q series, MELSEC-FX series, MELSEC-A series, and so on.

- **Omron Corporation**: Omron is also a Japanese company in Kyoto, Japan. They are also a specialist in PLCs. The brand name of Omron PLCs is SYSMAC. Their compact PLC series includes CPM1A, CPM2A, CP1E, CP1L, CP1H, and more. Their modular PLC series includes CJ2M, CJ2H, and CJ2H, while their rack PLC series includes CS1H, CS2G, and CS1D.

Other leading PLC manufacturers, and the country in which their headquarters are located, include the following:

- Schneider Electric: France

- Emerson (General Electric): United States

- Keyence: Japan

- B & R Industrial Automation: Australia

- ABB: Switzerland

- Delta Electronics: Taiwan

In this section, you have learned about the top manufacturers of PLCs, which include Siemens, Rockwell Automation, Mitsubishi Electric, Omron Corporation, and so on. There are other top manufacturers that are not mentioned here. The PLC market is big and fast growing. With the rise of automation in industries, the demand for PLCs in industries is getting higher. More PLC manufacturers will evolve to join the market.

Let's proceed by learning about PLC wiring in the subsequent sections of this chapter.

PLC wiring (1) – Wiring of switches, lights, and actuators to a PLC

Here, in this section, we will look into the connection (wiring) of input devices (switches) and output devices (lights and actuators) to the PLC. **Sinking** and **sourcing** are two important terms in PLC wiring. They are used to describe the direction of the conventional current flow between two circuits or devices. A conventional current always flows from positive potential (+) to negative potential (-). Anytime you have a current flowing between two devices, one will be sourcing while the other will be sinking.

The sourcing device is the device that provides the current (positive), while the sinking device is the device that absorbs the current (that is, the device connected to negative or ground).

In PLC wiring, the two devices in question are input devices and input modules or output devices and output modules. PLC manufacturers usually produce input/output modules that you can either sink or source. There are PLCs where you can only sink their input/output module, and for some, you can sink or source it. It all depends on the manufacturer's design. You are advised to consult the manual of the input or output module before wiring, regardless of the information you have gathered in this book.

Many PLC manufacturers usually have a common terminal on their PLC's input or output module which must be connected to positive (e.g. if the module will be sourcing) or connected to negative (e.g. if the module will be sinking). When the common terminal of the input/output module is connected to the positive (+24V) of the power supply, we say the input/output module is sourcing, and when that common terminal is connected to the negative (- or 0V) of the power supply, we say the input/output module is sinking.

We are going to follow simple rules in wiring PLCs in this book:

- Sourcing input devices (switches and sensors) must be connected to a sinking input module. The following diagram shows the connection of sourcing input devices (**normally open (NO)** push button, **normally closed (NC)** push button, and limit and float switches) to a sinking input module:

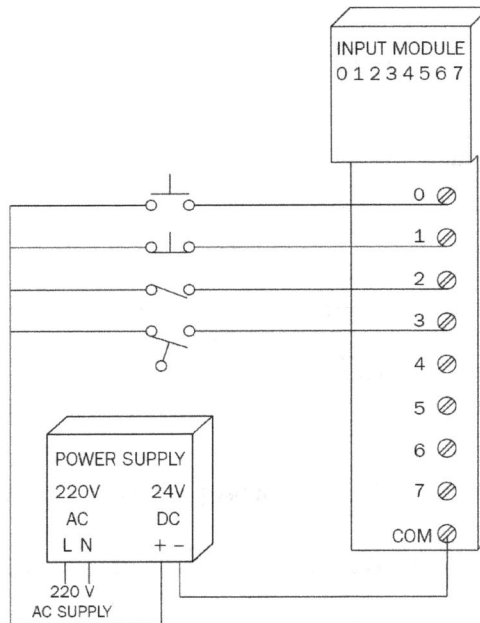

Figure 7.10 – Connection of sourcing input devices to a sinking input module

- Sinking input devices (switches and sensors) must be connected to a sourcing input module. The following circuit diagram shows the connection of sinking input devices (NO push button, NC push button, and limit switch to a sourcing input module:

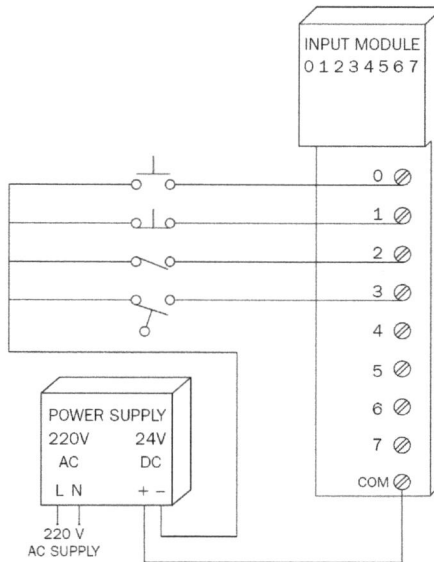

Figure 7.11 – Connection of sinking input devices to a sourcing input module

- Sourcing output devices (lights and actuators) must be connected to a sinking output module. The following circuit diagram shows the connection of sourcing output devices (pilot lamp, relay, contactor, and solenoid valve) to a sinking output module:

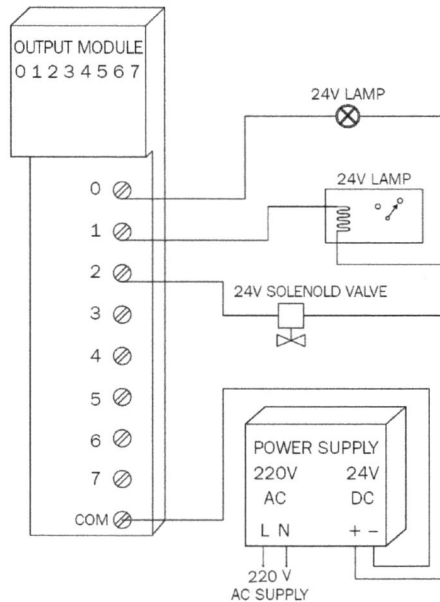

Figure 7.12 – Connection of sourcing output devices to a sinking output module

- Sinking output devices (lights and actuators) must be connected to a sourcing output module. The following circuit diagram shows the connection of sinking output devices (pilot lamp, relay, contactor, and solenoid valve) to a sourcing output module:

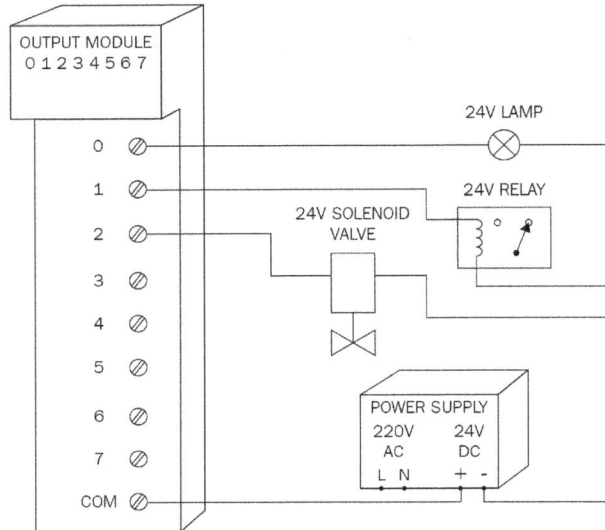

Figure 7.13 – Connection of sinking output devices to a sourcing output module

The preceding rules and wiring diagrams apply to PLCs whose input or output module is DC. The same concept goes for a PLC whose input or output module is AC, except that the terms *sinking* and *sourcing* will not be used. You can use **live** (**L**) and **neutral** (**N**) to describe the connection of the devices. For example, when an input device is connected to L, the input module must be connected to N, and vice versa. Also, when an output device is connected to L, the output module must be connected to N, and vice versa.

The following circuit diagram shows input devices (NO push button and NC push button) connected to L, while the input module is connected to N:

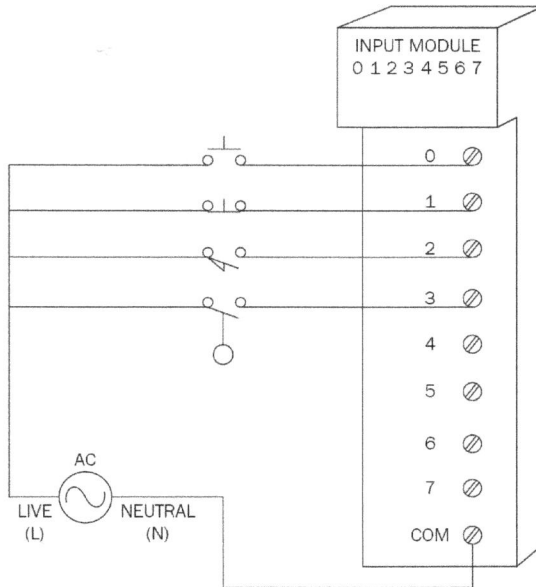

Figure 7.14 – Input devices connected to live (L), while the input module is connected to neutral (N)

The following circuit diagram shows input devices (NO push button and NC push button) connected to N, while the input module is connected to L:

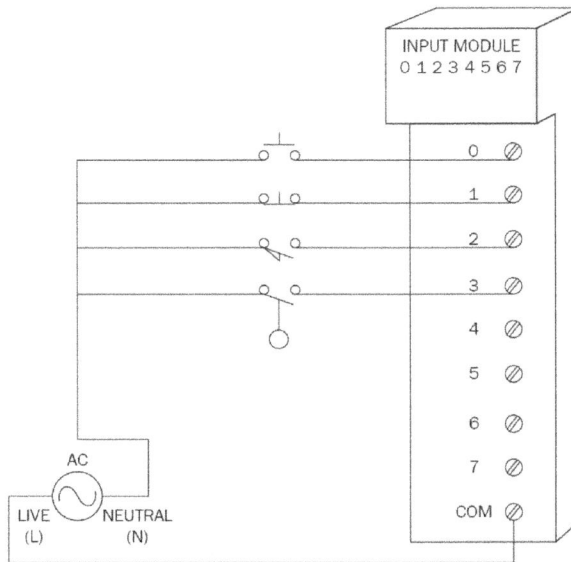

Figure 7.15 – Input devices connected to neutral (N), while the input module is connected to live (L)

The following circuit diagram shows output devices (pilot lamp, relay, contactor, and solenoid valve) connected to L, while the output module is connected to N:

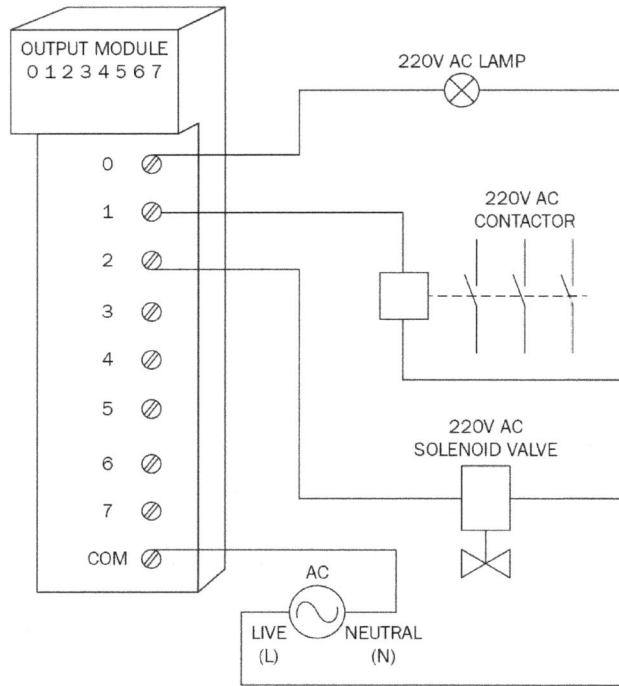

Figure 7.16 – Output devices connected to live (L), while the output module is connected to neutral (N)

The following circuit diagram shows output devices (pilot lamp, relay, contactor, and solenoid valve) connected to N, while the output module is connected to L:

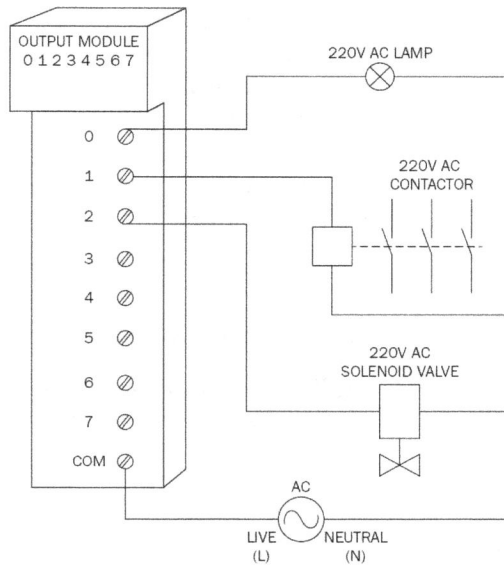

Figure 7.17 – Output devices connected to neutral (N), while the output module is connected to live (L)

There are cases in which a PLC output module will be DC and you need to wire an AC load to it. For example, you might have a PLC with DC output and you need to control a three-phase AC motor. The following circuit diagram shows the wiring of a DC output module that's wired to a 220V AC contactor and also to a three-phase motor AC motor:

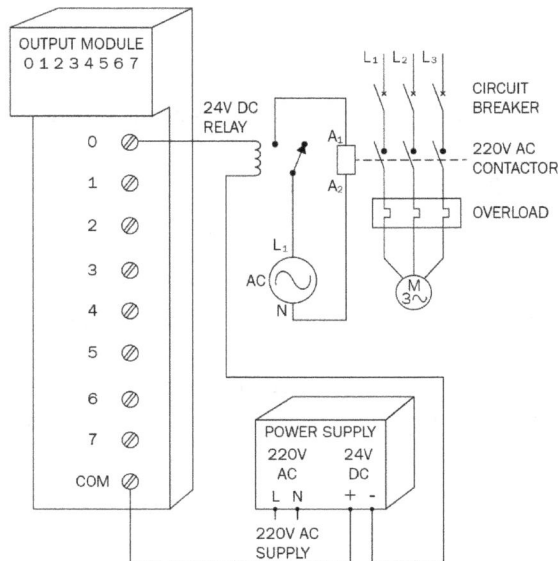

Figure 7.18 – Wiring of a DC output module that's wired to a 220V
AC contactor and also to a three-phase AC motor

In this section, we discussed various connections (wiring) that can be carried out on a PLC. The terms *sinking* and *sourcing* were thoroughly explained. You should now be able to connect sourcing input devices to a sinking input module, connect sinking input devices to a sourcing input module, connect sourcing output devices to a sinking output module, and connect sinking output devices to a sourcing output module. This section also explained how to connect input/output devices to input/output modules that use AC, and we finally showed a diagram that will help in understanding how to connect a DC output module to an AC load (*Figure 7.20*).

Let's now proceed to the next section to advance our knowledge of PLC wiring.

PLC wiring (2) – Wiring of proximity sensors (3-wire) to a PLC

In *Chapter 2*, *Switches and Sensors – Working Principles, Applications, and Wiring*, we learned about proximity sensors (capacitive and inductive). We will learn how to connect them to a PLC in this section.

Capacitive and inductive proximity sensors can be PNP or NPN. Either NPN or PNP type can come as NO or NC.

The following circuit diagram shows a 3-wire PNP (NO) and NPN (NO) proximity sensor wiring diagram:

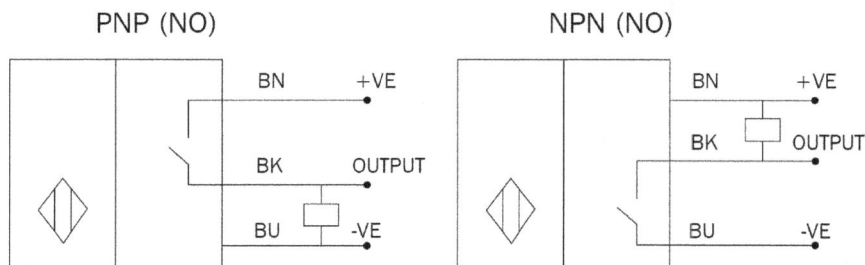

Figure 7.19 – 3-wire PNP (NO) and NPN (NO) proximity sensor wiring diagram

Now, let's differentiate between PNP and NPN sensors.

PNP sensor

A PNP sensor can be referred to as a sourcing sensor. Recall that a sourcing input device provides the current (+ve), so, a PNP sensor will provide positive (+ve). It switches to positive when an object is detected. It is usually used with a sinking input module (that is, an input module that has its common connection to negative or ground). The following circuit diagram shows a PNP sensor (NO) connection to a PLC:

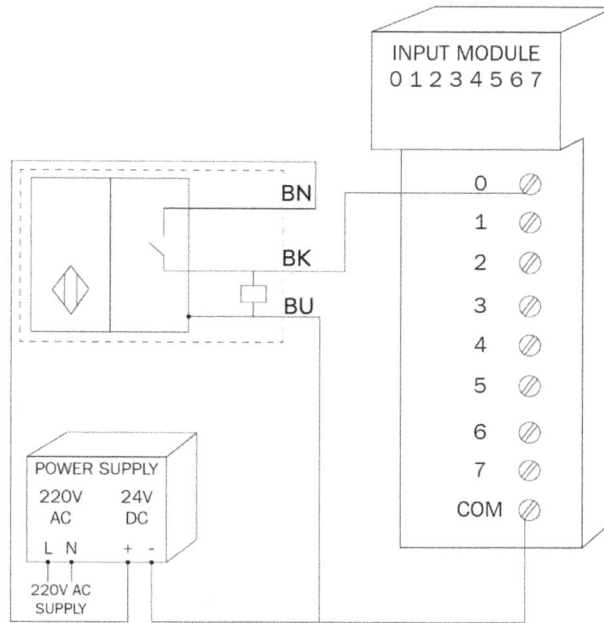

Figure 7.20 – PNP sensor (NO) connection to a PLC

Let's check the NPN sensor.

NPN sensor

An NPN sensor, on the other hand, can be referred to as a sinking sensor. Recall that a sinking input device absorbs current, so an NPN sensor switches to negative when an object is detected. It is used with a sourcing input module (that is, a module that is connected to positive). The following circuit diagram shows an NPN sensor (NO) connection to a PLC:

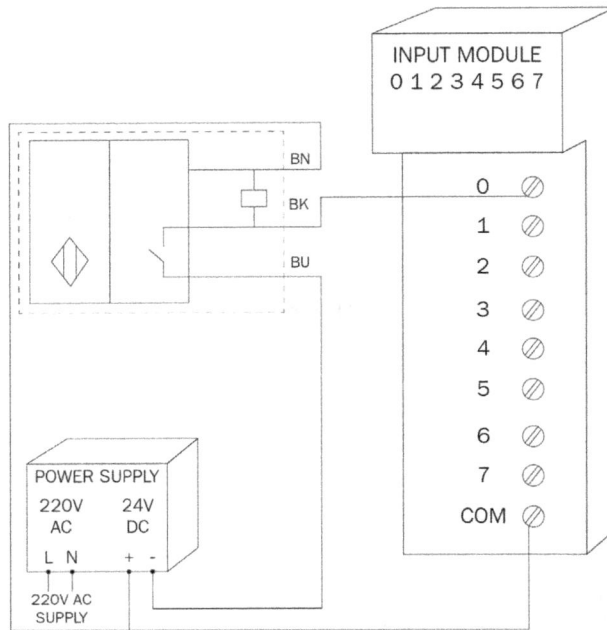

Figure 7.21 – NPN sensor (NO) connection to PLC

We have now been able to differentiate between PNP and NPN proximity sensors and how to wire each. Let's have a look at the pictorial diagram of the NPN sensor connection to a digital input module for better understanding:

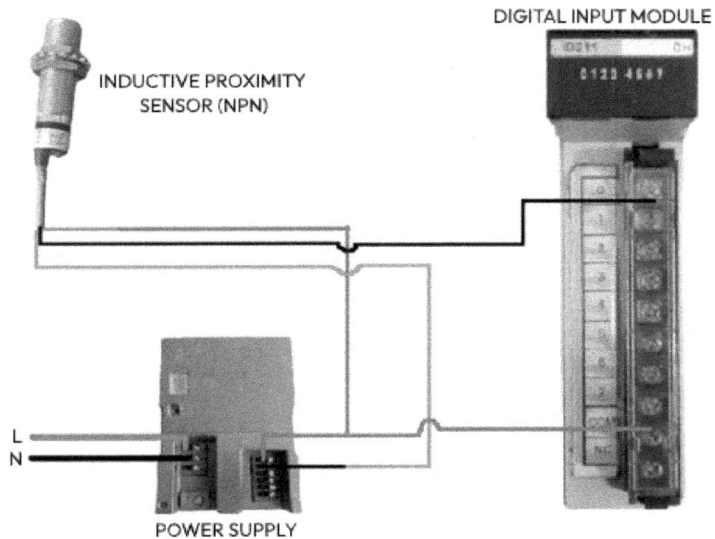

Figure 7.22 – Pictorial diagram of an NPN sensor connection to a digital input module

The credit for this figure goes to *Industrial Design Simulation Automation Services Ltd.* and *Showlight Technologies Ltd.*

In this section, you have learned about the basic connection (wiring) of proximity sensors (3-wire) to a PLC. You should now be able to connect a PNP or NPN proximity sensor to a PLC.

We will now proceed to the next section to learn about the connection (wiring) of a photoelectric sensor to a PLC.

PLC wiring (3) – Wiring of photoelectric sensors (retro-reflective)

We also discussed photoelectric sensors in *Chapter 2, Switches and Sensors – Working Principles, Applications, and Wiring*; in this section, we will learn how to wire a photoelectric sensor to an input module.

Figure 2.49 in *Chapter 2* shows the wiring diagram of a retro-reflective photoelectric sensor.

The retro-reflective photoelectric sensor described in *Chapter 2* can be made to switch to positive or negative depending on whether the input module to be used with it is sinking or sourcing. If the **White (WH)** is connected to positive, the **Black (BK)** will have an output (+ve) when an object is detected, and if the WH is connected to negative, the BK will have negative when an object is detected.

The following circuit diagram shows the wiring of a photoelectric sensor connection to a sinking input module:

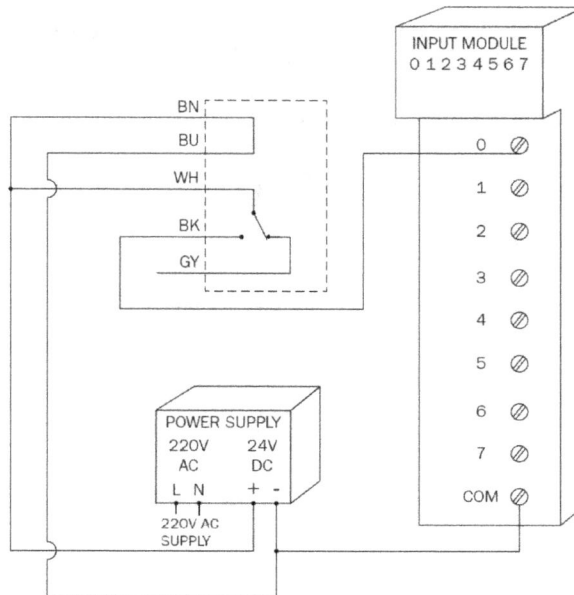

Figure 7.23 – Photoelectric sensor connection to a sinking input module

The following circuit diagram shows a photoelectric sensor connection to a sourcing input module:

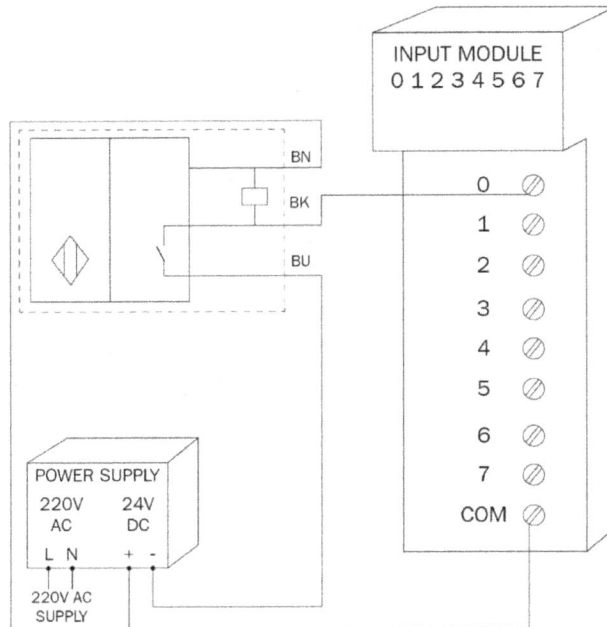

Figure 7.24 – Photoelectric sensor connection to a sourcing input module

In this section, you have learned about the basic connection of a photoelectric sensor to a PLC. You should now be able to connect a photoelectric sensor to a sinking or sourcing input module of a PLC.

Let's proceed to the next section to learn how to wire a PLC from one of the top PLC manufacturers.

Wiring of the Siemens S7-1200 PLC (CPU 1211C AC/DC/ Relay)

The PLC wiring explained in this chapter provides the basic skill needed to get PLCs wired to sensors, lights, and actuators. It is advisable to consult the manual of any PLC you want to wire and follow the wiring/guides provided in the manual. The following diagram shows the wiring of the Siemens S7-1200 PLC (CPU 1211C AC/DC/Relay) in the manual:

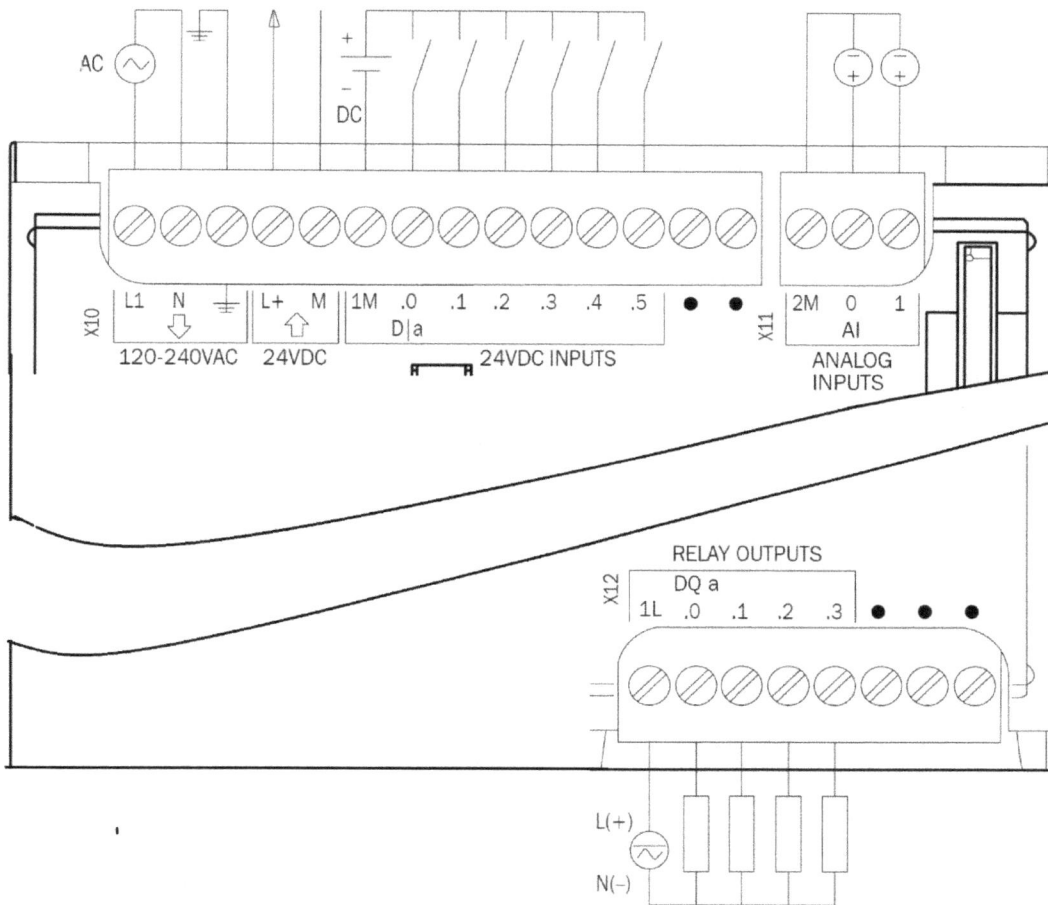

Figure 7.25 – Siemens S7 1200 PLC wiring (CPU 1211C AC/DC/Relay)

The pictorial diagram shown as follows will give a better understanding of the wiring of the Siemens S7-1200 PLC (CPU 1211C AC/DC/Relay):

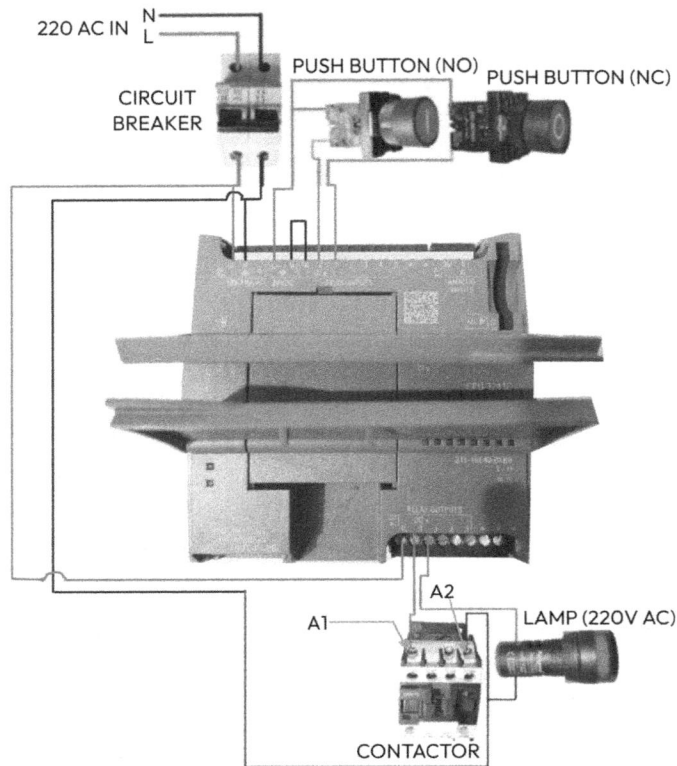

Figure 7.26 – Pictorial diagram of the Siemens S7-1200 PLC wiring (CPU 1211C AC/DC/Relay)

The credit for this figure goes to *Showlight Technologies Ltd.* (www.showlight.com.ng).

Explanation of the wiring

In *Figure 7.26*, a circuit breaker connects to the AC supply (220V AC) and is used to supply AC power to the PLC at the 120-220V input of the PLC. The PLC produces a 24VDC at the 24VDC output of the PLC. The 24VDC output is used to send a signal to the input of the PLC. The negative (M) of the 24VDC output is connected to the common (M) of the input module part of the PLC (which means the input module part of the PLC is sinking). The positive (L+) of the 24VDC output of the PLC is connected to one terminal of the push button (NO), while the other terminal is connected to the first input point of the PLC (I0.0). Similarly, the positive (L+) of the 24VDC output is also connected to one terminal of the push button (NC), while the other terminal is connected to the second input point on the PLC (I0.1). Live (L) from the circuit breaker is connected to the common (L) of the output module part of the PLC. One terminal (A1) of the contactor is connected to the first output point (Q0.0) of the PLC, while the other terminal of the contactor (A2) is connected to neutral (N) on the breaker. Similarly, one terminal of the lamp is connected to the second output point (Q0.1) of the PLC, while the other terminal of the lamp is connected to neutral (N) on the breaker.

Operation of the wiring

When the circuit breaker is switched on, AC power is supplied to the PLC, and a 24VDC output will be produced. The indicator light for the input point I0.0 will be OFF since a NO push button is used, and the button is not pressed to complete the circuit (that is, no signal is sent to the input point I0.0). The indicator light for the input point (I0.1) will be ON since an NC push button is used, and the button is not pressed (that is, the circuit is already completed).

When the NO push button connected to I0.0 is pressed, the corresponding indicator light (I0.0) will be ON, and when the button is released, the corresponding indicator light (I0.0) will be OFF.

When the NC push button connected to I0.1 is pressed, the corresponding indicator light (I0.1) will be OFF, and when the button is released, the corresponding indicator light (I0.1) will be ON.

The operation of the connected output devices (contactor and lamp) is a function of the program written into the PLC, which is covered in *Chapter 8, Understanding PLC Software and Programming with TIA Portal*.

When the written program is executed and the result of the logic energizes or activates Q0.0, the corresponding indicator light on the PLC (Q0.0) will be ON and the contactor will be energized. Similarly, when the result of the logic energizes or activates Q0.1, the corresponding indicator light on the PLC (Q0.1) will be ON and the lamp will be ON.

PLC wiring and programming work together. You need to understand the PLC wiring covered in this chapter to have a better understanding of the programming covered in *Chapter 8*.

You have just completed the PLC wiring part of this book. You should now be able to connect push buttons, limit switches, proximity sensors, lights, relays, contactors, solenoid valves, and other input or output devices to a PLC.

Summary

Congratulations! You have successfully completed *Chapter 7* of this book. Good job! The chapter gives a good explanation of what you need to know about PLCs starting from the basic to the more advanced part involving wiring. You should now have a good understanding of what a PLC really is, the basic types available, PLC modules, and various wiring that you may need to carry out on PLCs as an industrial automation engineer. This chapter focuses on the hardware part of PLC.

The next chapter focuses on the software part, where you will learn how to program PLCs to perform automated tasks.

Questions

The following are questions to test your understanding of this chapter. Ensure you have read and understood the topics in this chapter before attempting the questions:

1. _____ is an industrial computer that consists of both hardware and software used for automating industrial processes.

2. PLC is an acronym for _____.

3. The basic modules that make up a PLC are: _____, _____, _____, and _____.

4. _____ can be regarded as the brain of the PLC system.

5. The module or part that connects all the input devices (switches and sensors) to the PLC is called _____.

6. The module or part that connects the PLC to the output devices or actuators is called _____.

7. _____ is the cycle in which the PLC reads the inputs, runs the PLC programs, performs diagnostic and communication tasks, and updates the output.

8. The time taken by the PLC to complete one scan cycle is referred to as _____.

9. _____ and _____ are terms used to describe the direction of conventional current flow between two circuits or devices.

Understanding PLC Software and Programming with TIA Portal

In the previous chapter, we learned about the hardware part of the **Programmable Logic Controller** (**PLC**), which includes wiring the PLC to sensors and actuators. A PLC wired to switches, sensors, and actuators will do nothing unless there is a program written into it. A program (a set of instructions written using a programming language) is required for the CPU of the PLC to make decisions based on the inputs and carry out the required control function. It is important for an industrial automation engineer to have knowledge of both the hardware part of the PLC, including the wiring, and the software part, which includes programming.

This chapter focuses on the software aspect of the PLC. You will learn the basic things you need to start programming PLCs with the most common PLC programming language, **Ladder Diagram** (**LD**). You will learn how to download, install, and use **Totally Integrated Automation Portal** (**TIA Portal**), powerful software for programming Siemens PLCs.

We are going to cover the following main topics in this chapter:

- Understanding software/program in PLCs
- Introducing PLC programming languages
- Introducing PLC programming devices
- Understanding the basics of LD
- Downloading and installing TIA Portal V13 Professional and PLCSIM
- Creating a project and writing a program with Siemens programming software (TIA Portal)

Technical requirements

While every part of this book is valuable, *Chapter 2, Switches and Sensors – Working Principles, Applications, and Wiring, Chapter 3, Actuators and Their Applications in Industrial Automation*, and *Chapter 7, Understanding PLC Hardware and Wiring*, are very important to have a better understanding of this chapter.

Understanding software/program in PLCs

Before looking into PLC programming languages, let's get some knowledge of the software/program in PLCs.

In computing, generally, software is a set of instructions written in a language the computer understands to perform a task. Software can also be referred to as a program.

A PLC requires two types of software that must be executed by the CPU for the PLC to function. They are as follows:

- The operating system (firmware)
- User program

The operating system is the program written by the PLC manufacturers and designed to run automatically in the PLC once powered. It is responsible for executing the user program, establishing communication between devices, memory management, and updating the output.

A PLC user program is a set of instructions in either textual or graphical form that represents the control function that will be carried out for a specific industrial task or application. It is the program that the user or PLC programmer writes to dictate the required automation or control task to the PLC.

A programming language is usually used to convey instructions to a PLC or any other computing device to perform a specific task. Let's have a look at PLC programming languages in the next section.

Introducing PLC programming languages

According to **International Electrotechnical Commission** (**IEC**) standards, there are five programming languages, stated in part 3 of IEC 61131, for writing a program for PLCs to control and automate a task. These programming languages are the following:

- **Ladder Diagram** (**LD**)
- **Function Block Diagram** (**FBD**)
- **Sequential Function Chart** (**SFC**)
- **Instruction List** (**IL**)
- **Structured Text** (**ST**)

IEC 61131-3 can be purchased at their web store at the following link. However, there is an abstract you can read and also a preview that can be downloaded for free:

`https://webstore.iec.ch/publication/4552`

> **Note**
>
> IEC is an international standards organization responsible for preparing and publishing international standards for all electrical, electronics, and other related technologies.

Let's discuss each of these programming languages in detail.

Ladder Diagram (LD)

This is most common among several PLC programming languages. It is a graphical programming language that's easy to understand by most plant technicians because it's similar to the relay diagrams that they are familiar with. **Normally Open (NO)** or **Normally Closed (NC)** contacts are used as input. Specific outputs are activated depending on the logic performed on the input.

The following diagram shows a simple AND logic program using LD. Output 1 will be activated only when input 1 and input 2 are true (ON):

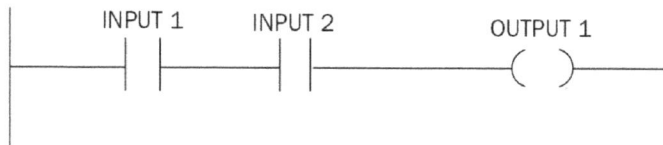

Figure 8.1 – A simple AND logic program using LD

The following diagram shows a simple OR logic program using LD. Output 1 will be activated when either input 1 or input 2 or both input 1 and input 2 are true (ON):

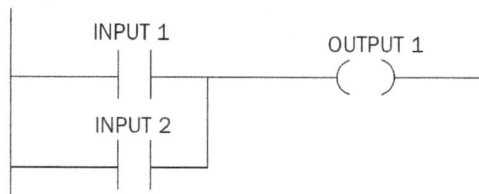

Figure 8.2 – A simple OR logic program using LD

The following diagram shows a program that combines both AND and OR logic. If input 1 OR input 2 is true AND input 3 are true (ON), output 1 will be activated:

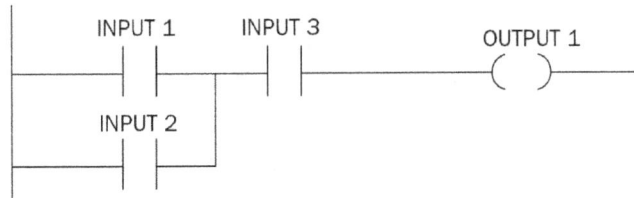

Figure 8.3 – A program that combines both AND and OR logic

Let's have an idea of how FBD looks.

Function Block Diagram (FBD)

This is a graphical language in which program elements are represented in the form of blocks and are connected together. A function with many lines of code is represented as a block and several blocks can be connected together in a program.

The following diagram shows a simple AND logic program using FBD. Output 1 will be activated only when input 1 and input 2 are true (ON):

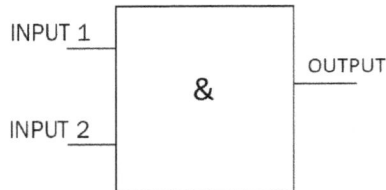

Figure 8.4 – A simple AND logic program using FBD

The following diagram shows a simple OR logic program using FBD. Output 1 will be activated when either input 1 or input 2 or both input 1 and input 2 are true (ON):

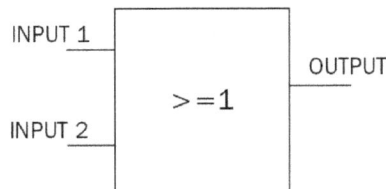

Figure 8.5 – A simple OR logic program using FBD

The following diagram shows a program that combines both AND and OR logic. If input 1 OR input 2 is true AND input 3 are true (ON), output 1 will be activated:

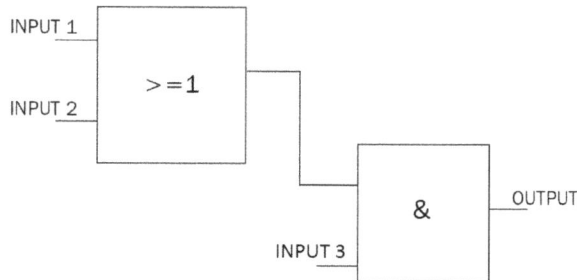

Figure 8.6 – A program that combines both AND and OR logic

We will now briefly discuss the next programming language in our list.

Instruction List (IL)

This is a text-based language that resembles assembly language. A series of instructions is listed just as assembly language. A program written using IL is processed faster than other programming languages defined by IEC for programming PLCs. One disadvantage of this programming language is that it's difficult to debug a large or complex program.

The following is a sample program written using IL. If input A or input B is true (ON) and input C is true (ON), output Y will be activated:

```
LD    A
OR    B
AND   C
OUT   Y
END
```

Let's now move on to the next language.

Structured Text (ST)

ST is also a text-based language, similar to traditional programming languages, such as C, C++, Python, and Java, in structure. It gives the possibility of working with inputs and outputs using statements such as if, while, and for, which are common in traditional programming languages. A complex algorithm can more easily be implemented in ST than LD or some other programming languages. One drawback of ST is the difficulty a programmer may experience in debugging. The following code block shows a simple program written using ST:

```
VAR
        PushButton1: BOOL;
```

```
        PilotLamp1: BOOL;
END_VAR
IF PushButton1 = TRUE THEN
        PilotLamp1:= TRUE
        ELSE
            PilotLamp1;
        END_IF
END_IF
```

Finally, from our list of programming languages, we will look at SFC.

Sequential Function Chart (SFC)

This is also a graphical programming language similar to a regular flow chart used in computer science. Steps/states and transitions are used. Separate steps (rectangular boxes) describe the operation or action to execute and are connected sequentially by a vertical line. A transition condition (a logical statement that must be true to proceed) is present between each step, as shown in the following diagram. A complex sequential task can easily be visualized and designed using SFC:

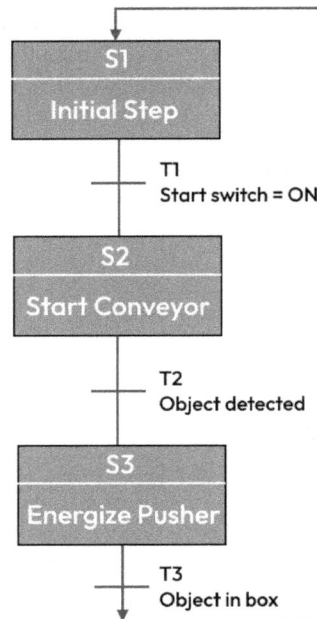

Figure 8.7 – A sample program using SFC

In the preceding figure, **S1**, **S2**, and **S3** represent the steps, while **T1**, **T2**, and **T3** represent the transitions.

We should now be able to differentiate between PLC programming languages (LD, FBD, ST, IL, and SFC). The just-concluded section, that is, *Introducing PLC programming languages*, is not meant to make you a PLC programmer using all five languages; rather, its purpose is to give you a brief idea of the programming languages and allow you to be able to differentiate between them. As mentioned earlier in this chapter, we will focus on LD, which will be covered in subsequent sections. We will first quickly look at PLC programming devices in the next section.

Introducing PLC programming devices

These are devices or tools used to write, edit, and transfer programs to the PLC. They can also be used to read programs from the PLC. Hence, a programming device can be used to download (write) or upload (read) a program. The term download or write means to transfer a program from a programming device to a PLC while the term upload or read means to transfer a program from the PLC to a programming device. Programming devices can also help in troubleshooting PLCs. There are two common programming devices:

- Handheld devices
- **Personal Computers (PCs)**

Handheld devices are usually connected to the PLC via a cable. They consist of keys for entering a program, editing a program, downloading a program to the PLC or uploading a program from the PLC. They also consist of a small display for viewing instructions. They are small in size and have limited capability.

A **PC** is the most popular programming device. A programming software from the PLC manufacturer can be installed on a PC for writing a program, editing a program, downloading a program to the PLC or uploading a program from the PLC. A (communication) cable is also required between the PC and PLC. We will focus on using a PC as a programming device in this book.

Let's proceed to learn about PLC programming software.

Looking at the different PLC programming software

In *Chapter 7, Understanding PLC Hardware and Wiring*, we learned about PLC manufacturers, which included Siemens, Rockwell Automation, and Mitsubishi Electric. Each of these manufacturers has its own software dedicated to programming and configuring their PLCs. The following are examples of programming software from different PLC manufacturers:

- **For Siemens PLCs**: MicroWIN, STEP 7 Manager, and TIA Portal
- **For Allen-Bradley PLCs by Rockwell Automation**: RSLogix 500 and RSLogix 5000 or Studio 5000
- **For Mitsubishi PLCs**: GX Developer and GX Works

Our focus in this book is TIA Portal for programming and configuring Siemens PLCs (S7-300, S7-400, S7-1200, and S7-1500). We will also focus on one of the Siemens S7-1200 PLCs (SIMATIC S7-1200, CPU 1211C, AC/DC/Rly) in this chapter. Refer to *Chapter 7, Understanding PLC Hardware and Wiring*, to learn about the wiring of Siemens S7-1200 PLC (CPU 1211C AC/DC/Relay). TIA Portal comes with different versions, including V13, V14, and V15. We will be using V13 in this book.

Let's proceed to learn the basics of the most common PLC programming language (LD) in the next section.

Understanding the basics of LD

LD is an easy-to-learn programing language for creating control programs in PLCs to automate or control machines in industries.

In *Chapter 1, Introduction to Industrial Automation*, it was mentioned that prior to the introduction of PLCs, there was automation but through the use of relays, timers, contactors, and so on. These were hardwired together to perform the required control or automation task. Relay logic circuits were used when designing such control and were also used for troubleshooting the system.

A relay logic circuit is an electrical circuit or drawing showing symbols of components, such as relays, switches, timers, and contactors, and their connections for the desired control or automation task. The circuit usually consists of two vertical lines (rails), one on the extreme left (supply voltage potential, live or positive) and the other on the right (zero potential, neutral or negative). One or several horizontal lines in a relay logic circuit are called rungs. A rung consists of inputs and an output. The inputs must be ON for an output to be enabled. The following circuit diagram shows a sample of a relay logic circuit:

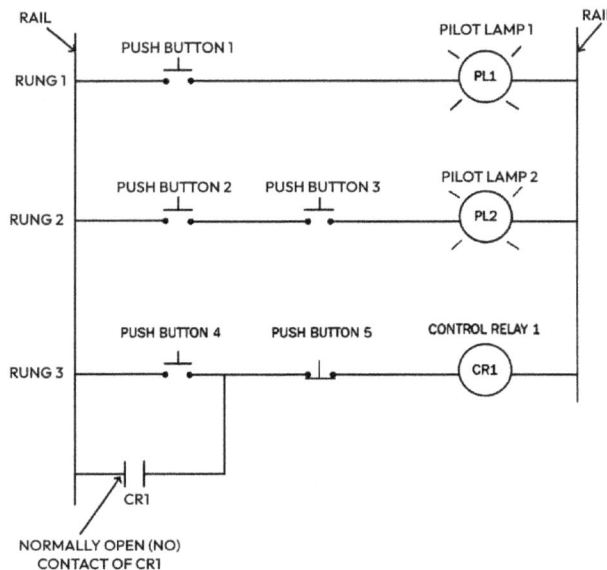

Figure 8.8 – A relay logic circuit

In the first rung, when push button 1 is pressed, lamp 1 will be ON, and when push button 1 is released, lamp 1 will be OFF.

In the second rung, push button 2 and push button 3 must be pressed for lamp 2 to be ON. When any of the buttons are released, the lamp will be OFF.

The third rung is a latching circuit that energizes a relay coil (CR1) when push button 4 is pressed. The coil remains energized even when the button is released. The reason for the coil remaining energized is due to the fact that the NO contact of the relay coil (CR1) is in parallel with push button 4. Hence, when the button is pressed, the coil (CR1) becomes energized and its contacts close. The closed contact of CR1 (which is in parallel with push button 4) serves as an alternative path for the current to flow to the coil. Hence, the coil remains energized. Push button 5 is a stop button (NC), which will cut the flow of current to the coil when pressed, and the coil will be de-energized until push button 4 is pressed.

LD, which we will learn about in this and subsequent sections, is similar to a relay logic circuit. This is the reason why most technicians find it easy to understand and work with LD. But unlike a relay logic circuit, which uses component symbols, as shown in the preceding diagram, LD uses symbolic notation to express logic operation. The following screenshot shows the LD of the preceding circuit using Siemens programming software:

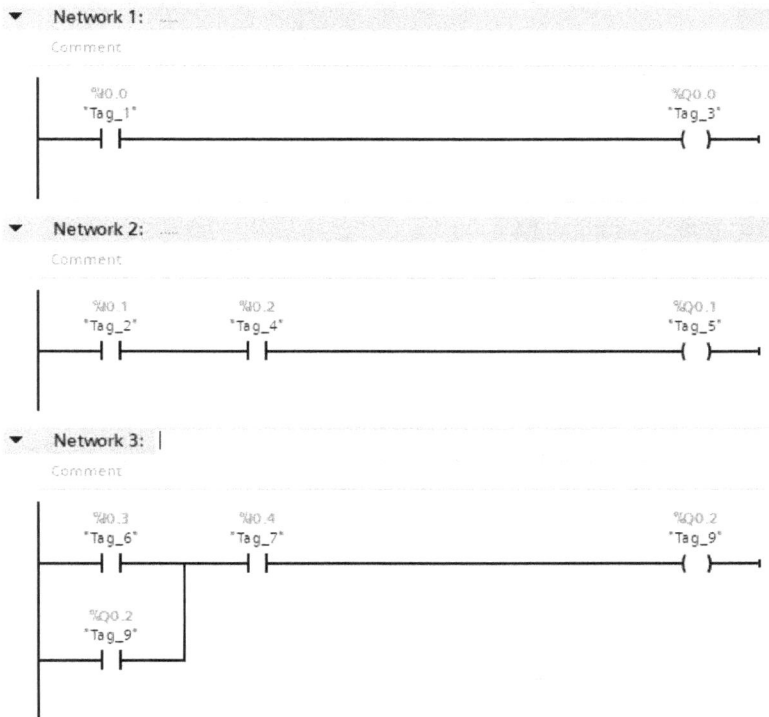

Figure 8.9 – LD using Siemens programming software

The following diagram shows how the push buttons will be wired to a PLC for the program to function as required. It is actually not economical to use a PLC and LD to implement the relay logic circuit in *Figure 8.8*. The reason for using a PLC and LD here to implement such simple logic is to aid understanding and help you to see how the relay logic circuit relates to the LD programming language:

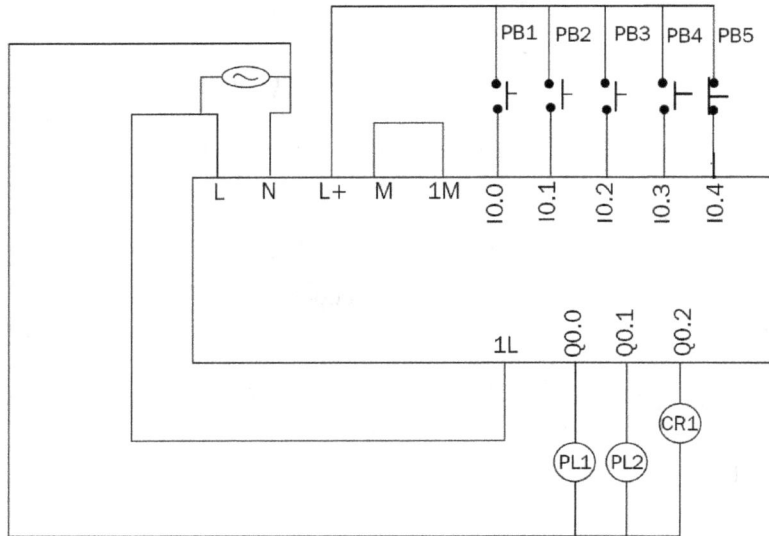

Figure 8.10 – Push buttons, pilot lamps, and control relay wired to a PLC (S7-1200, CPU1211C, AC/DC/Rly)

We should now have basic knowledge of LD. Subsequent sections will give further knowledge that will help you to start programming PLCs using LD.

Exploring the elements of LD (ladder logic) program

In order to write a program using LD. You need to understand the elements that make up a ladder program. The following elements make up an LD (ladder logic) program:

- **Rails**: These are the two vertical lines, one at the extreme left (hot rail) and the other at the extreme right. The vertical line at the extreme left is called the hot rail while the one at the extreme right can be referred to as the neutral.

- **Rungs**: These are horizontal lines that connect the two rails to complete a circuit. Power flows across the rung from the hot rail to the neutral rail. It connects the rails to the logic expression.

- **Inputs**: The inputs are the NO and NC contact symbols representing input devices, such as a push button, selector switch, limit switch, or float switch, that are hardwired to the PLC. The state of the input devices connected to the PLC determines whether the NO or NC contact symbol will open or close in the program.

- **Output**: The output is the coil symbol representing the output devices, such as a pilot lamp, relay, contactor, or solenoid valve, that are hardwired to the PLC.

- **Logic expression**: This is the combination of input and output symbols to formulate the desired control function.

- **Address notation and tag names**: The PLC uses an address to identify each input and output point where external devices (input and output devices) are connected in the PLC. Each point you connect an input device or output device to has an address that the PLC uses to identify it. The PLC also uses an address to identify internal/software-created devices, such as relays, timers, and counters. The addresses used by each manufacturer to identify inputs/outputs (inputs/output points) and software-created devices in their PLCs differ. A programmer needs to know the addressing or address notation used by a particular PLC manufacturer or model of PLC before programming it.

Siemens, for instance, uses I to represent input and Q to represent output. We can basically have inputs as I0.0, I0.1, I0.2, and so on and outputs as Q0.0, Q0.1, Q0.2, Q0.3, and so on. Software-created relays/auxiliary relays in Siemens are M0.0, M0.1, M0.2, and so on. Mitsubishi uses X to represent inputs and Y to represent outputs. We can basically have inputs as X000, X001, X002, and so on and outputs as Y000, Y001, Y002, and so on. Software-created relays/auxiliary relays in Mitsubishi PLCs are M0, M1, M2, and so on. However, our focus in this book will be Siemens PLC (S7 1200 CPU 1211C AC/DC/RLY). Address notation is simply a description of each input/output (input or output point) and software-created devices, such as relays, timers, and counters, while a tag name is a description of an address.

S7 1200 (CPU 1211C AC/DC/RLY) has six digital inputs and four digital outputs, which can be addressed in a program as follows:

- **Inputs**: I0.0, I0.1, I0.2, I0.3, I0.4, I0.5

- **Outputs**: Q0.0, Q0.1, Q0.2, Q0.3

You can find the input/output addressing of a PLC or model of a PLC in its manual. Always read the documentation/manual of any PLC you want to program.

- **Comments**: Comments are used to describe the logical operation that a rung or group of rungs will perform in an LD. They are usually displayed at the start of each rung. They make the program easy to understand.

The following diagram shows sample LD with labels:

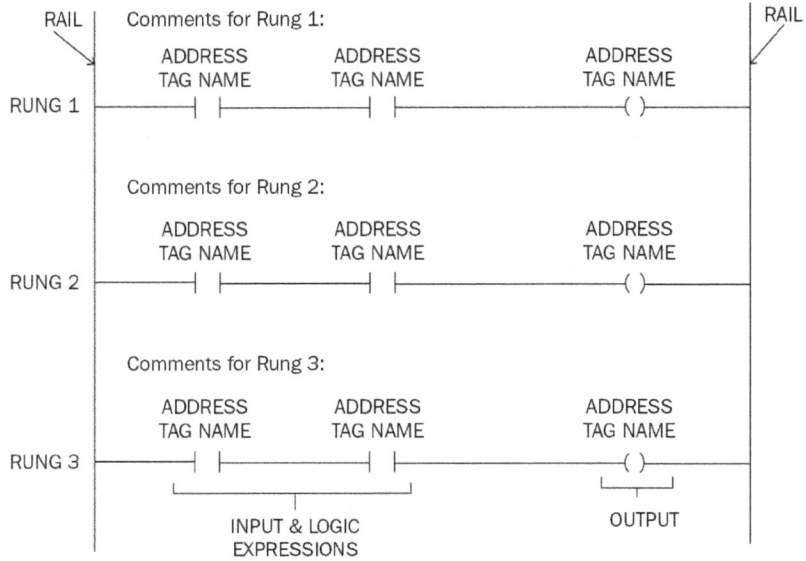

Figure 8.11 – LD with labels

We will now learn the rules for programming PLCs with LD.

Rules for LD programming

The following rules will help in writing a good program using LD:

- Inputs (NO contact and NC contact) can be connected in parallel or series in LD:

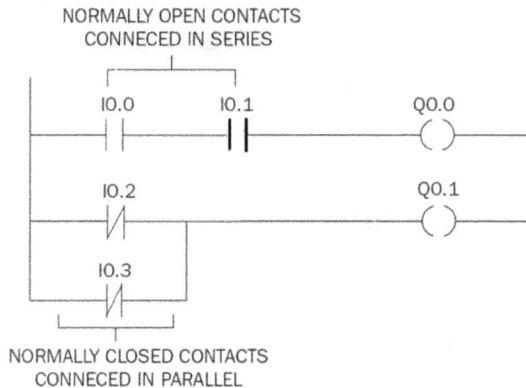

Figure 8.12 – Inputs of rung 1 connected in series and inputs of rung 2 connected in parallel

- Outputs (coils) cannot be connected in series; they can only be connected in parallel:

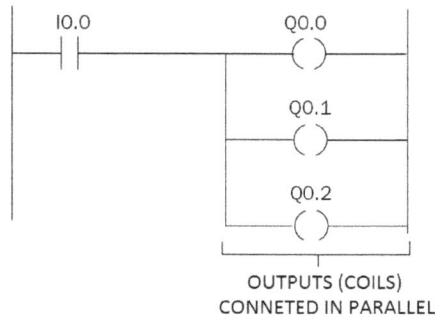

Figure 8.13 – Outputs (coils) connected in parallel

- One input can be used several times in one program; that is, a single input can be used in a different rung:

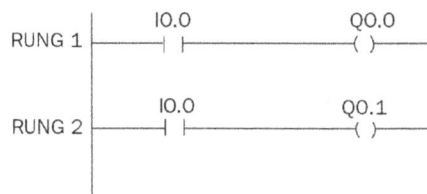

Figure 8.14 – Same input is used in rung 1 and rung 2

- One output cannot be used several times in one program except in set/reset and latch/unlatched program:

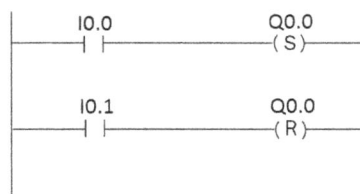

Figure 8.15 – One output used twice for set/reset in a Siemens PLC program

The following diagram shows a simple latch and unlatch program in an Allen-Bradley PLC program:

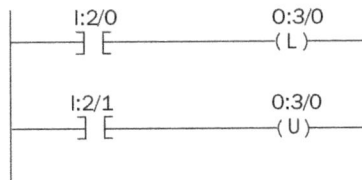

Figure 8.16 – One output used twice for latch/unlatch in an Allen-Bradley PLC program

- The address of an output (coil) can be used for input (NC or NO contact). You can use the address of an output (coil) for an input (NC or NO contact), which makes the input have the same logic state (1 or 0) as the output; that is, if the output is ON (1), the input will be HIGH (1), and if the output is OFF, the input will be LOW (0):

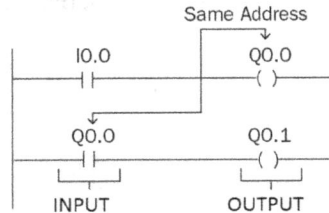

Figure 8.17 – Sample program showing output address used as input address

- The address of an input cannot be used for output. Don't make the mistake of using an input address for output (coil).

We should now know the rules for LD programming. These rules will be applied while writing programs using LD in subsequent sections. We will proceed to the next section to learn how to download and install programming software and also a simulator.

Downloading and installing TIA Portal V13 Professional and PLCSIM

Siemens offers a 21-day free trial of their programming software (TIA Portal). There is also a simulator that allows you to simulate your program. Hence, you can write a program and see how it runs even when you don't have a real PLC by using the simulator (PLCSIM). You can download both TIA Portal V13 Professional and PLCSIM free online and use it to learn PLC programming or purchase a license for TIA Portal to avoid any restrictions.

Downloading TIA Portal V13 and PLCSIM

The following steps should be followed to download TIA Portal V13 and PLCSIM:

1. The first step is to visit the download page. The following link should take you to the download page:

    ```
    https://support.industry.siemens.com/cs/document/109745155/
    simatic-step-7-including-plcsim-v13-sp2-trial-
    download?dti=0&lc=en-WW
    ```

    The following screenshot shows the page of the preceding link:

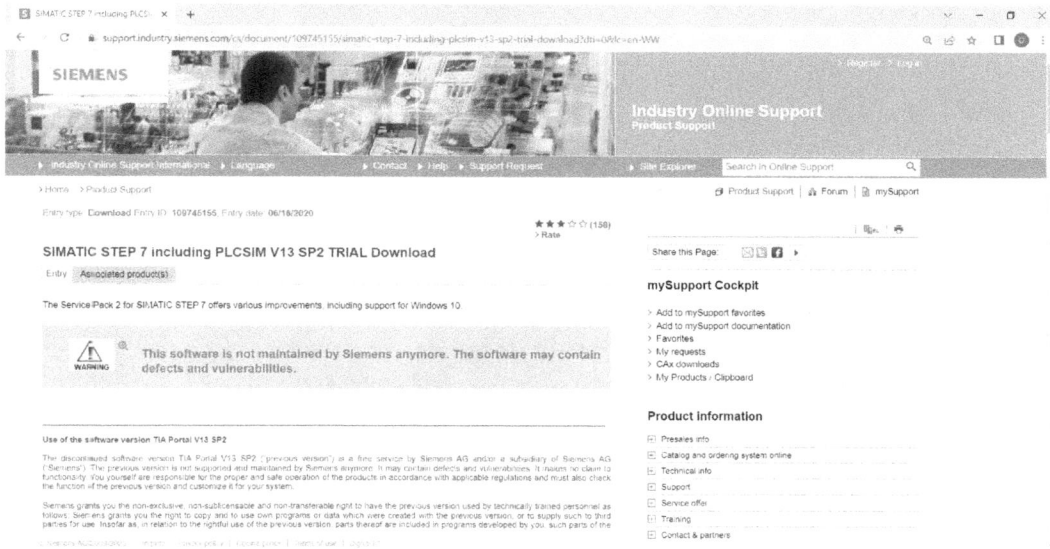

Figure 8.18 – Download page of Siemens TIA Portal V13 Professional and PLCSIM (1)

2. Scroll to the **STEP 7 Professional V13 SP2** section, as shown in the following screenshot, and download all the files indicated in the same folder:

DVD 1: (STEP 7 Prof. V13 SP2)

⬜ ♂ SIMATIC_STEP_7_Professional_V13_SP2_Upd4.001 (650.0 MB)

⬜ ♂ SIMATIC_STEP_7_Professional_V13_SP2_Upd4.002 (650.0 MB)

⬜ ♂ SIMATIC_STEP_7_Professional_V13_SP2_Upd4.003 (650.0 MB)

⬜ ♂ SIMATIC_STEP_7_Professional_V13_SP2_Upd4.004 (650.0 MB)

⬜ ♂ SIMATIC_STEP_7_Professional_V13_SP2_Upd4.005 (650.0 MB)

⬜ ♂ SIMATIC_STEP_7_Professional_V13_SP2_Upd4.006 (650.0 MB)

⬜ ♂ SIMATIC_STEP_7_Professional_V13_SP2_Upd4.007 (650.0 MB)

⬜ ♂ SIMATIC_STEP_7_Professional_V13_SP2_Upd4.008 (650.0 MB)

⬜ ♂ SIMATIC_STEP_7_Professional_V13_SP2_Upd4.009 (650.0 MB)

⬜ ♂ SIMATIC_STEP_7_Professional_V13_SP2_Upd4.010 (650.0 MB)

⬜ ♂ SIMATIC_STEP_7_Professional_V13_SP2_Upd4.011 (26.0 MB)

⬜ ♂ SIMATIC_STEP_7_Professional_V13_SP2_Upd4.exe (2.8 MB)

DVD 2: (Hardware Support Packages, Open Source Software, Tools)

⬜ ♂ SIMATIC_STEP_7_Professional_V13_SP2_Upd4_2.001 (650.0 MB)

⬜ ♂ SIMATIC_STEP_7_Professional_V13_SP2_Upd4_2.002 (650.0 MB)

⬜ ♂ SIMATIC_STEP_7_Professional_V13_SP2_Upd4_2.003 (650.0 MB)

⬜ ♂ SIMATIC_STEP_7_Professional_V13_SP2_Upd4_2.004 (650.0 MB)

⬜ ♂ SIMATIC_STEP_7_Professional_V13_SP2_Upd4_2.005 (501.3 MB)

⬜ ♂ SIMATIC_STEP_7_Professional_V13_SP2_Upd4_2.exe (2.8 MB)

SHA-256 checksum:

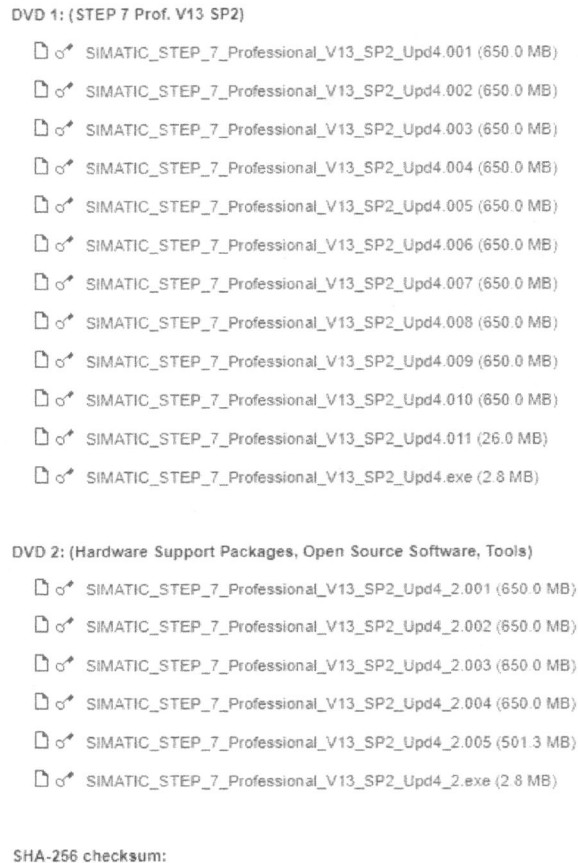

Figure 8.19 – Download page of Siemens TIA Portal V13 Professional and PLCSIM (2)

Note

A login screen will appear as soon as you click on any of the files to download. You will need to login with your Siemens Industry Online Support username and password if you already have a username and password and if not, you will have to click **Yes, I would like to register now**, follow the instructions to complete your registration and then login to download the files.

---

3. Also, scroll to **SIMATIC STEP 7 PLCSIM V13 SP2 for STEP 7 Basic and STEP 7 Professional** and download all the files indicated in the same folder, as shown in the following screenshot:

SIMATIC STEP 7 PLCSIM V13 SP2 for STEP 7 Basic and STEP 7 Professional

   □ ♂ SIMATIC_S7_PLCSIM_V13_SP2.001 (650.0 MB)

   □ ♂ SIMATIC_S7_PLCSIM_V13_SP2.002 (613.7 MB)

   □ ☑E SIMATIC_S7_PLCSIM_V13_SP2.exe (2.5 MB)

  SHA-256 checksum:

   □ ♂ SIMATIC_S7-PLCSim_V13_SP2.txt (1 KB) ↗ Information on SHA-256

Security Information

In order to protect technical infrastructures, systems, machines and networks against cyber threats, it is necessary to implement – and continuously maintain – a holistic, state-of-the-art IT security concept. Siemens' products and solutions constitute one element of such a concept. For more information about cyber security, please visit

https://www.siemens.com/cybersecurity#Ouraspiration.

Also available in the following languages:
> German
> Spanish
> Italian
> Chinese

Entry belongs to product tree folder(s):
> Automation Technology > Industry software > Automation software > TIA Portal > PLC programming > STEP 7 Professional (TIA Portal)

Rate entry

☆ ☆ ☆ ☆ ☆ no rating   Submit rating

Figure 8.20 – Download page of Siemens TIA Portal V13 Professional and PLCSIM (3)

Let's learn how to install TIA Portal V13 in the next section.

Installing TIA Portal V13

Take the following steps to install TIA Portal V13:

1. Open the folder where you have downloaded the files of TIA Portal V13 and double-click the setup file (SIMATIC_STEP_7_Professional_V13_SP2_Upd4.exe).

    You should see the following screen if you get it right:

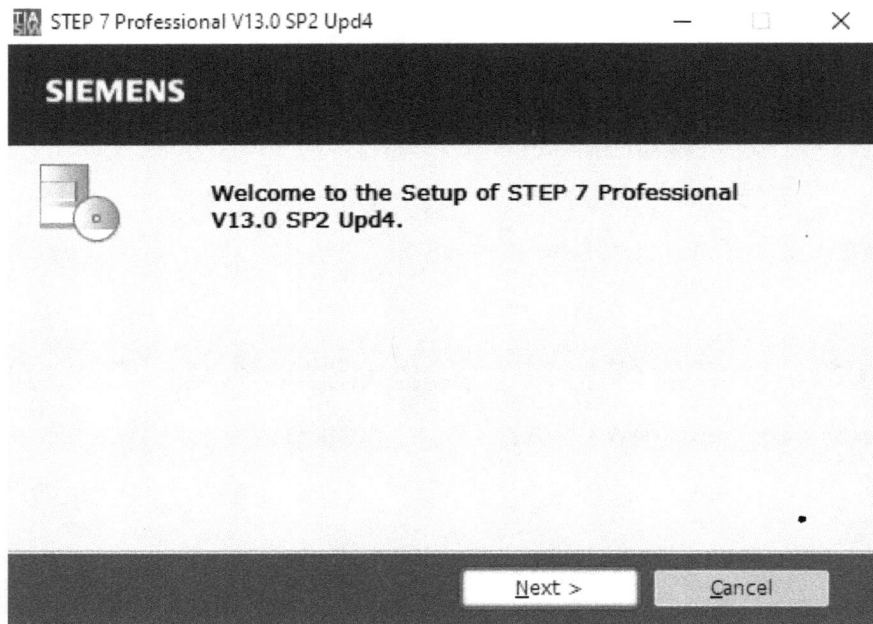

Figure 8.21 – Installing TIA Portal V13

2. Click **Next** to continue the setup. On the **License Transfer** screen, click on **Skip license transfer**. Follow the instructions on screen to complete the setup and restart the PC or laptop when prompted.

Let's now learn how to install PLCSIM in the next section.

Installing PLCSIM

The following steps should be taken to install PLCSIM:

1. Open the folder where you have downloaded the files of PLCSIM and double-click the setup file (SIMATIC_S7_PLCSIM_V13_SP2.exe).

    You should see the following screen if you get it right:

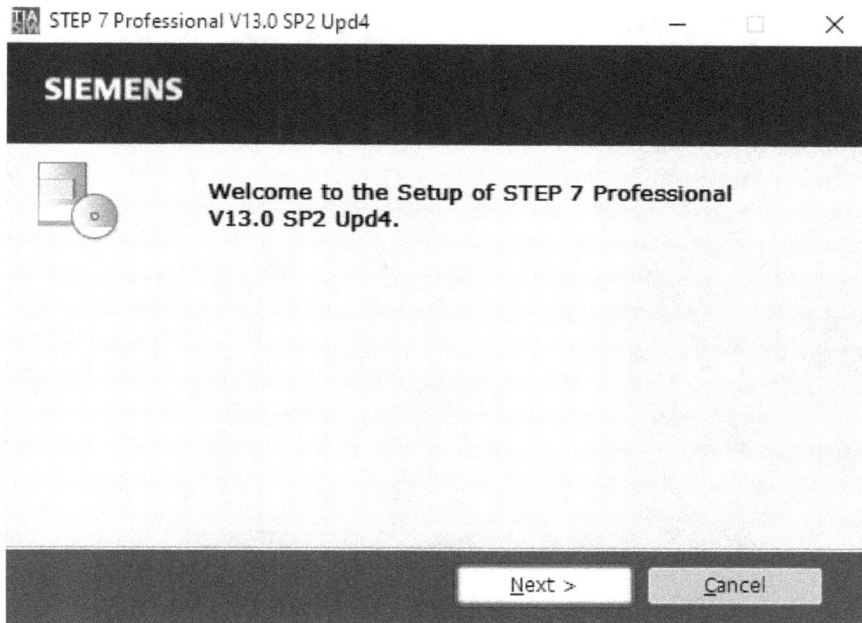

Figure 8.22 – Installing PLCSIM

2. Follow the instructions on the screen to complete the setup. Restart the PC or laptop when prompted.

We have now been able to download and install TIA Portal V13 and PLCSIM on our PC. We will be using this software to practice in subsequent sections of this chapter and also in the next chapter. We will begin by creating a project and learning how to use the software in the next section.

Creating a project and writing a program with Siemens programming software (TIA Portal)

We will now start TIA Portal, create a project, and write a simple program using the following steps:

1. Start TIA Portal V13 from the **Start** menu or double click on the TIA Portal V13 icon on your desktop.

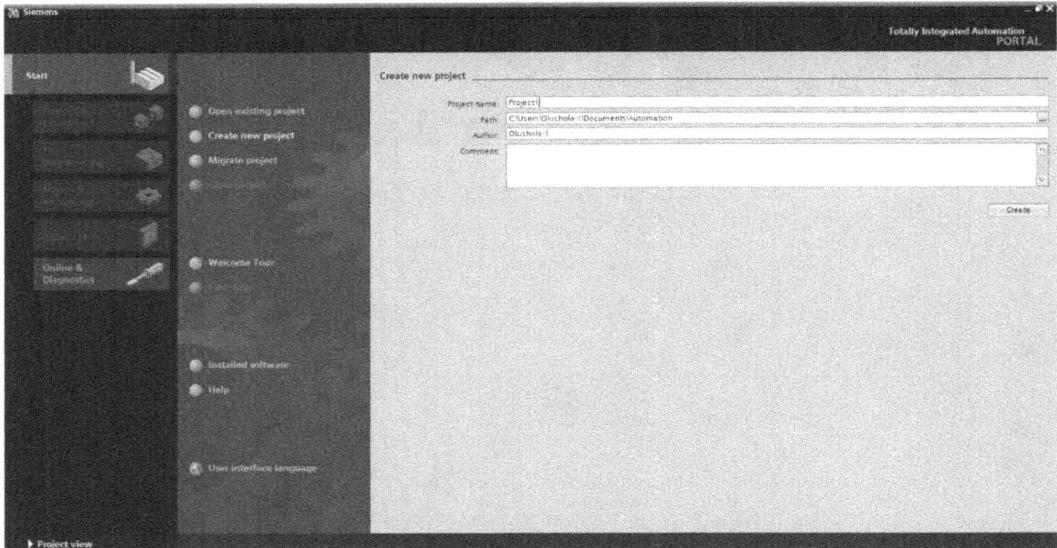

Figure 8.23 – Creating a project in TIA Portal V13

Click **Create new project**.

2.   Type the project name and click **Create**.

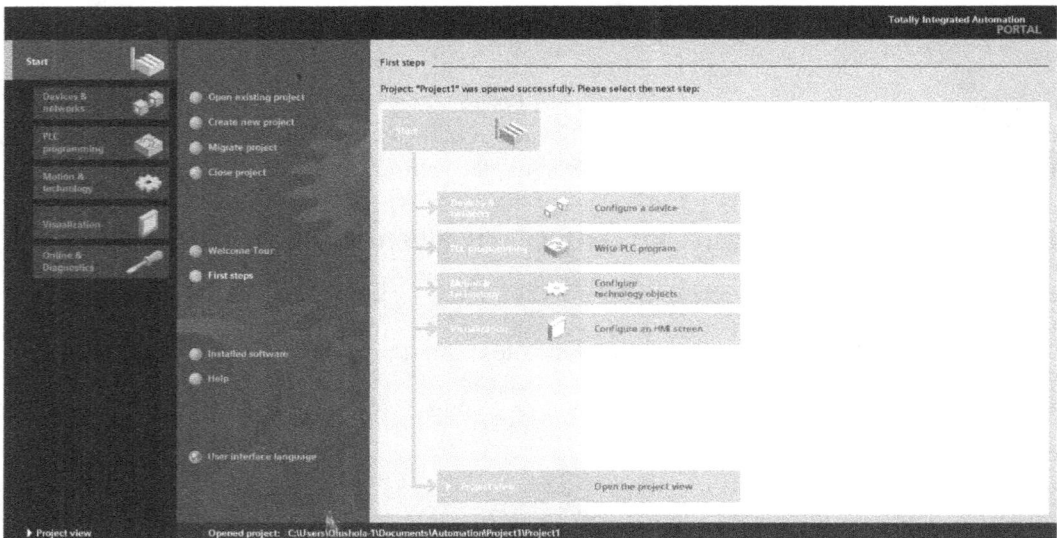

Figure 8.24 – Configuring a device in TIA Portal V13 (1)

3. Click on **Devices & networks**.

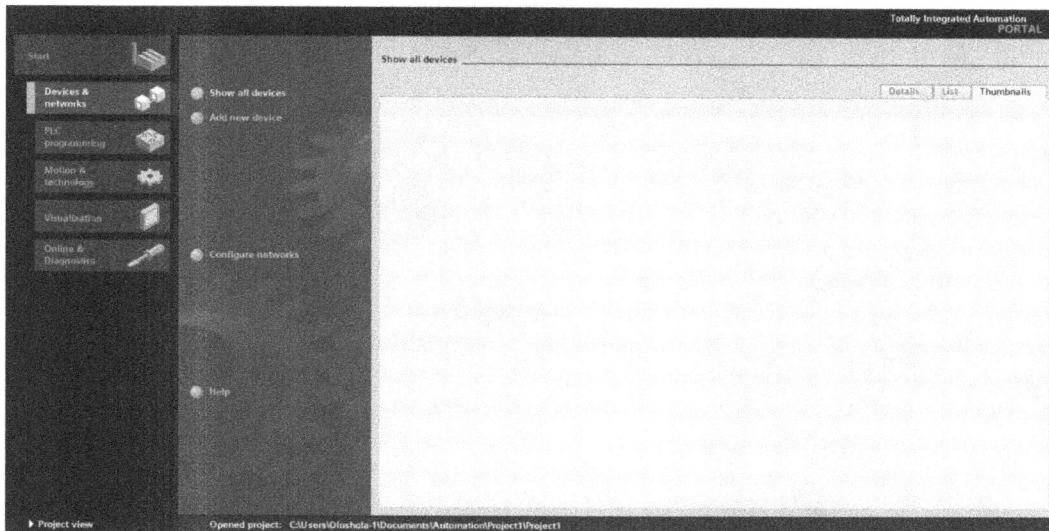

Figure 8.25 – Configuring a device in TIA Portal V13 (2)

4. Click **Add new device** and select **Controllers**.

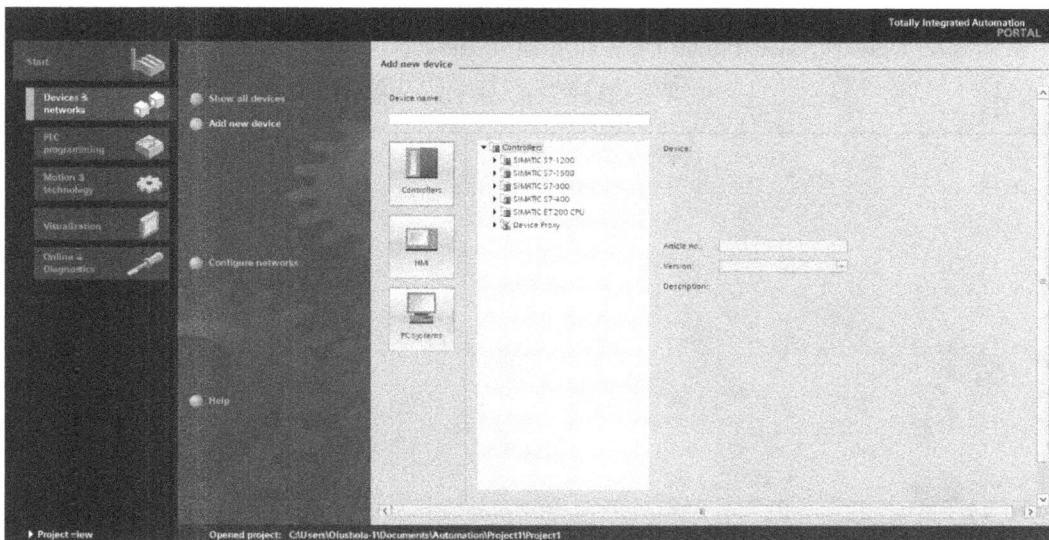

Figure 8.26 – Configuring a device in TIA Portal V13 (3)

5.   Expand **SIMATIC S7 1200 | CPU | CPU1211C AC/DC/Rly** and click **STEP 7 Professional** in the **Automation License Management - STEP 7 Basic** dialog box, as shown in the following screenshot:

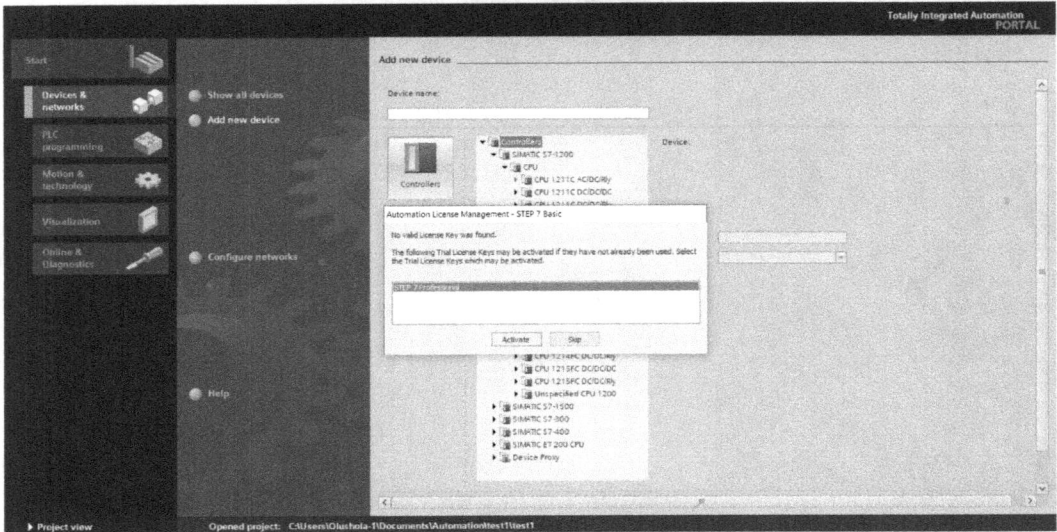

Figure 8.27 – Configuring a device in TIA Portal V13 (4)

6.   Click **Activate** to activate the 21-day trial license and select the article number of the PLC, for example, **6ES7 211-1BE40-0XB0**, from the list, as shown:

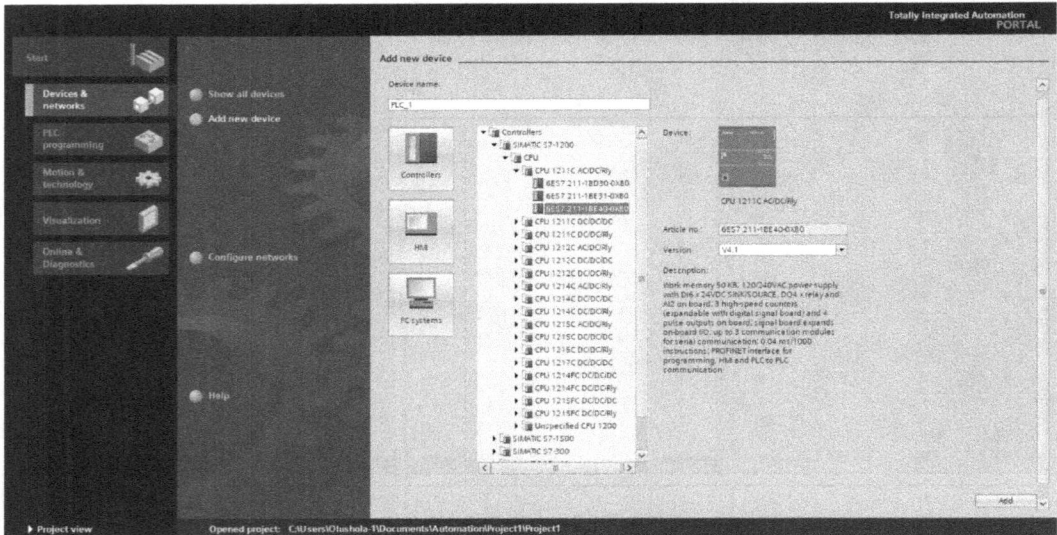

Figure 8.28 – Configuring a device in TIA Portal V13 (4)

7.   Click **Add** in the preceding screenshot. You should see the following screen:

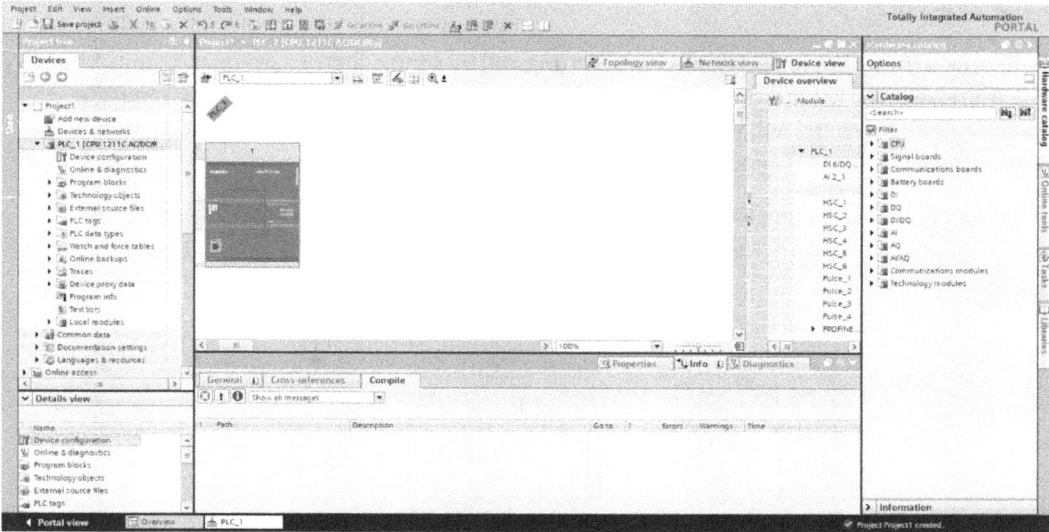

Figure 8.29 – Writing a program in Siemens TIA Portal (1)

8.   In the project tree on the left-hand side of the screen, expand **Programming blocks** and double-click on **Main [OB1]** to get to the programming environment:

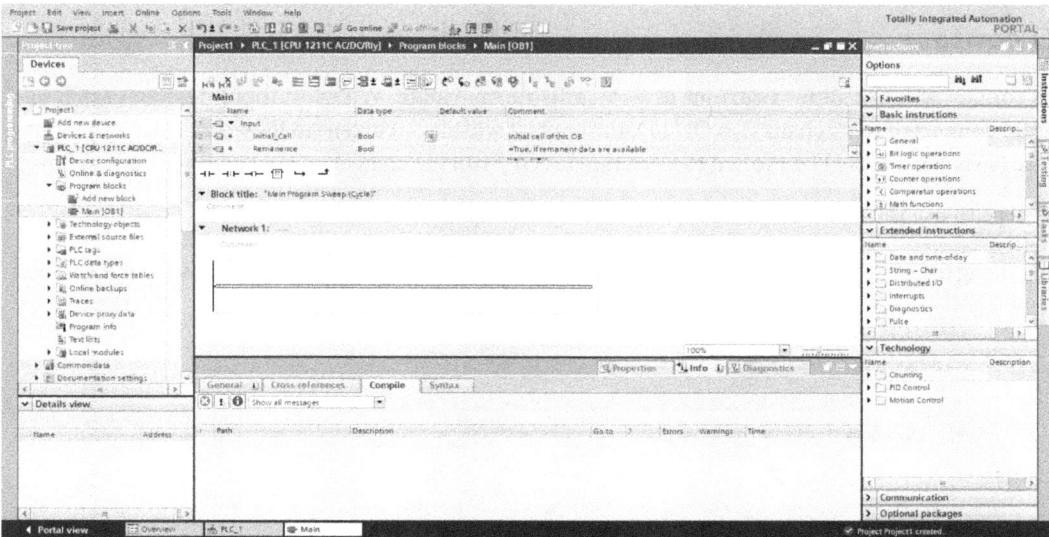

Figure 8.30 – Writing a program in Siemens TIA Portal (2)

Congratulations, you have now arrived at the interface for writing, editing, and downloading a program to the PLC.

We will now write a simple program. Our first rung will be simple logic to energize a lamp when a button is pressed. The second rung will be simple AND logic in which two NO contacts must be ON for the output to be energized, while the third rung will be simple OR logic in which either of the contacts can be ON for the output to be energized.

Let's have a brief look at some basic instructions to note before writing the preceding program:

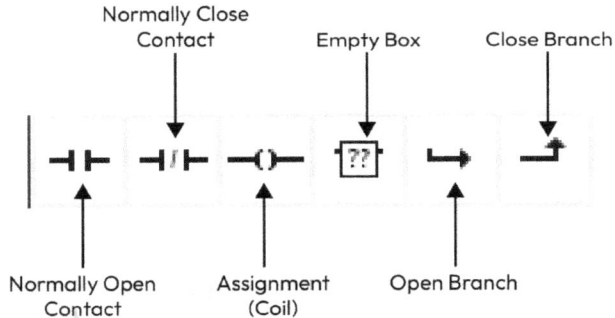

Figure 8.31 – Instructions and tools for writing a program in TIA Portal V13

We will start by learning how to use the NO contact.

NO contact

This contact will not allow a current to flow in its normal state. When an input device (switch or sensor) connected to its channel is ON (1), the contact will allow current to flow, and when the input device (switch or sensor) connected to its channel is OFF (0), the contact will not allow current to flow.

The steps to use the NO contact in TIA Portal are as follows:

1. Drag the NO contact to the line in the network, as indicated by the arrow in the following screenshot:

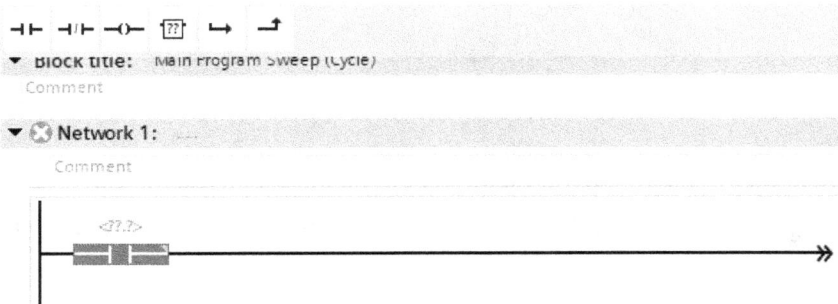

Figure 8.32 – Using NO contact instruction (1)

2. Double-click the question marks (**??.?**) and type the address of the input channel where a switch, push button, or sensor is connected physically, for example, I0 . 0.

3. Press *Enter* twice to have what is shown in the following screenshot:

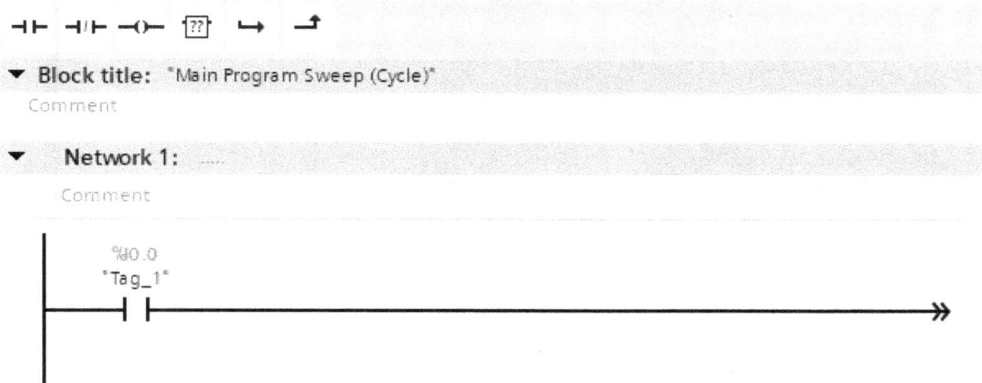

Figure 8.33 – Using NO contact instruction (2)

Let's learn how to use the NC contact.

NC contact

This contact will allow a current to flow in its normal state. When an input device (switch or sensor) connected to its channel is ON (1), the contact will not allow current to flow, and when the input device (switch or sensor) connected to its channel is OFF (0), the contact will allow current to flow.

The steps to use the NC contact in TIA Portal are as follows:

1. Drag the NC contact to the line in the network, as indicated by the arrow in the following screenshot:

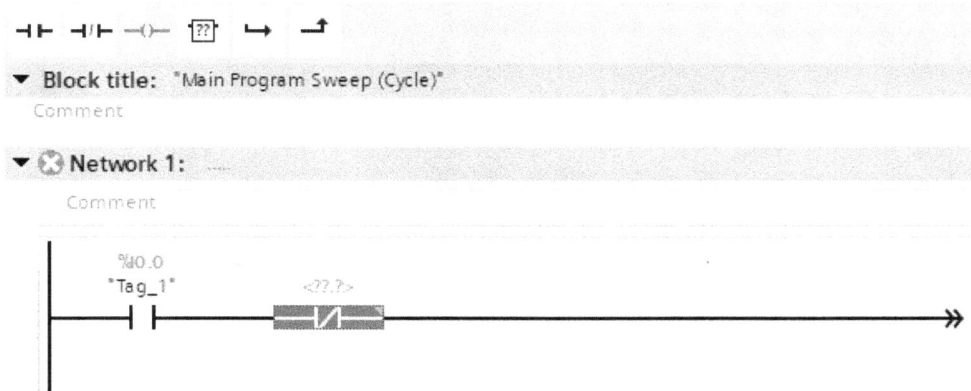

Figure 8.34 – Using NC contact instruction (1)

2.   Double-click the question marks (**??.?**) and type the address of the input channel where a switch, push button, or sensor is connected physically, for example, I0.1.

3.   Press *Enter* twice to have what is shown in the following screenshot:

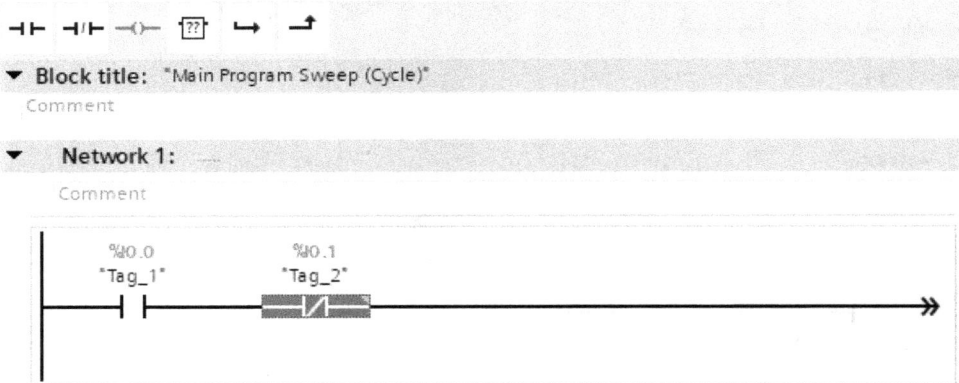

Figure 8.35 – Using NC contact instruction (2)

Let's proceed to learn about coil/assignment.

Coil/assignment

This represents a relay that will be energized when power flows to it. It will be de-energized when power does not flow to it. A load (light or other actuators) connected to its channel physically will turn on when it is energized.

The steps to use the coil/assignment in TIA Portal are as follows:

1.   Drag the coil/assignment to the line in the network, as indicated by the arrow in the following screenshot:

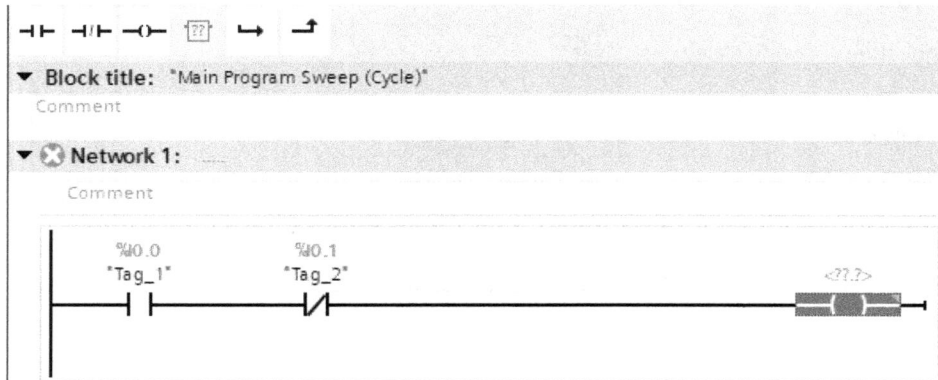

Figure 8.36 – Using coil/assignment instruction (1)

2. Double-click the question marks (**??.?**) and type the address of the output channel where a lamp or actuator is connected physically, for example, Q0 . 0.

3. Press *Enter* to have what is shown in the following screenshot:

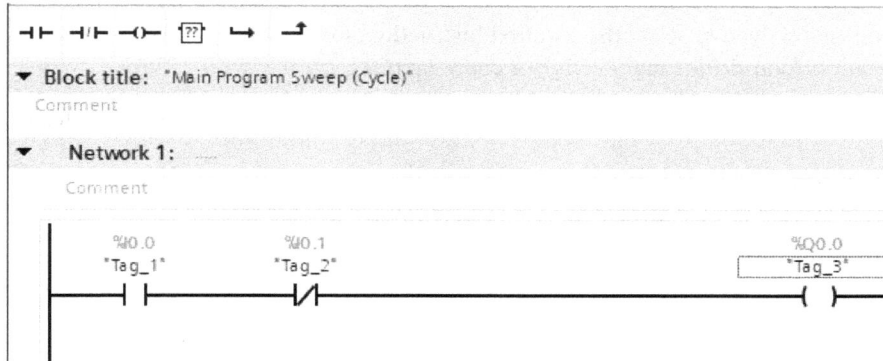

Figure 8.37 – Using coil/assignment instruction (2)

Let's now discuss how the preceding program will work.

Explanation of the program

Wiring and programming work together in a PLC. Let's consider the wiring in *Figure 8.38* to explain the program. NO push buttons (PB1 and PB2) are connected to I0.0 and I0.1, respectively, and a pilot lamp (PL1) is connected to Q0.0.

When the NO push button (PB1) connected to I0.0 is pressed, the NO contact with address I0.0 in the program will close to allow the signal to flow through it. Since the second input in the program with address I0.1 is NC and a NO push button (PB2) is connected to I0.1, the output (Q0.0) will be energized and the lamp will be ON as long as PB2 is not pressed. If PB2 is pressed, the NC contact with address I0.1 in the program will open and the output (Q0.0) will be de-energized:

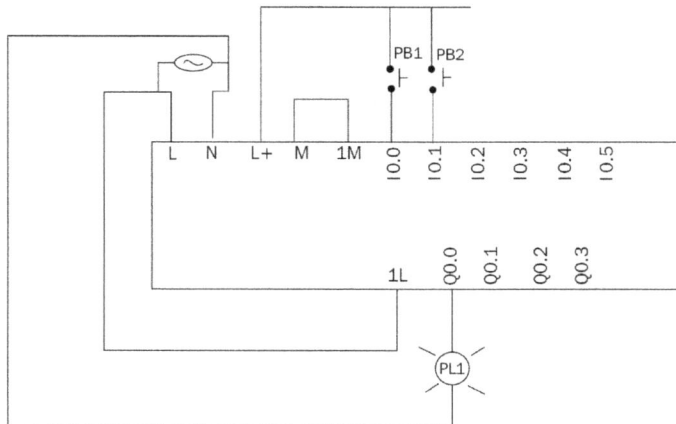

Figure 8.38 – Wiring two push buttons and a pilot lamp to S71200 CPU 1211C AC/DC/RLY

Let's learn about the empty box.

Empty box

The empty box allows you to select the required instruction, for example, NO contact, NC contact, coil/assignment (=), on-delay timer (TON), or count up (CTU).

Let's learn how to use the empty box to select the required instruction in the following steps:

1. Drag the empty box to the line in network 2, as shown:

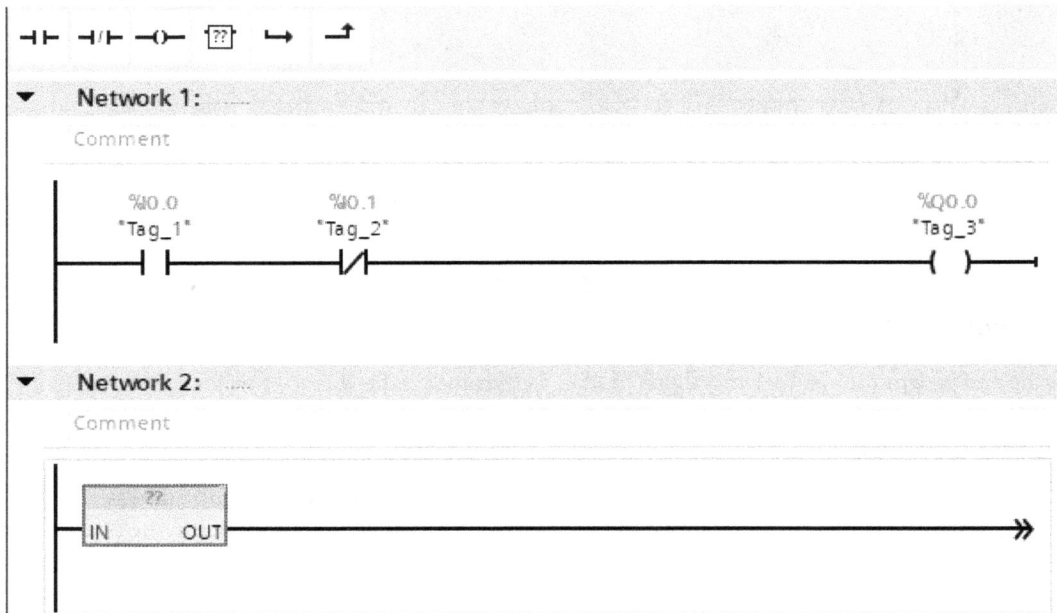

Figure 8.39 – Using empty box (1)

2. Double-click the two question marks (**??**) and click the icon at the right of the box to see the list of instructions that can be selected, as shown in the following screenshot:

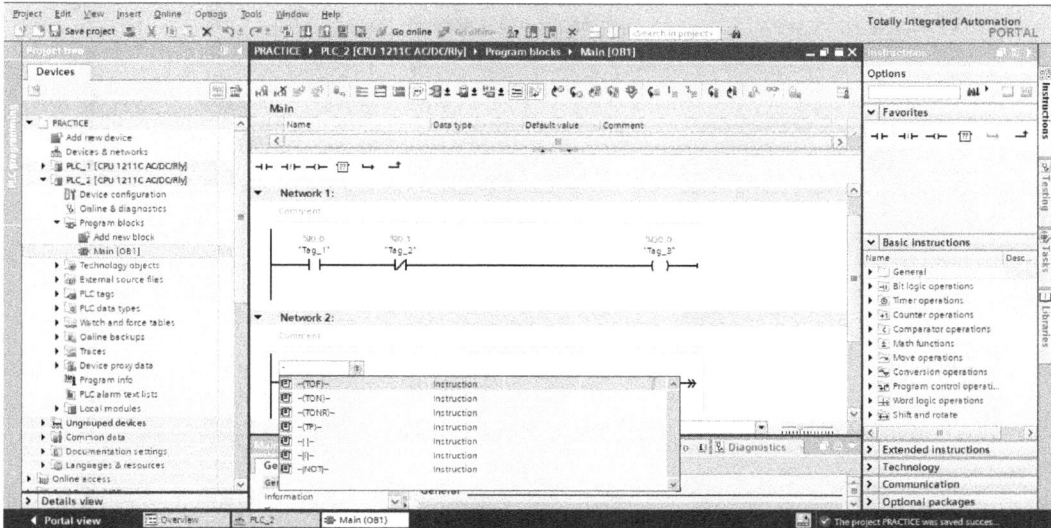

Figure 8.40 – Using empty box (2)

3. Select the NO instruction, for example, and press *Enter* twice. You should have what is shown in the following screenshot:

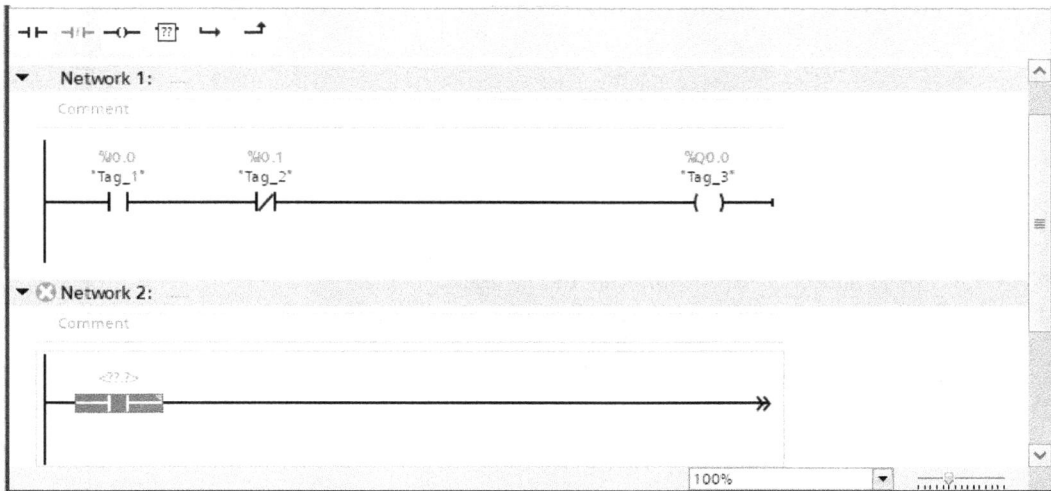

Figure 8.41 – Using empty box (3)

4. Double-click the question marks (**??.?**) and type the address of the input channel where a pushbutton, switch, or sensor is connected physically, for example, I0.2.

5. Press *Enter* to have what is shown in the following screenshot:

Comment

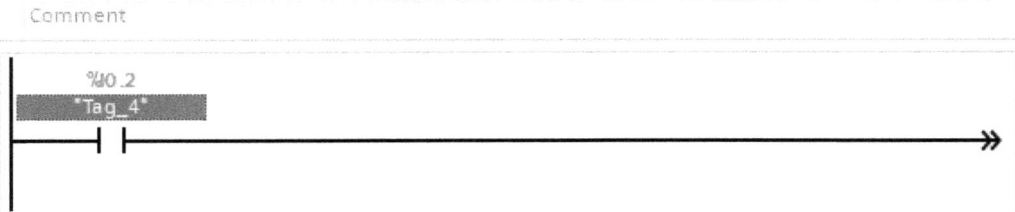

Figure 8.42 – Using empty box (4)

6. Repeat *steps 1-5* to add a coil/assignment (=) instruction and give it an address, Q0.1 which refers to the address of the output channel where a lamp or an actuator is connected physically. You should have what is shown in the following screenshot:

▼   **Network 2:**

Comment

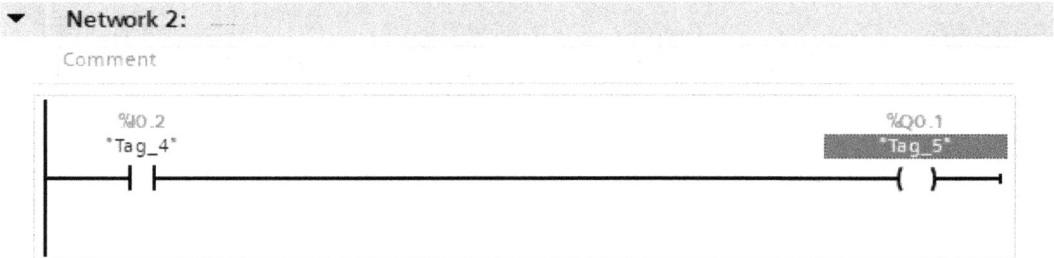

Figure 8.43 – Adding coil/assignment to the rung

Let's discuss how the preceding program will work.

Explanation of the program

Let's say a NO push button (PB3) is connected to I0.2 and a pilot lamp (PL2) is connected to Q0.1, as shown in *Figure 8.44*. When PB3 is pressed, the NO contact, I0.2, in the program will close, signal will flow, and the coil with address Q0.1 will be energized. Pilot lamp PL2 connected to Q0.1 will be ON. When PB3 is released, the NO contact with address I0.2 will open and the coil with address Q0.1 will be de-energized, making the pilot lamp (PL2) connected to Q0.1 be OFF:

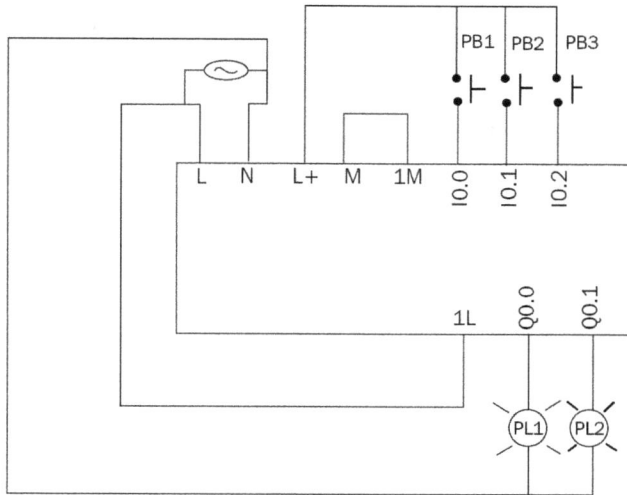

Figure 8.44 – Wiring three push buttons and two pilot lamps to S71200 CPU 1211C AC/DC/RLY

Let's move further to learn how to use the open branch.

Open branch

As the name implies, this opens a branch. It is used when instructions need to be connected in parallel.

Let's now open a branch that will be used to add an instruction in parallel to the preceding program in netwcrk 2 by following these steps:

1. Drag the open branch to the line in network 2, as indicated by the arrow in the following screenshot:

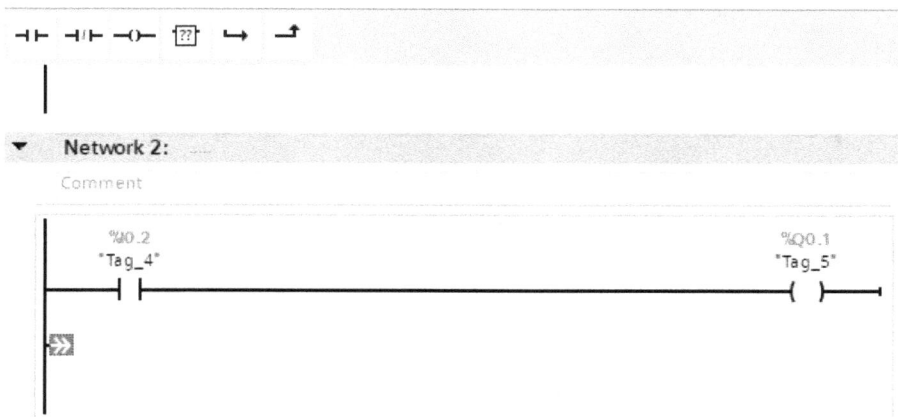

Figure 8.45 – Using open branch

2. Drag an instruction to the new branch, as indicated by the arrow in the following screenshot:

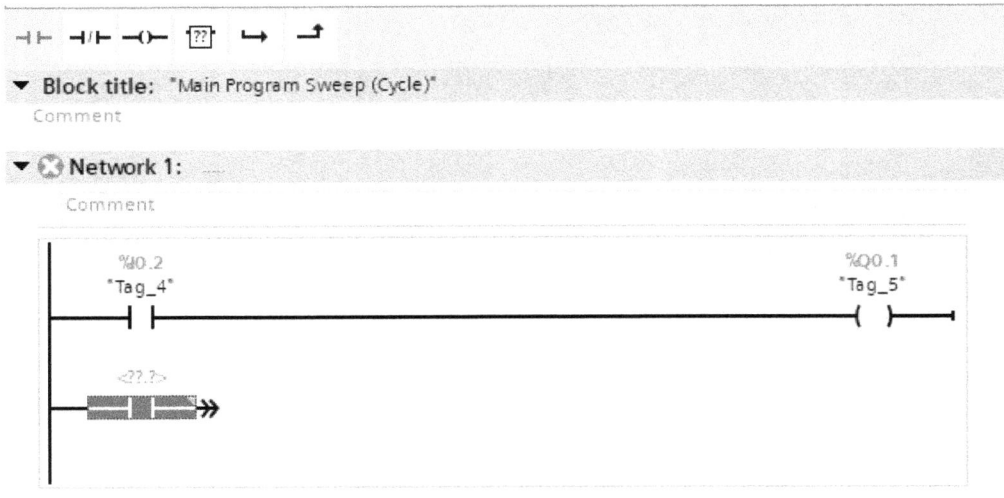

Figure 8.46 – Adding instruction to a branch (1)

3. Double-click the question marks (**??.?**) and type the address of the input channel where a switch, push button, or sensor is connected physically, for example, I0.3.

4. Press *Enter* twice to have what is shown in the following screenshot:

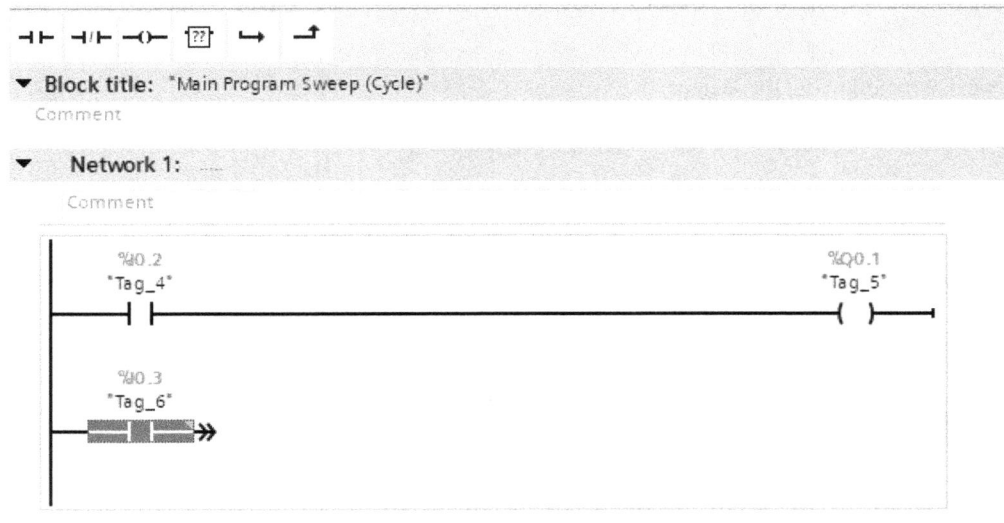

Figure 8.47 – Adding instruction to a branch (2)

Let's now learn about the close branch.

Close branch

As the name implies, this closes a branch that is open. Let's now close the branch to complete the program in network 2. Drag the close branch to the point indicated by the arrow in the following screenshot:

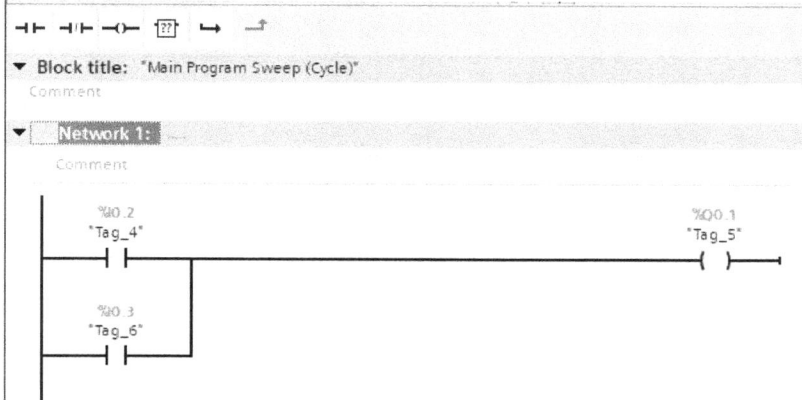

Figure 8.48 – Using close branch

The preceding is a simple OR logic circuit. Let's discuss the program briefly.

Explanation of the program

Let's say two NO push buttons (PB3 and PB4) are connected to I0.2 and I0.3, respectively, and a pilot lamp, PL2, is connected to Q0.1, as shown in *Figure 8.49*. Q0.1 will be energized if either PB3 or PB4 or both PB3 and PB4 are pressed, and the pilot lamp (PL2) connected to Q0.1 will be ON. When both buttons are released, there will be no signal flowing to the coil with address Q0.1 and the pilot lamp (PL2) connected to Q0.1 will be OFF:

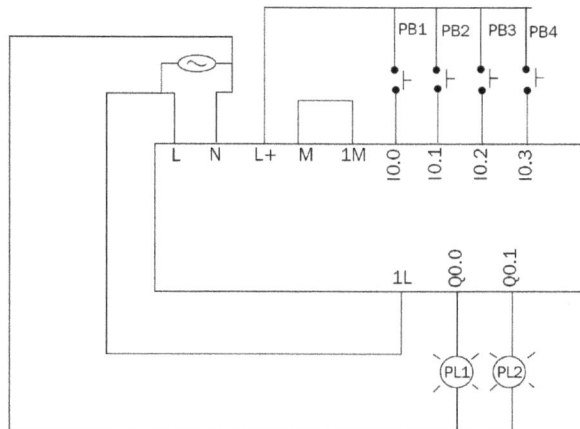

Figure 8.49 – Wiring four push buttons and two pilot lamps to S71200 CPU 1211C AC/DC/RLY

> **Note**
> Always remember to save changes to your project. Click **Project | Save**.

You have just learned how to create a project and write a simple program in TIA Portal. You should now understand how to use basic instructions and tools, such as NO contact, NC contact, and empty box, to write a simple ladder logic program.

Summary

Congratulations! You have successfully completed *Chapter 8* of this book. The chapter explained the software part of PLC. You should now be able to differentiate between various programming languages. You should also be able to explain programming devices, programming software, the basics of LD, and so on. An industrial automation engineer needs to be aware of the elements of LD and the rules for LD programming, which was well explained in this chapter. You should now be familiar with Siemens programming software (TIA Portal) and be able to use it to write some basic PLC programs using the most common PLC programming language (LD).

The next chapter will further discuss PLC programming. You will learn how to simulate your program, how to download program to a live PLC and also how to write programs that are more advanced than the ones learned in this chapter.

Questions

The following are questions to test your understanding of this chapter. Ensure you have read and understood the topics in this chapter before attempting the questions:

1. _____ is a set of instructions written in a language the computer understands to perform a task.

2. _____ is a set of instructions in either textual or graphical form that represents the control function that will be carried out for a specific industrial task or application.

3. According to IEC standards, the five programming languages stated in part 3 of IEC 61131 for writing a program for PLCs to control and automate tasks are _____, _____, _____, _____, and _____.

4. _____ is the most common among the several PLC programming languages. It is a graphical programming language that's easy to understand for most plant technicians because it's similar to the relay diagrams that they are familiar with.

5. IEC is an acronym for _____.

6. _____ is a graphical language in which program elements are represented in the form of blocks and are connected together.

7. _____ is a text-based language that resembles assembly language.

8. The two common programming devices are _____ and _____.

9. _____ are the horizontal lines that connect the two rails to complete a circuit.

9

Deep Dive into PLC Programming with TIA Portal

In the previous chapter, we learned the fundamentals of PLC programming, which include PLC programming languages, PLC programming devices, PLC programming software, Ladder Diagram basics, elements of Ladder Diagrams (ladder logic programs), rules for Ladder Diagram programming, downloading and installing TIA Portal V13 Professional and PLCSIM, and how to create a project and write some programs with the programming software (TIA Portal).

In this chapter, you will dive deeper by practicing how to use the programming software you downloaded and installed in the previous chapter. This chapter also includes a simulation of programs, which will enable you to see the result of the programs you have written right there on your PC or laptop even when you don't have the real PLC to test or practice.

We are going to cover the following main topics in this chapter:

- Opening a saved program
- Simulating programs with Siemens TIA Portal using PLCSIM
- Latching and unlatching in PLC programming
- Using an output address as an input in a program
- Using the SET and RESET instructions
- Using the timer instruction
- Using the counter instruction
- Using the move instruction
- Using the compare instructions

- Level control using PLC
- Automated filling, capping, and wrapping system using PLC

Technical requirements

While every part of this book is valuable, *Chapter 2, Switches and Sensors – Working Principles, Applications, and Wiring, Chapter 3, Actuators and Their Applications in Industrial Automation,* and *Chapter 7, Understanding PLC Hardware and Wiring,* as well as *Chapter 8, Understanding PLC Software and Programming with TIA Portal,* are very important to have a better understanding of this chapter.

This chapter is actually a continuation of the previous chapter. Hence, you must have familiarized yourself with the necessary terms and the programming software (TIA Portal) and also practiced the programs explained using the necessary instructions before reading this chapter.

Opening a saved program

We will continue learning the programming of PLC here by opening the project/program we saved in the previous chapter:

1. Start the TIA Portal application.
2. Click on **Open existing project**, select the project, and then click **Open**:

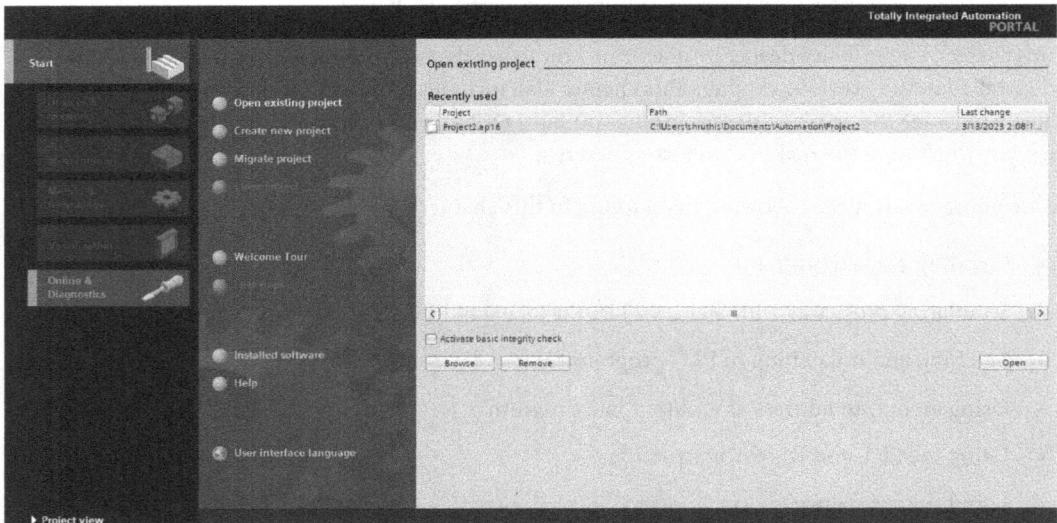

Figure 9.1 – Opening a saved program – recently used project

3.   Upon clicking **Open**, you should see the following screen. Click on **Open the project view**:

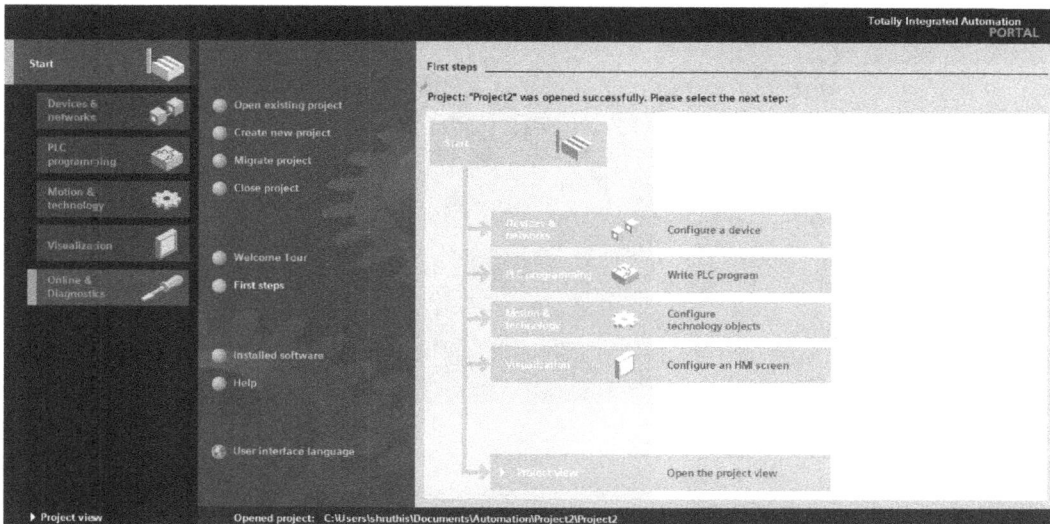

Figure 9.2 – Opening a saved program – Open the project view

4.   Double-click the PLC folder (**PLC_1**) on the left-hand side, and then double-click **Program blocks**:

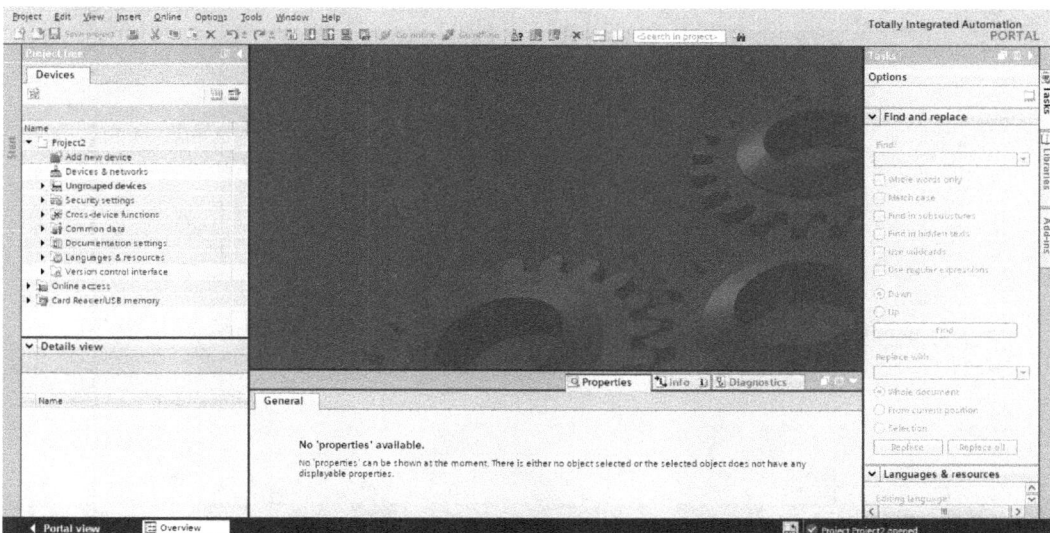

Figure 9.3 – Opening a saved program – PLC_1 [CPU 1211C AC/DC/Rly]

5.   Double-click **Main [OB1]**:

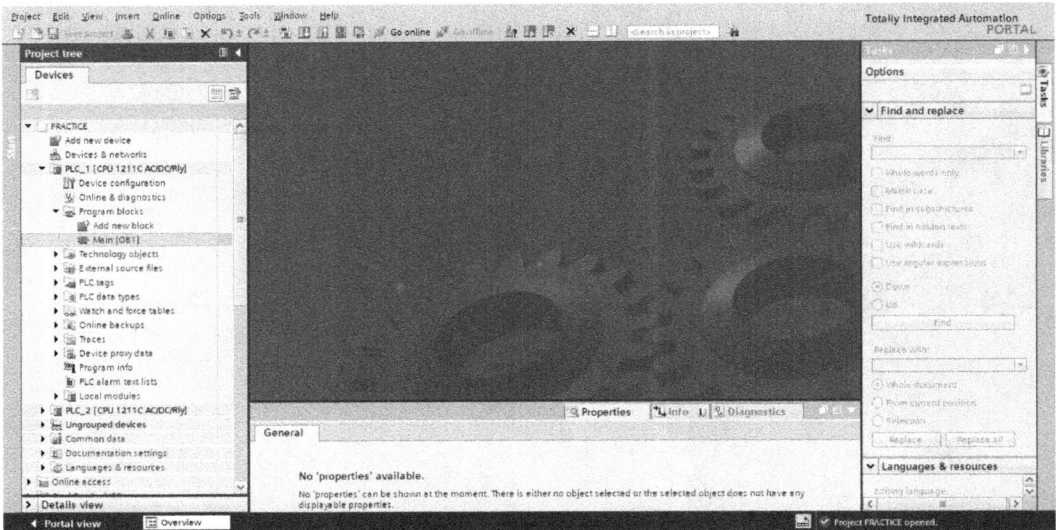

Figure 9.4 – Opening a saved program – Main [OB1]

The program will be opened showing the ladder diagram, as shown in the following screenshot:

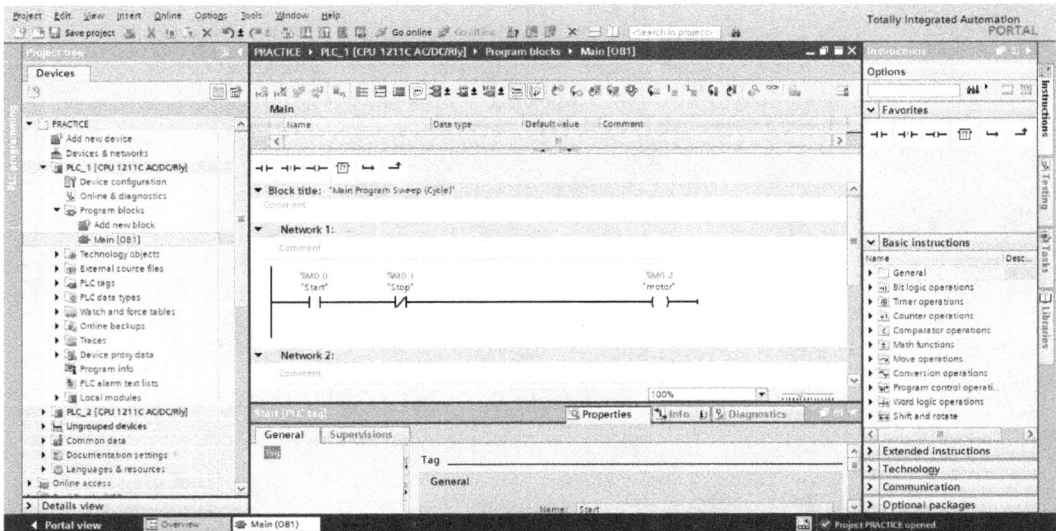

Figure 9.5 – Program opened showing the ladder diagram

We have just learned how to open a saved program in TIA Portal. We will now move on and learn how to simulate a program in the next section.

Simulating programs with Siemens TIA Portal using PLCSIM

We will now learn how to simulate a program using Siemens TIA Portal. We will use the program written in the previous chapter as an example:

1. Open the project/program using the steps in the previous section. You should have what looks like the following screenshot. Click on the **Compile** icon and ensure there is no error in your program.

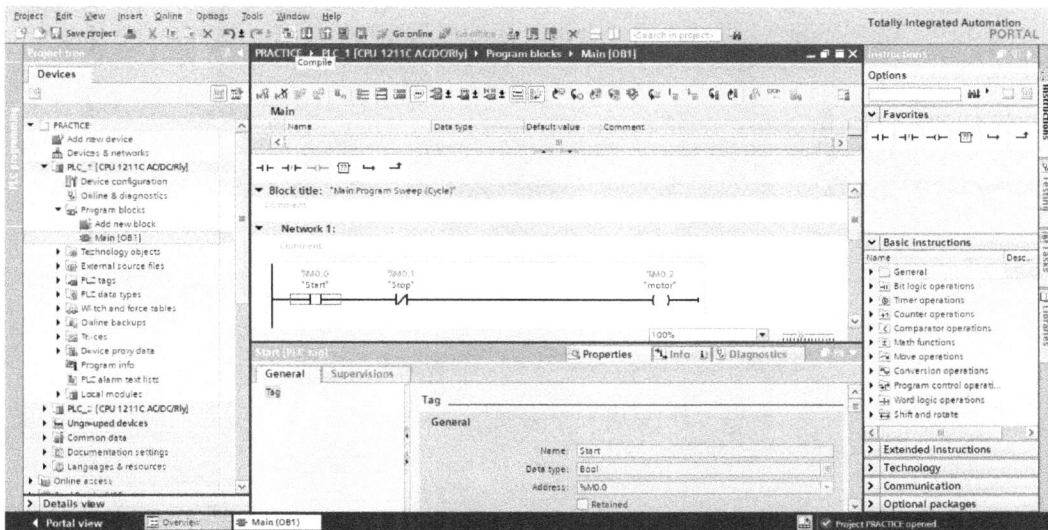

Figure 9.6 – Opened project showing the Compile button (red arrow)

2. Click on the **Start simulation** icon, as indicated in the following screenshot:

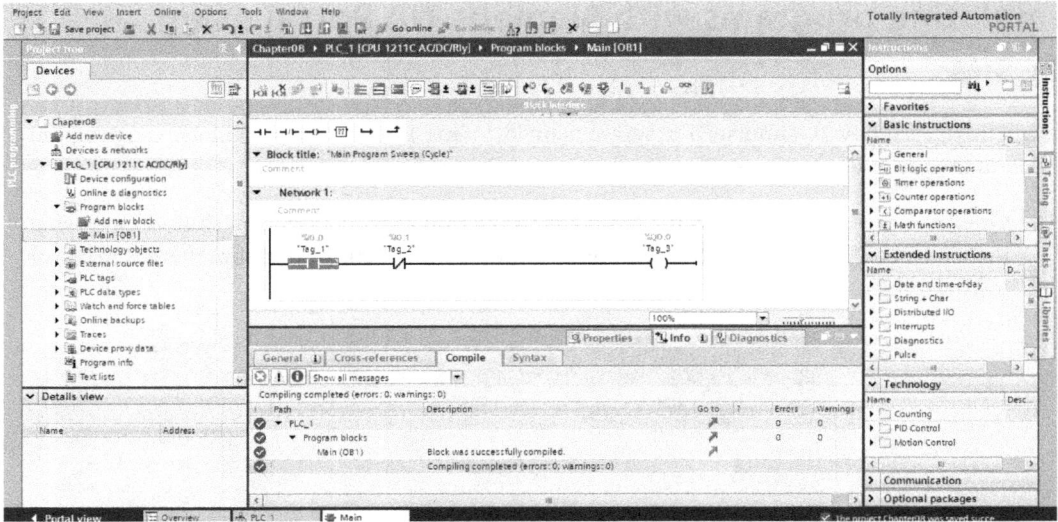

Figure 9.7 – Compiled program showing no error, that is 0 errors

3.   A warning message shown as follows will appear. Click **OK**:

Figure 9.8 – Simulation – Warning message

You should have what looks similar to the following screenshot:

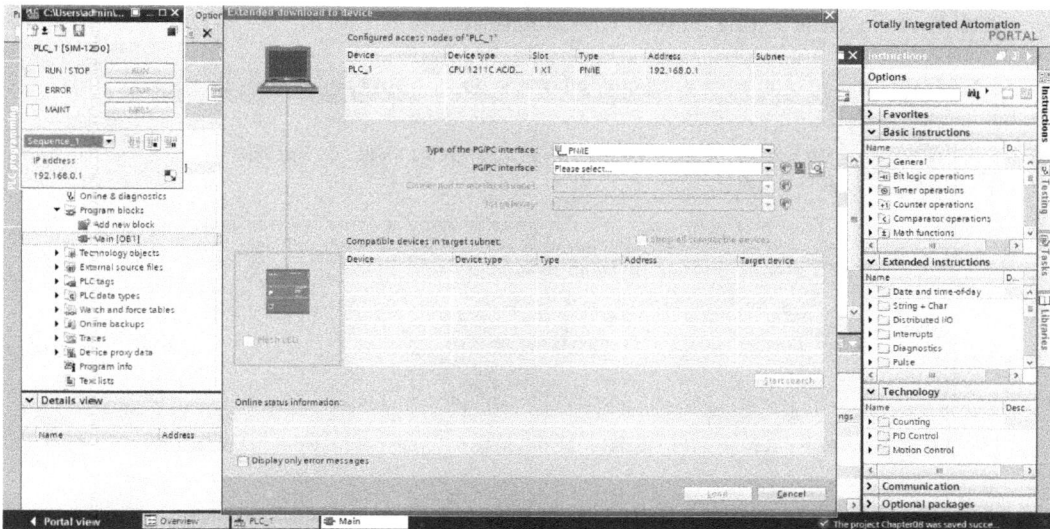

Figure 9.9 – Simulation – Extended download to device dialog box

4.   Select **PLCSIM S7-1200/S7-1500** in the **PG/PC interface** list and then click **Start search**. Select **CPUcommon** with the address `192.168.0.1` and then click **Load**:

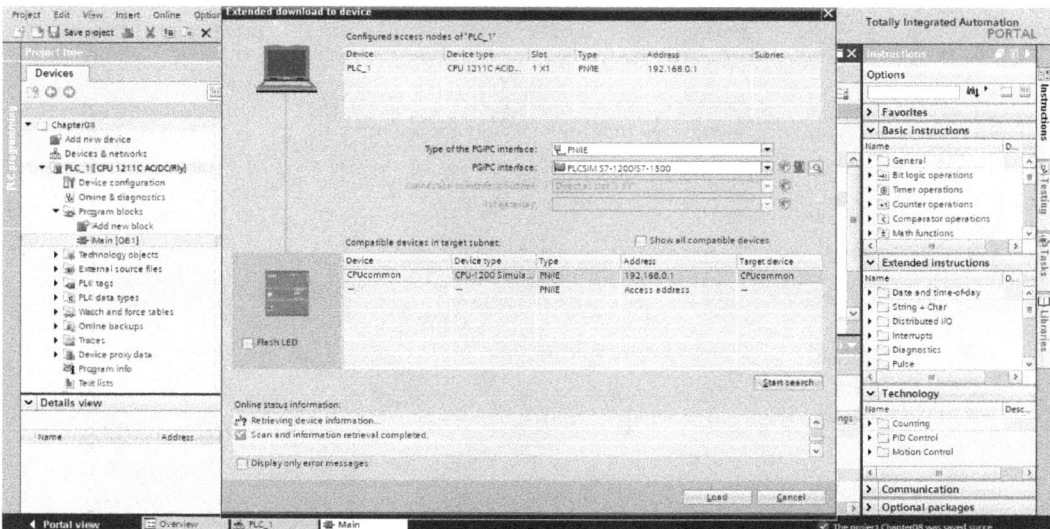

Figure 9.10 – Simulation – Search result

5.   Click **Yes** to save the setting for the PG/PC interface:

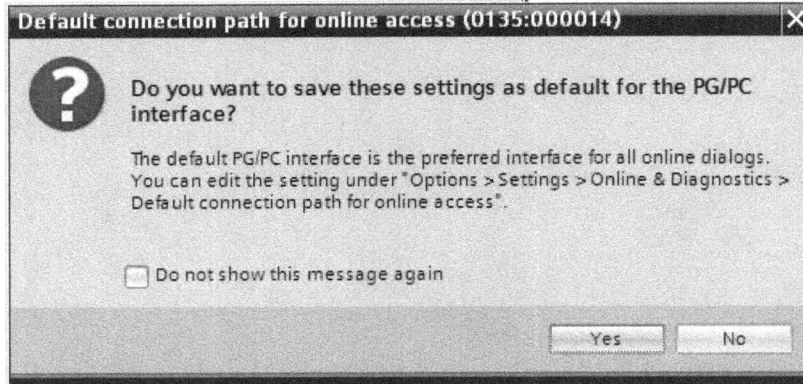

Figure 9.11 – Simulation – Default connection path for online access message

6.   Click **Load**:

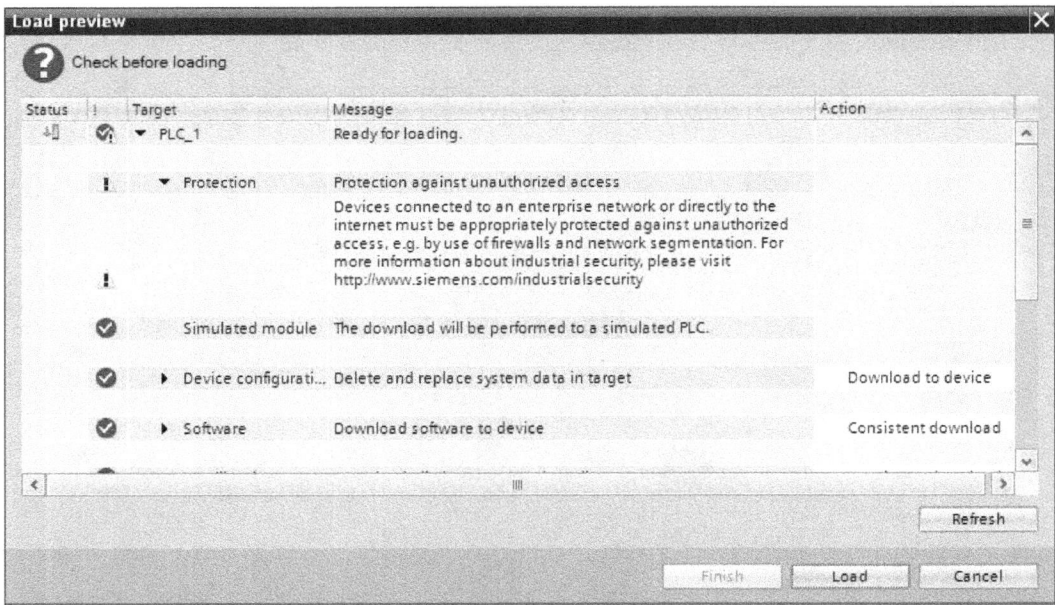

Figure 9.12 – Simulation – Summary screen

7.   Mark **Start all** and then click **Finish**:

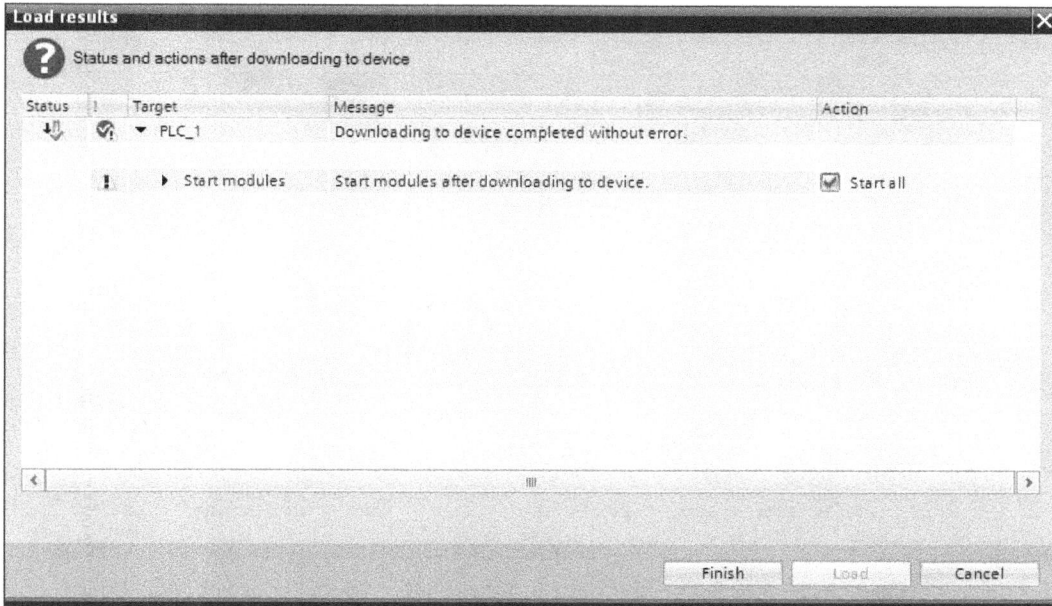

Figure 9.13 – Simulation – Load results

You should have what looks like the following screenshot:

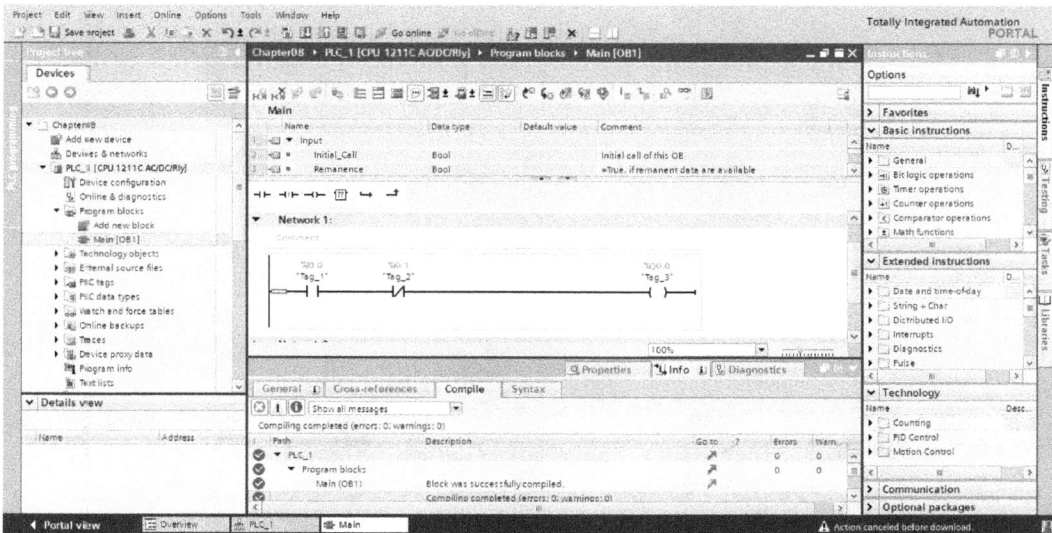

Figure 9.14 – Simulation – Program ready for simulation; PLCSIM on the taskbar

8. Click **PLCSIM** on the taskbar to have the simulator on your screen, as shown in the following screenshot. Click on the **Switch to project view** icon:

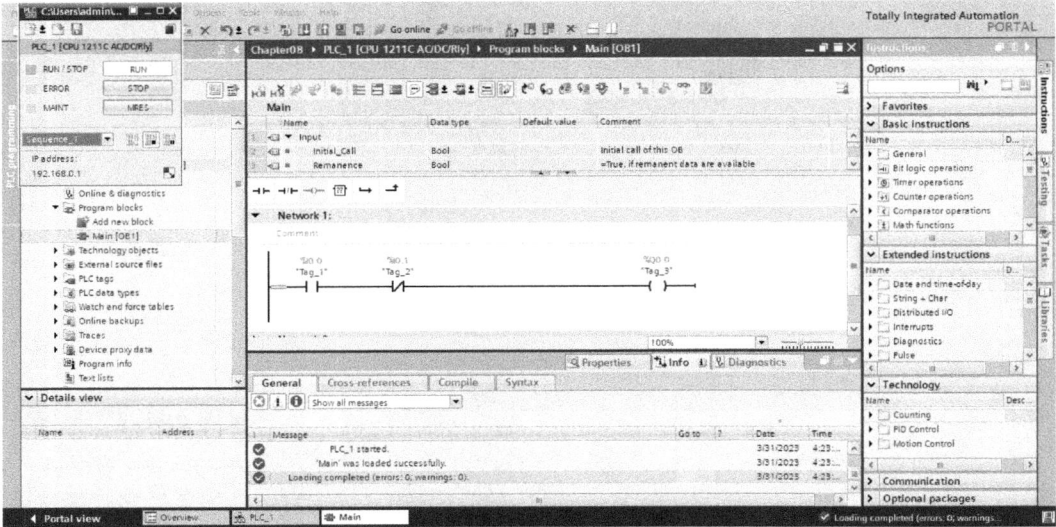

Figure 9.15 – Simulation – Simulator in the top-left corner of the screen

Your screen should look like the following screenshot:

Figure 9.16 – Simulation – Project view of the simulator

9. Expand the left pane to have something that looks like the following screenshot:

Figure 9.17 – Simulation – Left pane expanded

10. Expand **SIM tables**:

Figure 9.18 – Simulation – SIM tables

11. Double-click **SIM table_1**:

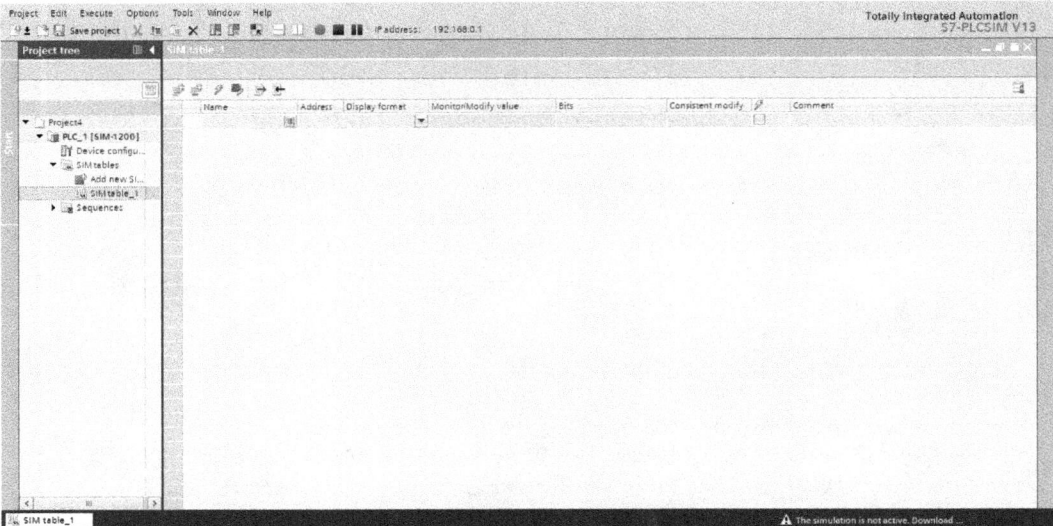

Figure 9.19 – Simulation – SIM table_1

12. Add all the I/O addresses you have used in the **Address** column, as shown in the following screenshot:

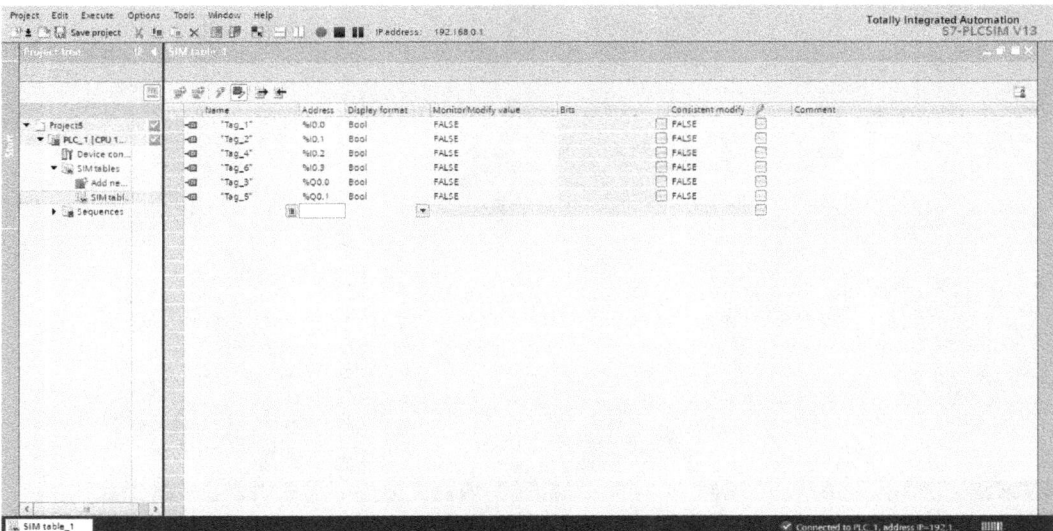

Figure 9.20 – Simulation – Project view of the simulator with I/O addresses added

13. Return to the programming environment by clicking **TIA V13** on the taskbar and then clicking on the **Monitoring on/off** icon to switch on the monitor mode.

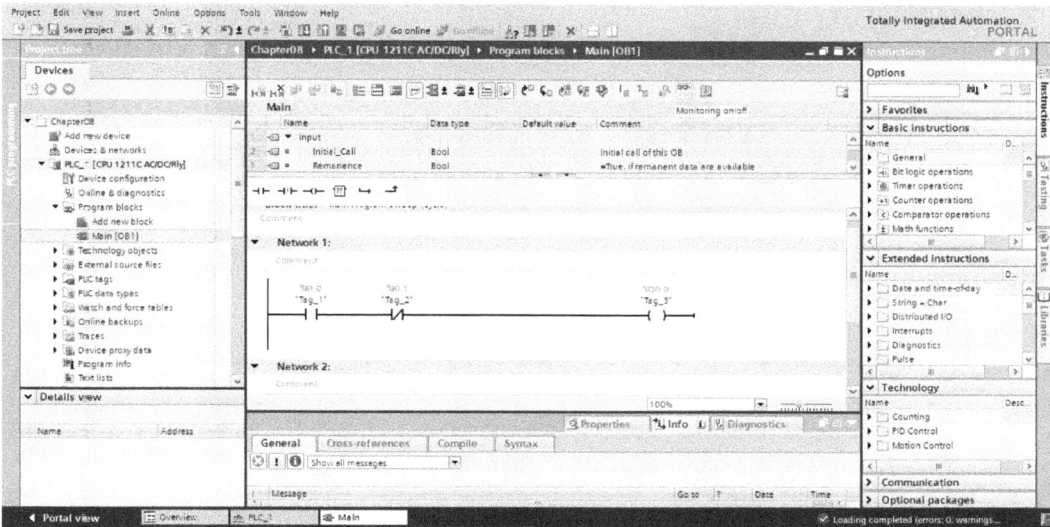

Figure 9.21 – Simulation – Monitoring on/off

14. Mark or unmark the bit of the input address you want to turn ON or OFF in the simulator.

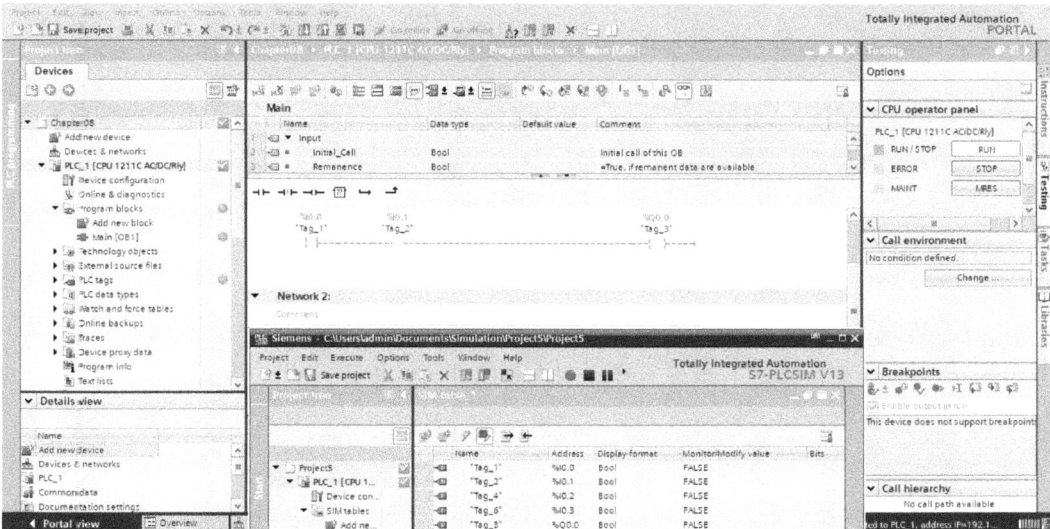

Figure 9.22 – Simulation – Screen rearranged to show both the program
and the simulator's project view with I/O addresses added

Now, let's simulate network 1. In the preceding screenshot, it can be seen that the bit for I0.0 is not marked, while the output Q0.0 remains OFF because a signal cannot flow to it.

Mark the bit for I0.0. The coil with the address Q0.0 will come ON, as shown in the following screenshot:

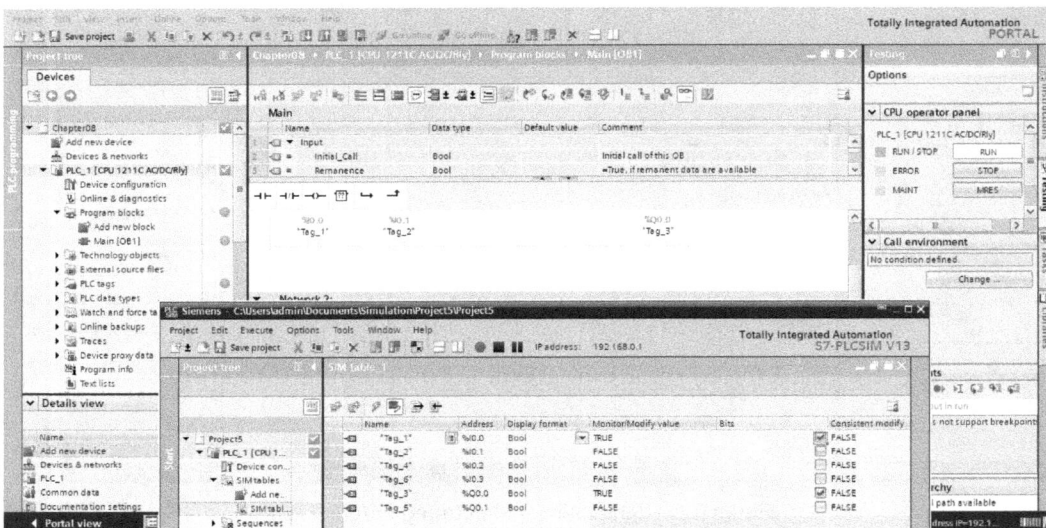

Figure 9.23 – Simulation – Simulation result of network 1

You can mark or unmark the bit for any input address you want to turn ON or OFF in the simulator and see how the output responds to the inputs.

Let's now simulate network 2. Mark the bit for I0.2. Q0.1 will turn ON, as shown in the following screenshot:

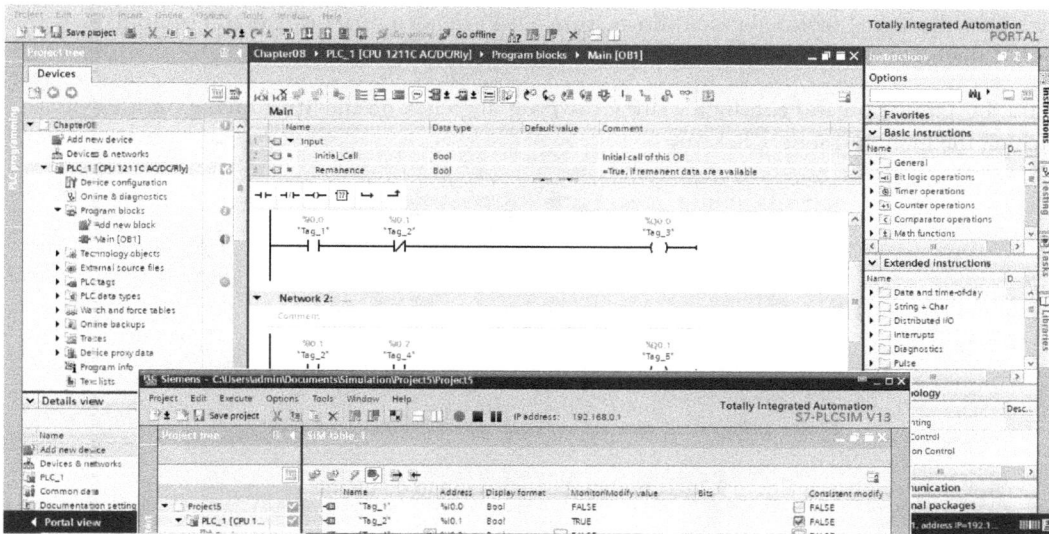

Figure 9.24 – Simulation – Simulation result of network 2 when I0.2 is marked (ON)

Unmark the bit for I0.2 and mark the bit for I0.3. Q0.1 will still turn ON, as shown in the following screenshot:

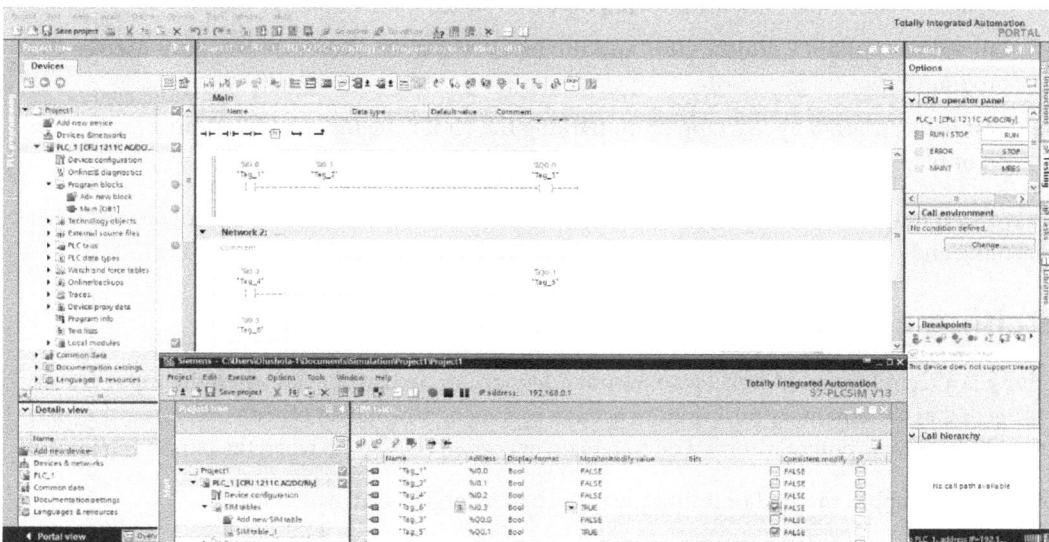

Figure 9.25 – Simulation – Simulation result of network 2 when I0.2 is unmarked and I0.3 is marked

Watch how the signal (colored green) flows from left to right to turn ON the coil (output) as you turn ON the input.

To end the simulation and continue writing the program, click on the **Monitoring on/off** icon to switch off the monitor mode. Click **Yes**:

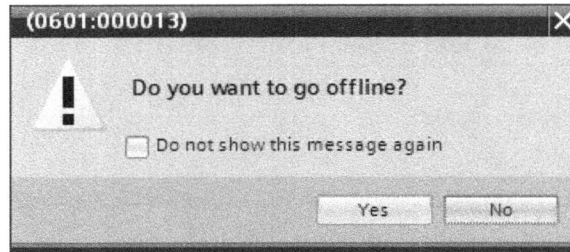

Figure 9.26 – Confirmation message to go offline

You can continue writing or editing the program using the instructions (normally open contact, normally closed contact, and assignment/coil) explained in the previous section.

Perform the following steps to simulate other changes you have made to your program:

1. Click **Compile** and then click **Download to device**.
2. Click **Load** and then click **Monitoring on/off** to switch on the monitor mode.
3. On your simulator's project view, add the input or output addresses you have used in the program as we did previously.
4. Test your program by marking or unmarking the bit for the input address you want to turn on or off.

We have seen how to simulate programs using Siemens TIA Portal. Next, we will understand what the latching and unlatching techniques are in PLC programming.

Latching and unlatching in PLC programming

Latching is a technique used to keep an output energized or activated even when the input ceases. Latching makes a momentary push-button act as a maintained switch; in other words, after the push button is pressed, the output is turned on to remain energized (ON) even when the button is released.

Unlatching is simply a method or technique used to de-energize a latched output.

A latch and unlatch program using normally open push buttons for both starting and stopping

Write the simple program in network 3 to see how latching and unlatching works with a normally open push button for starting and a normally open push button for stopping:

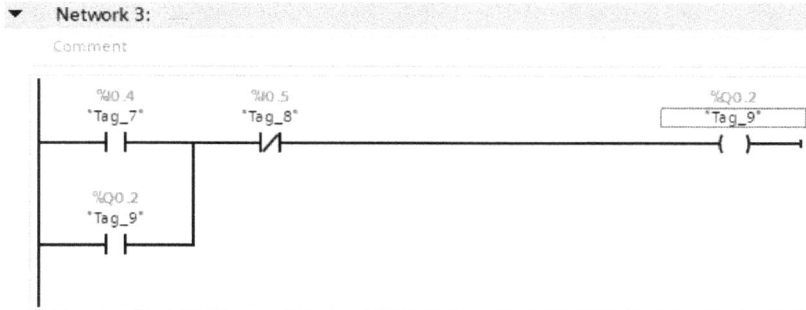

Figure 9.27 – A latch and unlatch program using normally open
push buttons for both starting and stopping

The PLC wiring for the program is as follows:

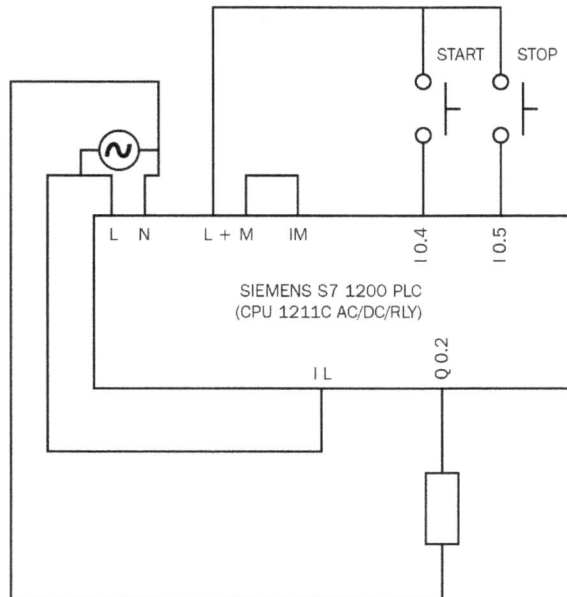

Figure 9.28 – Wiring to implement a latch and unlatch program using
normally open push buttons for both starting and stopping

The preceding wiring diagram shows the start push button (normally open) connected to I0.4, and the stop push button (normally open) connected to I0.5. A load, which can be the coil of a contactor, is connected to Q0.2.

When the start button is pressed, the normally open contact with the address I0.4 in the program will be closed. Power will flow to energize the coil with the address Q0.2 because I0.5 in the program is a normally closed contact. When the start button is released, the normally open contact with the address I0.4 in the program opens, but the output with the address Q0.2 remains energized due to the normally open contact with the same address as the output (Q0.2) that is parallel to I0.4. Coil Q0.2, which is energized (ON), closes the normally open contact with the address Q0.2 in parallel to I0.4. This creates an alternative path for the signal to flow to Q0.2. Hence, the output coil with the address Q0.2 remains ON even when the button is released. This is referred to as latching.

When the stop push button is pressed, the normally closed contact with the address I0.5 is opened. Power will no longer flow to the coil. The output coil with the address Q0.2 will become de-energized (OFF). This is referred to as unlatching.

The stop push button used here is unsafe because when the push button fails, the output will remain energized and the machine will keep running. This is not preferred and is unsafe for industrial use.

A latch and unlatch program using a normally open push button for starting and a normally closed push button for stopping

Edit the program in network 3 such that the normally closed contact used for I0.5 will be a normally open contact, as shown in the following screenshot:

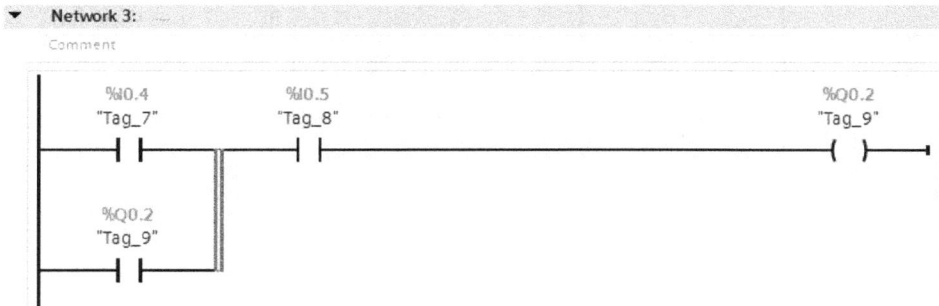

Figure 9.29 – A latch and unlatch program using a normally open push button
for starting and a normally closed push button for stopping

The PLC wiring for the program is as follows:

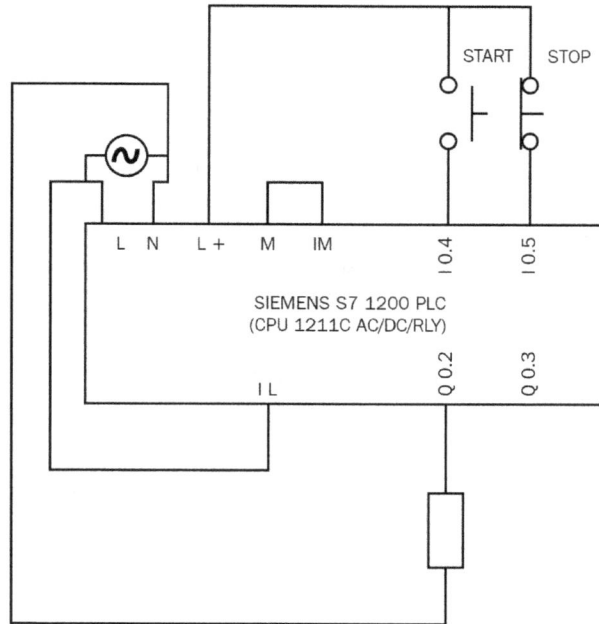

Figure 9.30 – Wiring to implement a latch and unlatch program with a normally open
push button for starting and a normally closed push button for stopping

The preceding wiring diagram shows the start push button (normally open) connected to I0.4, and the stop push button (normally closed) connected to I0.5. A load, which can be the coil of a contactor, is connected to Q0.2.

When the start button is pressed, the normally open contact with the address I0.4 in the program will be closed. The stop push button (normally closed push button) will close the normally open contact with the address I0.5 in the program. Power will flow to energize the coil with the address Q0.2 because address I0.5 in the program is closed due to the normally closed push button used in the wiring. When the start button is released, the normally open contact with the address I0.4 in the program is opened, but the output with the address Q0.2 remains energized due to the normally open contact with the same address as the output (Q0.2) that is parallel to I0.4 in the program. Coil Q0.2, which is energized (ON), closes the normally open contact with the address Q0.2 that is parallel to I0.4 in the program.. This creates an alternative path for the signal to flow to Q0.2. Hence, the output coil with the address Q0.2 remains ON even when the button is released. This is referred to as latching.

When the stop push button is pressed, the normally closed contact with the address I0.5 is opened. Power will no longer flow to the coil. The output coil with the address Q0.2 will become de-energized (OFF). This is referred to as unlatching.

The stop push button used here is safe because when the push button fails, the output will be de-energized and the machine that the stop push button is connected to will not run. This is preferred and safe for industrial use.

> **Note**
>
> If the output (Q0.2) will be used to power on an induction motor, an overload relay is necessary between the contactor and the motor to ensure the motor stops when there is an overload. In this case, the normally closed contact (95-96) will be wired as input to the PLC, and a normally open contact will be used for it in the program (in series with the normally open contact for the stop) such that, when there is an overload, the normally open contact (95-96) opens and the motor will stop. This idea is used to start and stop an induction motor using PLC.

We have now learned about latching and unlatching. The two terms are important in PLC programming. We will now proceed to learn how to use an output address for an input in a program in the next section.

Using an output address as an input in a program

In the previous chapter, we learned that an output address can be used as an input address (NC or NO contact) in the rules for ladder diagram programming.

The following example shows an output address (Q0.2) being used for an input. You can write the program in network 4 and simulate the program to see how it works.

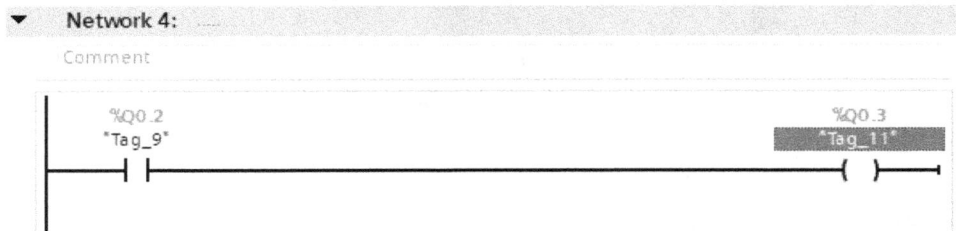

Figure 9.31 – Output address (Q0.2) used for a normally open (NO) contact at the input

The wiring diagram for the program is as follows:

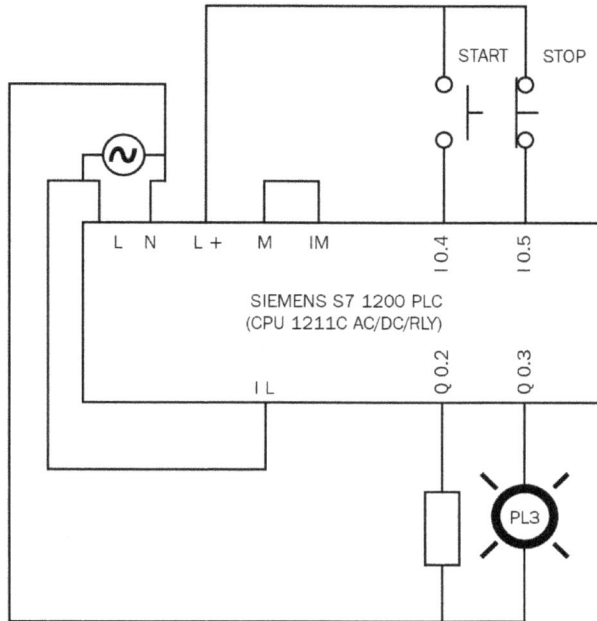

Figure 9.32 – Wiring diagram to implement a program using an output address for input

Let's say a pilot lamp (PL3) is connected to Q0.3, as shown in the preceding diagram.

Anytime Q0.2 is ON, the normally open contact with the address Q0.2 will be closed since the coil with the address Q0.2 is energized and PL3 connected to Q0.3 will be ON.

If we edit the program by replacing the normally open contact with the address Q0.2 with a normally closed contact (NC), as shown in the following screenshot, the program will behave differently, as explained next.

Anytime Q0.2 is ON, the normally closed contact with the address Q0.2 will be opened since the coil with the address Q0.2 is energized, and PL3 connected to Q0.3 will be OFF.

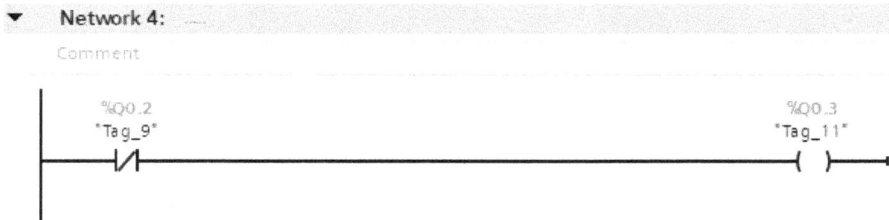

Figure 9.33 – Output address (Q0.2) used for a normally closed (NC) contact at the input

We've just learned how an output address can be used as input. This is another important technique in PLC programming. Let's proceed to learn the SET and RESET instructions in the next section.

Using the SET and RESET instructions

The SET and RESET instructions are two important instructions used in ladder diagram programming. They are both output (coil/assignment) instructions.

The SET coil is energized when power is supplied to it via an input (switch, sensor, or push button) and it remains SET (ON) even when the switch, sensor, or push button that supplied power to it is OFF until it is reset.

The RESET coil is de-energized (OFF) when power is supplied to it via an input (switch, sensor, or push button) and it remains OFF even when the switch, sensor, or push button that supplied power to it is OFF until it is SET.

The sample program in the following screenshot shows the SET and RESET instructions in use:

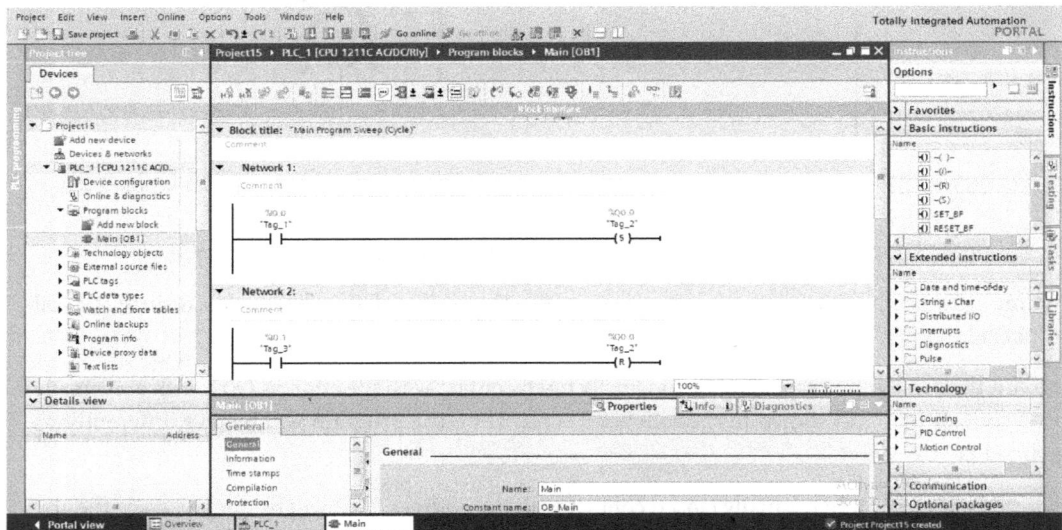

Figure 9.34 – SET and RESET program

> **Note**
>
> The set/reset instruction can be added to your rung using the following steps:
>
> On the right-hand side of the programming environment, click **Instructions | Bit Logic operation**, drag the set output instruction or reset output instruction to the rung/network, type the output address, for example, Q0 . 0, and then press *Enter*.

The wiring diagram for the program is as follows:

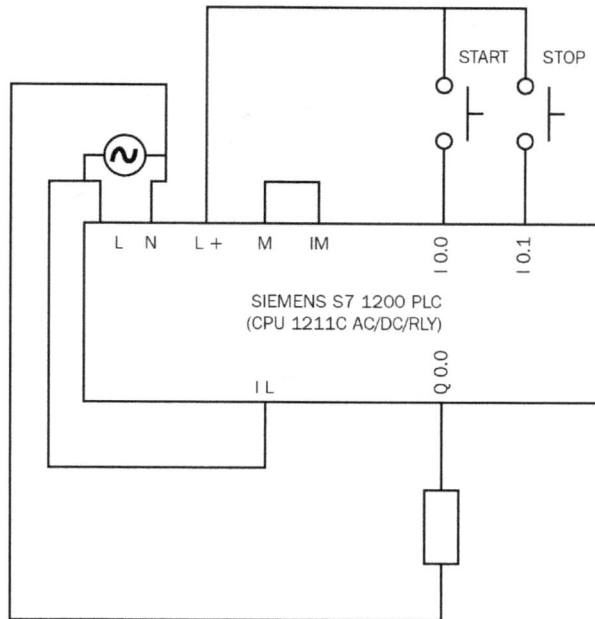

Figure 9.35 – PLC wiring to implement the SET and RESET instructions

Considering the wiring diagram, when the push button connected to I0.0 is pressed, the output Q0.0 will be energized (ON) and it will remain ON even when the push button is released.

When the push button connected to I0.1 is pressed, the output Q0.0 will be de-energized (OFF) and it will remain OFF even when the push button is released.

You should now be able to use SET and RESET instructions in your PLC program. We will proceed to learn about the timer instruction in the next section.

Using the timer instruction

The timer instruction is a delay instruction. It can be used to delay an operation. Several timers exist; the common ones are the ON-delay timer (TON) and the OFF-delay timer (TOF).

An **ON-delay timer (TON)** activates an output when the input is ON for a specified amount of time.

Let's write a simple program to demonstrate the ON-delay timer using the following steps:

1. Cn a new network, add a normally open instruction and specify an address, for example, I0.0.

2. On the right-hand side of the programming environment, click **Instructions** | **Timer operation**, drag the **TON** (ON-delay timer) instruction to the rung/network, and type a name, for example, timer1, as shown in the following screenshot. Click **OK**:

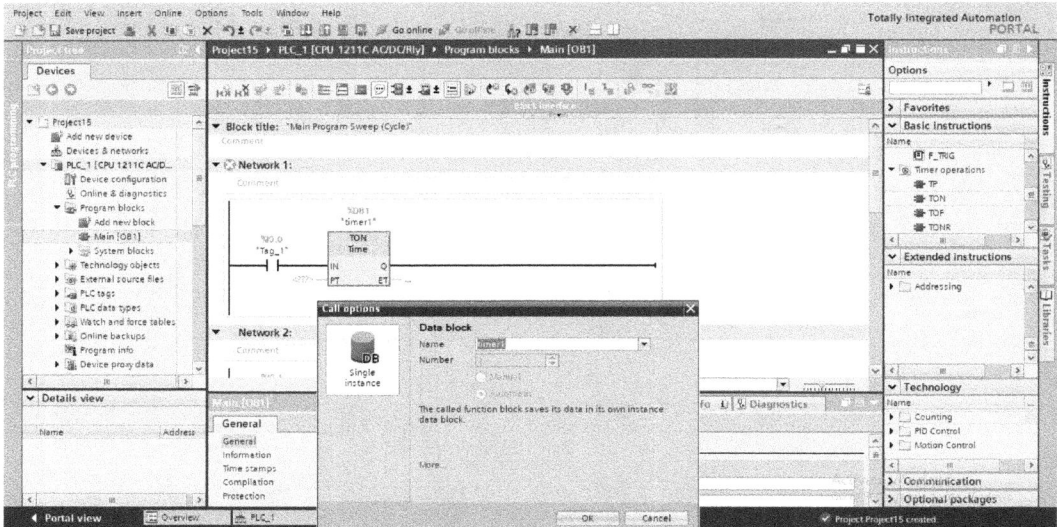

Figure 9.36 – How to use the timer instruction – Call options dialog box

3. Double-click the question marks (**???**) at **PT** (programmed time) and type the time you want, Q (output), to become energized, for example, 25s or **T#25s**.

4. Press *Enter* twice.

5. Drag a coil/assignment instruction to the output side of the rung/network and type the output address (Q0.0), as shown:

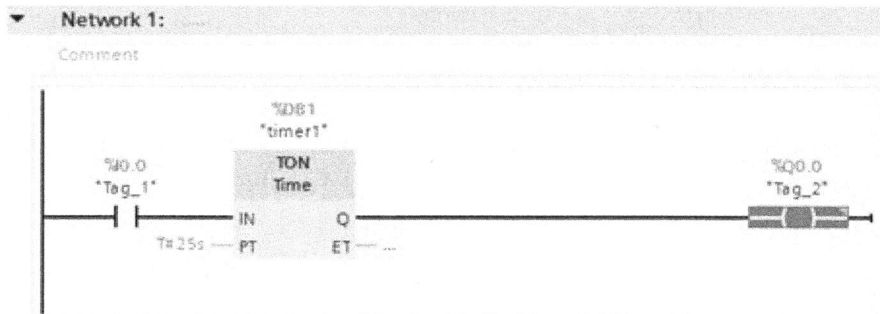

Figure 9.37 – How to use the timer instruction – Complete timer program

The wiring diagram for the program is as follows:

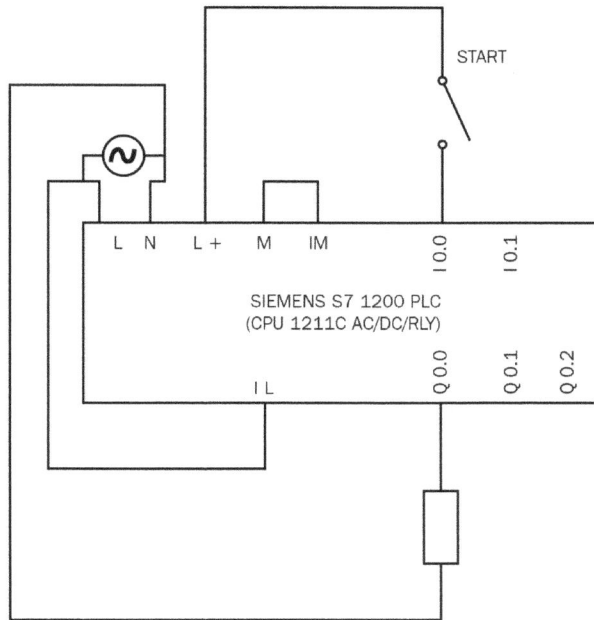

Figure 9.38 – PLC wiring to implement the program

Considering the wiring diagram, when the switch connected to I0.0 is pressed (ON), I0.0 in the program will be closed and Q0.0 will be energized (ON) after 25s according to the program. The input has to be ON for the timer to complete the time and energize the output. Hence, a switch that will remain ON when pressed is preferred to be used to test the preceding program rather than a push button.

The timer's output can be used to activate or deactivate any other output in the program. In network 2, add a normally closed contact, input "timer1".Q, and also add another output (coil/assignment) with the address Q0.1, as shown in the following screenshot:

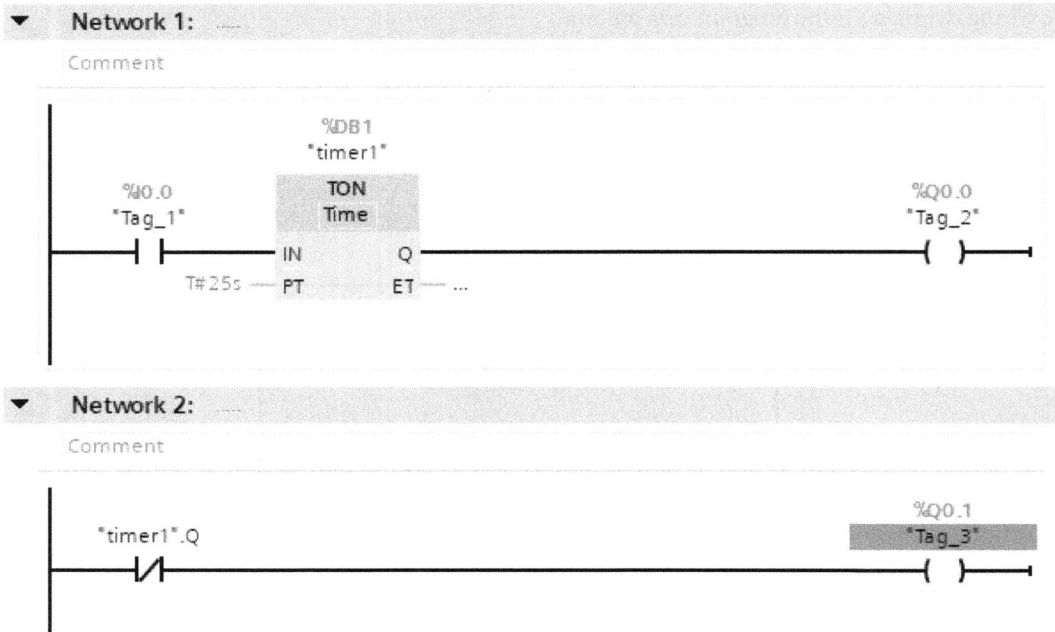

Figure 9.39 – How to use the timer instruction – the timer's output
used to activate or deactivate another output

Also, connect an output (load) to channel (Q0.1) in the wiring, shown in *Figure 9.38*.

When the switch at I0.0 in the wiring is not ON, channel I0.0 is not ON, the timer will not start, and Q0.1 will be energized (ON) since a normally closed contact is used at the input side of the network (network 2) in the program. When I0.0 is ON, the timer will start and, after 25 seconds, Q0.0 will be ON while Q0.1 will be OFF because the timer's output is energized (high) and that will open the normally closed contact used in network 2.

An **OFF-delay timer (TOF)** activates an output when the input is OFF for a specified amount of time. Here, when the input is ON, the output will be energized, but the timer will not start. The timing function starts as soon as the input to it is OFF. Unlike ON-delay timer, the output will be de-energized when the timing has elapsed.

Write the program (as shown in the following screenshot) in network 3 of the program to demonstrate the TOF.

Use the OFF-delay timer (TOF) in place of the ON-delay timer (TON) used in the previous section.

Also, connect a switch to channel I0.1 and an output (load) to channel (Q0.2) in the wiring in *Figure 9.38* to demonstrate the program.

Network 3:

Comment

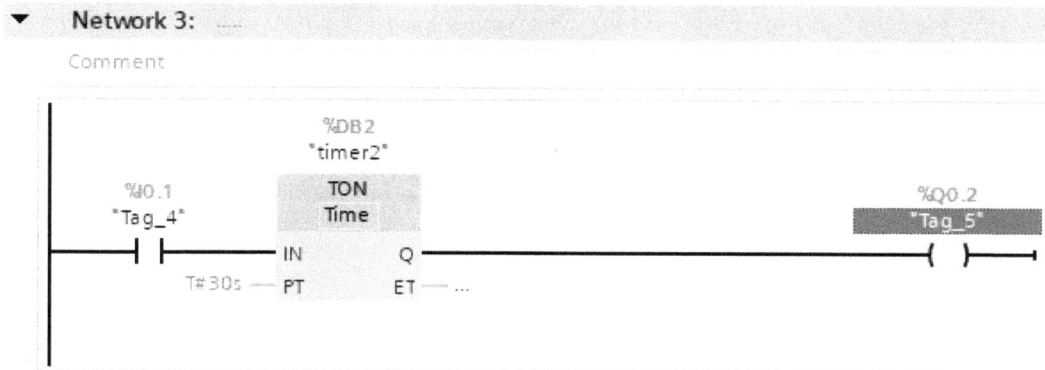

Figure 9.40 – OFF-delay timer program

When the switch connected to I0.1 is pressed (ON), the output Q will be energized, but the timer will not start. The timer will start counting as soon as the input (I0.1) is OFF, while the output will remain energized until the time has elapsed. Try to write and simulate the program to have a better understanding.

We've just learned about the timer instruction and how to use it to activate or deactivate another output. Let's now move on to learn about the counter instruction In the next section.

Using the counter instruction

This is an instruction that can either count up or down when its input is triggered, and an output is activated when the set value is reached. It can be used to determine the number of times an event has happened in a process. In this section, we will be looking at the common counter instructions, including counting up and counting down.

Count up (CTU)

The count up instruction will count up or increase by 1 each time its input is triggered (0 to 1); for example, when a button is pressed and released or when a sensor is activated. The counter's output will be energized when the number of times the input is triggered (current value of the counter) is equal to, or greater than, the pre-set value (PV).

In this example, a pilot lamp (PL1) is connected to Q0.0. A PNP proximity sensor (NO) is connected to I0.0 and a normally open push button is connected to I0.1, as shown in the following wiring diagram:

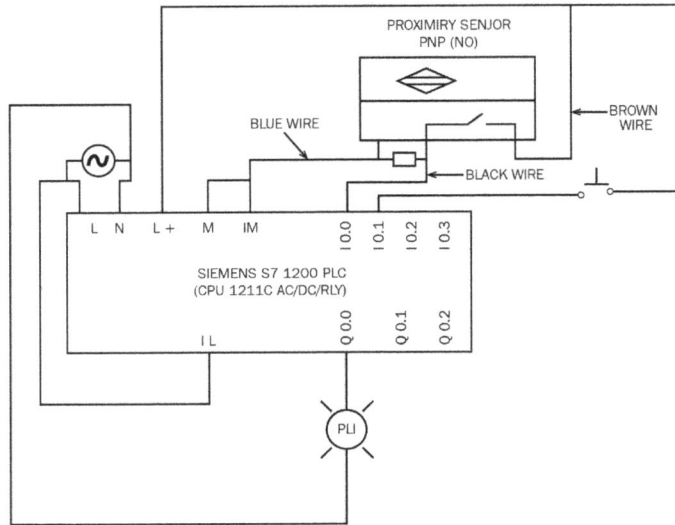

Figure 9.41 – PLC wiring for using the counter instruction with a proximity sensor

Perform the following steps to write a simple program to learn how to use the counter instruction.

1. On a new network, add a normally open instruction and specify an address, for example, I0.0.

2. On the right-hand side of the programming environment, click **Instructions | Counter operation**, drag the **CTU** (count up counter) instruction to the rung, and type a name, for example, counter1, as shown in the following screenshot. Click **OK**:

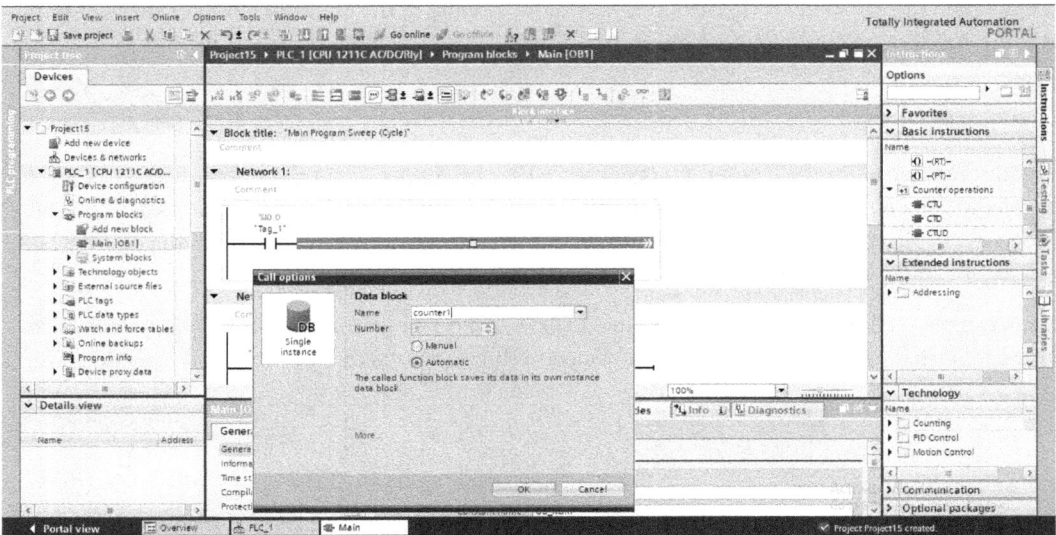

Figure 9.42 – Using the counter instruction – Call options dialog box

3. Double-click the question marks (**???**) at **PV** (preset value) and type the number of times the input will be triggered to get Q (output) energized, for example, 5.

4. Press *Enter* twice.

5. Drag a normally open contact to the R (reset) and type an address, I0.1.

6. Drag a coil/assignment instruction to the output side of the network and type the output address, Q0.0, as shown in the following screenshot:

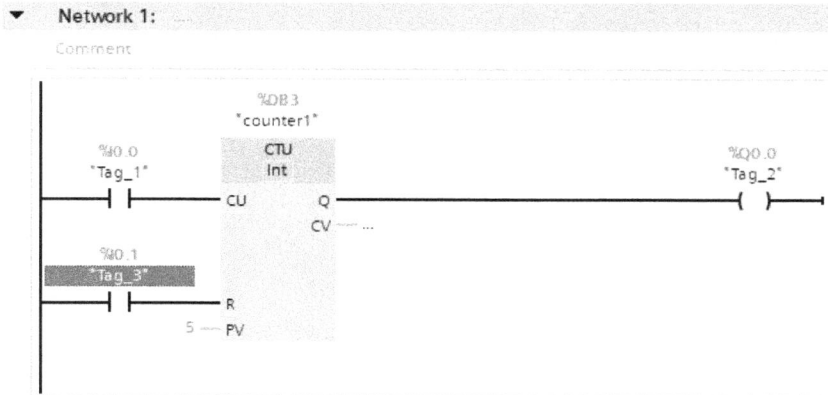

Figure 9.43 – Using the counter instruction – Complete counter program

From the wiring diagram in figure 9.41, when the proximity sensor detects an object, I0.0 will be high (ON). It will be low (OFF) when the object is no longer detected.

In the program in figure 9.43, when the input I0.0 switches between ON and OFF five times, Q0.0 will be energized. When the push button connected to I0.1 is pressed and released, the counter will reset.

Similar to a timer, the counter's output can be used to activate or deactivate any other output in the program.

Try the program in network 2 by performing the following steps:

1. Add a normally closed (NC) contact to network 2, type "counter1".QU at the ???, and press *Enter* twice.

2. Add an output (coil/assignment) and specify an address, Q0.1, as shown in the following screenshot:

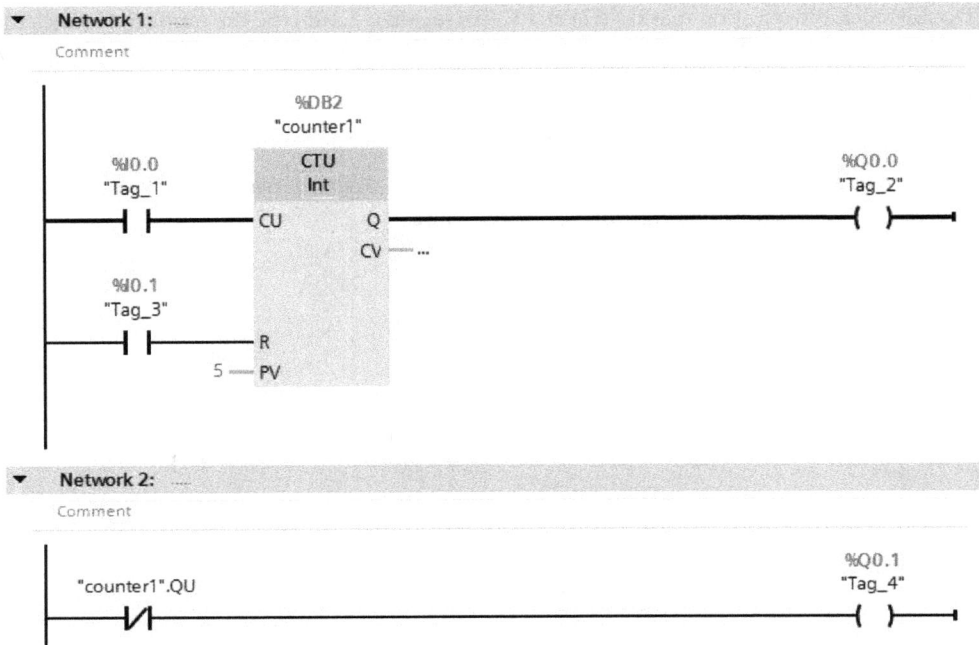

Figure 9.44 – Using the counter instruction – the counter's output
used to activate or deactivate another output

Also, add a load (output) to channel Q0.1 in the wiring in *Figure 9.41* to demonstrate the program.

In the preceding program (*Figure 9.44*), when I0.0 is not triggered, the counter will not count. Q0.0 will be de-energized and Q0.1 will be energized (ON) since a normally closed contact is used at the input side of the network (network 2). When I0.0 is triggered five times, Q0.0 will be ON, while Q0.1 will be OFF because the counter's output is energized (high) and that will open the normally closed contact used in network 2.

Count down (CTD)

The count down instruction will count down or decrease by 1 each time its input is triggered (0 to 1); for example, when a button is pressed and released or when a sensor is activated and deactivated. The counter's output will be energized when the counter's current value equals zero (counting down from the preset value).

In the following example, a count down (CTD) instruction is used. The preset value used is 7.

Add a load (output) to channel Q0.2, add a push button to channel I0.2, and another push button to channel I0.3 in the wiring in *Figure 9.41* to demonstrate the succeeding program (*Figure 9.45*). I0.2 is used for the trigger input, while I0.3 is used for LD (load) input. Q0.2 is used as the output.

When the push button connected to I0.3 is pressed and released, **7** will be loaded. The value will begin to decrease by 1 as the push button connected to I0.2 is pressed and released. The output, **Q**, will turn ON when the count reaches zero and Q0.2 will be energized.

Try to write and simulate the program to further your understanding.

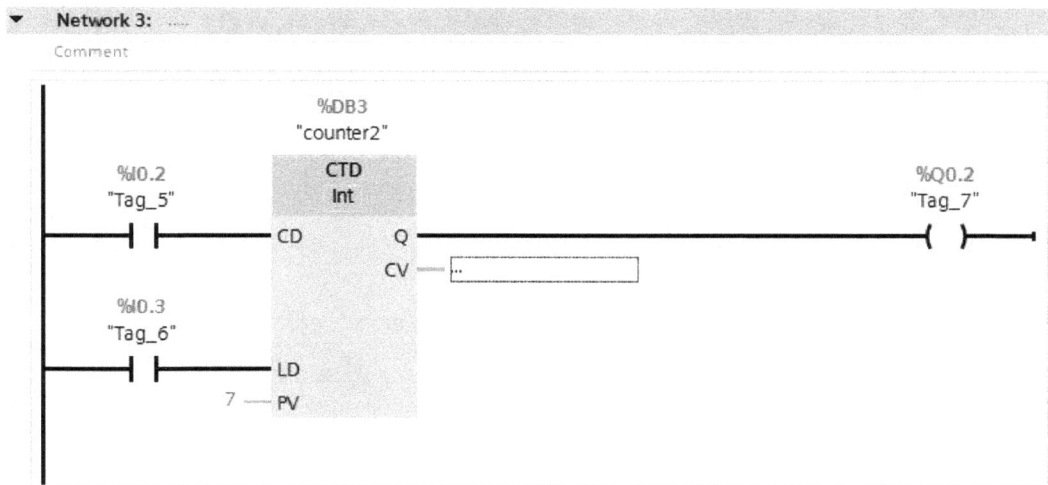

Figure 9.45 – Using the countdown counter

Now that we have learned how to use the counter instruction, we will move on to the next section and learn how to use the move instruction.

Using the move instruction

The move instruction moves a value from a source to a destination. There are various move operations in TIA Portal. In this book, we will learn how to use MOVE (the move value), which is the common move instruction.

Perform the following steps to write a simple program to move the value 20 to a register, MW0, when a push button connected to I0.0 is pressed:

1. On a new network, add a normally open instruction and specify an address, I0.0.

2. On the right-hand side of the programming environment, click **Instructions | Move operation**, and drag the MOVE (move value) instruction to the rung/network.

3. Double-click **???** for **IN** and specify a value, for example, 20.

4. Double-click **???** for **OUT1** and specify the destination address, for example, MW0, as shown in the following screenshot:

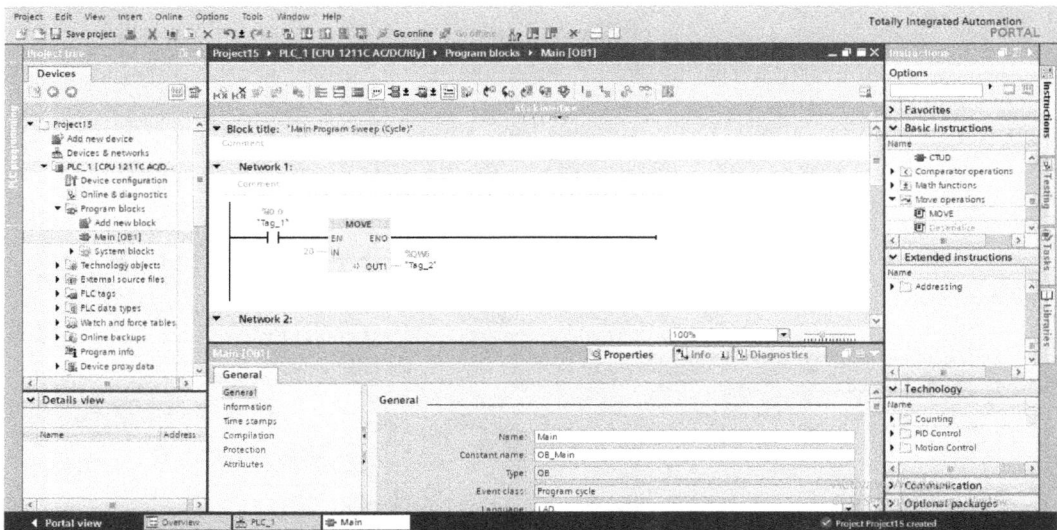

Figure 9.46 – Using the move instruction

When the push button connected to I0.0 is pressed, the normally open contact I0.0 in the program will be closed and the value 20 will be available at the register MW0.

> **Note**
> MW0 is a memory address for the word data type (16 bits) in Siemens PLC.

We've just learned about the move instruction. Now, we will learn about the compare instructions in the next section.

Using the compare instructions

The compare instructions are used to test the relationship between two values from two sources.

The following table shows the compare instructions available in S7 1200:

Compare Instruction	Meaning
CMP ==	Equals
CMP <>	Not equal
CMP >=	Greater than or equal to
CMP <=	Lesser than or equal to
CMP <	Less than
CMP >	Greater than

Table 9.1 – The compare instructions and their meaning

We will learn how to use CMP > in this book. The same concept applies to other compare instructions.

Perform the following steps to write a simple program to compare a value in register MW0 and 5. We will continue with the program from the previous section:

1. On the right-hand side of the programming environment, click **Instructions | Comparator operation**, and drag the CMP > (greater than) instruction to the rung (network 2).
2. Double-click **???** at the top and input MW0
3. Double-click **???** at the bottom and input 5.
4. Drag the coil/assignment instruction to the output side of the rung and specify an output address (Q0.0).

You should have what looks like the following screenshot if everything is correct:

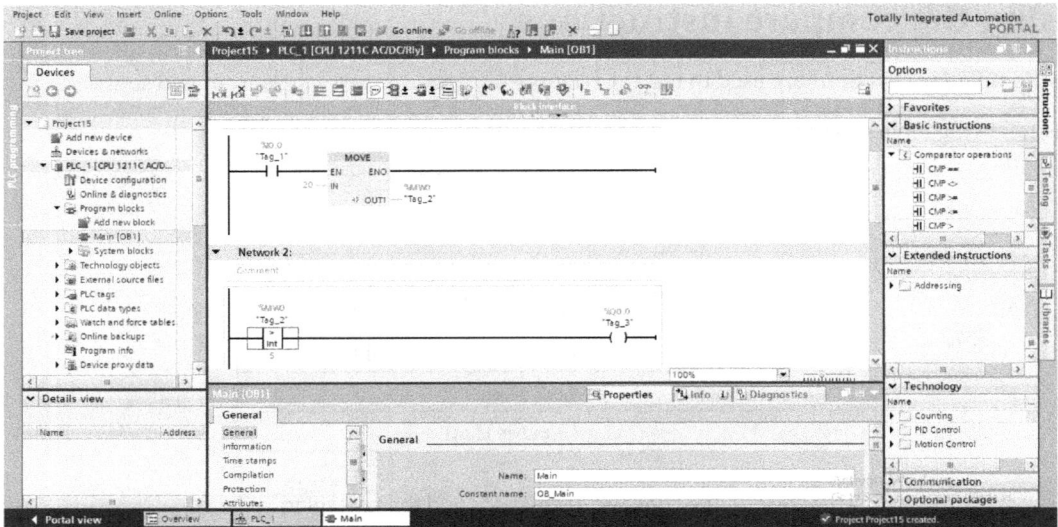

Figure 9.47 – Using the compare instruction

When the push button connected to I0.0 is not pressed, MW0 is 0, and since 0 is less than 5, the output Q0.0 will be de-energized. When the push button is pressed, MW0 will be 20, and since 20 is greater than 5, the output Q0.0 will be energized.

When you simulate the program, you will find out that when I0.0 is not marked (low), the value of MW0 is 0, and since 0 is less than 5, the output Q0.0 is de-energized. When I0.0 is marked (high), MW0 becomes 20, and since 20 is greater than 5, the output Q0.0 becomes energized.

The following screenshot shows the simulation result with the output (Q0.0) energized when I0.0 is marked (high):

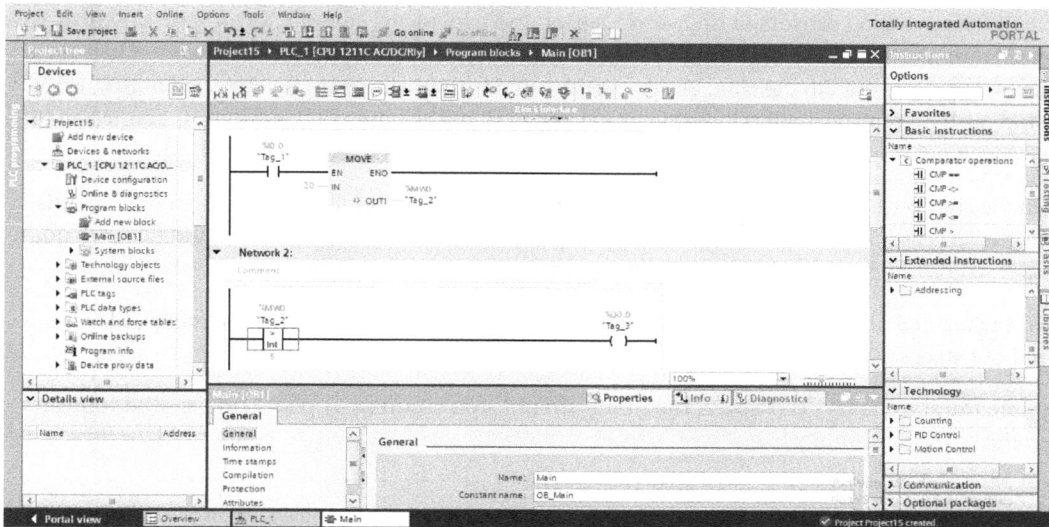

Figure 9.48 – Simulation result of the compare program

We have now learned how to use the compare instruction in Siemens TIA Portal. Now that we understand various PLC programming instructions, in subsequent sections, we will be looking at some real-life industrial applications of PLC using some of the instructions we have learned.

Level control using PLC

The following diagram shows a level control that can be used in the industry:

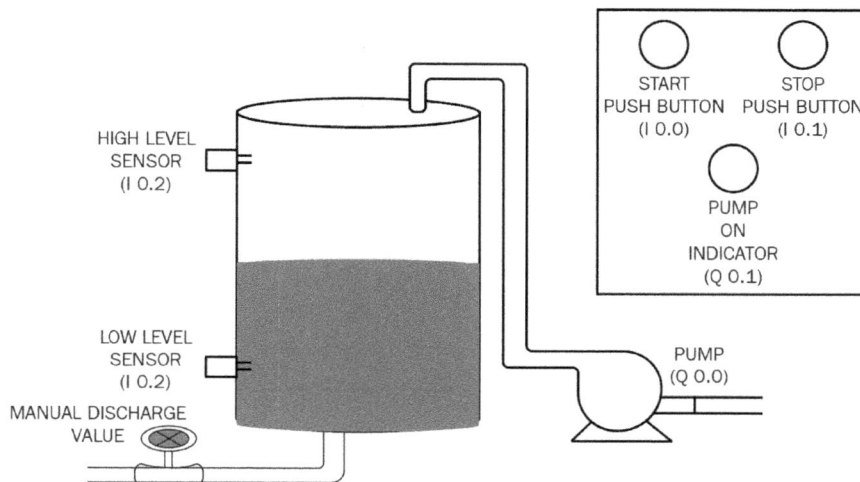

Figure 9.49 – Level control

The tank has two level sensors (a high-level sensor and a low-level sensor). The low-level sensor is connected to I0.2, while the high-level sensor is connected to I0.3. A pump connected to Q0.0 via a contactor supplies liquid to the tank. The start push button (normally open) is connected to I0.0, while the stop push button (normally closed) is connected to I0.1. The pump ON indicator is connected to Q0.1. A manual discharge valve is used to discharge the liquid when required.

When the start push button is pressed and released and there is no liquid in the tank, in other words, the low-level sensor cannot detect liquid, the pump should start running until the liquid fills the tank, in other words, the high-level sensor detects liquid.

When the manual discharge valve is opened and the liquid falls below the high-level sensor, the pump should not start until the liquid falls below the low-level sensor.

Anytime the stop button is pressed, the pump should stop running. A pilot lamp should indicate when the pump is running.

> **Note**
>
> The level sensor used here is a vibrating fork-level sensor. It's a type of level sensor in which a tuning fork vibrates or oscillates at its natural frequency when not in contact with a material (liquid, free-flowing solids, granules, or powders), its frequency changes when in contact with a material and it gives an output signal that can be fed to a PLC as input.

The program can be written as follows:

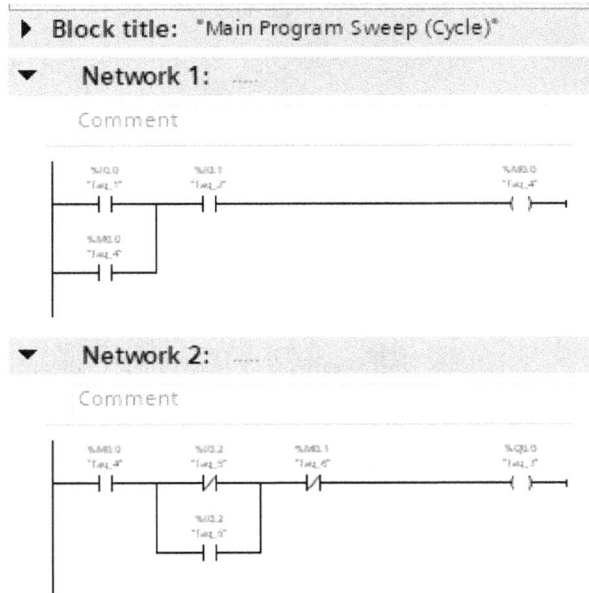

(a)

Network 3:

Comment

Network 4:

Comment

(b)

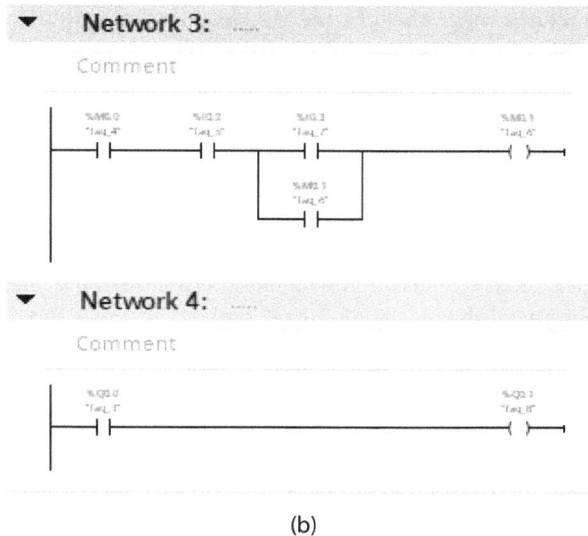

Figure 9.50 – Level control program using a ladder diagram

The different networks are explained as follows:

- **Network 1**: When the start push button connected to I0.0 is pressed, I0.0 in the program will be closed, and since the stop push button connected to I0.1 is normally closed, I0.1 in the program will be closed and this energizes the coil M0.0. When the start push button is released, the coil M0.0 will remain energized due to a normally open push button with the same address as M0.0 that's in parallel with I0.0. When the stop push button is pressed, the normally closed contact with the address I0.1 in the program will be opened and the coil with the address M0.0 will be de-energized.

- **Network 2**: When M0.0 is energized (due to the start button being pressed and released) and the low-level sensor is not detecting any liquid, the normally closed contact with the address I0.2 is closed, and since M0.1 is a normally closed contact, Q0.0 will be energized and the pump will start running to supply water to the tank. The normally open contact with the address I0.2 that's in parallel with the normally closed contact with the address I0.2 creates an alternative path for the signal to flow to the coil (Q0.0) to keep the pump running even when the low-level sensor detects liquid and the high-level sensor has not yet detected any liquid.

- **Network 3**: When M0.0 is energized (due to the start button being pressed and released) and the low-level and high-level sensors have not detected liquid, the normally open contact with the addresses I0.2 and I0.3 will be opened in the program, M0.1 will be de-energized and the pump keep running. M0.1 is used in rung 2 (network 2) to cause the pump to stop when energized. The normally open contact with the address M0.1 in rung 2 (network 2) will become open when the address M0.1 in rung 3 (network 3) is energized and this will cause the pump to stop running. This will happen when the tank is full.

The normally open contact with the address M0.1 in parallel with I0.3 is used to create an alternative path for the signal to flow to M0.1 when the manual discharge valve is open and the water level in the tank decreases gradually. As the water decreases, it will get to a point where it will fall below the high-level sensor, causing the normally open contact with the address I0.3 (which was closed when water was full) to open, and since we don't want M0.1 to be de-energized to cause the pump to start immediately (in other words, we want it to start when the liquid level falls below the high- and low-level sensors), the address M0.1 is used to keep the coil M0.1 energized so that the pump can remain off until the liquid falls below the low-level sensor with the address I0.2. Recall from the operation that when the manual discharge valve is opened and the liquid falls below the high-level sensor, the pump should not start until the liquid falls below the low-level sensor.

- **Network 4**: This rung is responsible for the pump ON indicator. A normally open contact with the same address as the coil for the pump, in other words, Q0.0, is used as input, while a coil with the address Q0.1 is used in the output section of the rung. Anytime Q0.0 is energized (which must happen in order for the pump to run), the normally open contact with the address Q0.0 will close and Q0.1 will be energized. Hence, the indicator lamp connected to Q0.1 will turn on anytime the pump is running.

We've just learned a simple level control that can be used in the industry. This knowledge can help in developing more advanced level controls. We will see an automated filling, capping, and wrapping system using PLC in the next section.

Automated filling, capping, and wrapping system using PLC

The following diagram shows another industrial application of PLC:

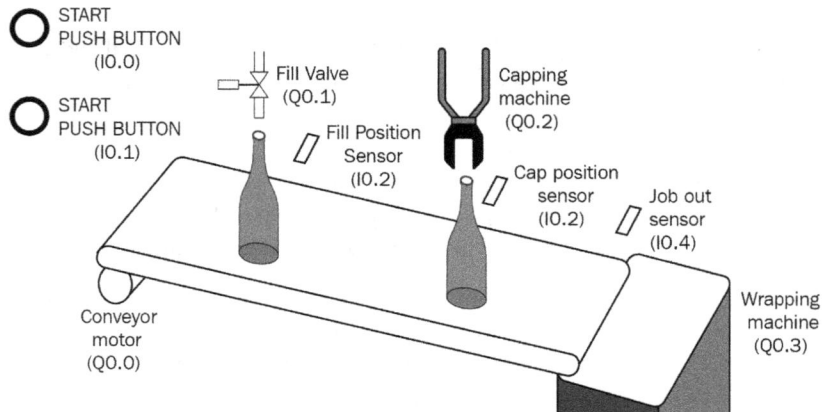

Figure 9.51 – Filling, capping, and wrapping

The start push button (normally open) is connected to I0.0, while the stop push button (normally closed) is connected to I0.1. A conveyor motor is connected to Q0.0, a fill valve is connected to Q0.1, a capping machine is connected to Q0.2, and a wrapping machine is connected to Q0.3. A fill position sensor is connected to I0.2 and a cap position sensor is connected to I0.3, while a job out sensor is connected to I0.4, as indicated in the preceding diagram.

When the start push button is pressed and released, the conveyor motor should start running until the bottle gets to the fill position sensor; in other words, the fill position sensor detects the bottle and the conveyor motor stops.

The fill valve should open as soon as the conveyor motor stops at the fill sensor for 10 seconds to fill the bottle. After 10 seconds, the valve should close and the conveyor motor should continue running until the bottle gets to the cap position sensor, in other words, the cap position sensor detects the bottle and the conveyor motor stops.

The capping machine should start as soon as the conveyor motor stops at the cap sensor for 4 seconds to cap the bottle. After 4 seconds, the capping machine should stop and the conveyor motor should continue running.

When the bottle gets to the job out sensor, a counter should increment by one and the bottle should drop to the wrapping machine. When the number of bottles detected by the job out sensor reaches 6, in other words, when the counter increments to 6, the wrapping machine should start to wrap the 6 bottles and stop after 5 seconds.

The program can be written as follows:

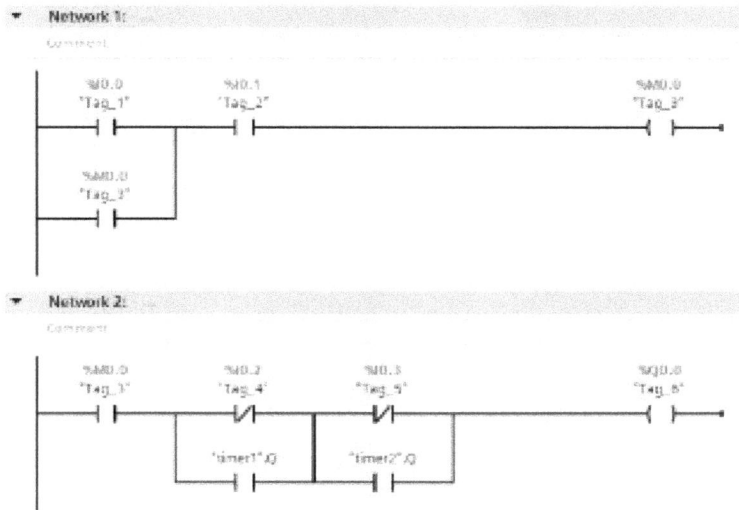

(a)

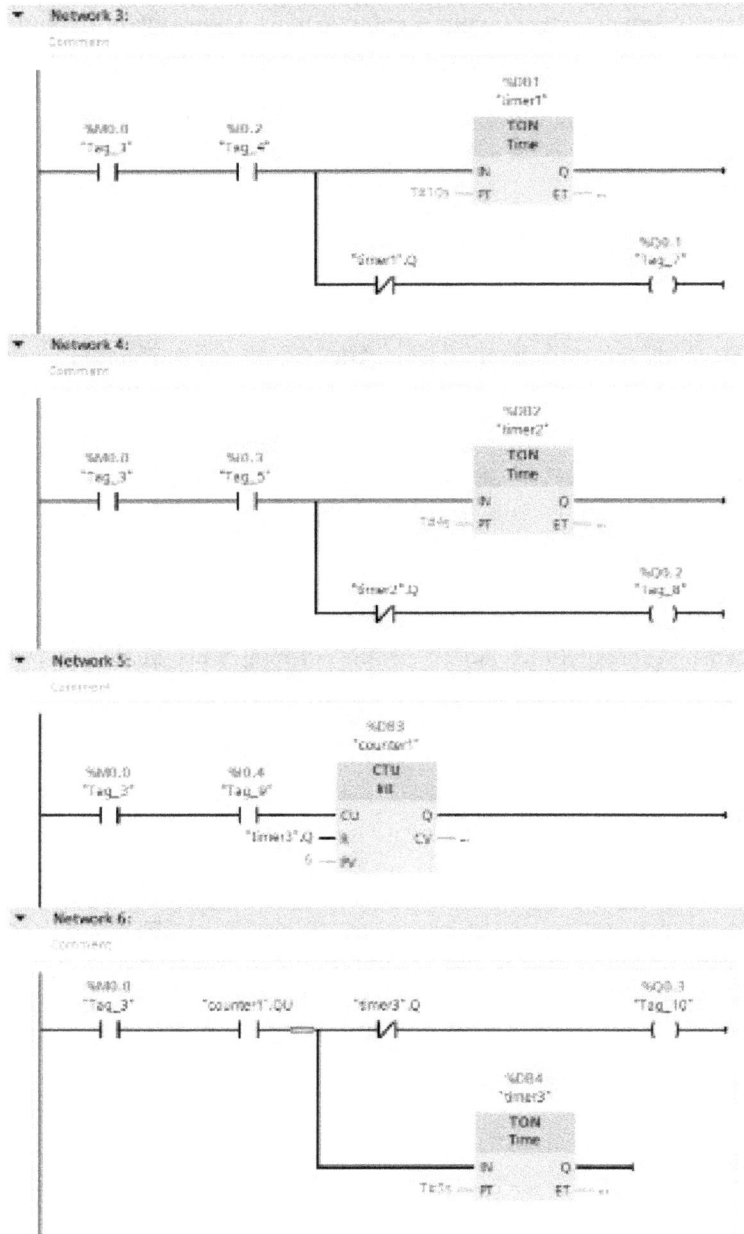

(b)

Figure 9.52 – Filling, capping, and wrapping program

The different networks are explained as follows:

- **Network 1**: This is a simple start and stop program to start and stop the system. Refer to the explanation of network 1 in the previous section to understand how it works. When the start push button is pressed and released, M0.0 will be energized, and it will remain energized until the stop push button at I0.1 is pressed.

- **Network 2**: M0.0, which will be energized in network 1 when the start button is pressed and released, will cause the normally open contact with the address M0.0 to close, and since I0.2 (address of the fill position sensor) and I0.2 (address of the cap position sensor) are normally closed, Q0.0 (address of the conveyor motor) will be energized and the conveyor motor will start running. When the fill position sensor detects a bottle, it will turn ON and this will make I0.2 to open and cause the motor to stop at that position until timer1 (filling time) elapses and close the normally open contact (timer1) in parallel to I0.2 to create an alternative path for the signal to flow to coil Q0.0 to keep the conveyor running. When the cap position sensor detects a bottle, it will turn ON, this will make I0.3 to open and cause the motor to stop at that position until timer2 (capping time) elapses and close the normally open contact (timer2) in parallel to I0.3 to create an alternative path for the signal to flow to coil Q0.0 to keep the conveyor running.

- **Network 3**: When the normally open contact with the address M0.0 is closed due to the start button being pressed and released, and the normally open contact with the address I0.2 is closed due to the fill position sensor that detects a bottle, a 10-second timer (timer1) will start and, at the same time, an output with the address Q0.1 will be energized. Hence, the valve connected to Q0.1 will open to fill the bottle. The valve will close after 10 seconds due to the normally closed contact (timer1) that is in series with the coil with the address Q0.1 opening when timer1 is energized after 10 seconds.

- **Network 4**: When the normally open contact with the address M0.0 is closed due to the start button being pressed and released, and the normally open contact with the address I0.3 is closed due to the cap position sensor that detects a bottle, a 4-second timer (timer2) will start and, at the same time, an output with the address Q0.2 will be energized. Hence, the capping machine connected to Q0.2 will start to cap the bottle. The capping machine will stop after 4 seconds due to the normally closed contact (timer2) that is in series with the coil with the address Q0.2 opening when timer2 is energized after 4 seconds.

- **Network 5**: When the normally open contact with the address M0.0 is closed due to the start button being pressed and released, and the normally open contact with the address I0.4 is closed due to the job out sensor detecting a bottle, a counter (counter1) will start. The counter will increment each time a bottle passes.

- **Network 6**: When the normally open contact with the address M0.0 is closed due to the start button being pressed and released, and the counter in network 5 has incremented to 6, the normally open contact (counter1) will close and since timer3 is a normally closed contact, the coil with the address Q0.3 will be energized and a wrapping machine connected to Q0.3 will start to wrap the 6 bottles. A 5-second timer (timer3) also starts when counter1 is incremented to 6.

After 5 seconds, the normally closed contact (timer3) will open and the wrapping machine connected to Q0.3 will stop. The counter (counter1) in network 5 will also reset after the 5 seconds of timer3 have elapsed so that the next bottle will be counted as 1 and keep incrementing up to 6, when the wrapping machine will start to wrap again.

The program for an automated filling, capping, and wrapping system learned in this section covers almost all the instructions we have learned so far in this book. Try to write and simulate the program on your own to gain more knowledge regarding the practical application of the instructions we have learned in the previous chapter and also in this chapter for programming PLCs with ladder diagram programming language.

Summary

Congratulations! You have successfully completed another chapter of this book. Good job! PLC is the brain behind most modern automated machines and programming is key when automating a machine with PLC. You should now understand latching and unlatching, know how to use an output address as input, and be familiar with timer, counter, move, and compare instructions. You should also be able to write programs using the techniques and instructions explained in the chapter. Simulation, which is key, was also discussed. You should be able to simulate your program after writing to know whether it will work. The final two sections of this chapter were the most interesting, where we looked into some real-life applications of some of the techniques learned in the chapter (that is, level control using PLC and an automated filling, capping, and wrapping system using PLC). Try to get your hands dirty with them.

In the next chapter, we will learn about the **Human Machine Interface** (**HMI**) and how to use it with a PLC. HMI is another interesting and powerful device that an industrial automation engineer needs to be familiar with. Don't miss it!

Questions

The following are questions to test your understanding of this chapter. Ensure you have read and understood the topics in this chapter before attempting the questions:

1. _____ is a technique used to keep an output energized or activated even when the input ceases.

2. A _____ instruction can be used to delay an operation.

3. A _____ instruction can be used to determine the number of times an event has happened in a process.

4. _____ instructions are used to test the relationship between two values from two sources.

10
Understanding Human Machine Interfaces (HMIs)

The previous three chapters covered a great deal on **Programmable Logic Controllers** (**PLCs**), which are the brains behind most automated machines. We learned how to automate industrial processes using PLCs, sensors, and actuators. However, in most cases, PLCs, sensors, and actuators alone cannot give us the functionality required in a real-life industrial process. A **Human Machine Interface** (**HMI**) is usually integrated into a manufacturing line and other industrial processes to give users easy control of the machines and to give them feedback on machine statuses.

In the previous chapter, we learned how to control machines using PLCs, whereby the user or operator presses a push button to start the machine and a pilot light gives an indication of whether the machine is running or not. Integrating HMIs into such systems makes them more user-friendly. HMIs allow you to start or stop your machine and even get feedback or status updates for the machine through a graphical interface or touchscreen.

In this chapter, you will learn about HMIs and how to interface them with PLCs to give commands to machines and also get feedback from machines. You will use TIA Portal to create an HMI screen for monitoring and control. You will learn how to simulate your HMI program (a good way to test your HMI program without a real HMI panel) and you will also learn how to download a designed HMI graphic to a real HMI panel (Siemens HMI - (KTP400) for a real-life application.

We are going to cover the following topics in this chapter:

- Introducing HMIs
- Exploring the applications of HMIs
- Understanding HMI programming and development
- Understanding HMI programming software
- Exploring HMI manufacturers and their programming software
- Interfacing PLCs and HMIs
- Downloading programs to PLCs and HMIs

Technical requirements

You will need an understanding of PLCs, sensors, and actuators from the previous three chapters.

Ensure you have downloaded and installed TIA Portal V13 on your PC or laptop. Please refer to *Chapter 8, Understanding PLC Software and Programming with TIA Portal.*

Other devices that may be required include Siemens HMI (KTP400) and Siemens PLC (S7 1200, CPU 1211C, AC/DC/RLY).

Introducing HMIs

HMIs can be seen as graphical interfaces that allow humans and machines to interact. We have learned that PLCs can control and automate industrial machines. HMIs, on the other hand, allow humans to interact or communicate with machines. No matter how we automate a process, we still need a human (an operator) to start or initiate the process, perform other functions, and get feedback on the machine's operation.

HMIs allow humans to give commands to PLCs to control a process and also receive feedback from PLCs about the process. With HMIs, the human operator can interact with, monitor, and control machines via the PLC. The HMI is usually connected to the PLC via a communication cable, depending on the protocol being used (Ethernet/IP, Profinet, Profibus, or Modbus). This chapter does not discuss protocols. We will look into protocols in *Chapter 13, Industrial Network and Communication Protocols Fundamentals.* However, we will use Profinet as the protocol to establish communication between the HMI and PLC. Basically, the PLC connects to the machine or sensors and actuators via cables, as we learned in *Chapter 7, Understanding Programmable Logic Controllers – Part 1*, where we discussed PLC wiring. The following diagram depicts the human operator, HMI, PLC, and machine, as well as the communication flow between them:

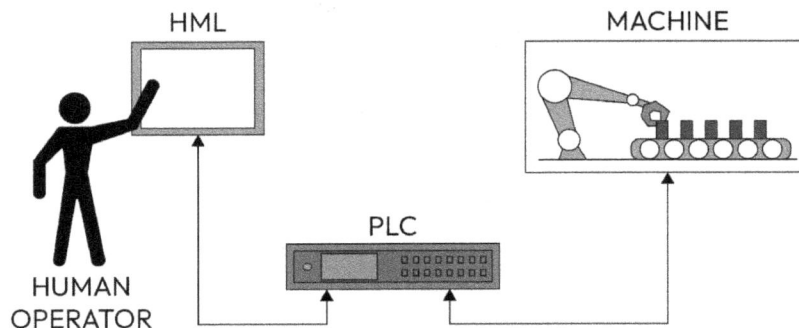

Figure 10.1 – Human operator, HMI, PLC, and machine

The HMI is not peculiar to manufacturing and other industrial settings alone. There are many HMIs that we have and even use regularly without knowing them as HMIs. For instance, some cars have a touchscreen interface through which the driver or passenger can control the air conditioning, heating, sound, and other systems. These touchscreens are HMIs. The driver or passenger is able to interact with the machine via the touchscreen-enabled interface. An HMI is simply an interface (usually graphical) through which a human can control a machine. It can also be referred to as a **Man-Machine Interface (MMI)**, **Operator Interface Terminal (OIT)**, or **User Interface (UI)**. HMIs are designed to ensure that operators can monitor or control machines easily. HMIs are also designed to ensure operators can monitor alarms and respond accordingly. Various colors can be used to represent the level of priority of the alarms. HMI alarms give information about important events and faults. Another good thing about HMIs is trend. Trend can be referred to as a graphical representations of data over a period of time. HMIs allow you to monitor or view the data from sensors over a period of time in a graphical interface. HMIs also support setpoint management. HMIs can be designed and programmed to allow operators to change setpoint values easily from a graphical interface.

HMIs comprise both hardware and software that allow communication between a human operator and machines.

The following figure shows an HMI that can be used by an operator (user):

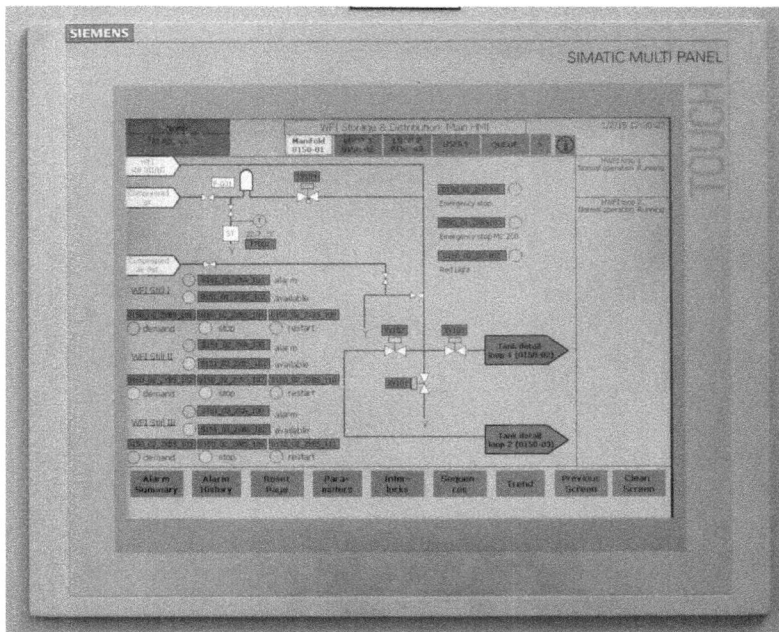

Figure 10.2 – HMI showing the GUI through which the user can interact with the machine or process

Credit for this image goes to Wikimedia Commons: https://creativecommons.org/licenses/by-sa/4.0/deed.en.

An HMI replaces switches and other forms of controls and status indicators connected to a PLC with a graphical interface, which is usually a touchscreen, as shown in the preceding figure. HMIs allow human operators to monitor a process, start or stop the process, adjust setpoint, and perform other control functions that require human attention.

Exploring the applications of HMIs

HMIs are used in almost all industries. Some industrial areas of application include the following:

- **Food and beverages**: HMIs can be used to easily monitor and control production processes to ensure quality in the food and beverages industry. HMIs allow recipes to be used in production. "Recipes" simply refers to data or values that can be selected by a user depending on the type of product that is to be produced. Recipes summarize production data or machine configurations for a particular product. The data is then transferred from the HMI to the PLC or controller at a given time to switch production to another product type. Recipes are used when multiple products are to be produced. Recipes are not only advantageous in the food industry; they are also of great benefit in other industries that produce multiple products.

- **Oil and gas**: HMIs provide a means by which drilling and rig operators can control and monitor their applications in the field in real time. HMIs also help in other areas of oil and gas, such as refining and pipeline integrity management.

- **Power**: HMIs can be used in substations and other areas of the power system to execute switching and other control operations easily.

- **Wastewater management systems**: HMIs are used in wastewater treatment to view all aspects of the treatment system, including pH, chemical usage, and tank levels.

- **Transportation**: The performance, efficiency, and safety of public transportation systems can be improved through interaction between humans and machines, which HMIs enable.

We have just learned about HMIs and their application in various industries. We will learn about HMI programming and development in the next section.

Understanding HMI programming and development

HMI programming involves the creation of the visual representation (graphic) of the machine operation, control commands, and writing codes to execute required functions.

The main purpose of an HMI is to allow the operator to control a machine or process. Hence, an HMI design must do the following:

- Be easy for the user to have access to the necessary information and control.

- Allow the user to navigate to different areas as quickly as possible.

- Allow the user or operator to see the current status of the machine. The operator must be able to tell whether the machine is running or has stopped through visual indication on the HMI. Other information that can help the operator to make the right decision can also be provided.

Let's proceed to learn about HMI programming software in the next section.

Understanding HMI programming software

HMI programming software is special software used for designing the graphical interface and writing some code where necessary for the interaction between the user (human) and the machine. The software is usually installed on the PC where programming will be done before it is downloaded to the HMI panel via a communication cable, just as we do for PLCs.

There are different kinds of programming software for different brands of HMI panels. The next section discusses various HMI manufacturers and their programming software.

Exploring HMI manufacturers and their programming software

Some of the popular HMI manufacturers are listed here:

- **Siemens**: Siemens is a German company, one of the top companies in the automation industry, producing SIMATIC HMIs. Some of their HMI models include KTP400, KTP600, and KTP700. Their HMI programming software is WinCC.

- **Rockwell Automation**: Rockwell Automation is one of the top companies in North America. Their HMI brand is called **PanelView**, which includes PanelView 600, Panel View 800, and PanelView+400. The programming software for their HMIs includes RSview32 and FactoryTalk View.

- **Schneider Electric**: Schneider Electric is one of the top companies in Europe. Their HMI brands include Mangelis and Harmony. Their HMI programming and configuration software includes Ecostructure Operator Terminal Expert and Vijeo Designer.

- **Mitsubishi Electric Corporation**: **Mitsubishi Electric Corporation** (**MELCO**) is a division of Mitsubishi. Their line of HMIs is called **Graphical Operator Terminal** (**GOT**), for example, the GOT 1000 series and GOT 2000 series. Their HMI programming software is GTworks2 (GOT HMI designer software).

We've just learned about some of the leading HMI manufacturers and their software. This book will focus on a Siemens HMI (KTP 400) and their software (WinCC). We will learn how to connect a Siemens HMI (KTP400) to a PLC and how to use WinCC in TIA Portal to develop an HMI screen in subsequent sections.

Interfacing PLCs and HMIs

In *Chapter 9, Deep Dive into PLC Programming with TIA Portal*, we learned how to use pushbuttons to start and stop and also how to use a lamp as an indicator that shows when a load is on or off.

As we discussed earlier, HMIs allows human operators to interact with machines via a PLC. This means that a PLC is required to communicate with the machine.

Here, in this section, we will learn how to interface a Siemens HMI (KTP 400) and a Siemens PLC using Profinet as a communication protocol. We will use TIA Portal to program the PLC and develop the HMI application using WinCC, which was installed when we installed TIA Portal in *Chapter 8, Understanding PLC Software and Programming with TIA Portal*.

By the end of this section, you should be able to interface PLCs and HMIs such that the HMI can be used to start and stop and get feedback on the status of a machine (for example, a motor).

We are already familiar with the Siemens S7 1200 PLC (CPU 1211C). The following photo shows a Siemens HMI (KTP 400):

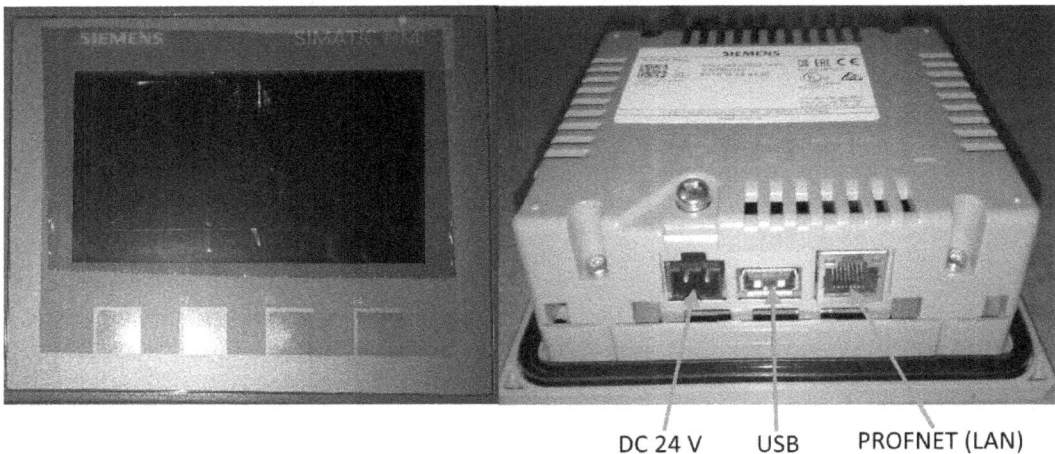

DC 24 V   USB   PROFNET (LAN)

Figure 10.3 – Siemens HMI panel (KTP 400). (Credit for this image goes to Showlight Technologies LTD. www.showlight.com.ng)

Let's get started:

1. Start TIA Portal, click **Create new project**, and type the project name (for example, PLC-HMI-PRACTICE), as shown in the following screenshot. Click **Create**:

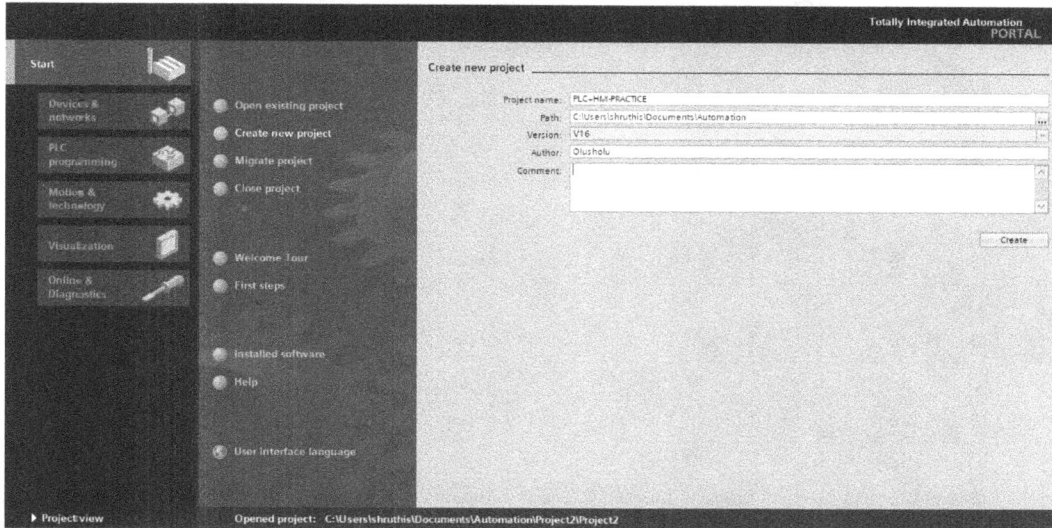

Figure 10.4 – Create a new project

2. Click **Configure a device**:

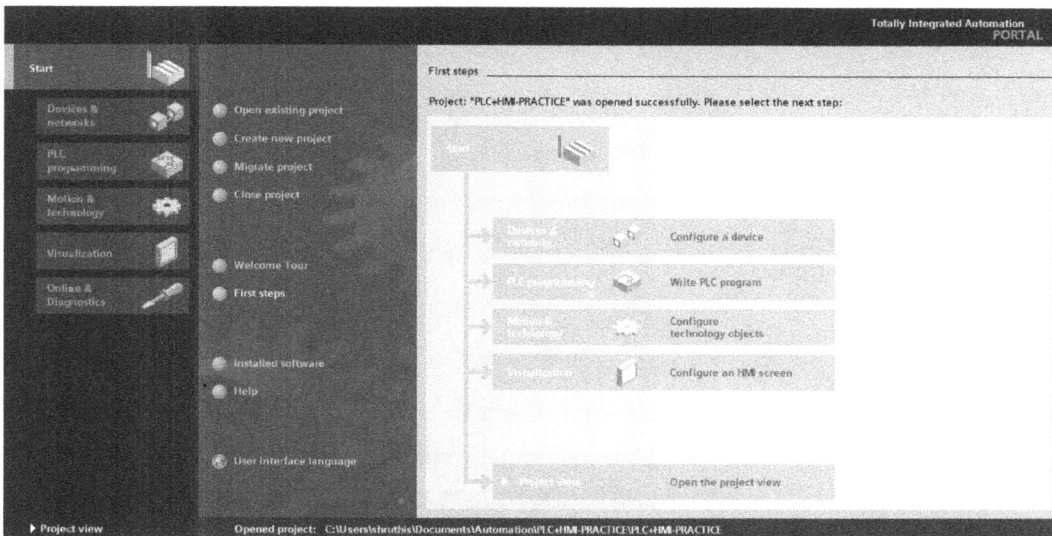

Figure 10.5 – New project created – portal view

3. Click **Add new device** and select **Controllers** from the device list. Expand **SIMATIC S7-1200 | CPU | CPU1211C AC/DC/Rly** and select the order number of your PLC, for example, **6ES7 211-1BE40-0XB0**, depending on the PLC you intend to use:

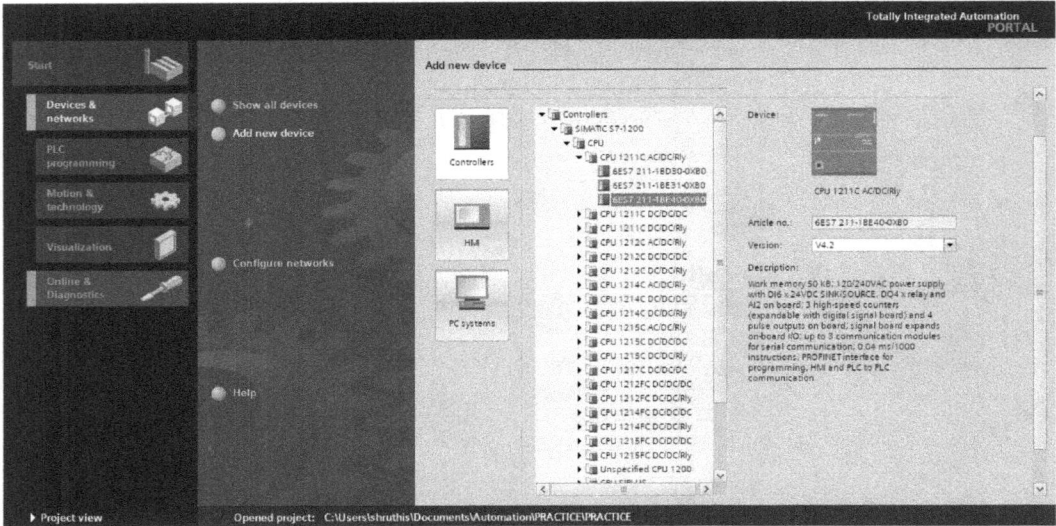

Figure 10.6 – Selecting your controller (PLC)

4. Click **Add** and you should see the following screen:

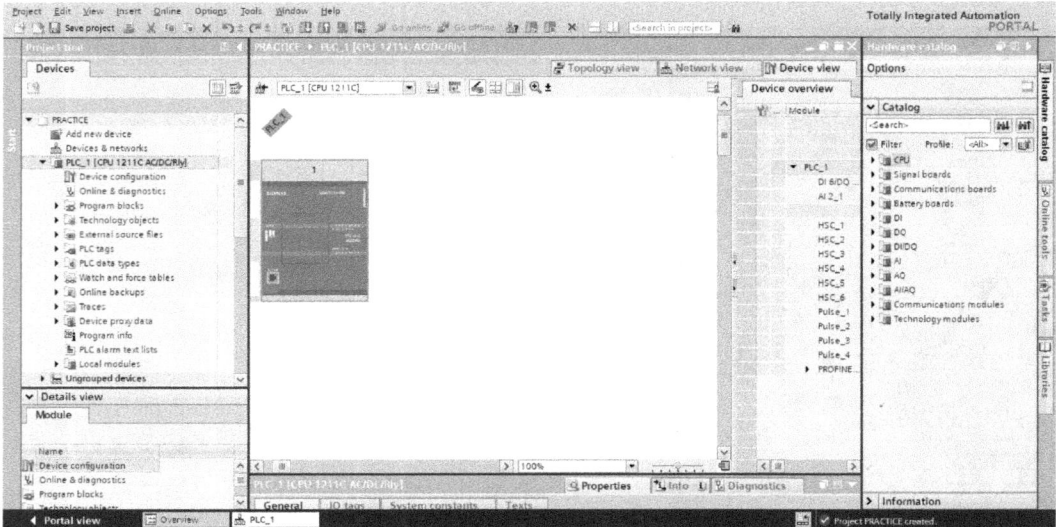

Figure 10.7 – PLC added to your project

5. Expand **Program blocks** and double-click **Main (OB1)** to create an environment to write your program in, as shown here:

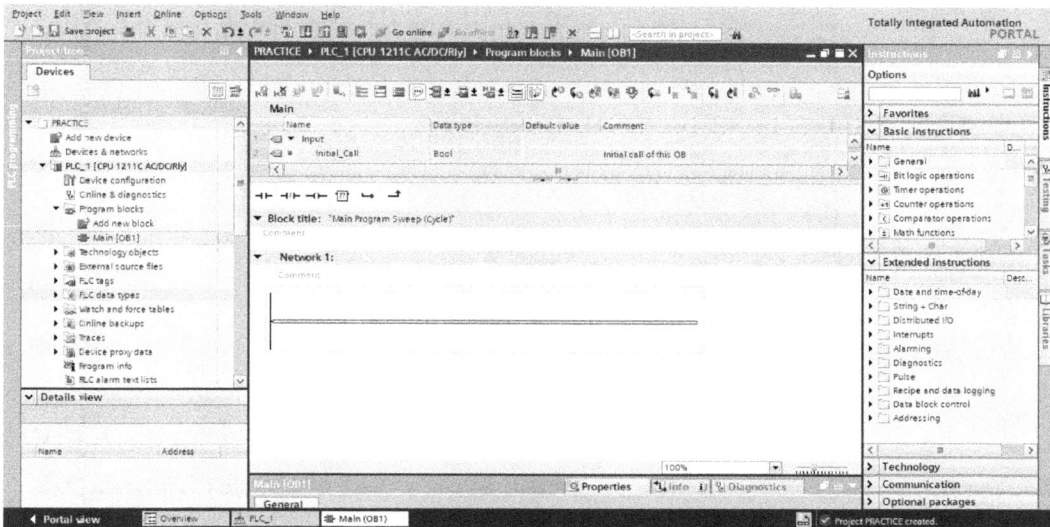

Figure 10.8 – Ladder logic programming environment

6. Write your program such that there will be no physical button attached to the PLC. Memory bits will be used, as shown here:

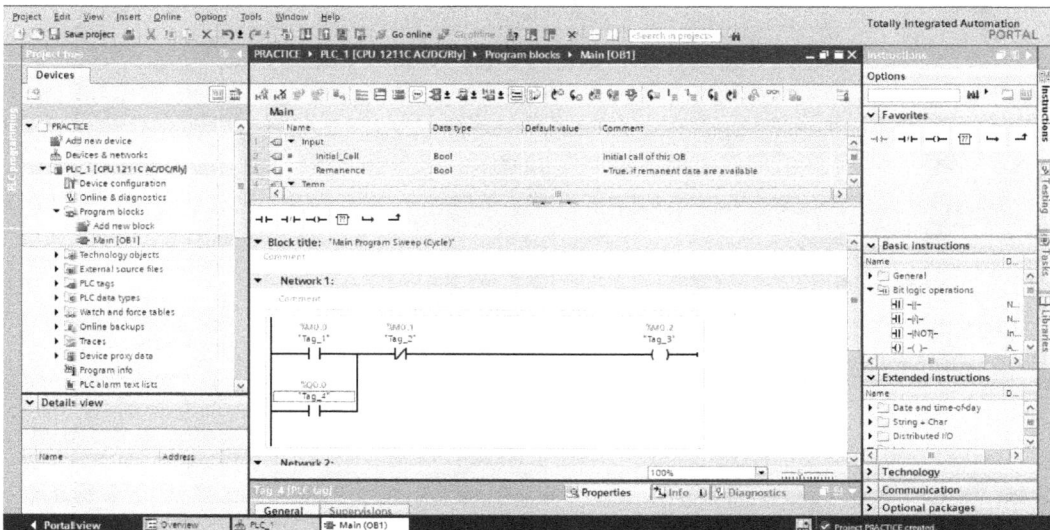

Figure 10.9 – Ladder logic program for start and stop using M0.0 and M0.1 as memory bits

7.   Right-click each tag and click **Rename tag**:

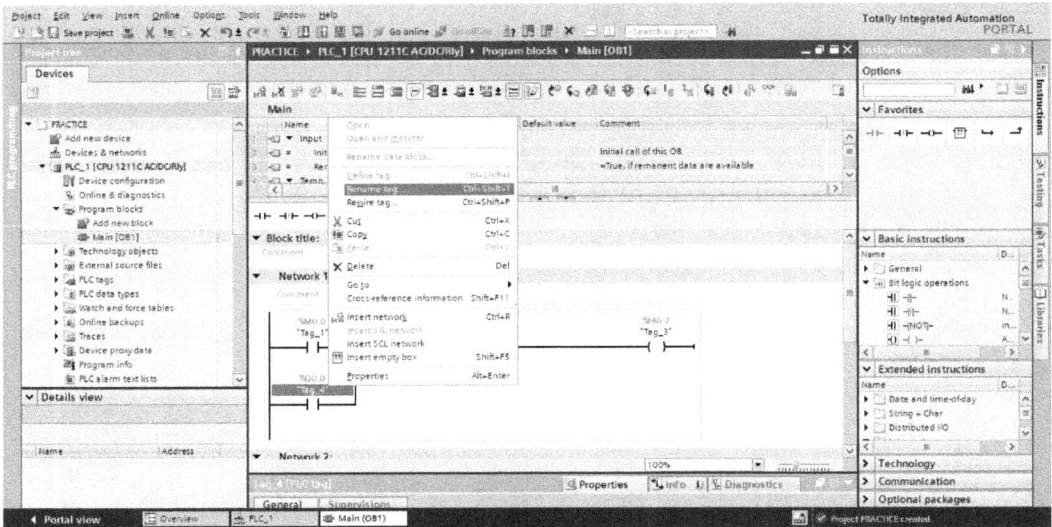

Figure 10.10 – Renaming the tags

8.   Type the new name and click **Change**. Do this for all tag names in your program to get the following:

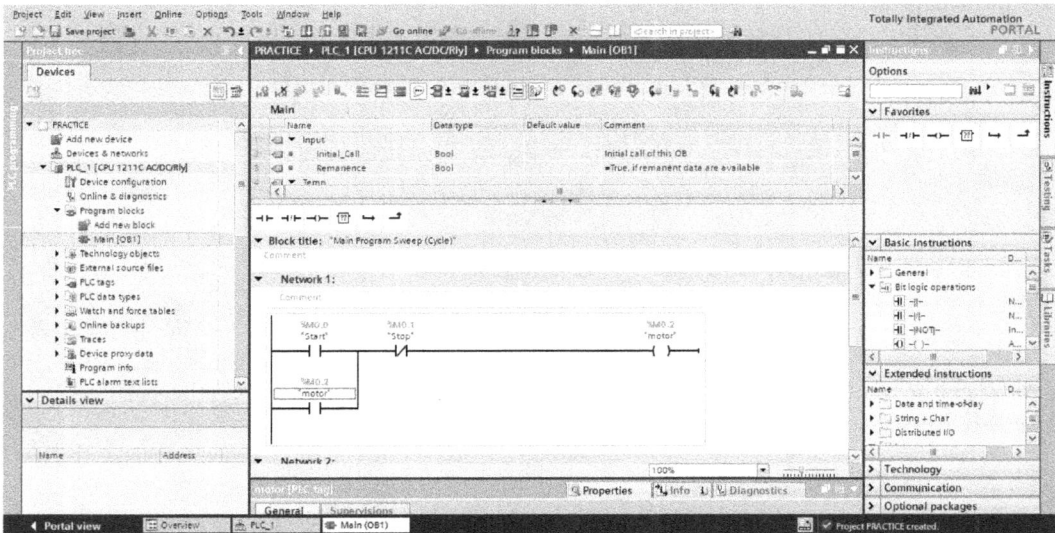

Figure 10.11 – All tags renamed

9.   Click **Save project** to save the project.

10. Click **Start simulator**:

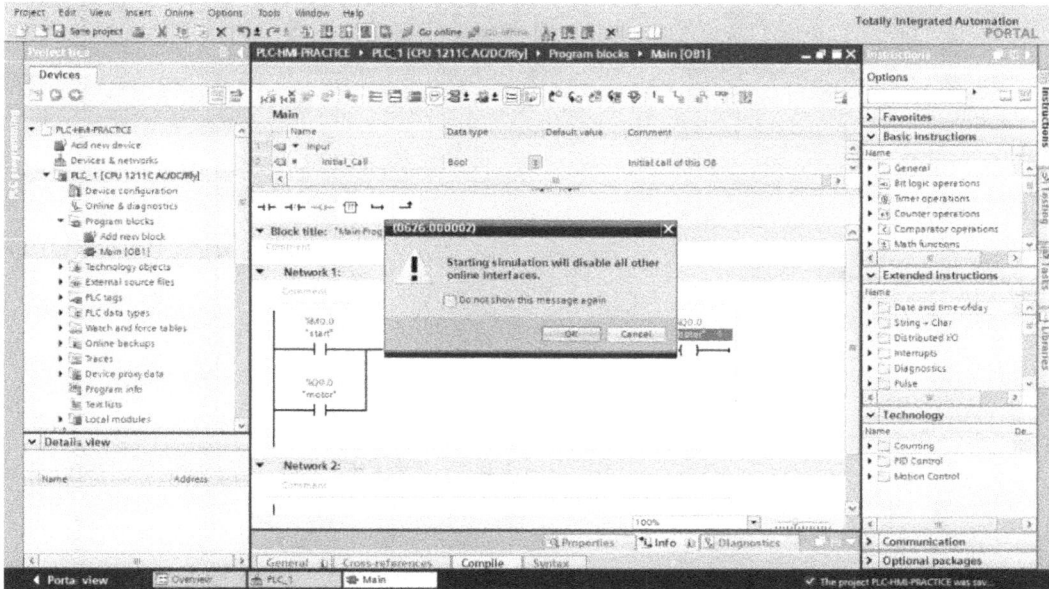

Figure 10.12 – Message that appears when starting the simulator

11. Click **OK**.

12. Select **PN/IE** and **PLCSIM1200/1500** as indicated in the screenshot, then click **Start search**:

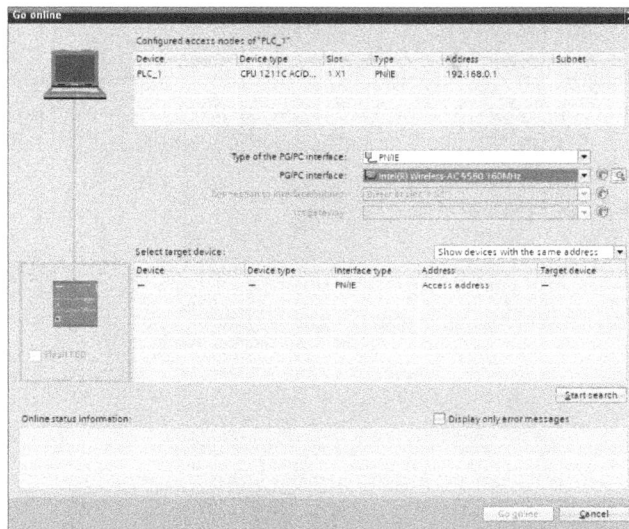

Figure 10.13 – Extended download to device dialog box

13. Select **CPUcommon** and click **Load** to have the program loaded to the virtual PLC:

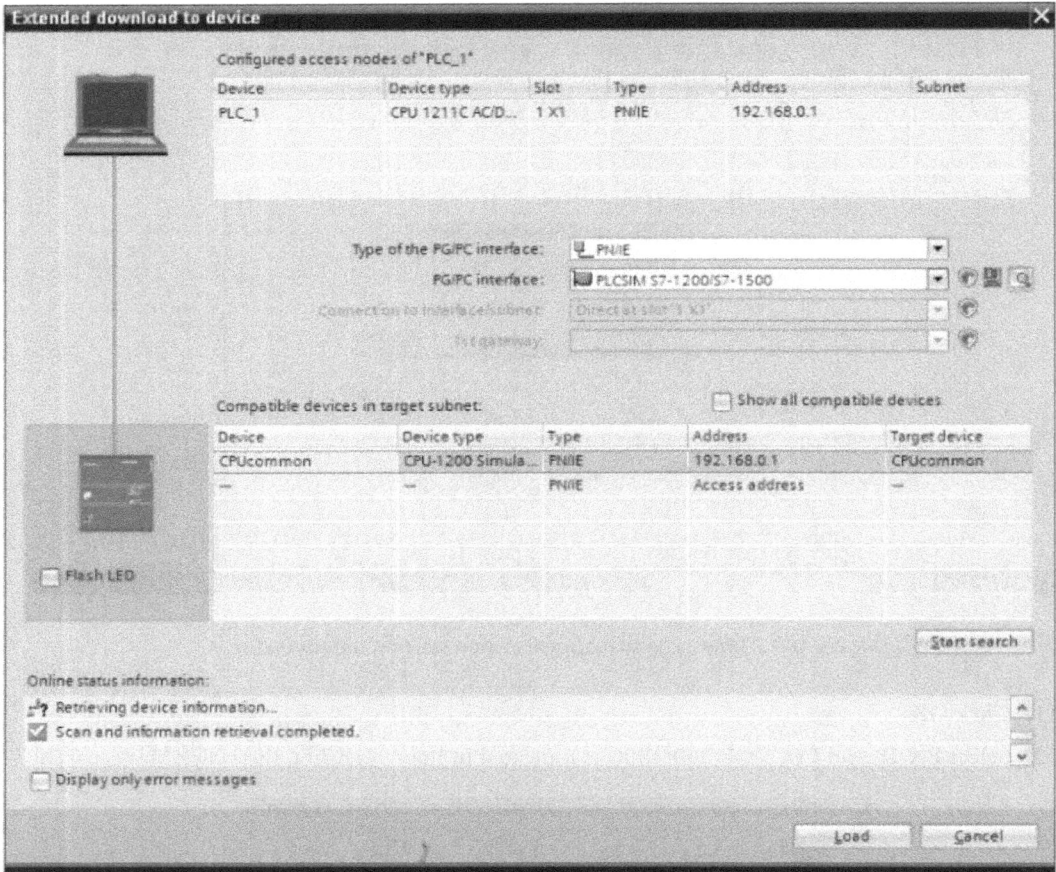

Figure 10.14 – Extended download to device dialog box – selecting
PLCSIM S7-1200/S7-1500 as the PG/PC interface

14. Click **Load** in the **Load preview** dialog box:

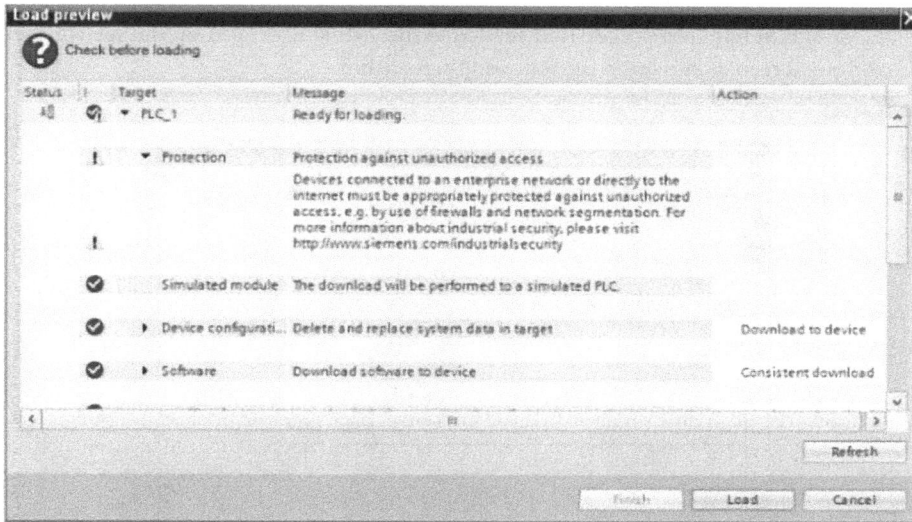

Figure 10.15 – Load preview

15. Mark **Start all** and click **Finish** in the **Load results** dialog box:

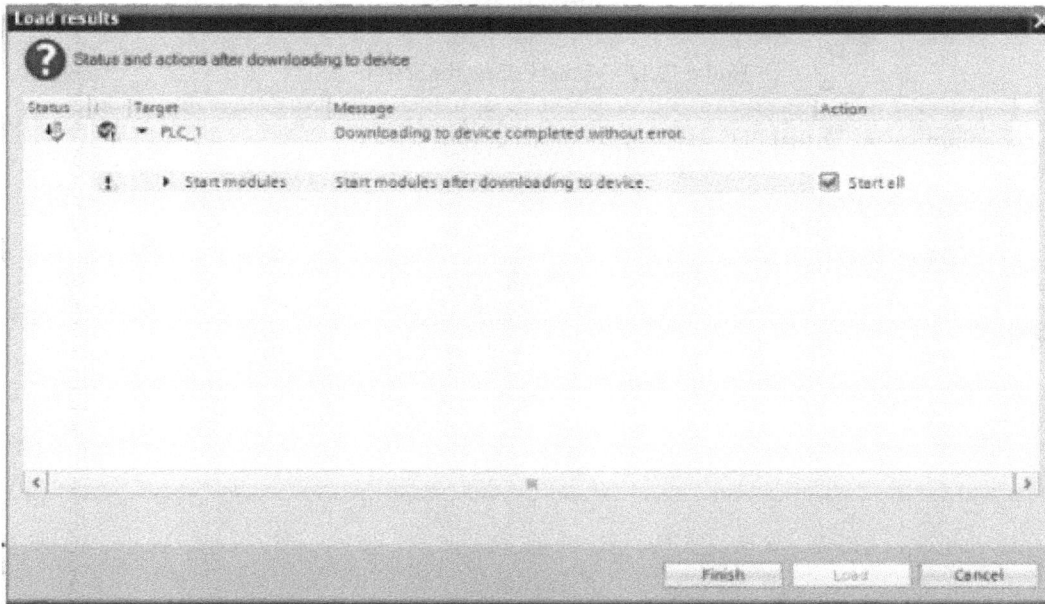

Figure 10.16 – Load results

16. The PLC program has been successfully loaded to the virtual PLC. You should see your virtual PLC on your screen as shown in the following screenshot:

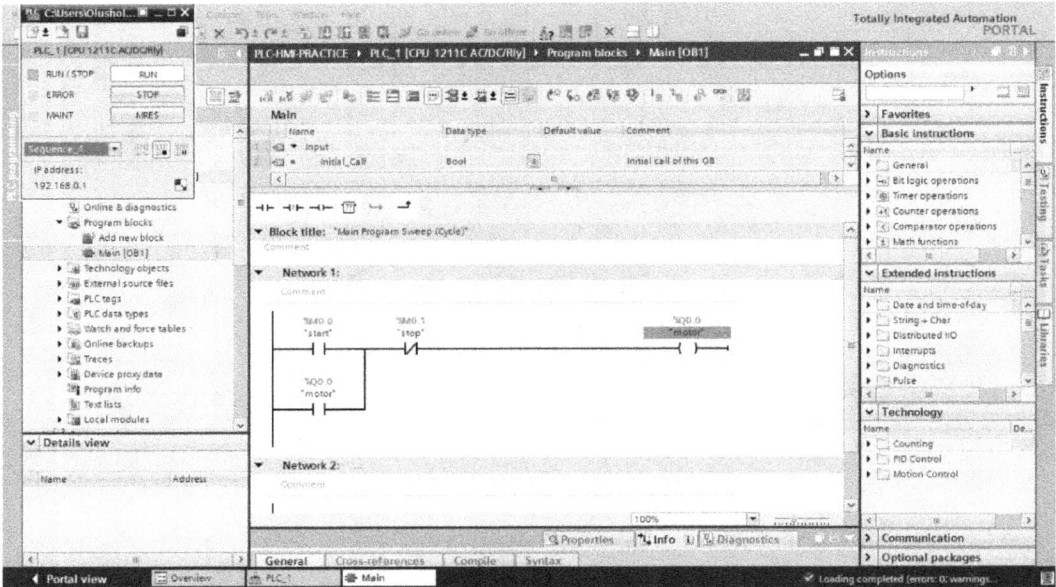

Figure 10.17 – Virtual PLC on the screen

You can minimize the virtual PLC so it will not obstruct your screen.

Let's now add an HMI to the project:

1. Click **View | Go to portal view**:

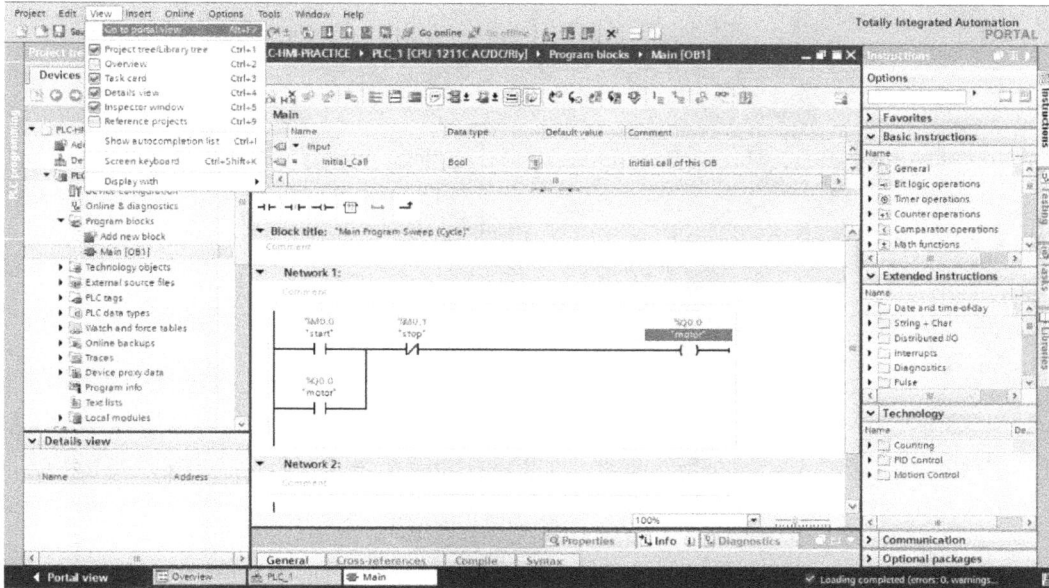

Figure 10.18 – View | Go to portal view

2. Click **Add new device** and select **HMI** from the device list.

3. Expand **SIMATIC Basic Panel | 4" Display | KTP400 Basic** and select the order number of your HMI, for example, **6AV2 123-2DB03-0AX0**, depending on the HMI you intend to use. Click **Add**:

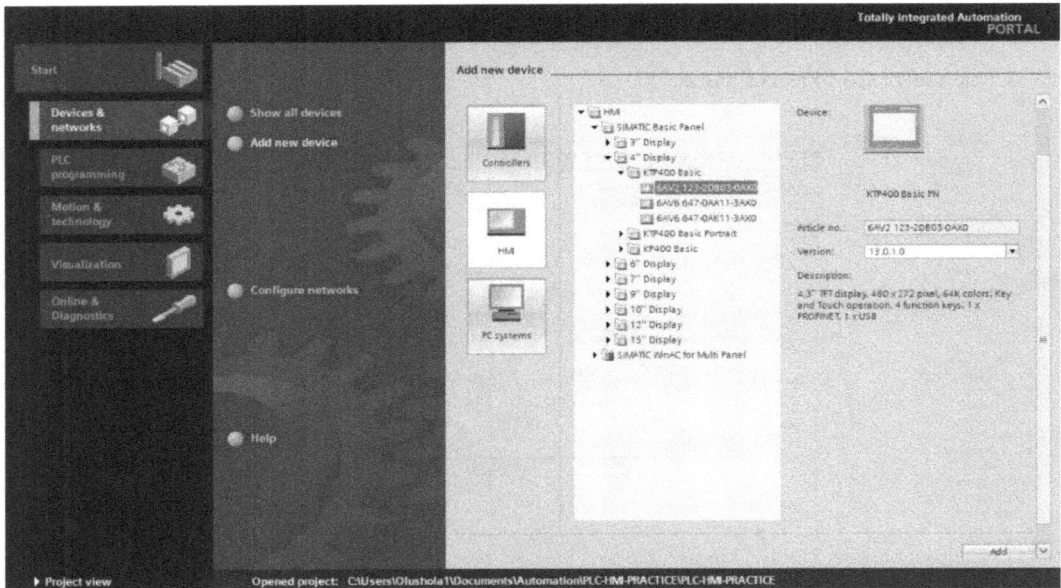

Figure 10.19 – Selecting your HMI

4.   Click **Cancel** to cancel the HMI device wizard because we want to do it without using the wizard:

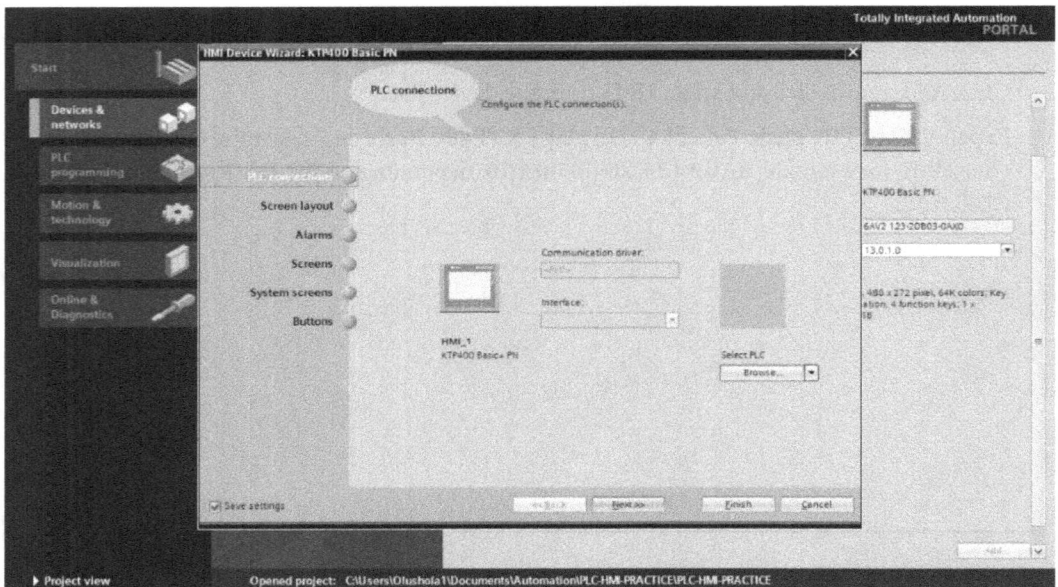

Figure 10.20 – HMI device wizard dialog box

> **Note**
>
> Using the wizard provides a faster way to create an HMI screen. However, it is more professional to create it from scratch as we will be doing here.

5. Click **OK**:

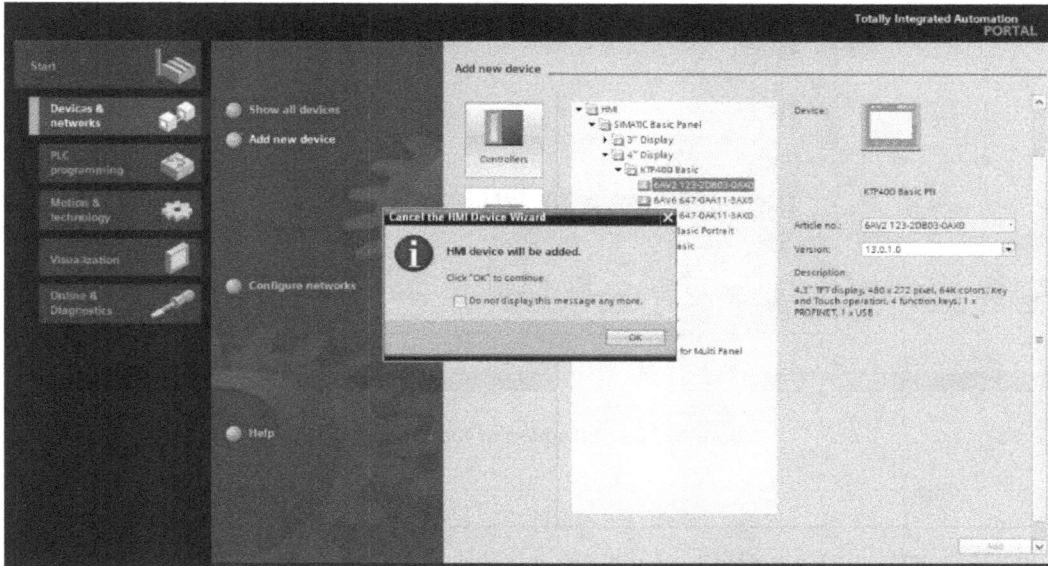

Figure 10.21 – Warning message for canceling the HMI device wizard

Once you click **OK**, you should get the following screen:

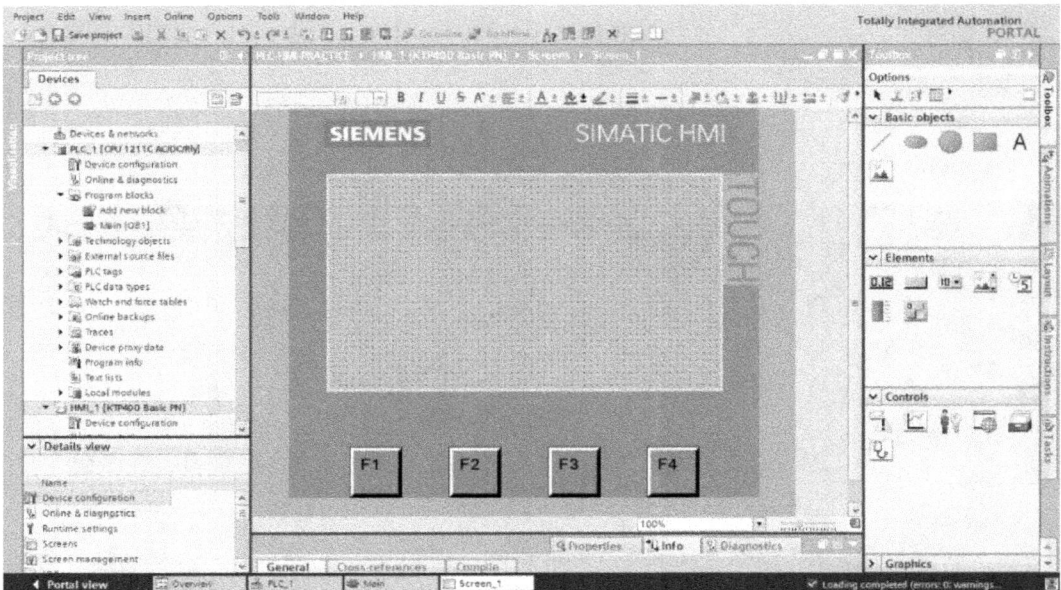

Figure 10.22 – HMI added to your project

> **Note**
>
> HMI_1 (KTP400) has now been added to our project. We now have two folders in the
> project (PLC_1 and HMI_1).

6.   Double-click **Devices & networks** in your project folder on the left of your screen:

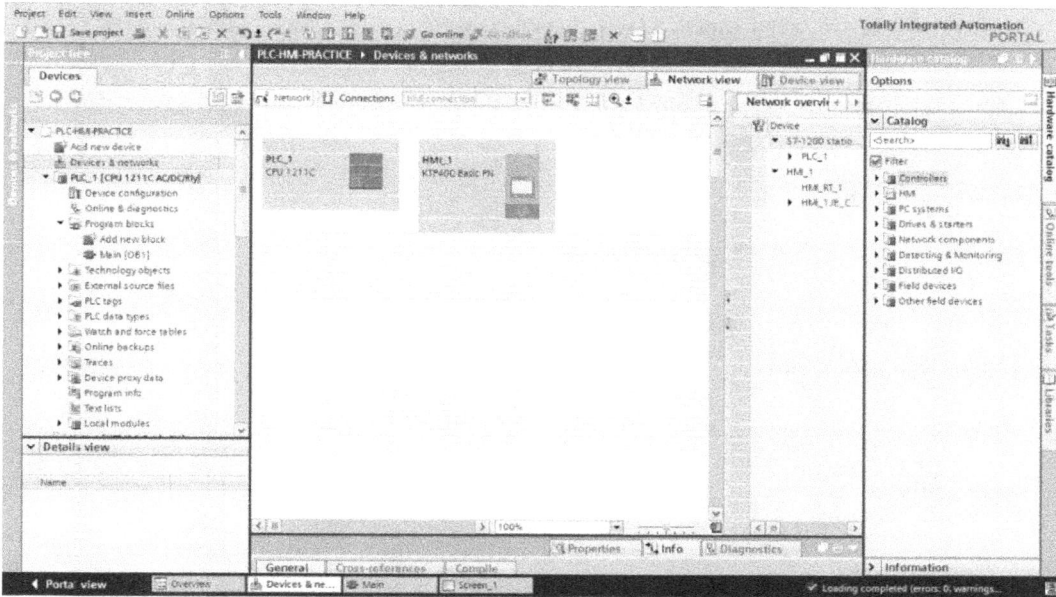

Figure 10.23 – Devices & networks

7.   Drag the network port from the PLC to the HMI to get the following:

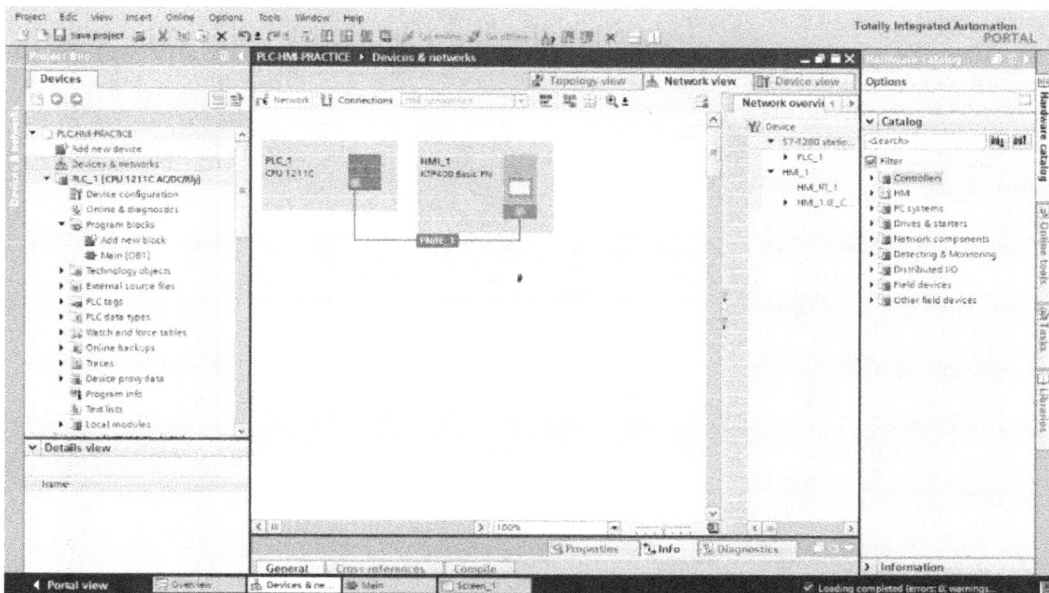

Figure 10.24 – Linking the PLC and HMI (Profinet)

Let's now design a screen, that is, a GUI, that an operator can use to start or stop a motor with an indicator to show the status of the motor.

8.   At the left-hand side, expand your HMI folder, **HMI_1 | Screens |** double-click **Screen_1**:

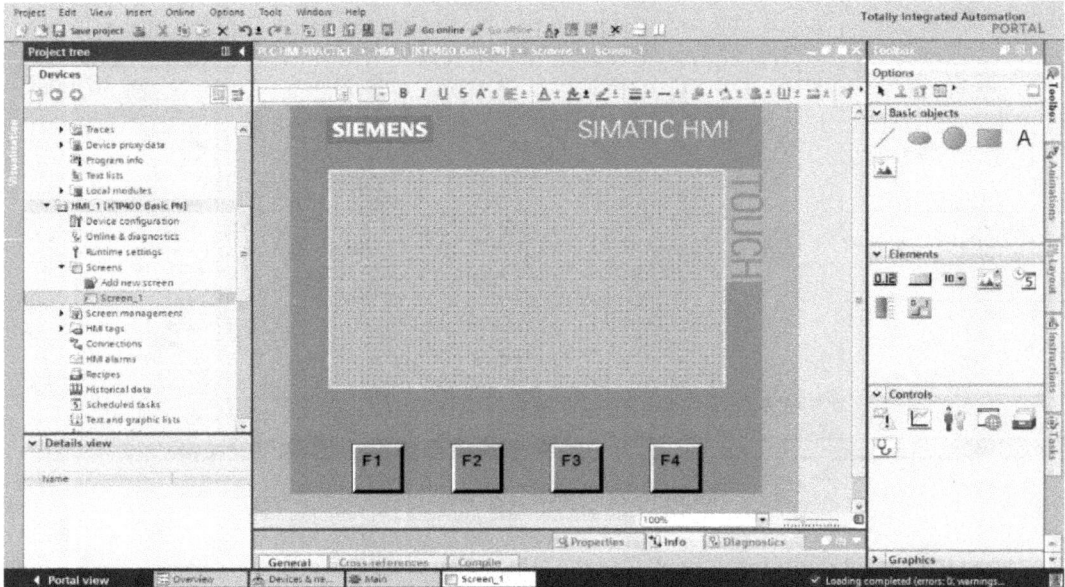

Figure 10.25 – HMI screen

9.   Insert a button by dragging it from the **Elements** tab to the screen, as shown here, and change the name to `Start`:

Figure 10.26 – Inserting a button

10. Insert one more button and change the name to Stop as shown here:

Figure 10.27 – Inserting an additional button

Let's change the appearance of our button.

11. You can select a button and change its appearance through the **Properties** tab as shown here:

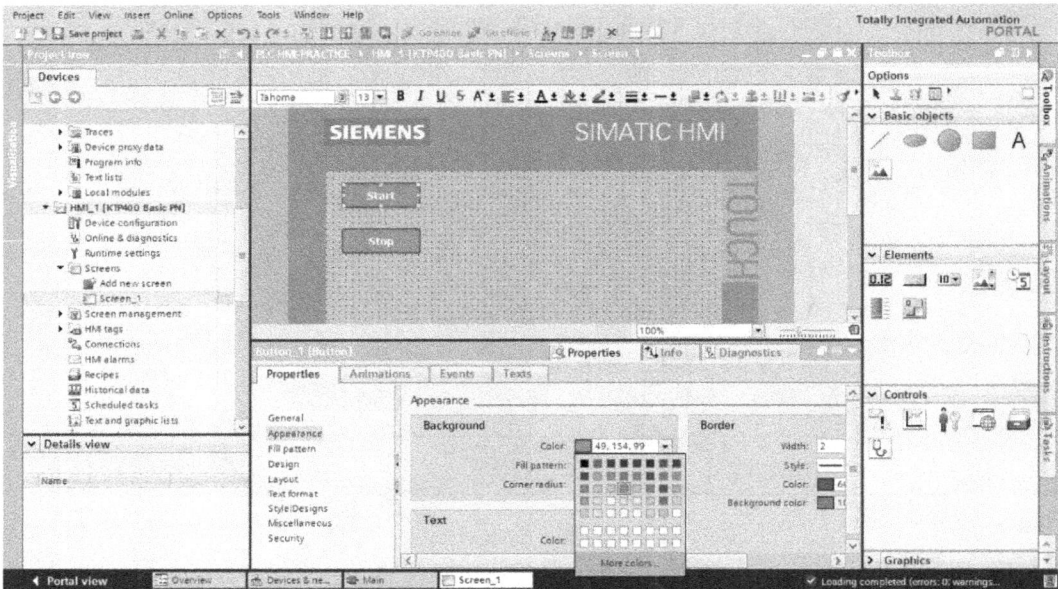

Figure 10.28 – Changing the appearance of a button

> **Note**
>
> In the preceding screenshot, **Solid** was selected as **Fill pattern** and a green color was selected. You can try to create something similar.

Let's now add functionality to our button.

12. Select the **Start** button and click the **Events** tab:

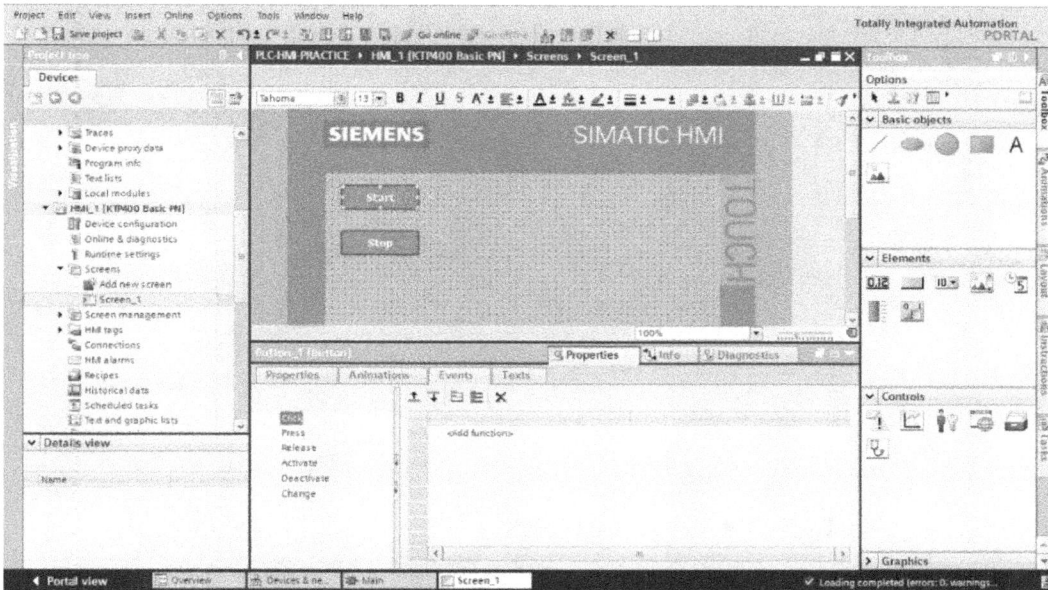

Figure 10.29 – Events tab

As you can see, various options are available (**Click**, **Press**, **Release**, and so on). You can specify a function you want the button to carry out when clicked, pressed, released, and so on. Here, we are going to use **Press**.

13. Select **Press** and click the **Add function** drop-down list:

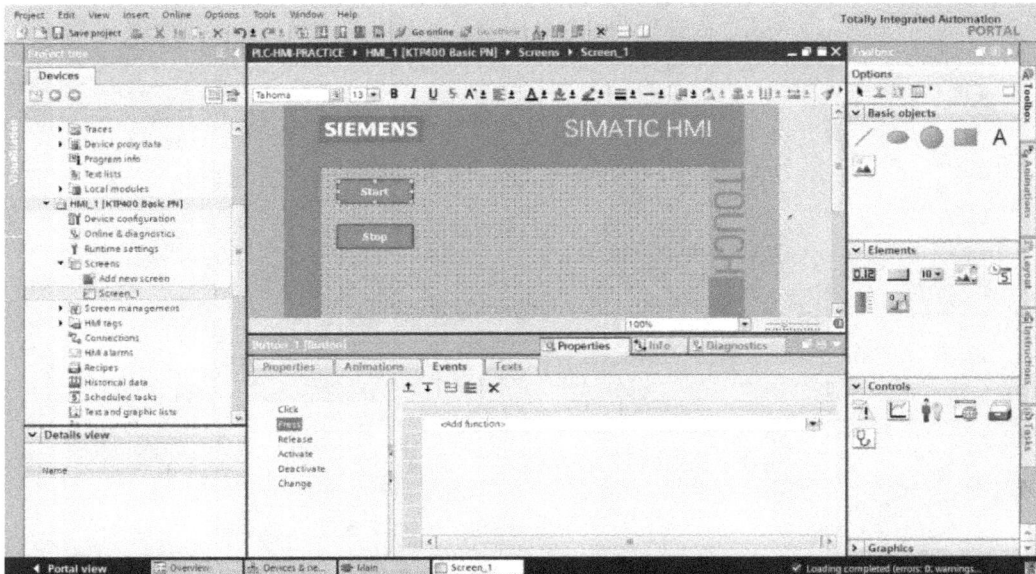

Figure 10.30 – Add function drop-down arrow

14. Search for `SetBitWhileKeyPressed` and select it:

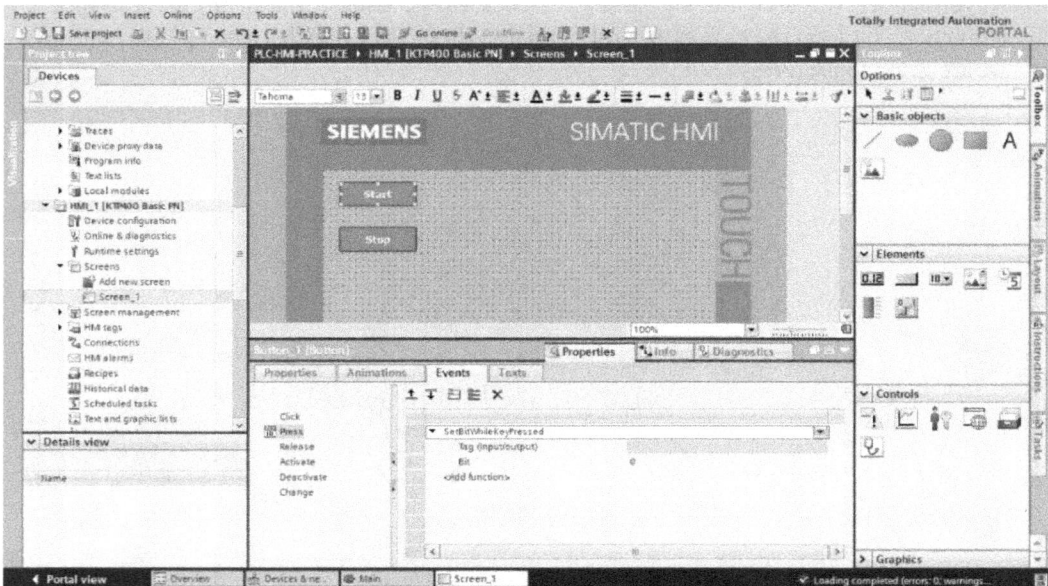

Figure 10.31 – SetBitWhileKeyPressed selected

15. Click inside the **Tag (Input/output)** box and click ...:

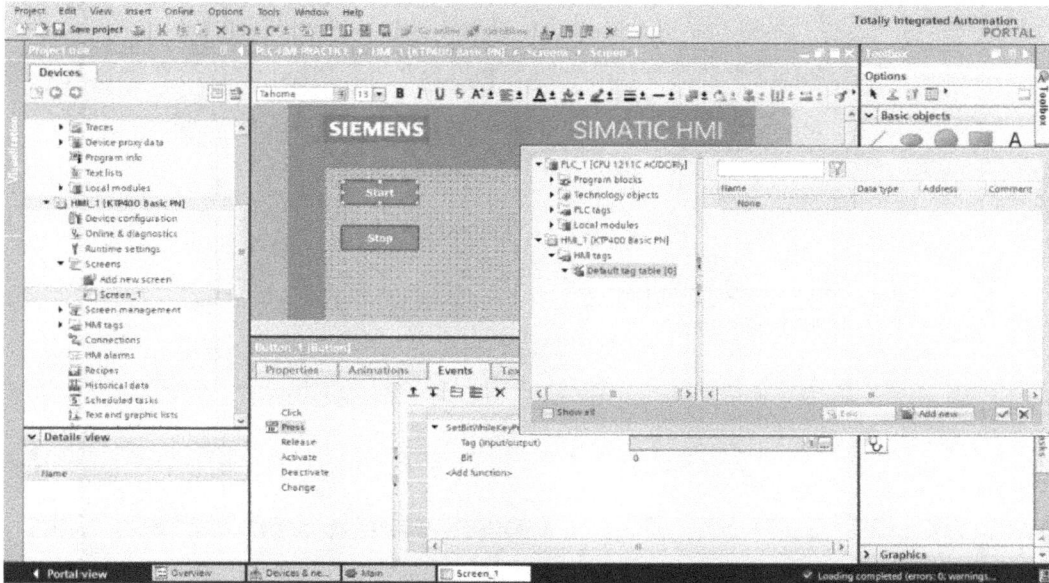

Figure 10.32 – Tag (input/output) options

16. Expand **PLC tags** and double-click **Default tag table(31)** to get what it looks like in the following screenshot.

17. Double-click **Start**:

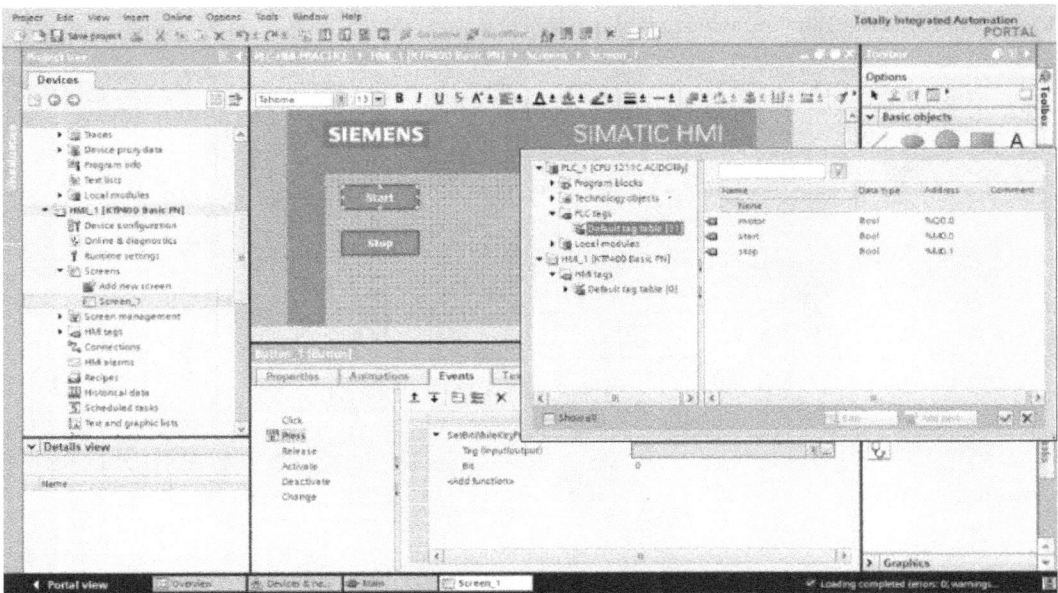

Figure 10.33 – Default tag table (31)

You should get the following:

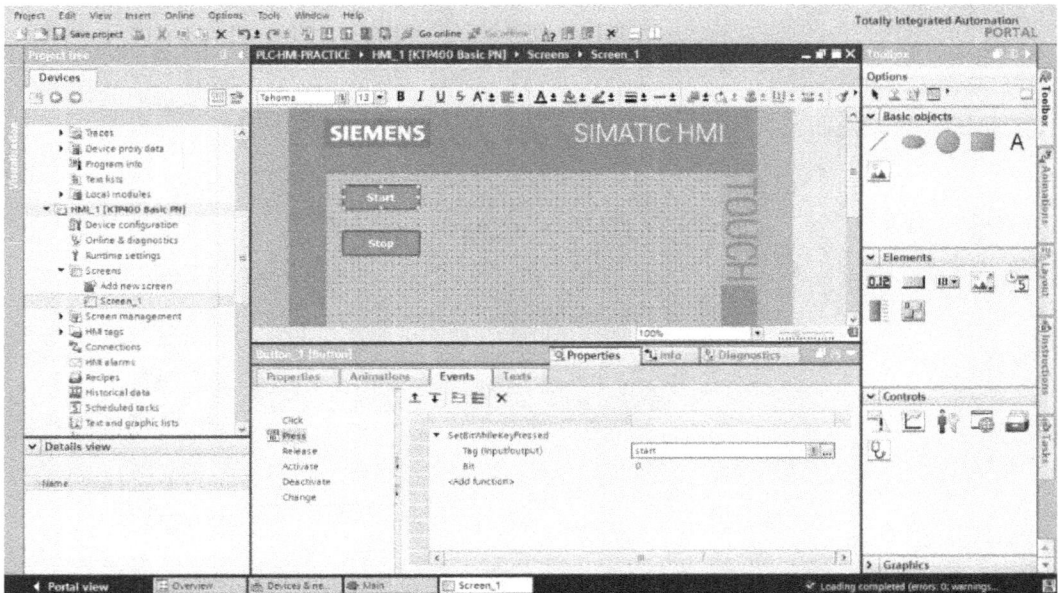

Figure 10.34 – PLC tag (start) selected for the Start button

18. Select the **Stop** pushbutton, click **Events**, and repeat *steps 13 to 16*.

19. Double-click **Stop**:

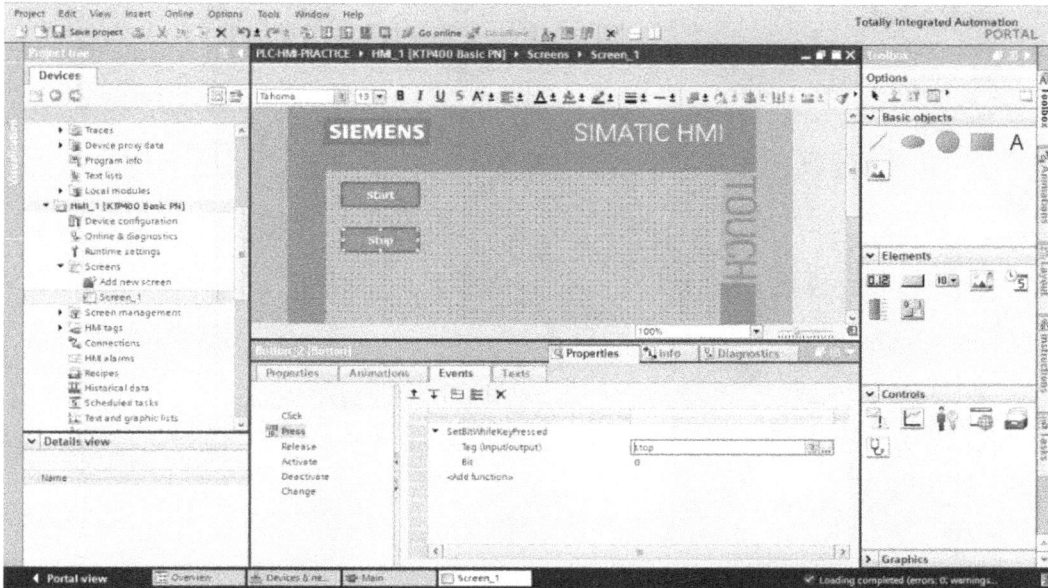

Figure 10.35 – PLC tag (stop) selected for the Stop button

We have successfully added functionality to our buttons.

Let's now draw a rectangle to represent our motor:

1. Insert a rectangle by dragging it from the **Basic objects** tab on the right-hand side to the screen as shown here:

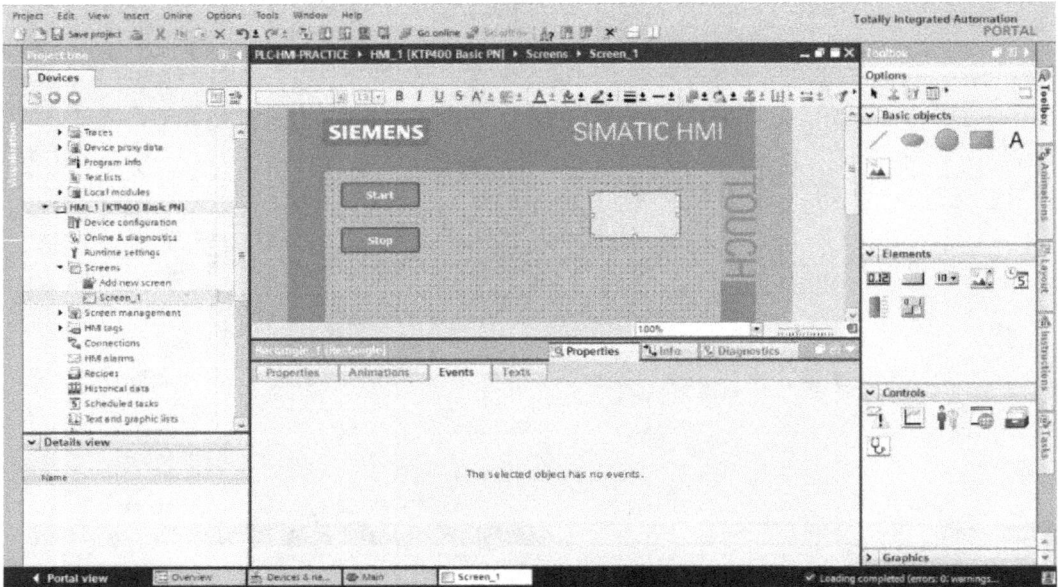

Figure 10.36 – Rectangle added to the screen

2. Click **Animations** and select **Dynamize colors and flashing**:

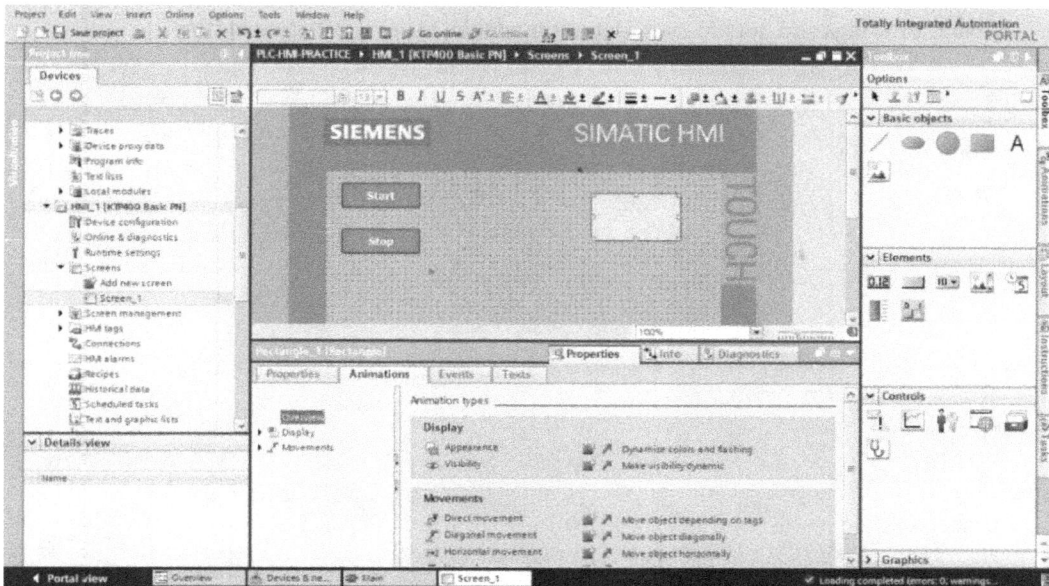

Figure 10.37 – Animation types

You should have the following:

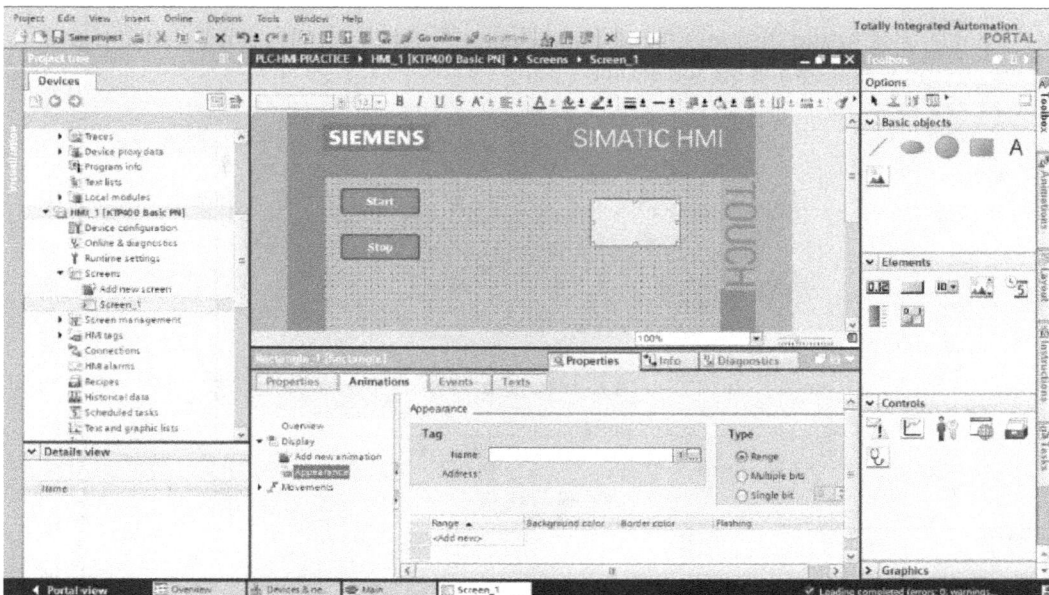

Figure 10.38 – Tag name box

3.   Click **…** on the **Tag | Name** box.

4.   Expand **PLC tags** and double-click **Default tag table[31]**. Double-click **motor**:

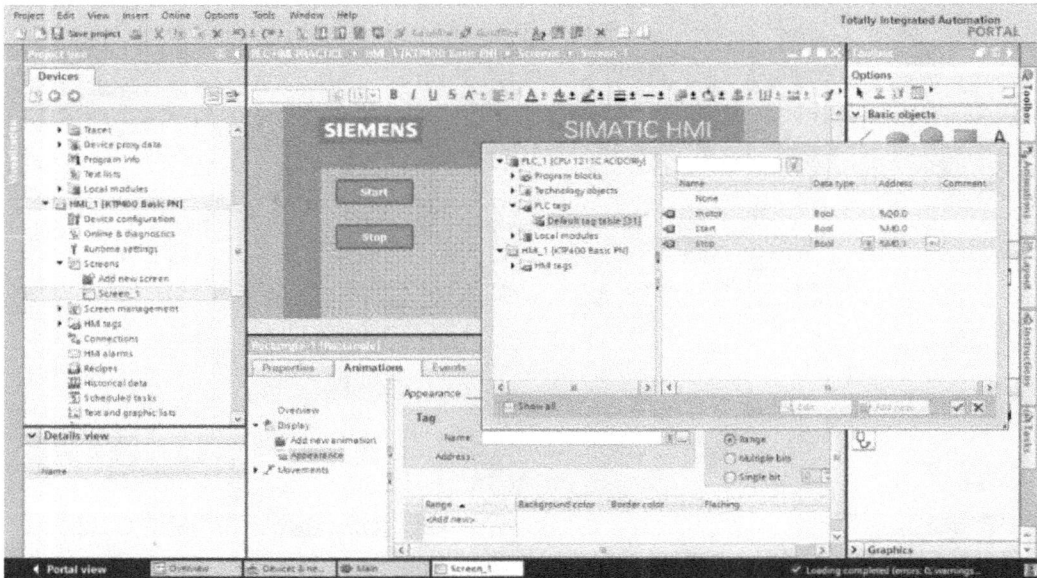

Figure 10.39 – PLC tag list

5.   Double-click to add the range **0** and **1** and change the background color to red and green respectively, as shown in the screenshot:

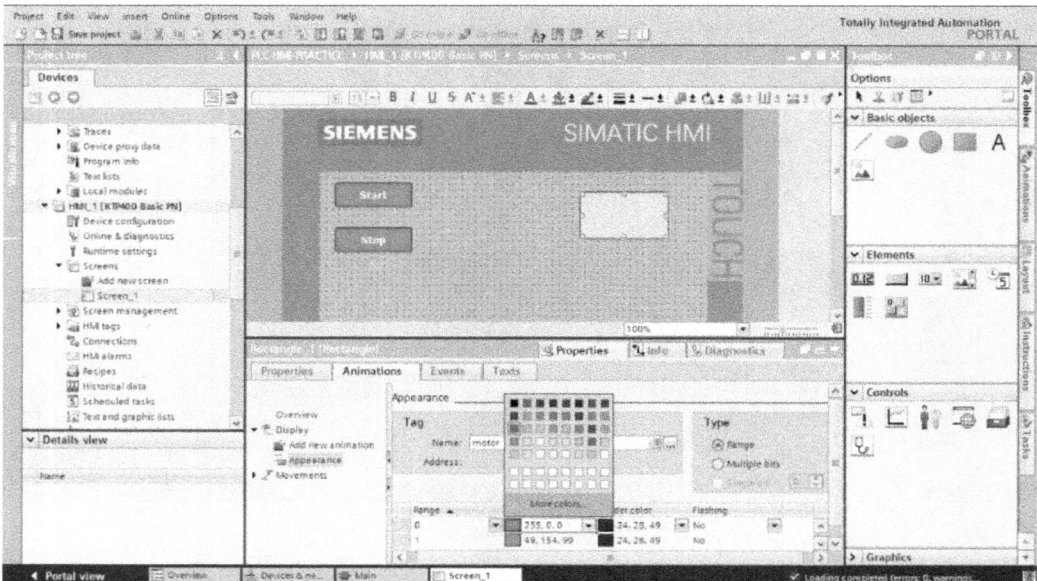

Figure 10.40 – Range 0 and 1 added for the motor tag

6.   Click **Save project**, click on your screen to deselect the rectangle, and click **Start simulator | OK**:

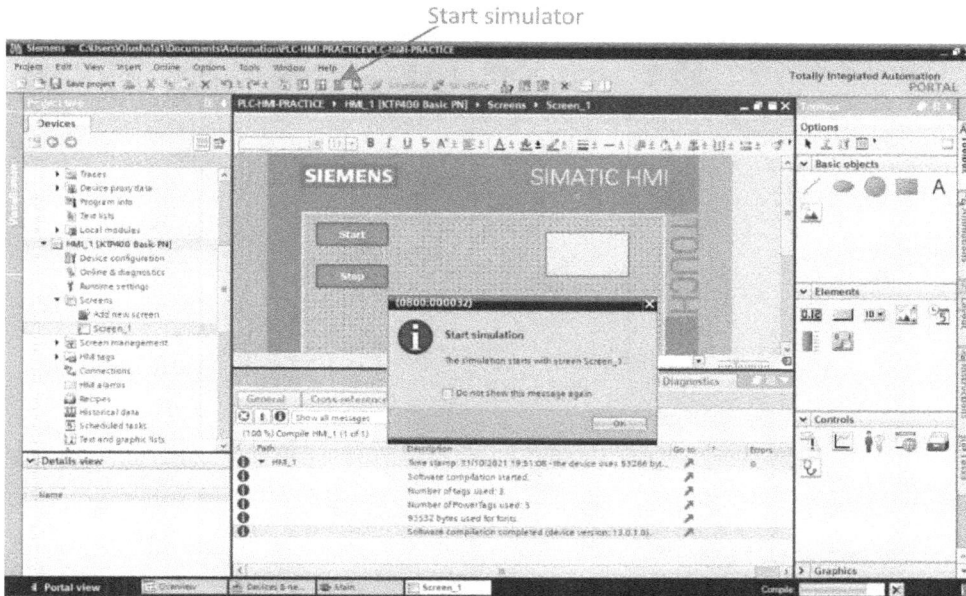

Figure 10.41 – Start simulation message

You should now see the simulator (**RT Simulator**) as shown in the following screenshot:

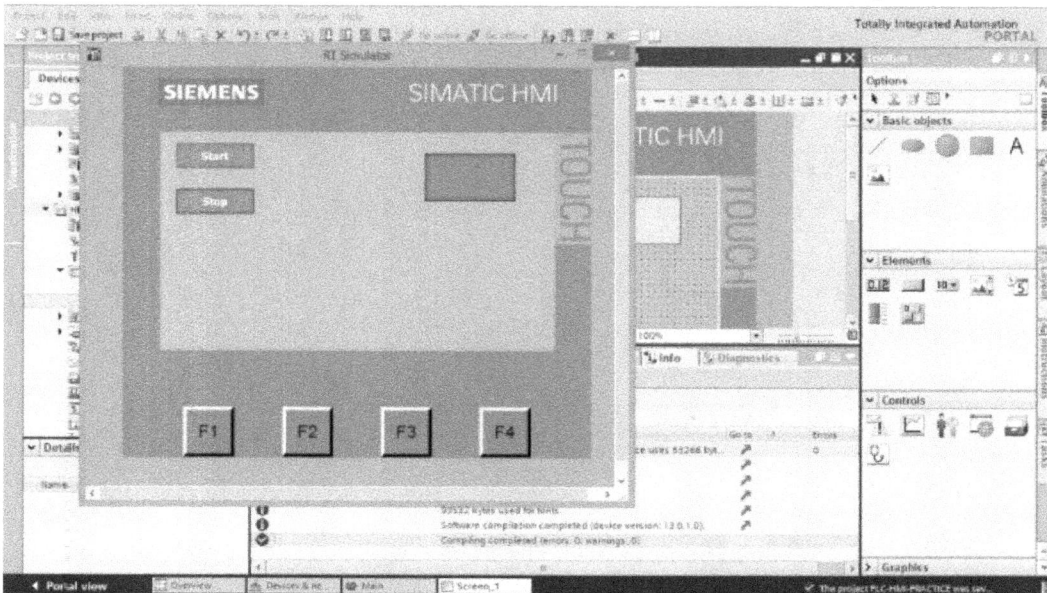

Figure 10.42 – HMI simulator

7. On the left-hand side, expand your PLC folder (PLC_1) and expand **Program blocks**. Double-click **Main [OB1]** to view your PLC program:

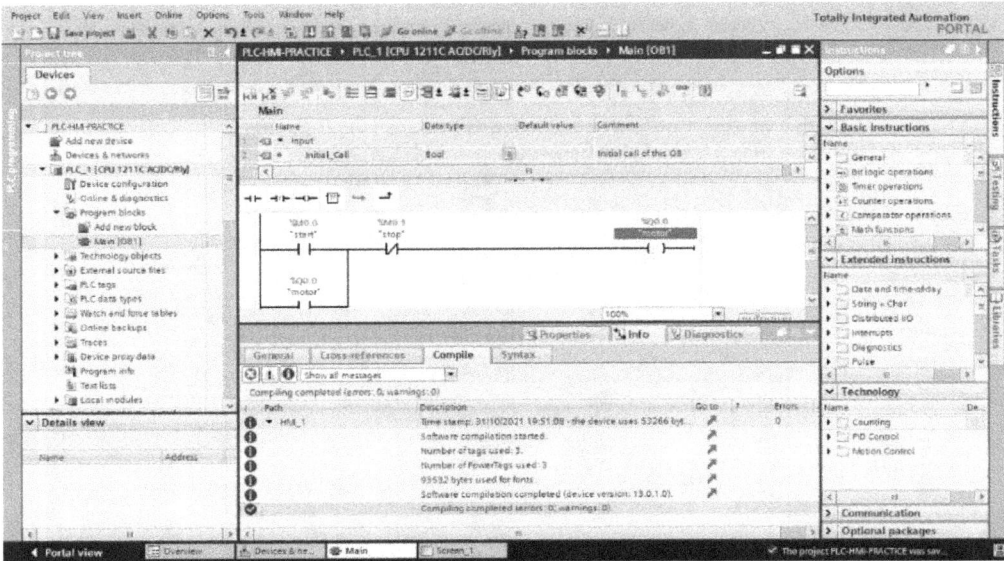

Figure 10.43 – PLC ladder logic program for the start and stop program

8. Click **Monitoring ON/OFF** to get the following:

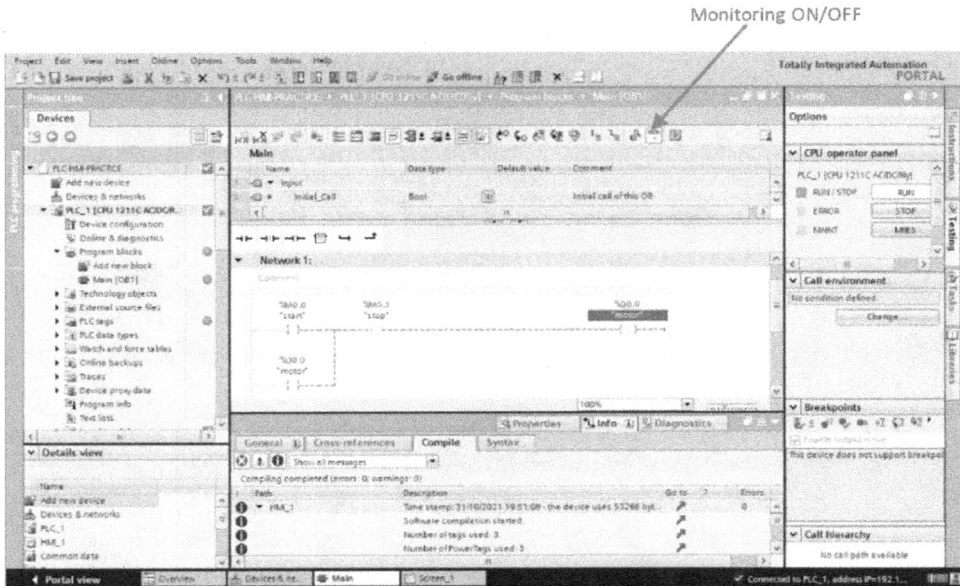

Figure 10.44 – Monitoring the ladder logic program

9. Arrange your windows to show both the PLC ladder logic program and the HMI screen, as shown here:

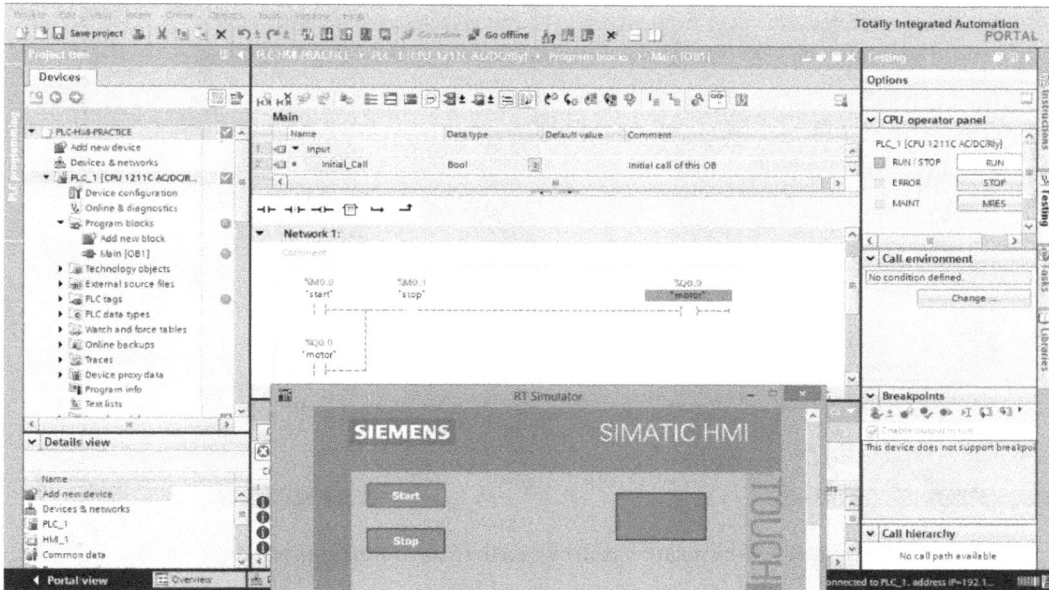

Figure 10.45 – PLC ladder logic program and HMI simulator

10. Press and release the **Start** button:

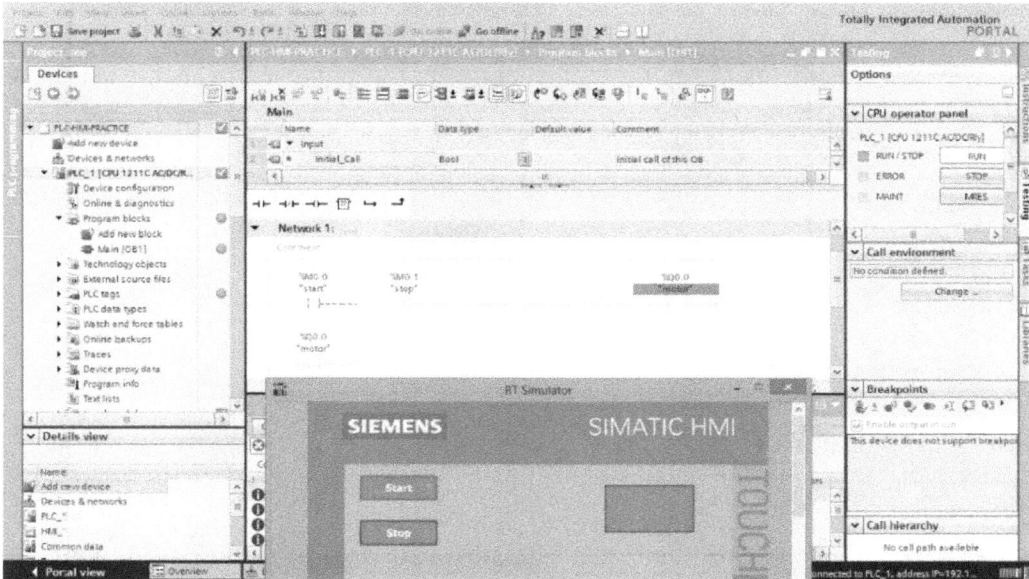

Figure 10.46 – Simulation result when the Start button is pressed

11. Press and release the **Stop** button:

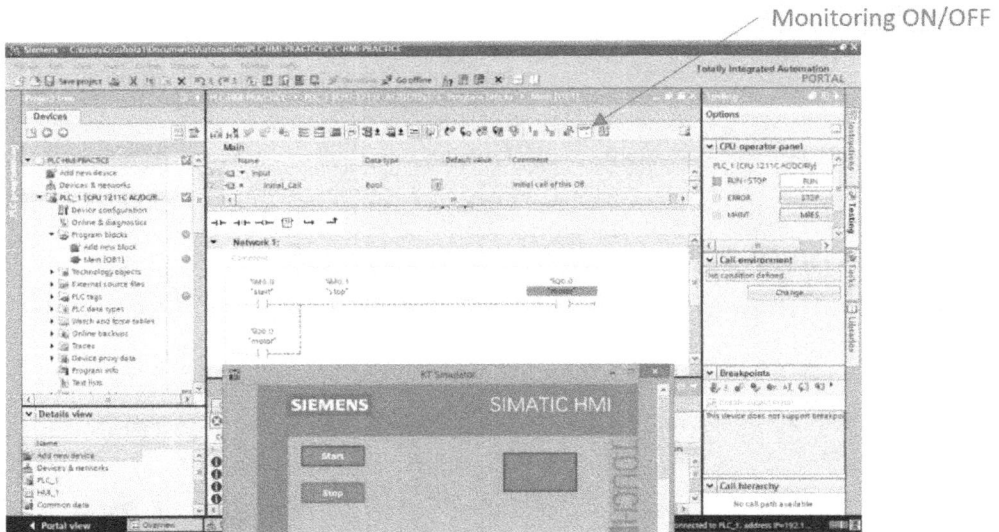

Figure 10.47 – Simulation result when the Stop button is pressed

12. Click **Monitoring ON/OFF** to stop the PLC program simulation:

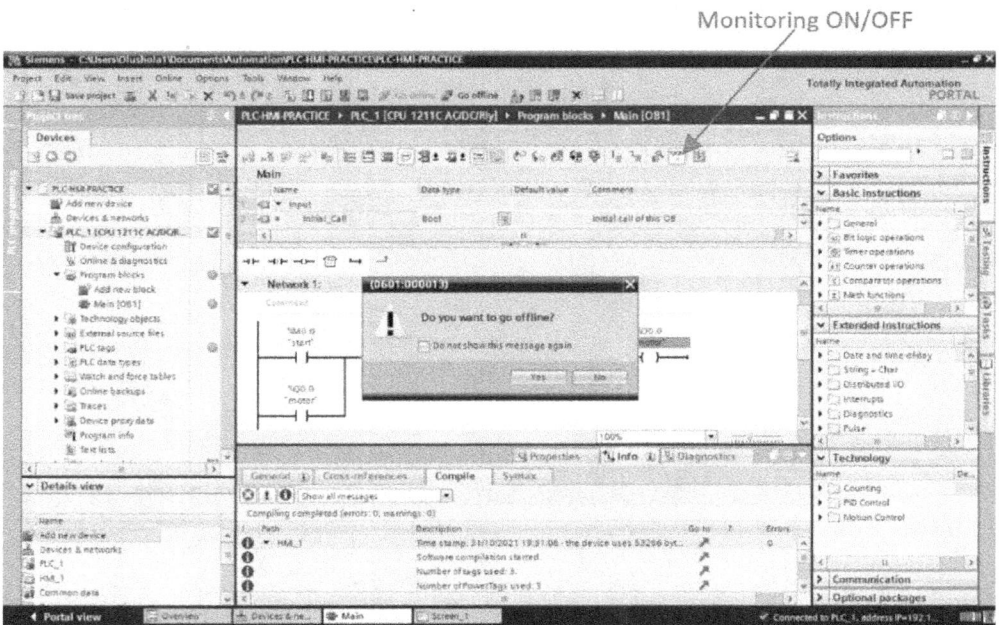

Figure 10.48 – Go offline confirmation message

13. Click **Yes** to go offline.

14. Close the HMI simulator. Click **Save project** to save changes to your project.

Let's advance our knowledge more by learning how to use switches and graphics:

1. Write this program in **Network 2** and give it an appropriate address (e.g. M0.1, Q0.2) and tag name as shown:

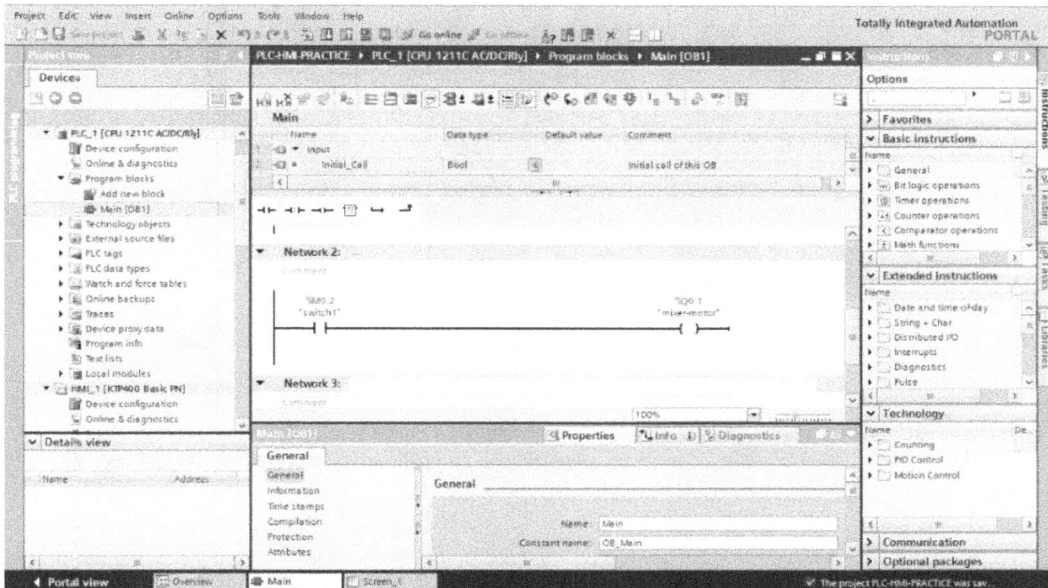

Figure 10.49 – Program to start and stop a mixer-motor

2. Click **Download to device** since our simulator is already running.

3. Click **Load | Finish**.

4. At the left-hand side, expand the HMI folder (**HMI_1**) and double-click **screen_1**.

5. Insert a switch by dragging it from the **Elements** tab at the right-hand side to the screen you are currently developing and resize as required. Your screen should look similar to the following:

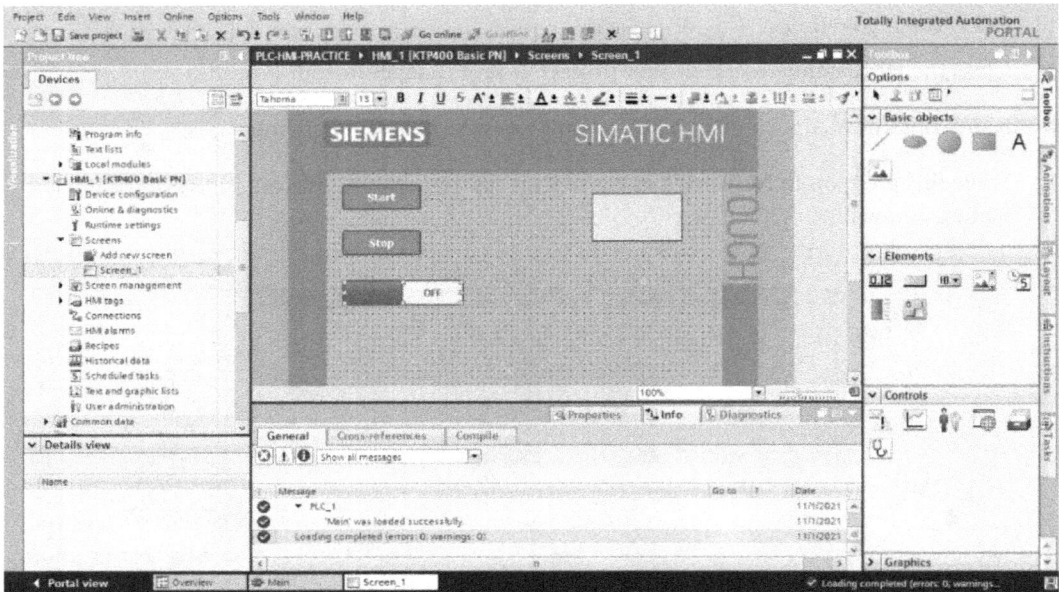

Figure 10.50 – Switch inserted and resized on the screen

6. Click on **Properties** and click **General**, as shown in the following screenshot. For **Mode**, select **Switch with text**:

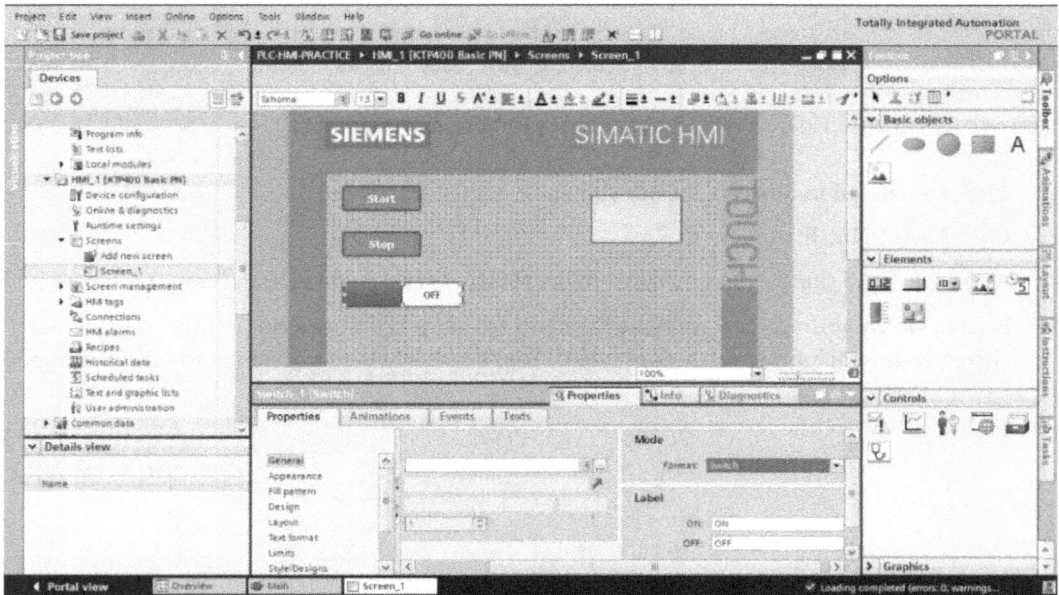

Figure 10.51 – General tab when the switch is selected

7.   In the label section, type some text for **ON** and some text for **OFF**, as shown in the following screenshot:

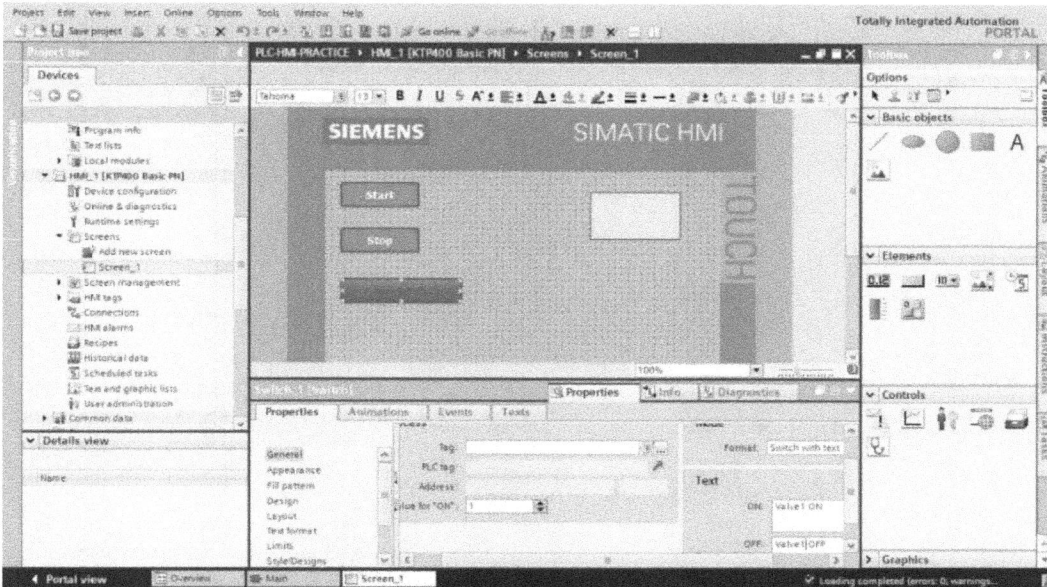

Figure 10.52 – General tab when switch with text is selected in the format drop down list for Mode

8.   Click **…** on **Tag**, expand **PLC tags**, and double-click **Default tag table**, as shown here:

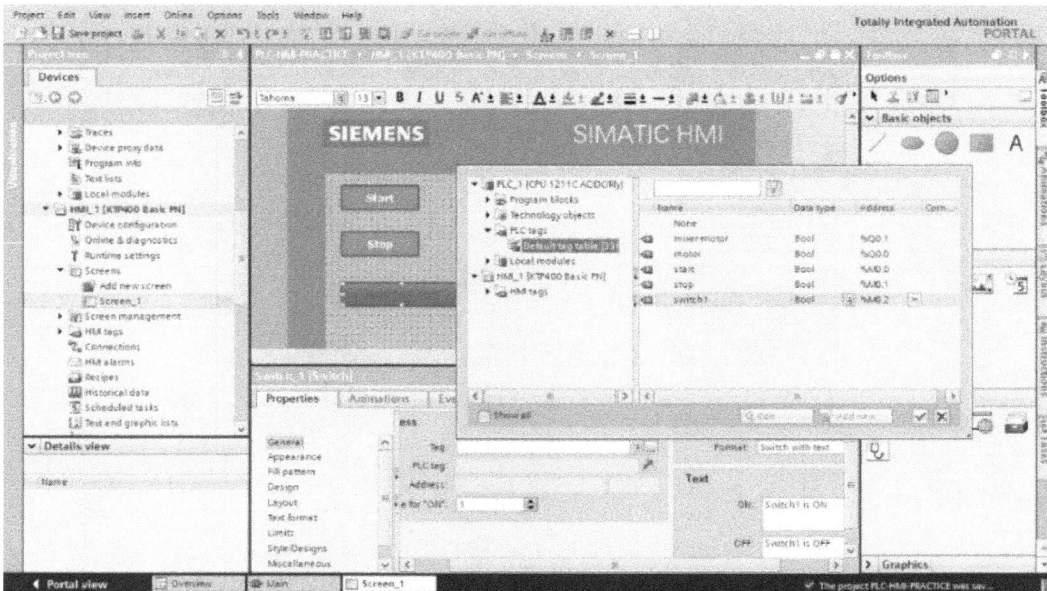

Figure 10.53 – Default tag table

9. Double-click **switch1** to select **switch1** and have your screen look as follows:

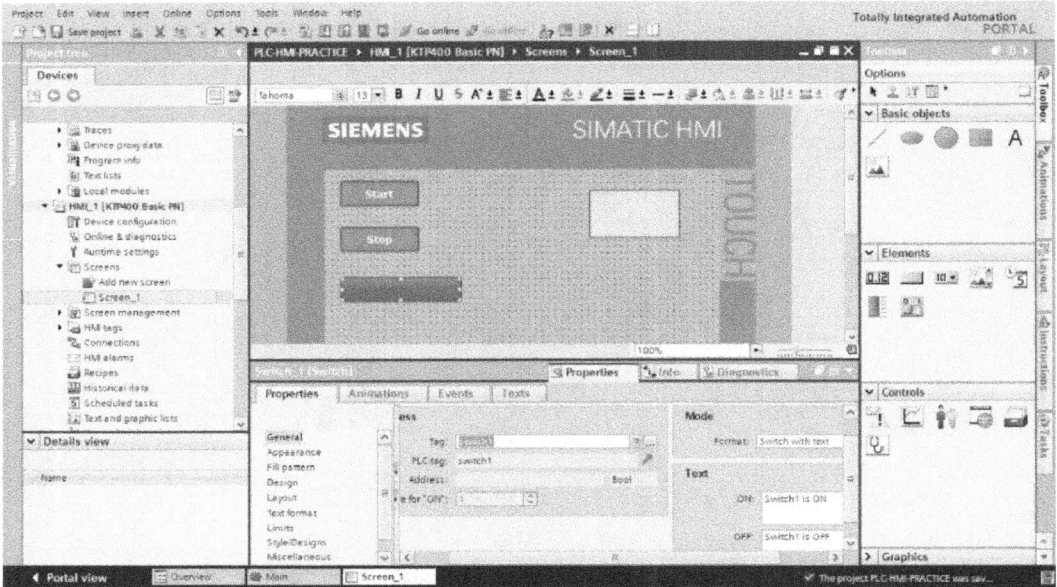

Figure 10.54 – switch1 selected as the tag

10. Click **Appearance** and change the text color to red as shown in the following screenshot:

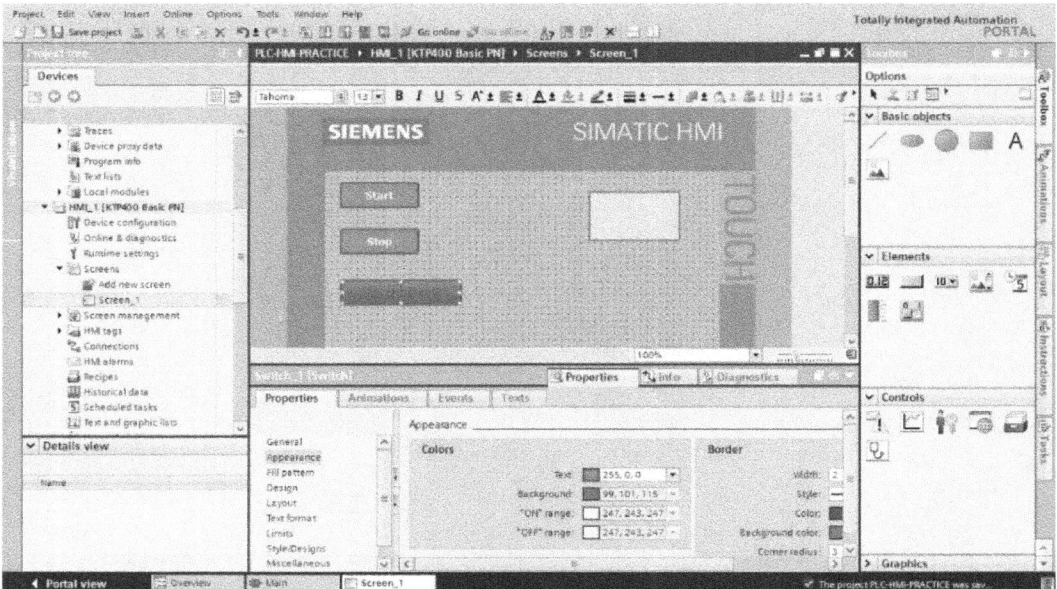

Figure 10.55 – Text color changed to red

11. Insert a motor by dragging it from the **Graphics** tab at the right-hand side onto the screen and resize as required. You will have to go to **Graphics** | **WinCC graphics** | **Automation equipment** | **Motors** | **True color**. You should see a list of motor graphics below your selection, as shown in the following screenshot:

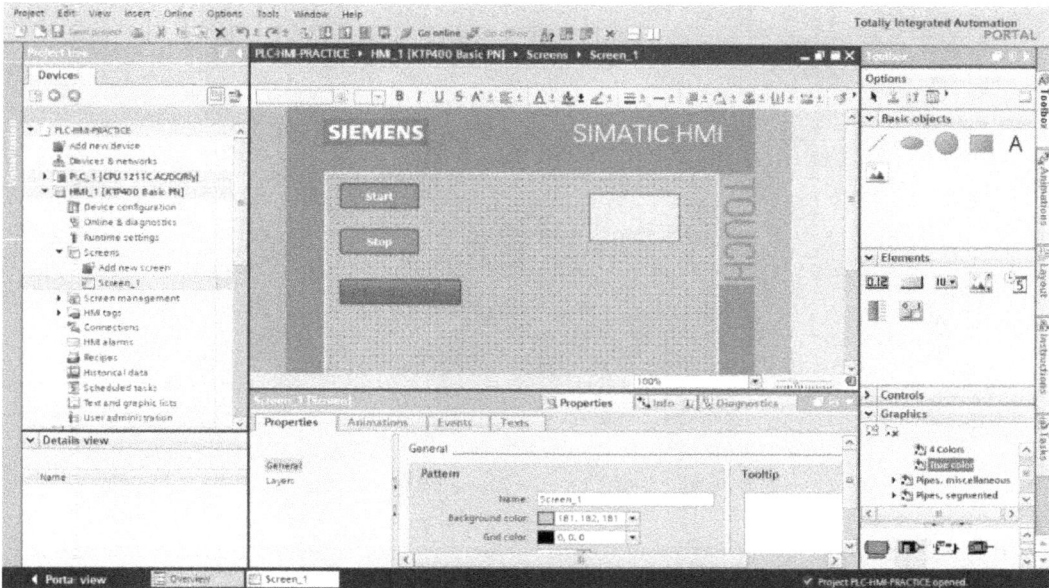

Figure 10.56 – List of motor graphics

12. Now drag the motor to your screen to get the following:

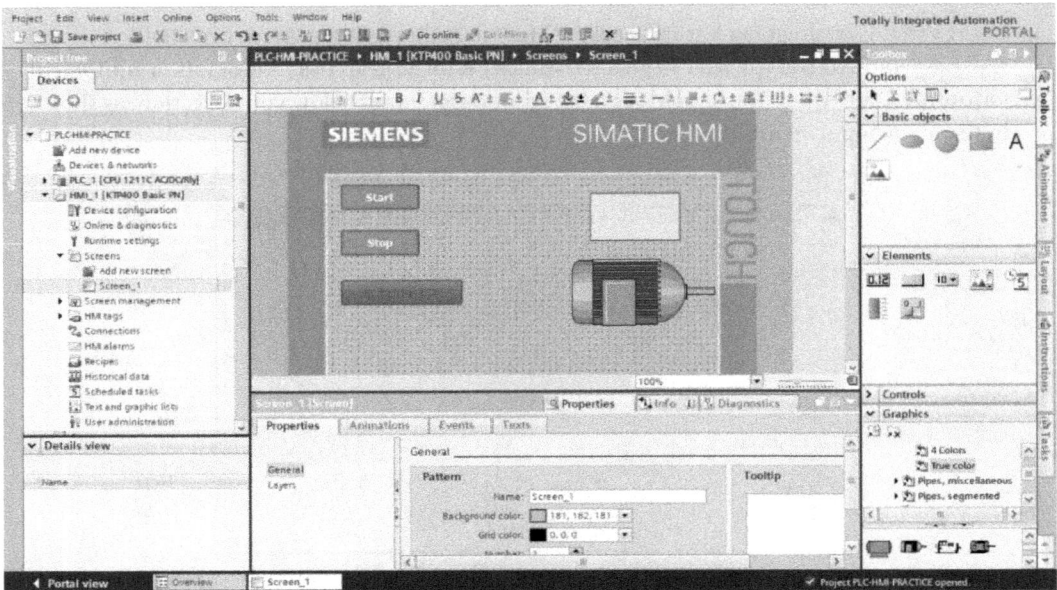

Figure 10.57 – Graphic of the motor on the screen

13. Insert a circle by dragging one from the **Basic objects** tab and carefully placing it on the motor, as shown in the following screenshot:

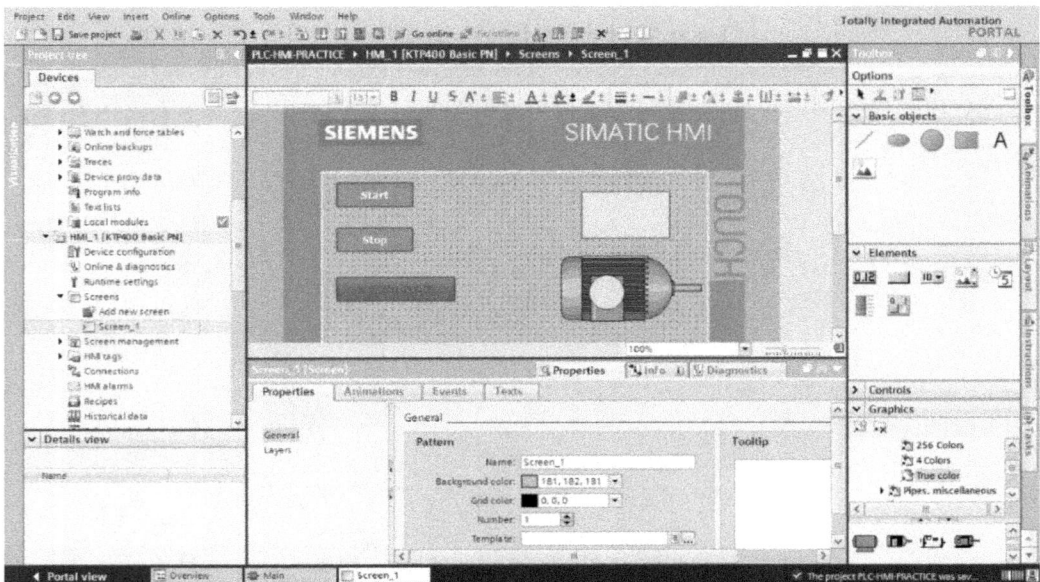

Figure 10.58 – Circle placed on the motor graphic

14. Select the circle on the motor and click **Animations | Dynamize color and flashing**, then follow the steps you took for the rectangle, but you should now see **mixer-motor** when you double-click **Default tag table[31]**, as shown in the following screenshot:

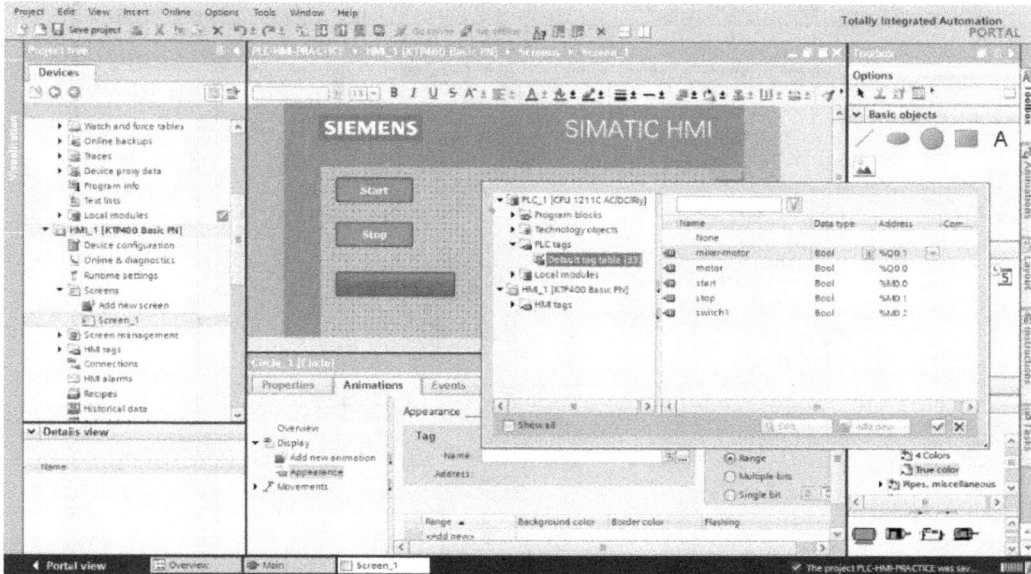

Figure 10.59 – Default tag table showing mixer-motor in the PLC tag list

15. Double-click **mixer-motor** to get the following:

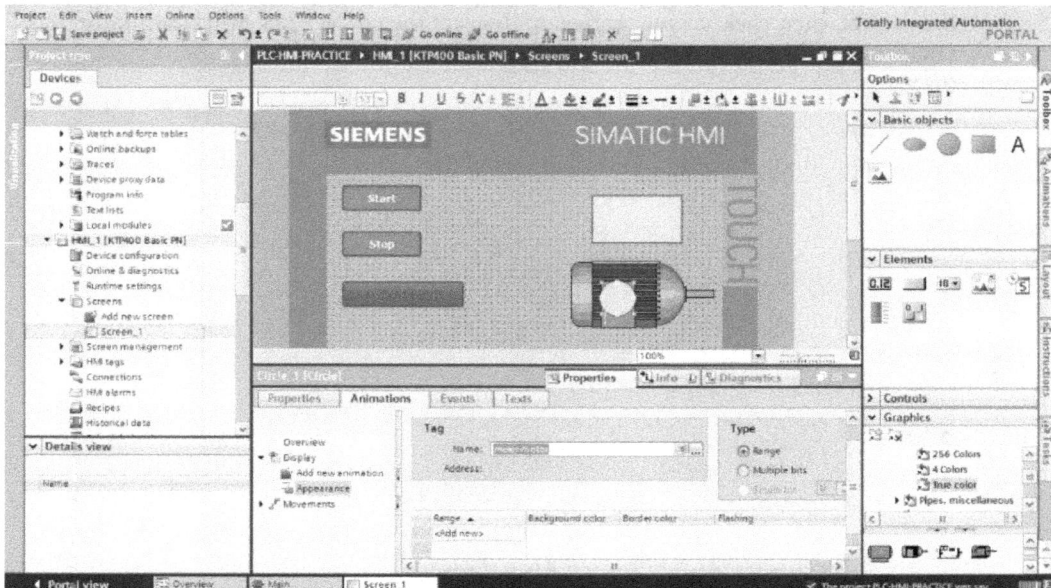

Figure 10.60 – mixer-motor selected

16. Double-click to add the range **0** and **1** and change the background color to red and green respectively, as shown in the following screenshot:

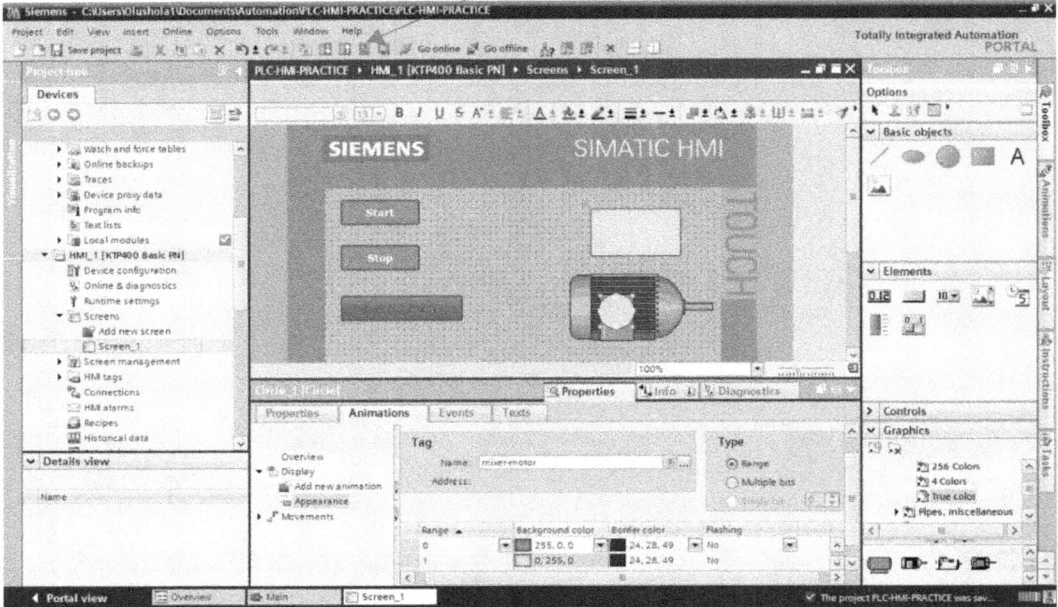

Figure 10.61 – Range and color selected

17. Save the project, then click on your screen to deselect the circle. Click **Start simulator** and click **OK** when the start simulation message appears:

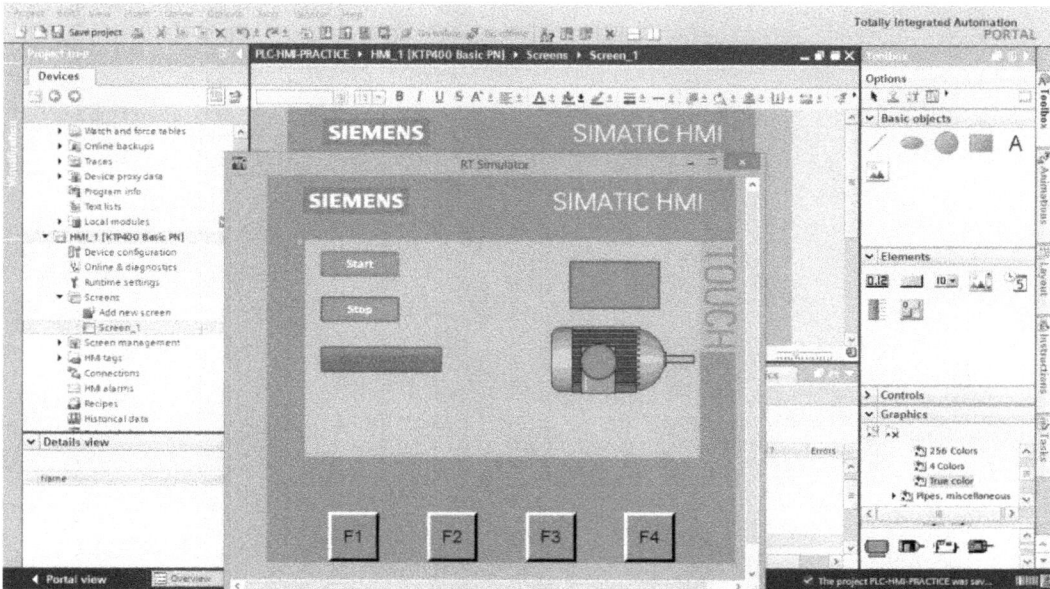

Figure 10.62 – HMI Simulator

18. Expand **PLC | Program blocks** and double-click **Main_0B1**.

19. Click **Monitoring ON/OFF** to get the following:

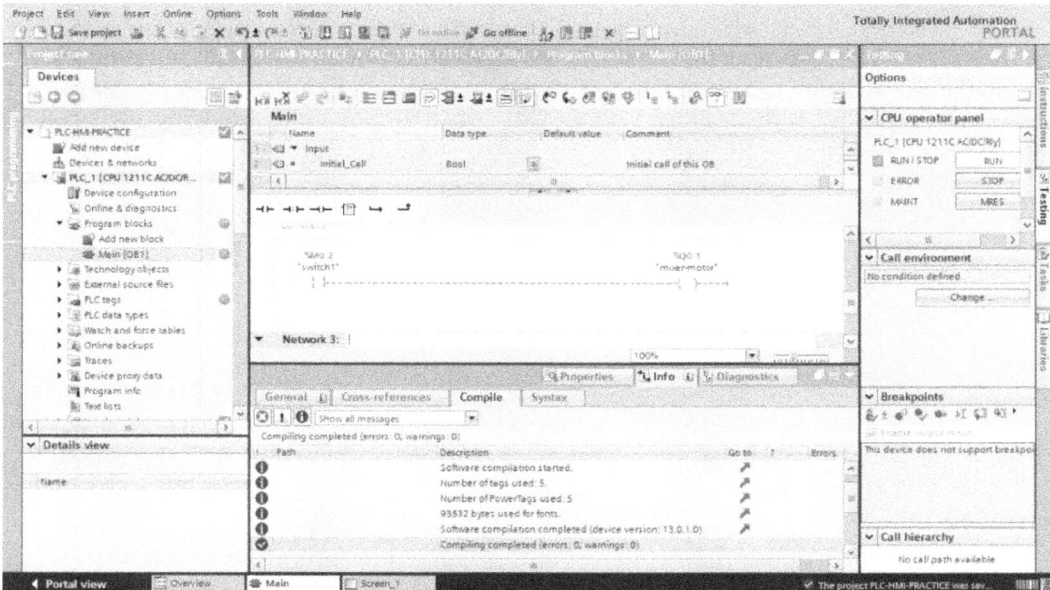

Figure 10.63 – Monitor mode for the ladder logic

20. Now arrange your screen to show both the PLC ladder simulation and the HMI simulator as shown here:

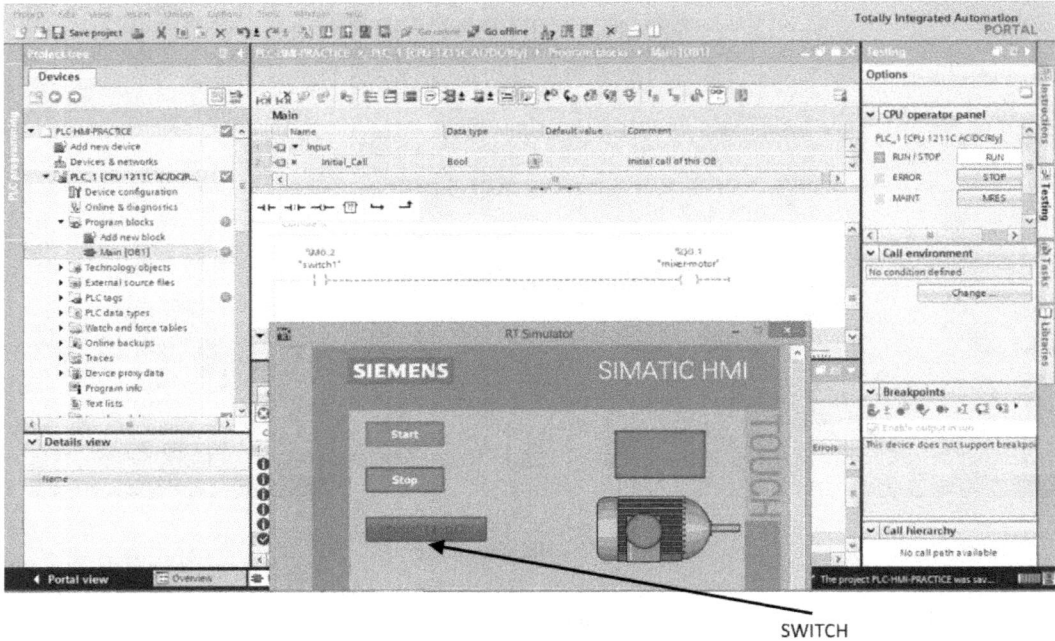

Figure 10.64 – HMI simulator and PLC ladder logic in monitor mode

21. Click the switch and check the result. The switch should show **Switch is ON** while the circle on the motor should turn green, as shown here:

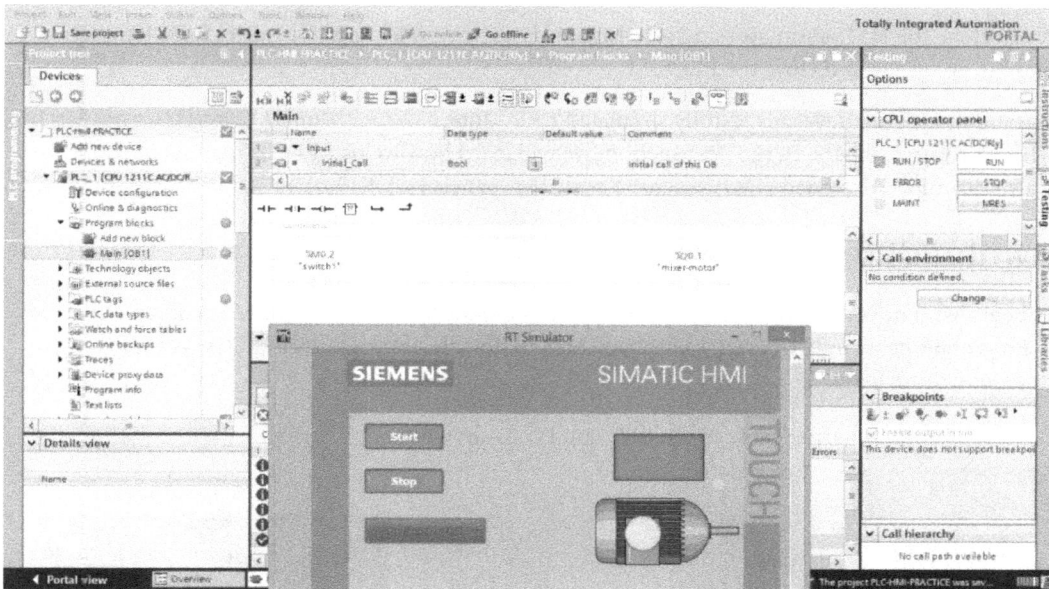

Figure 10.65 – Simulation result when the switch is clicked

22. Now click the switch again. The switch should show **Switch is OFF** while the circle on the motor should turn red, as shown in the following screenshot:

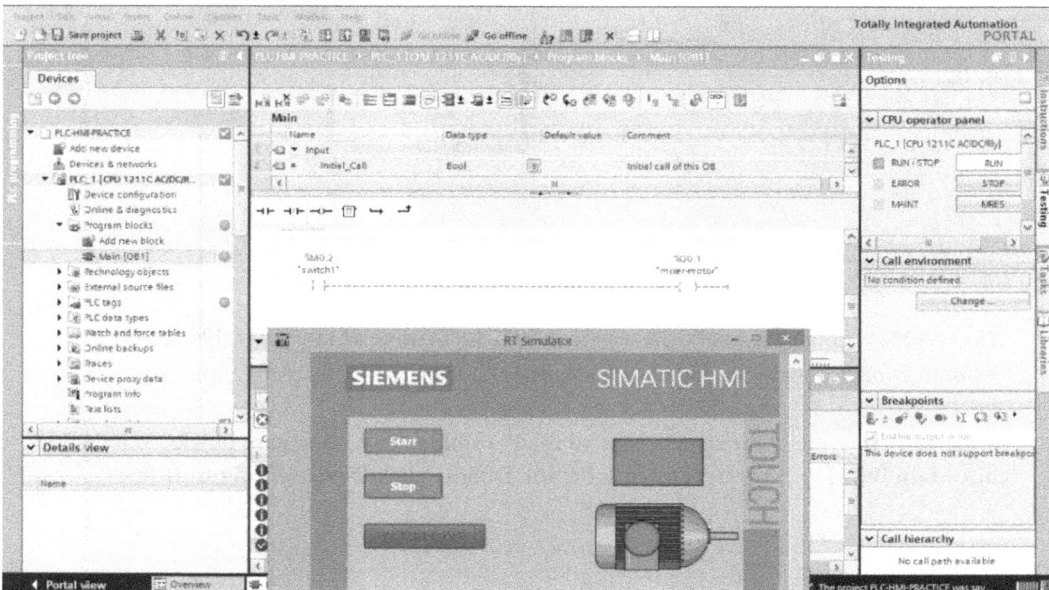

Figure 10.66 – Simulation result when the switch is clicked (second click)

You can close the HMI simulator and also turn monitoring off to stop the ladder logic simulation. Choose to go offline and close the PLCSIM.

Congratulations! You have successfully simulated a PLC interfaced with an HMI. You have now seen how PLCs relate to HMIs. In the next section, we will learn how to load PLC programs to real PLCs and also download designed HMI screens to real HMI panels (KTP 400).

Downloading programs to PLCs and HMIs

Let's begin with downloading a program to a PLC (Siemens S7-1200). In *Chapter 9, Deep Dive into PLC Programming with TIA Portal*, we learned how to download programs to PLCs. We are going to do the same thing here:

1. Connect an Ethernet cable to the PLC and PC as shown:

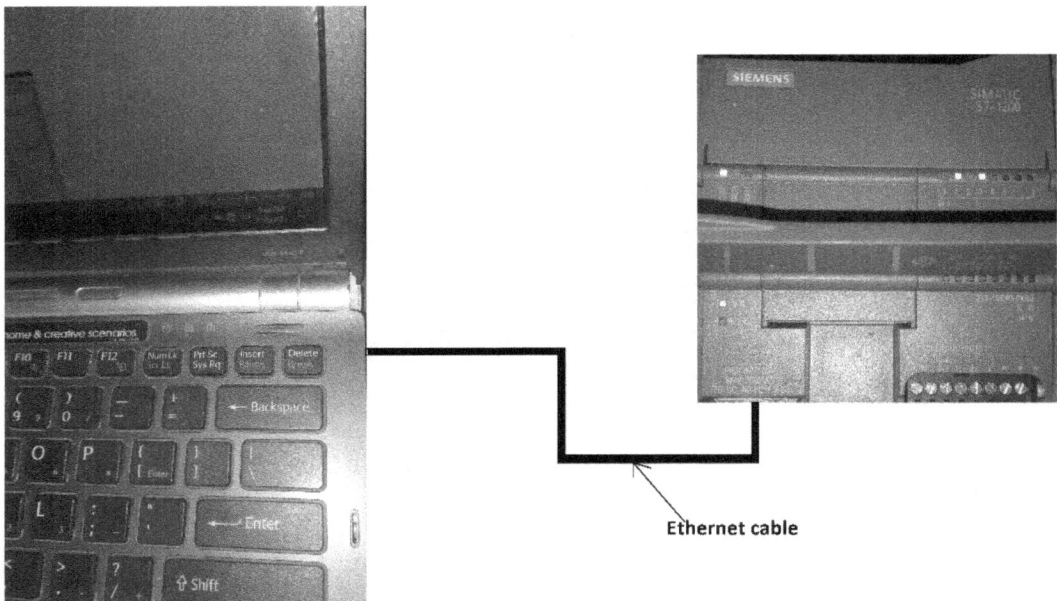

Figure 10.67 – Connection of PC (laptop) to PLC (S7 1200) using an Ethernet cable. (Credit for this image goes to Showlight Technologies LTD. www.showlight.com.ng)

2. At the left-hand side of the screen, expand the PLC folder, go to **Program blocks**, and double-click **Main [0B1]** to get your written program. Double-click **Device configuration**:

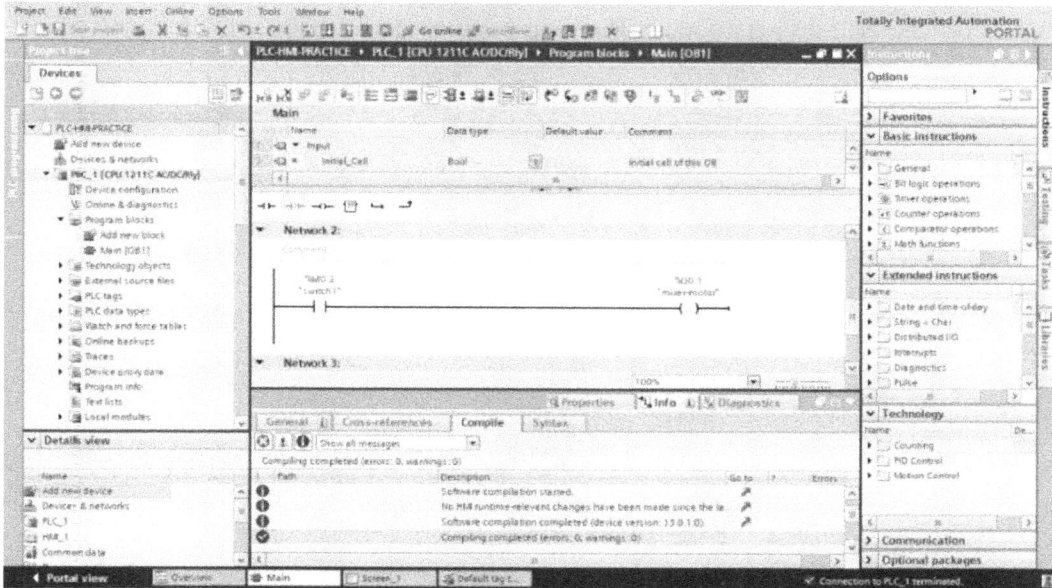

Figure 10.68 – Ladder logic program

3. Click **Properties** | **General** | **PROFINET Interface [X1]**:

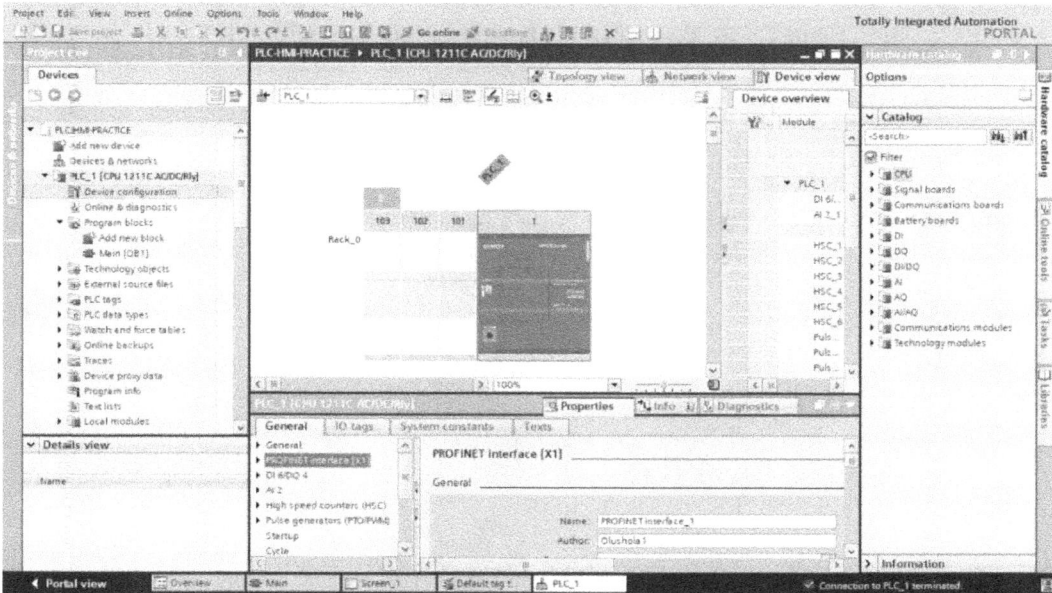

Figure 10.69 – Device configuration (PLC_1)

4.   Scroll down and enter an IP address and an appropriate subnet mask:

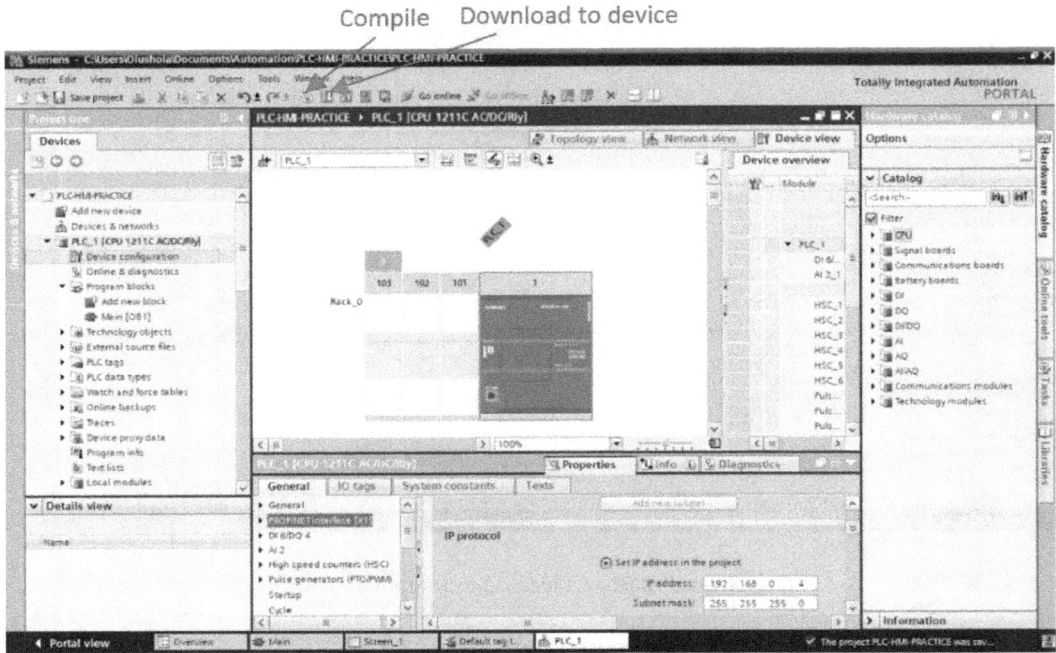

Figure 10.70 – IP address and subnet mask boxes for PLC_1

5.   Click **Save**, click **Compile**, and click **Download to device**:

> **Note**
> Ensure that the PLCSIM is not running before clicking **Download to device**.

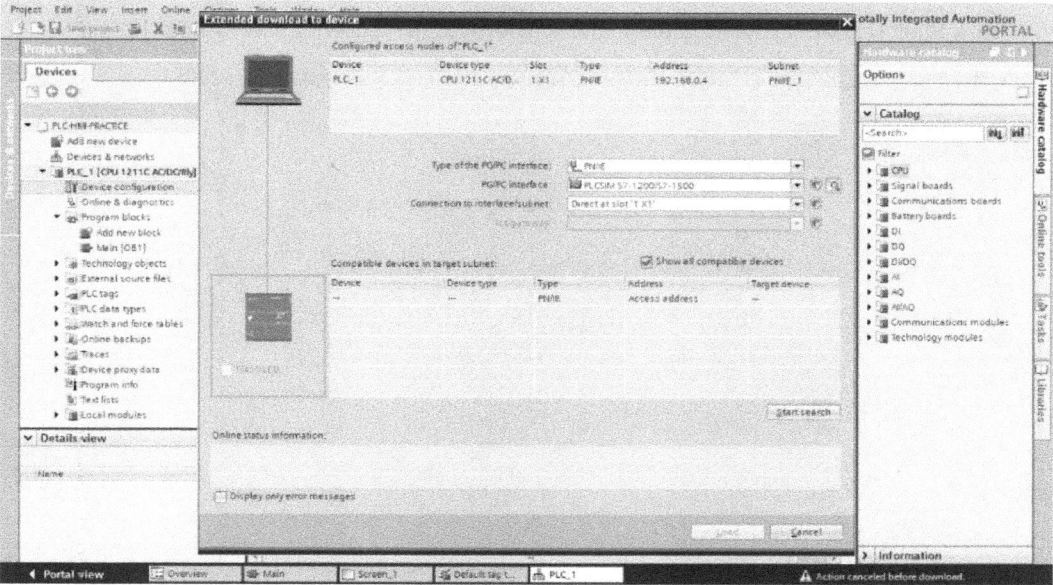

Figure 10.71 – Extended download to device dialog box

6. Choose **PN/IE** as the type of the **PG/PC** interface and choose your Ethernet card as the **PG/PC** interface:

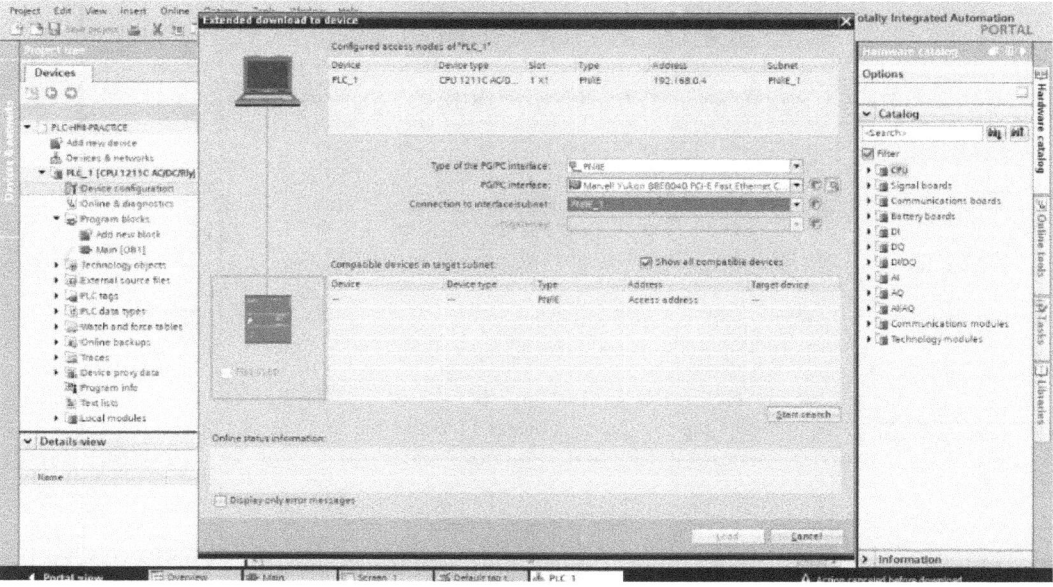

Figure 10.72 – PG/PC interface selected

7.   Click **Start search**. It should find the PLC as shown in the following screenshot:

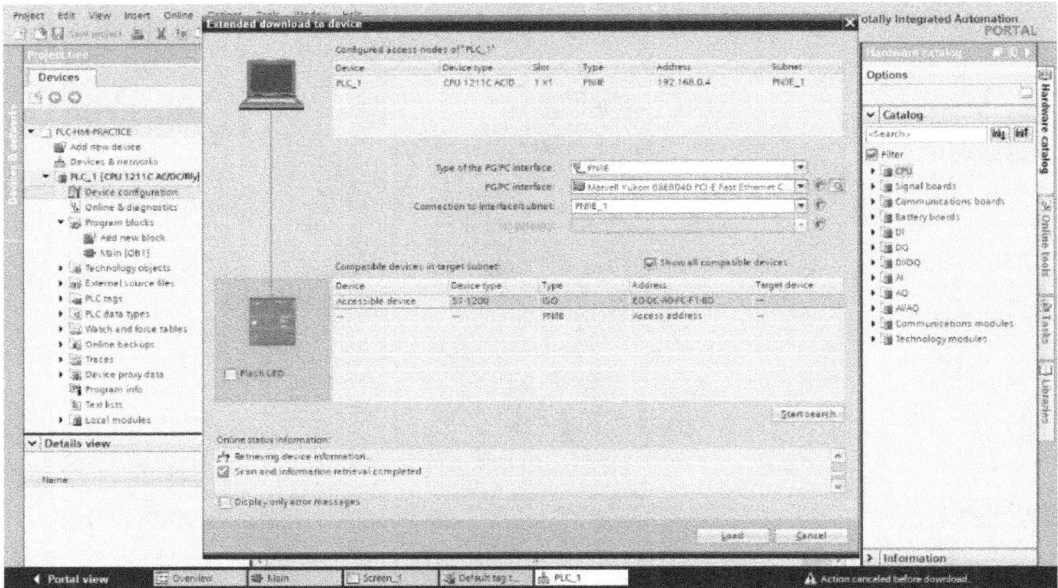

Figure 10.73 – PLC found after search

8.   Select the PLC and click **Load**:

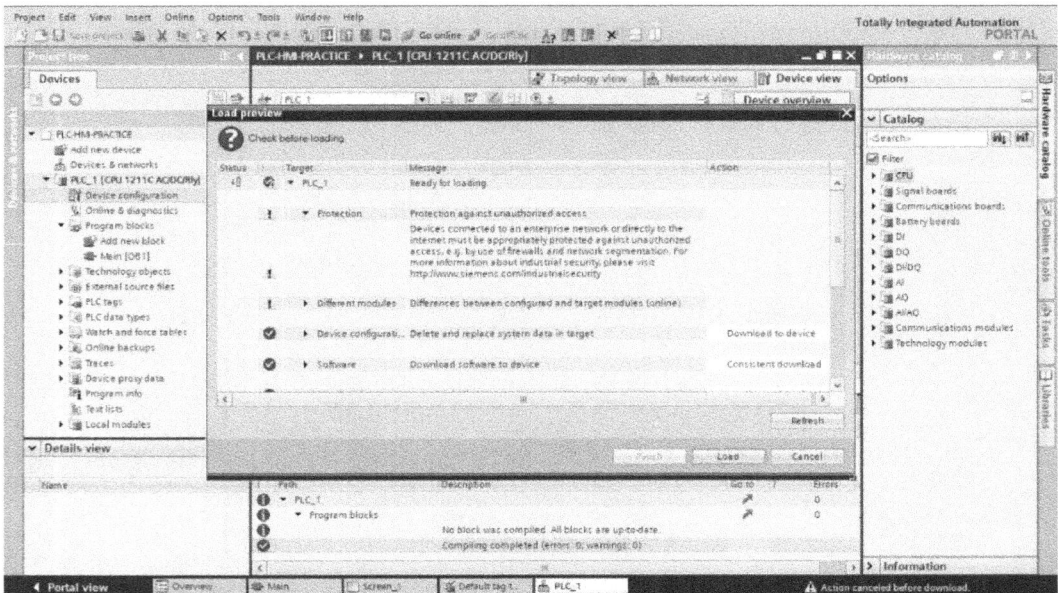

Figure 10.74 – Load preview dialog box

> **Note**
>
> Other screens/messages might appear before the preceding screen; read the messages and respond accordingly. It should take you to the load preview dialog box if everything is OK.

9. Mark **Start all** in the **Load results** dialog box. Click **Finish**:

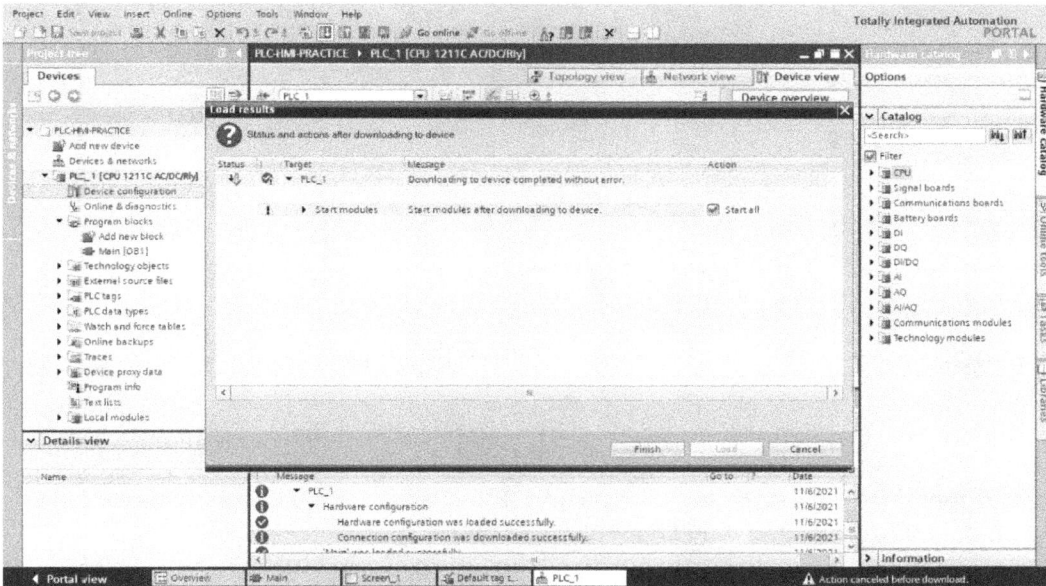

Figure 10.75 – Load result dialog box

10. If loading is successful, you will see a screen similar to the following, showing that the loading was completed in the **Message** section:

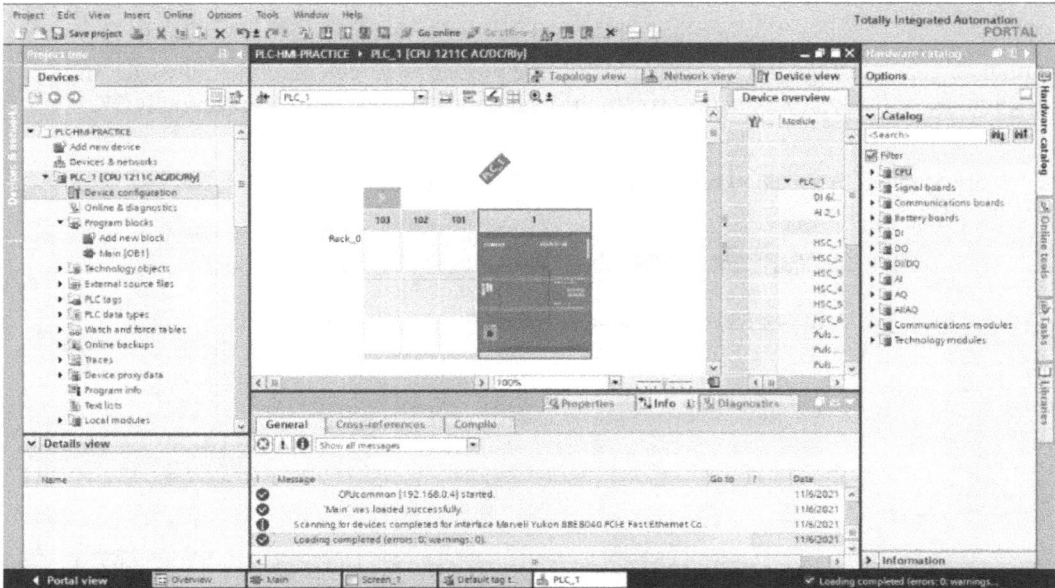

Figure 10.76 – Screen after loading showing loading completed in the Message section

Congratulations, you have successfully assigned an IP address to your PLC and loaded your PLC program and configuration to a real PLC (S71200, CPU 1211C).

You can confirm whether your PC is communicating with your PLC by pinging the IP address assigned to your PLC (192.168.0.4) from Command Prompt as shown in the following screenshot. You should see the reply shown in the screenshot if there is communication:

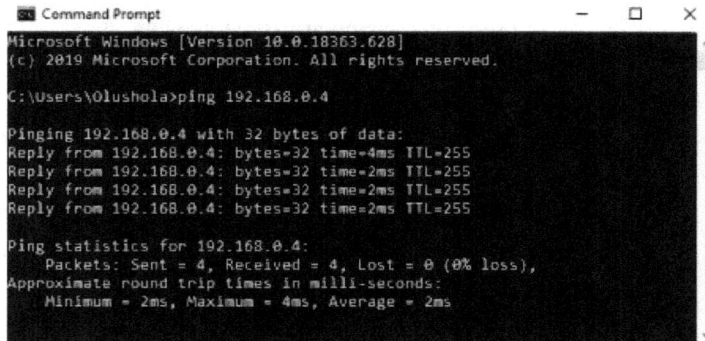

Figure 10.77 – Ping result for 192.168.0.4 (IP address assigned to the PLC)

Let's now proceed to download a program to the HMI:

1. Power on the HMI by connecting the $L+$ and M of a 24V power supply unit to the $L+$ and M of the HMI, as shown:

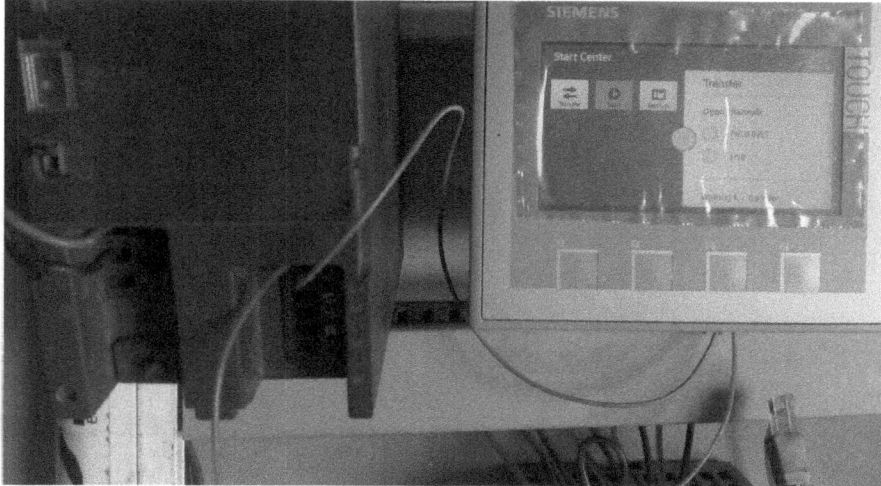

Figure 10.78 – Power supply and HMI connection. (Credit for this image goes to Showlight Technologies Ltd. www.showlight.com.ng)

2. Remove the Ethernet cable from the PLC and connect it to the HMI such that the Ethernet cable will be connected between the PC (laptop) and HMI.

3. On your PC (laptop), expand the HMI folder, double-click **Device configuration**, click on the HMI image, click **Properties**, click **PROFINET Interface**, and specify the IP address and subnet mask for the HMI, as shown in the following screenshot:

> **Note**
> The IP address of the PLC must be different from that of the HMI, while the subnet mask must be the same. Here, 192.168.0.4 is the IP address of the PLC and 192.168.0.5 is the IP address of the HMI. The PLC and HMI have the same subnet mask, 255.255.255.0.

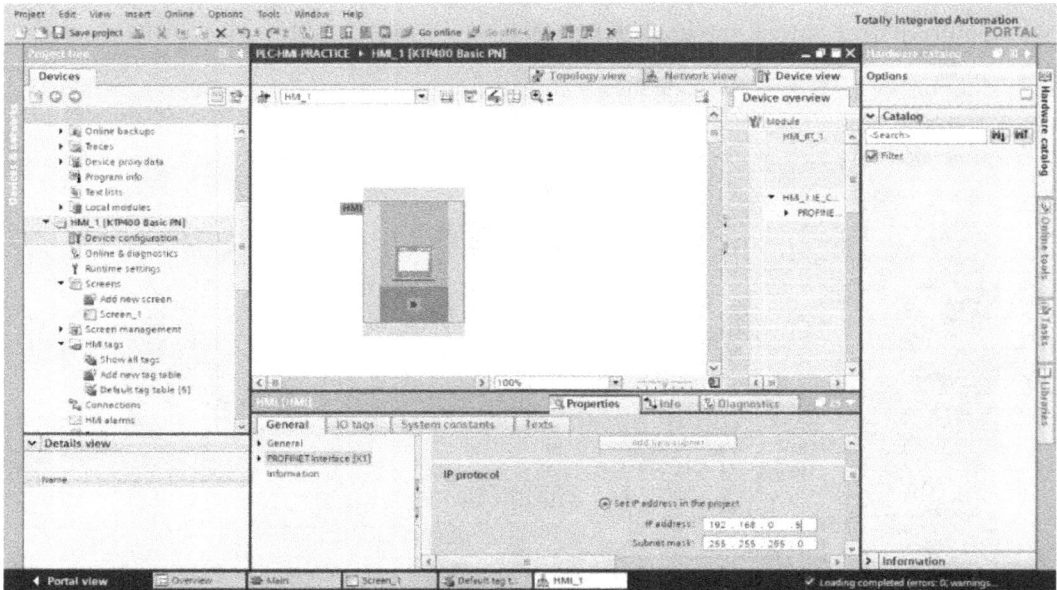

Figure 10.79 – IP address 192.168.0.5 with subnet mask 255.255.255.0 specified for the HMI

4.   Click **Save project** and click **Compile**. You should get the following:

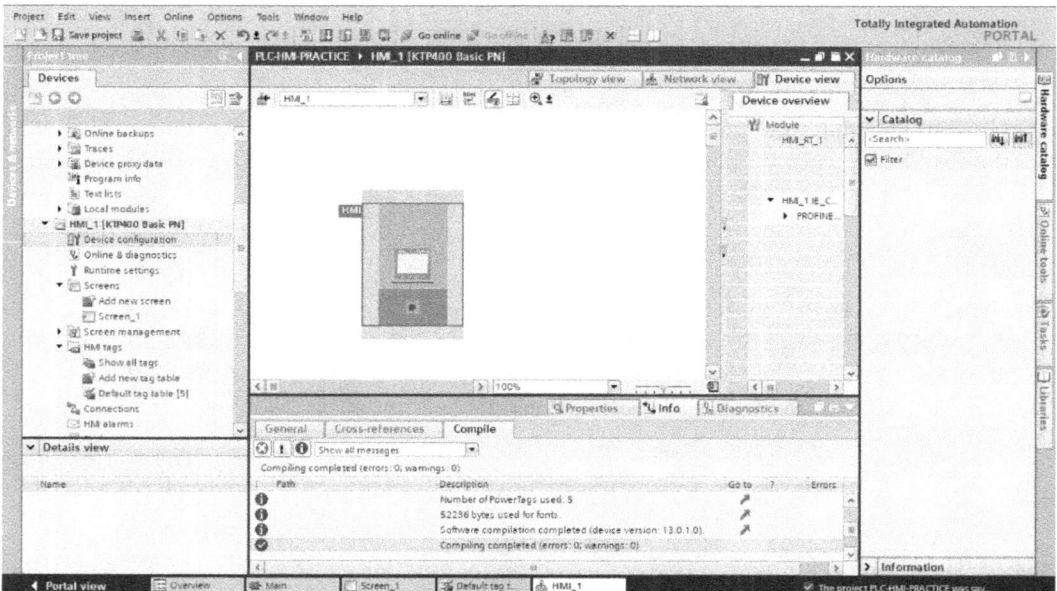

Figure 10.80 – Screen after saving and compiling

5. You can ping the IP address of your HMI in Command Prompt as we did for the PLC. You should get a reply as shown in the following screenshot if there is communication between your HMI and your PC:

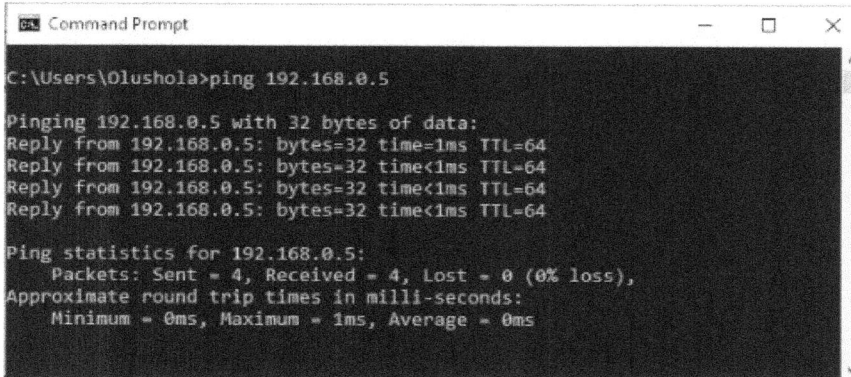

Figure 10.81 – Ping result of 192.168.0.5 (IP address assigned to the HMI)

6. Double-click **screen_1** to view your designed screen as shown:

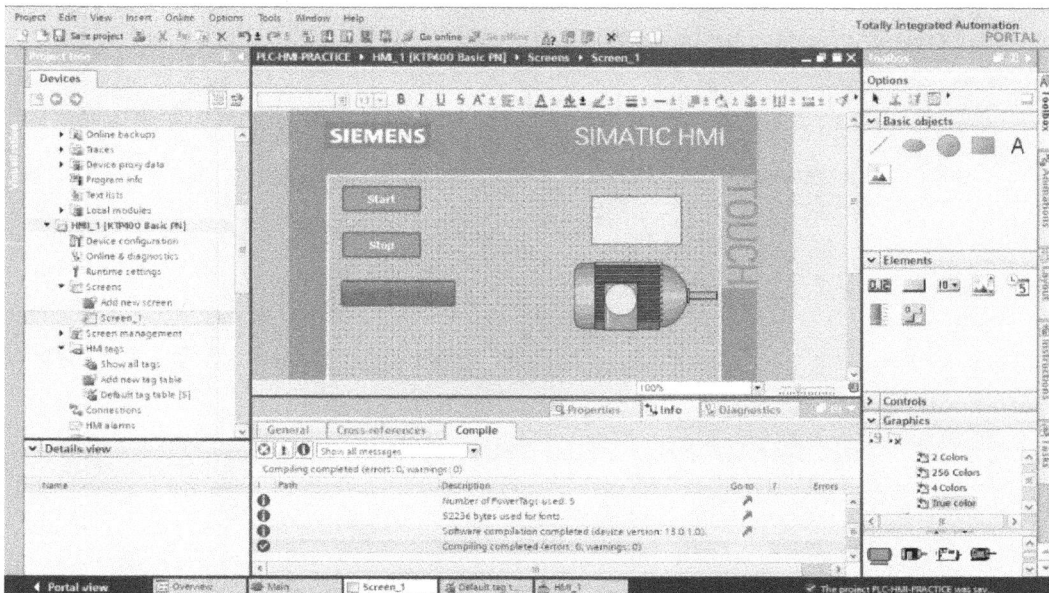

Figure 10.82 – HMI screen developed in the previous section

7. Click **Download to device**. Select **PN/IE** in the **Type of the PG/PC interface** box. Also, select your Ethernet card in the **PG/PC** interface box, click **Start search**, to search for the connected HMI, select the HMI found, and click **Load**:

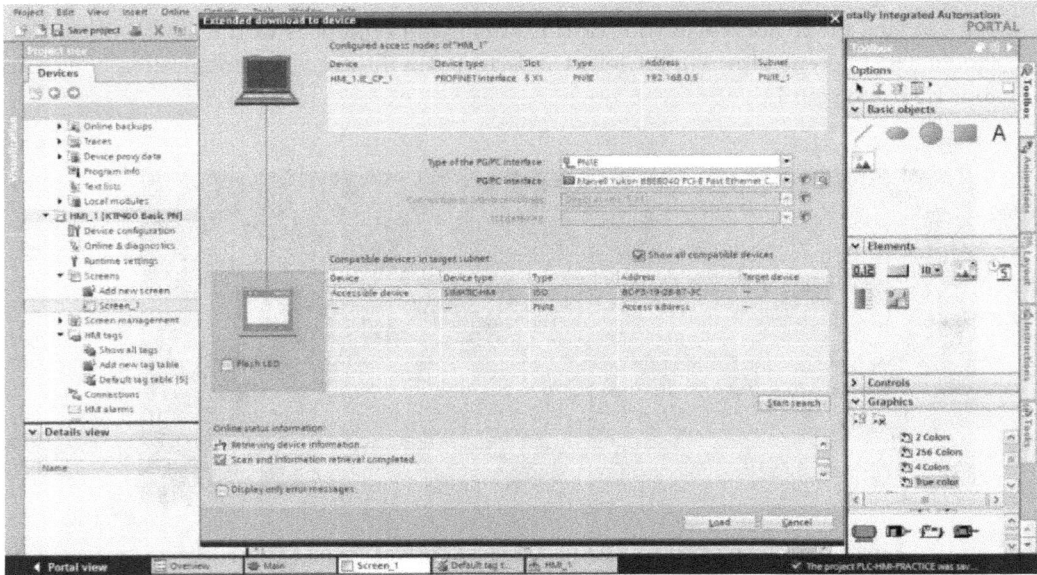

Figure 10.83 – Extended download to device dialog box showing the HMI found

8.   Mark **Overwrite all** and **Fit**, and click **Load**:

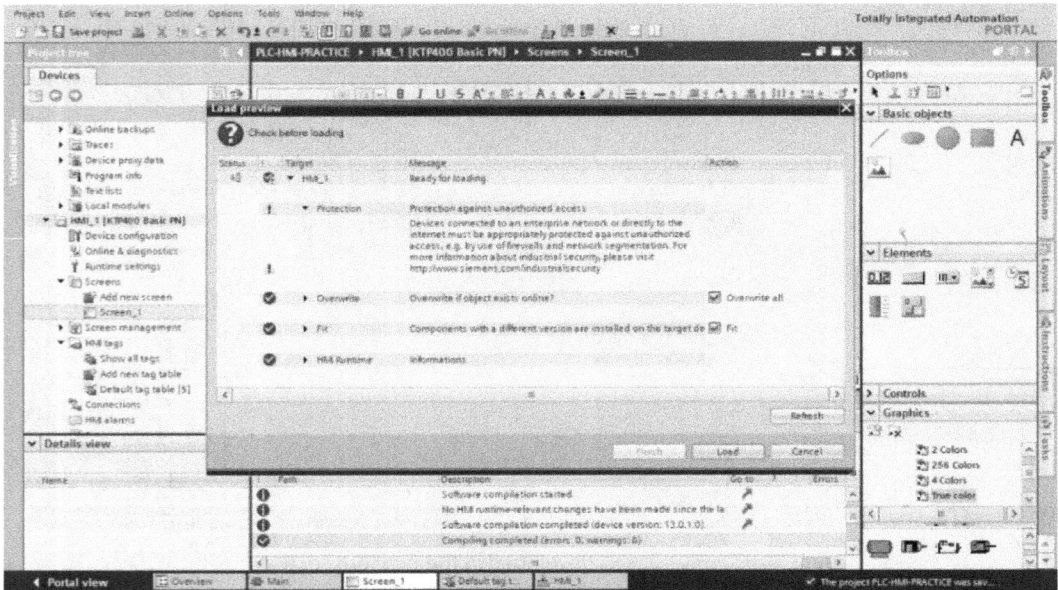

Figure 10.84 – Load preview dialog box

9. You might get the following error:

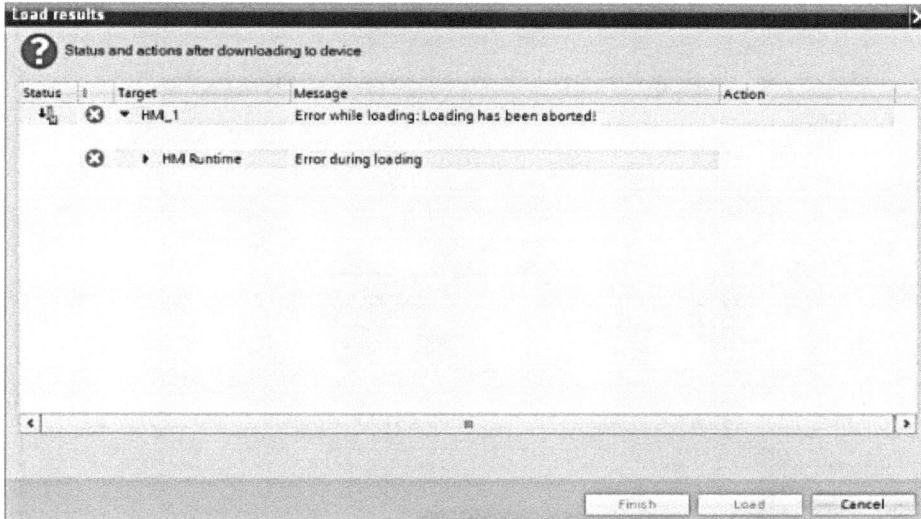

Figure 10.85 – Load result with error

If you see an error similar to the preceding one and the HMI panel displays the error shown in the following image, simply deactivate digital signature on your HMI panel by selecting **Setting | Transfer setting** and select **Off** for digital signature, then click **Transfer** to return to the transfer screen on your HMI panel. You can simply close the **Load results** error on your PC and click **Download to device**, select the HMI found, click **Load**, mark **Overwrite all**, and click **Load**:

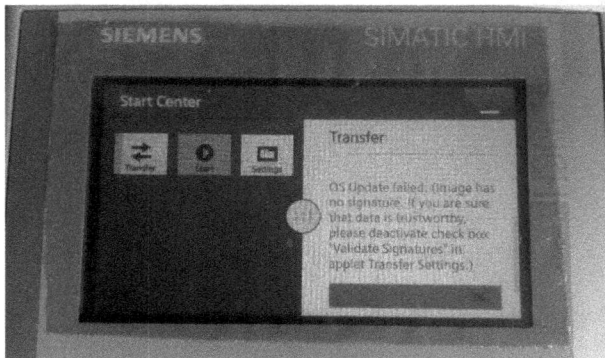

Figure 10.86 – HMI panel showing error during transfer. (Credit for this image goes to Showlight Technologies LTD. www.showlight.com.ng)

10. If everything is OK, it will start loading and display a **Loading completed** message, as shown in the following screenshot:

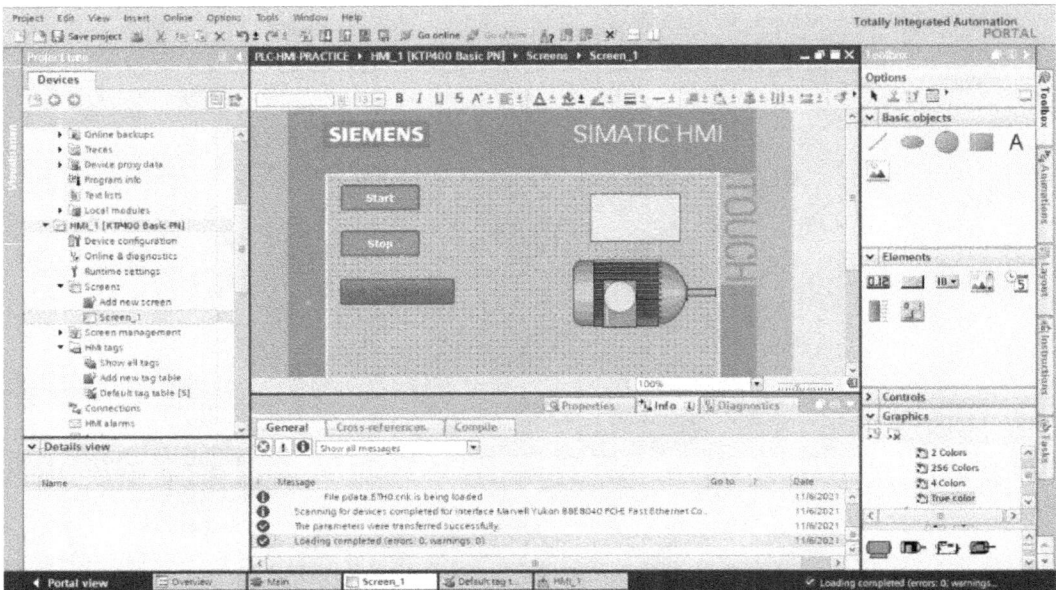

Figure 10.87 – Loading completed

The following figure shows the HMI panel already loaded with the designed screen:

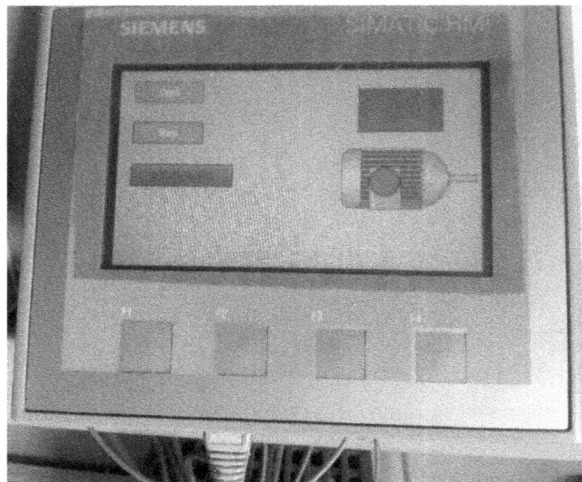

Figure 10.88 – HMI panel (KTP400) with the program loaded. (Credit for this image goes to Showlight Technologies LTD. www.showlight.com.ng)

11. Now remove the Ethernet cable from the PC and connect it to the PLC so that the Ethernet cable connects between the PLC and HMI:

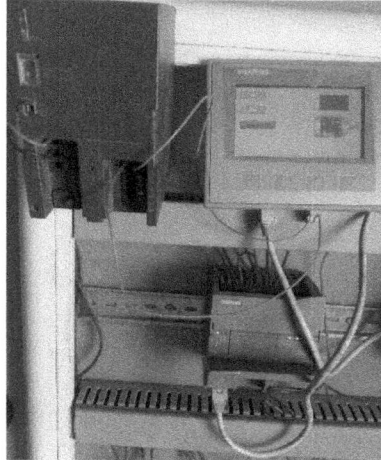

Figure 10.89 – The HMI and PLC connected via an Ethernet cable. (Credit for this image goes to Showlight Technologies LTD. www.showlight.com.ng)

Let's now do a test to check whether everything works fine:

1. Press **Start** on the HMI. The red rectangle should turn to green and *I0.0* on the PLC should turn on, as shown in the following photo:

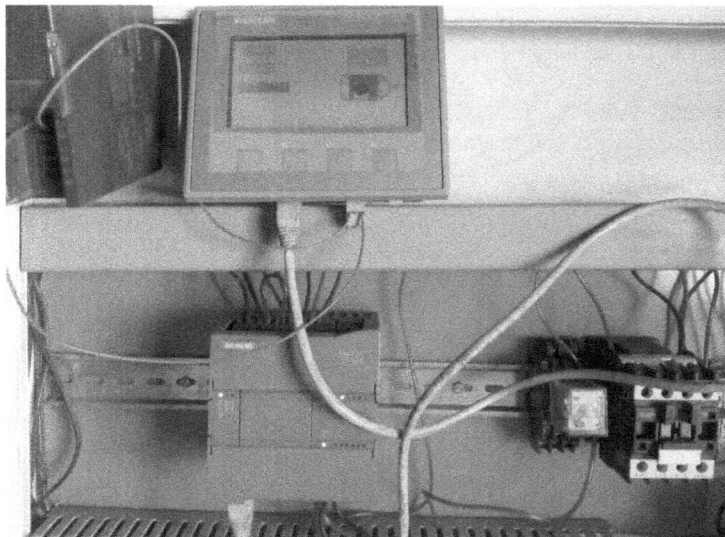

Figure 10.90 – Red rectangle turns green. I0.0 turns on. (Credit for this image goes to Showlight Technologies LTD. www.showlight.com.ng)

2. Press **Stop** on the HMI. The green rectangle should turn to red and *I0.0* on the PLC should turn off.

Good, we have now successfully downloaded a program to our PLC and HMI and have also established communication between them. Now, an operator can press **Start** or **Stop** on the HMI panel to start or stop the motor. The mixer-motor can also be switched on or off using the HMI panel and we can see the status of the motor and mixer-motor, with red indicating that it has stopped and green indicating that it is running.

Summary

Congratulations! You have successfully completed another chapter of this book. Good job! HMIs are used in conjunction with PLCs in various industries to give commands to machines and also to get feedback about a machine's status. The hands-on/practical part of this chapter showed how HMIs, interfaced with PLCs, can be used to control a machine and also get feedback from a machine. This chapter also included simulation. You should be able to simulate your PLC and HMI programs even when you don't have a real PLC and HMI panel to work with. The last section of this chapter explained how to download a program to a real PLC and HMI. Try to get your hands dirty with these tools.

In the next chapter, we will be learning about **Supervisory Control and Data Acquisition** (**SCADA**), which is another interesting topic that an industrial automation engineer needs to know about. Don't miss it!

Questions

The following are questions to test your understanding of this chapter. Ensure you have read and understood the topics in this chapter before attempting the questions:

1. HMI comprises both _____ and _____ that allow the communication between human operators and machines.
2. MMI is an acronym for _____.
3. OIT is an acronym for _____.
4. UI is an acronym for _____.

11
Exploring Supervisory Control And Data Acquisition (SCADA)

In the previous chapter, we learned how to monitor and control a machine through a **Human-Machine Interface (HMI)**. An HMI is always close to the machine or **Programmable Logic Controller (PLC)** – that is, an HMI is always local to the machine, and it is basically meant to control a single machine/PLC. Here in this chapter, we will be learning about **Supervisory Control And Data Acquisition (SCADA)**, which also provides a means of monitoring and controlling along with other functionalities, such as alarms, trending, and logging. However, the screen or PC for monitoring, controlling, or other functionalities can be in a remote place – that is, far away from the machine – and we can have one SCADA for more than one machine at different remote locations, unlike HMI. SCADA is an interesting and advanced aspect of industrial automation.

We are going to cover the following main topics in this chapter:

- Introducing SCADA
- Understanding the functions of SCADA
- Exploring the applications of SCADA
- Overviewing SCADA hardware
- Overviewing SCADA software
- Downloading and installing mySCADA
- Interfacing SCADA with an S7-1200 PLC using mySCADA software (a practical project)

Introducing SCADA

SCADA is a system consisting of various hardware and software that work together to monitor and control machines in an industrial process.

Supervisory control refers to a high level of overall control of several individual PLCs or multiple control loops.

Data acquisition refers to the gathering of information and data from a process that is analyzed in real time. Real-time data collection helps to reduce overhead costs, monitor an entire process, and increase efficiency.

SCADA system helps to collect, analyze, and visualize data from equipment or a machine and also provides a means of controlling the equipment or machine from a central location, which can be local or remote (that is, near the machine or far away from the machine).

SCADA is typically used to monitor and control geographically dispersed machines or devices. The following figure shows a simple SCADA system:

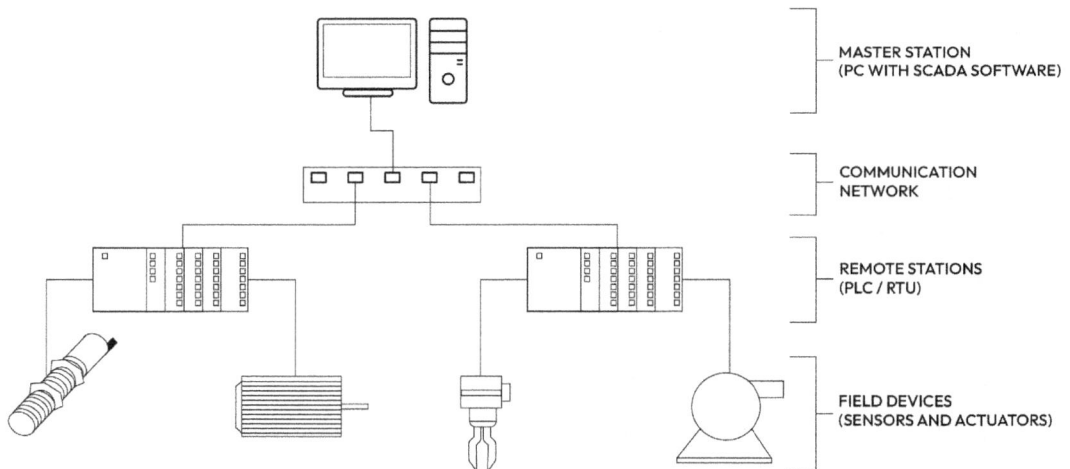

Figure 11.1 – A simple SCADA system showing SCADA hardware

One production machine can be located in one geographical area (location A) while the other machine can be in a different geographical area (location B). If a SCADA system is in place for the two production machines, we can monitor the status of the two machines from a central location and also control either of them. SCADA basically acquires data about a system or machine in order to control it. The system or machine you want to monitor and control can be a water treatment plant, power generation, a power distribution system, or anything else.

Understanding the functions of SCADA

The meaning of SCADA already implies its major functions. The following goes into detail about what a SCADA system does:

- It controls industrial processes either locally or remotely. Switching on or off a machine, increasing or decreasing setpoints, and so on are control functions that a SCADA system can be programmed to perform.

- It collects data (acquires real-time data) from a field or machine, which provides a status indicator to the operator or process engineer for monitoring.

- It enables a process engineer to have a bird's-eye view of an entire plant, which can consist of various machines/devices at different geographical locations.

- It generates alarms when abnormal conditions occur within a process. It can trigger alarms that can be in the form of audible sounds, lights, emails, SMS, or other forms when something is wrong in a process.

- It records events (data logging).

Next, let's take a look at the applications of SCADA.

Exploring the applications of SCADA

The following are a few applications of SCADA:

- **Water treatment plant**: SCADA helps in monitoring reservoir levels, pump operations, water flow, and lots more, and also helps in controlling pumps and other equipment in a plant.

- **Power generation and distribution**: A SCADA system helps to monitor and control power generation and distribution networks. SCADA can monitor various equipment in substations and also send a signal to control equipment in a remote location to switch ON or OFF a breaker.

- **Manufacturing**: SCADA controls industrial automation equipment used in the manufacturing of various goods. It also monitors manufacturing processes.

- **Building and facility management**: SCADA is used by facility managers to monitor and control lighting, **Heating Ventilation and Air Conditioning (HVAC)**, and so on.

- **Traffic light**: SCADA is used to monitor traffic lights such as failure detection, incorrect signals, and lots more. It is also used to control traffic flow.

In the next section, we will discuss the hardware components of SCADA.

Overviewing the SCADA hardware

The hardware components of a SCADA system include the following:

- **Field devices**: These are sensors, transmitters, and actuators that are directly connected to the machine or plant and generate a digital or analog signal for monitoring. Sensors or transmitters are for monitoring machine status or parameters, while actuators are for carrying out control action on equipment or a machine.

- **Remote station**: This can be a PLC or a **Remote Terminal Unit** (**RTU**). It is installed at a remote site where the machine or equipment to be monitored or controlled is located. Field devices (sensors, transmitters, and actuators) are connected to remote stations (a PLC or an RTU) to allow monitoring and control from the host computer at a remote site via a network. It gathers data through the sensors connected to it and transfers them to the host computer via a network. It also transfers an electrical control signal from the host computer to the actuators connected to it in order to perform the required control action.

- **Communication network**: The main purpose of a communication network in SCADA is to connect the remote stations (a PLC or an RTU) to the host computer (master station). It allows the flow of information in a SCADA system. The communication medium for a SCADA system can either be wired or wireless. A common wired medium is Ethernet. Various network topologies (star, bus, ring, and so on) are available for use in a SCADA system, and various network protocols (Profinet, Modbus, Profibus, and so on) are also available for use. More details will be provided on network topologies and network protocols in *Chapter 13, Industrial Network and Communication Protocols Fundamentals*.

- **Master station**: This can also be referred to as a supervisory station. It is the master computer that runs SCADA software that provides the graphical presentation of the system. It runs the HMI application that provides the graphical picture of the switches, sensors, transmitters, pumps, and so on for monitoring and control purposes. Alarms can be set up to activate at different predefined values. The entire control system at various plant sites can be monitored and controlled through the GUI on the master station. A single computer can be configured as a master station or networked to workstations or multiple servers. In a SCADA system that uses multiple servers as a master station, one of the servers can be dedicated to an **Alarm Management System** (**AMS**), and this can be referred to as an AMS console. An AMS console consists of AMS software that provides an overview of alarms from various areas in the plant, makes a quick response possible when there is an abnormal condition in the plant, and so on.

Basically, the master station collects information from the remote station (a PLC or an RTU), represents them on a GUI for easy interpretation, generates an alarm, and also provides an interface for various control actions that can be carried out at remote sites.

Figure 11.1 illustrated a simple SCADA system, showing the various SCADA hardware components that are explained in this section. Next, we will look at the different software that can be used to create the GUI for SCADA.

SCADA software

This is software that can be used to create the interface (GUI) required to monitor and control the industrial equipment and perform all necessary configurations. It is usually installed on a central host computer, where monitoring and control operations will be carried out.

Common SCADA software includes the following:

- WinCC by Siemens
- FactoryTalk View by Rockwell Automation
- InTouch by Wonderware
- iFix by General Electric
- Citect SCADA by Schneider
- mySCADA

We are going to use one of the preceding SCADA software (mySCADA) in this chapter to learn how SCADA works, and to understand its basic configuration and wiring for a simple system.

mySCADA is powerful SCADA software that runs on Microsoft Windows, macOS, and Linux. It supports various PLCs, including S7-1200, S7-1500, S7-300, S7-400, ControlLogix, CompactLogix, Micrologix 1200, Micrologix 1400, Micrologix 1500, SLC 500, PLC 5, Omron PLCs, and Melsec-Q.

It basically consists of two pieces of software (myDESIGNER and myPRO). myDESIGNER is a development platform that can be used to create screen/visualization and perform all the necessary configurations that suit your application, while myPRO allows visualization. You can view your SCADA to control and monitor your equipment through myPRO. It runs on common browsers, including Microsoft Internet Explorer IE (version 9 and above), Microsoft Edge, Chrome, Firefox, and Safari.

In the next section, we will see how to download and install the mySCADA software on your system.

Downloading and installing mySCADA software

Let's start by learning how to download the software, as it will be used in the SCADA project in the *Interfacing SCADA with an S7-1200 PLC using mySCADA software* section of this book.

Downloading mySCADA

We will perform the following steps to download the two pieces of software (myDESIGNER and myPRO) to our PC:

1. Visit www.myscada.org and click on **RESOURCES**:

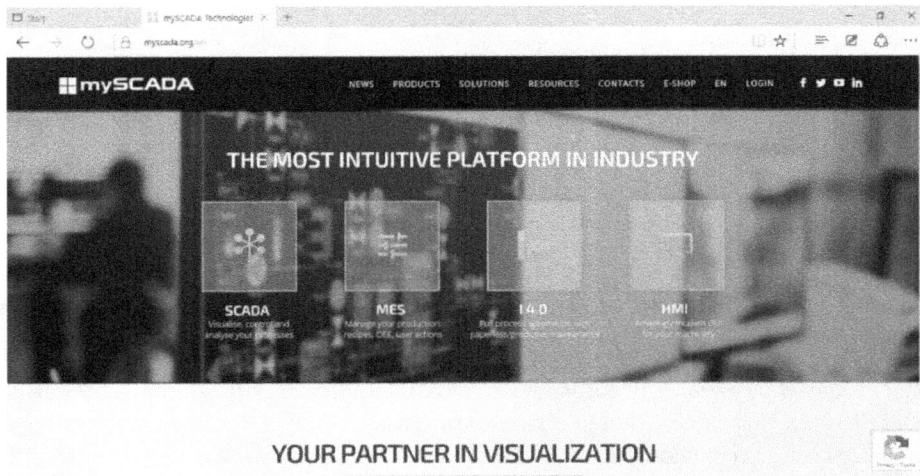

Figure 11.2 – The home page of mySCADA.org

2. Click on **Downloads**:

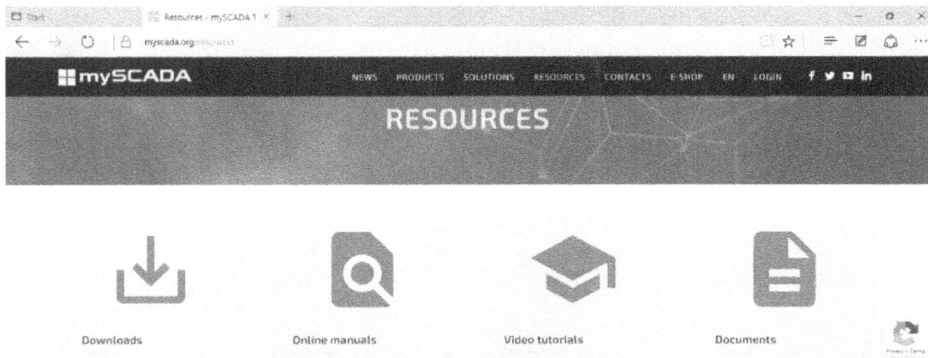

Figure 11.3 – The RESOURCES page

3. Click on **register or log in here**, fill in the necessary details, and click **Register**:

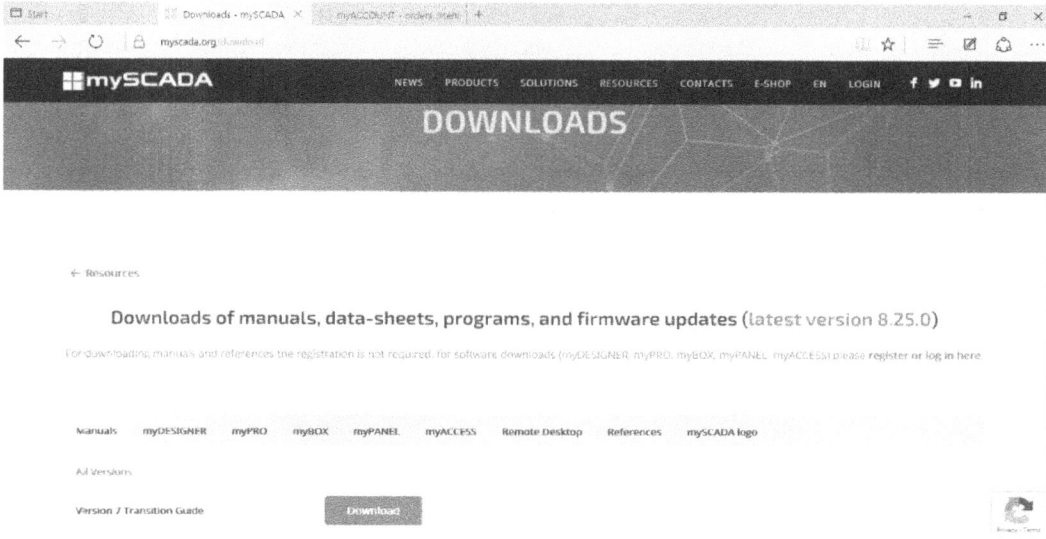

Figure 11.4 – The DOWNLOADS page

4. You should have the following screen if everything is okay:

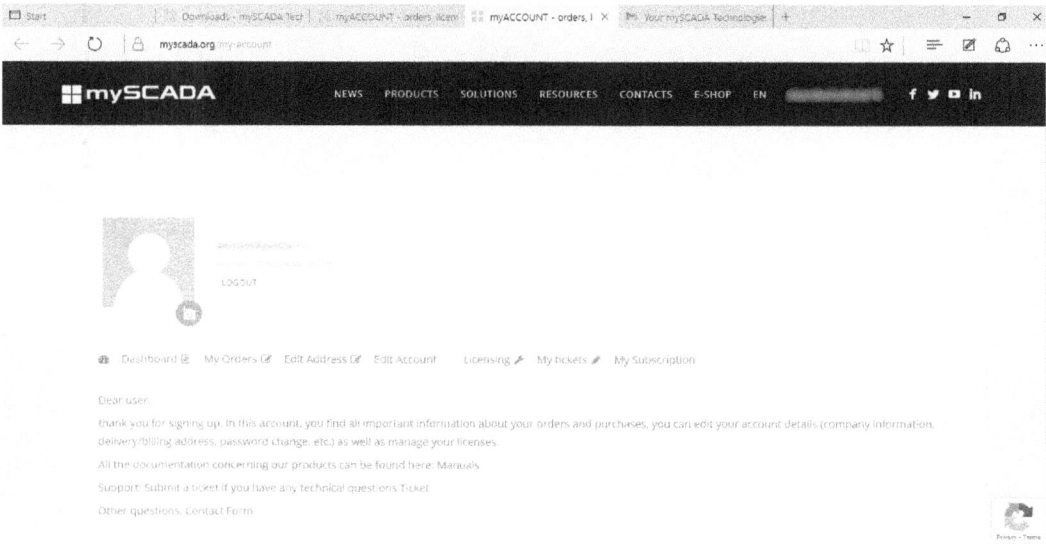

Figure 11.5 – The welcome message

5.   Check your email for a message from mySCADA, which contains the link to access your account or set a new password..

6.   You can log in anytime to download the necessary mySCADA software. Click on **Resources | Downloads** to reach the page shown in the following screenshot:

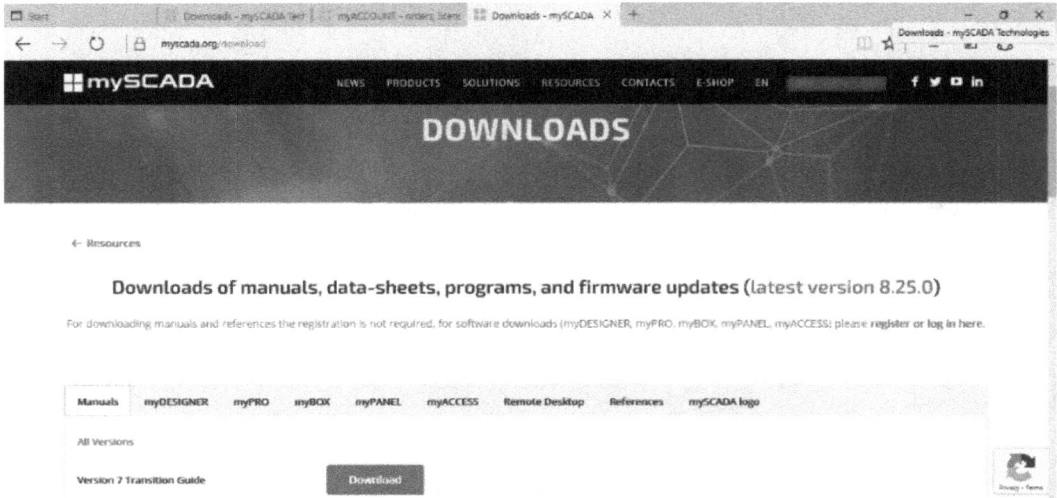

Figure 11.6 – The DOWNLOADS page after login

7.   Click on **myDESIGNER | Download** (for Windows):

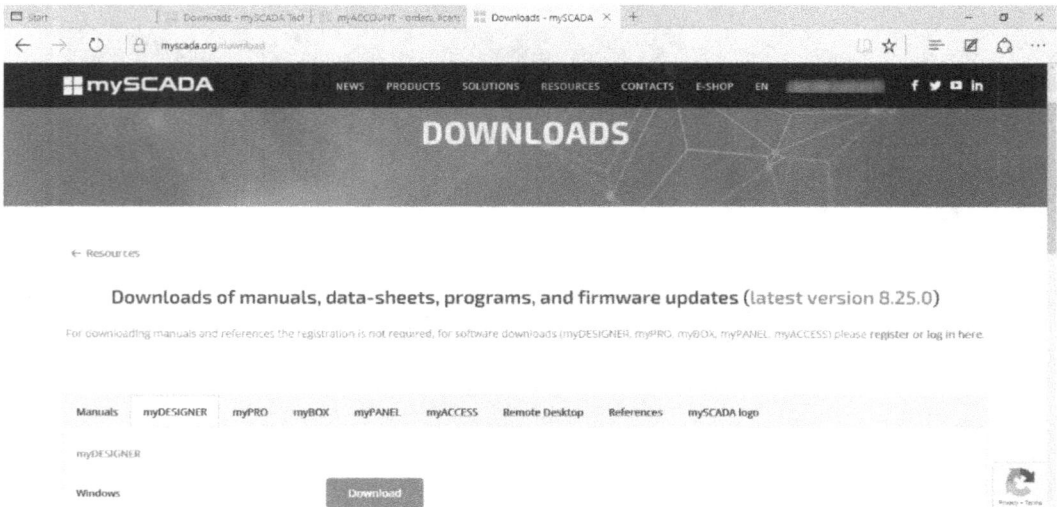

Figure 11.7 – The myDESIGNER DOWNLOADS page

8. Click on **myPRO | Download** (for Windows):

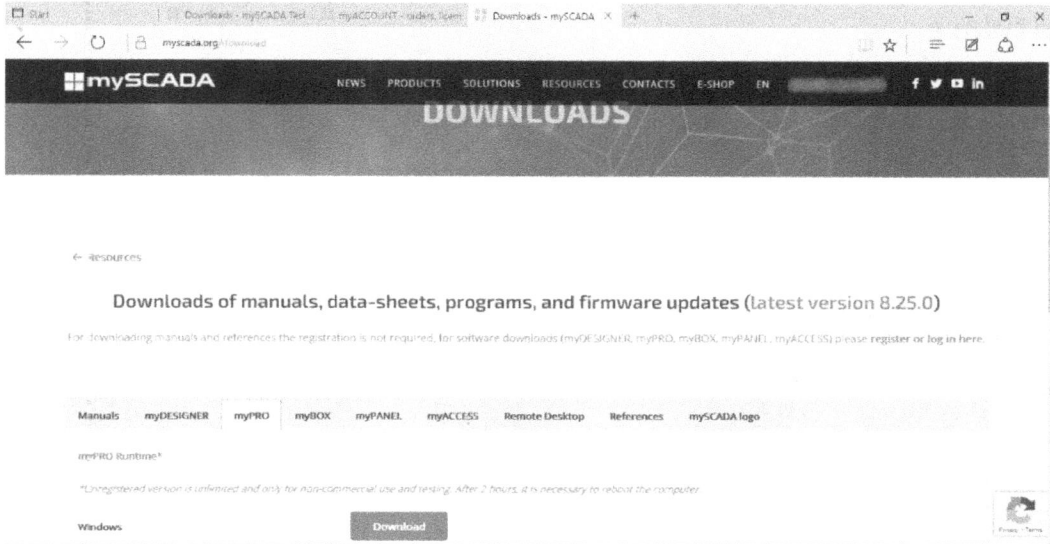

Figure 11.8 – The myPRO DOWNLOADS page

We now have the two setup files (myDESIGNER and myPRO) downloaded to our PC.

Installing mySCADA software (myDESIGNER and myPRO)

Perform the following steps to install myDESIGNER and myPRO on your PC:

1. Double-click the setup for myDESIGNER and follow the instructions on the screen.
2. Unmark **Launch myDESIGNER** and click on **Finish**.
3. Next, double-click the setup file for myPRO and follow the instructions on your screen.
4. Restart your PC when prompted.

We have successfully learned how to download and install mySCADA software. Follow the steps to have mySCADA on your laptop or PC, as it will be required in the next section.

Interfacing SCADA with an S7-1200 PLC using mySCADA software

We will learn how to configure a simple SCADA in this practical project. You will have a basic understanding of how to monitor and control a machine using SCADA.

We will be using Siemens S7-1200 as our remote station. Hence, TIA Portal will be used for the configuration and programming of the PLC (S7-1200).

Let's get started:

1. Start TIA Portal, click on **Create new project**, and type the project name (for example, SCADA-PRACTICE), as shown in the following screenshot. Click **Create**:

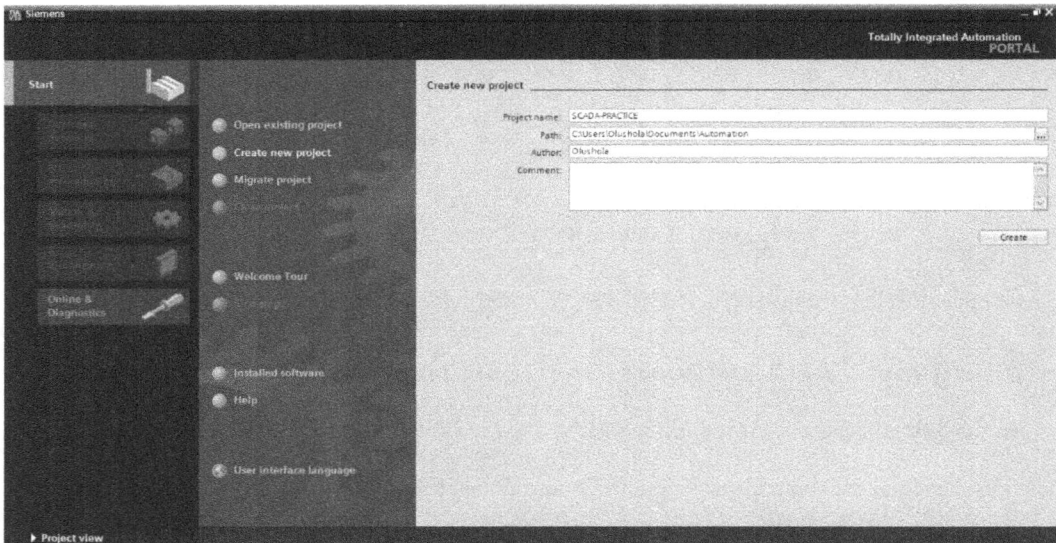

Figure 11.9 – Creating a new project

2. Click on **Configure a device**:

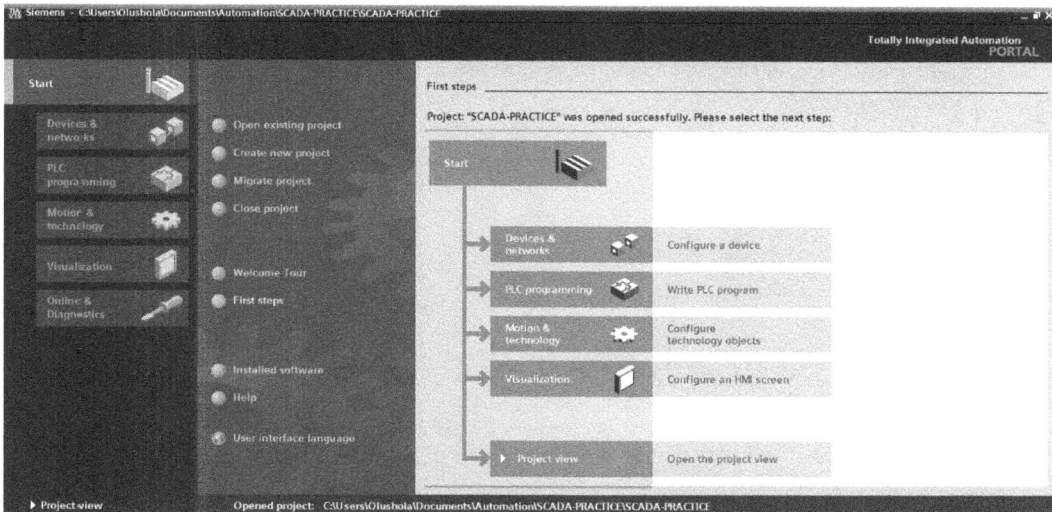

Figure 11.10 – A new project created in the PORTAL view

3. Click on **Add new device** and select **Controllers** from the device list. Expand SIMATIC S7-1200|CPU|CPU 1211C AC/DC/Rly and select the order number of your PLC – for example, 6ES7 211-1BE40-0XB0 – depending on the PLC you intend to use:

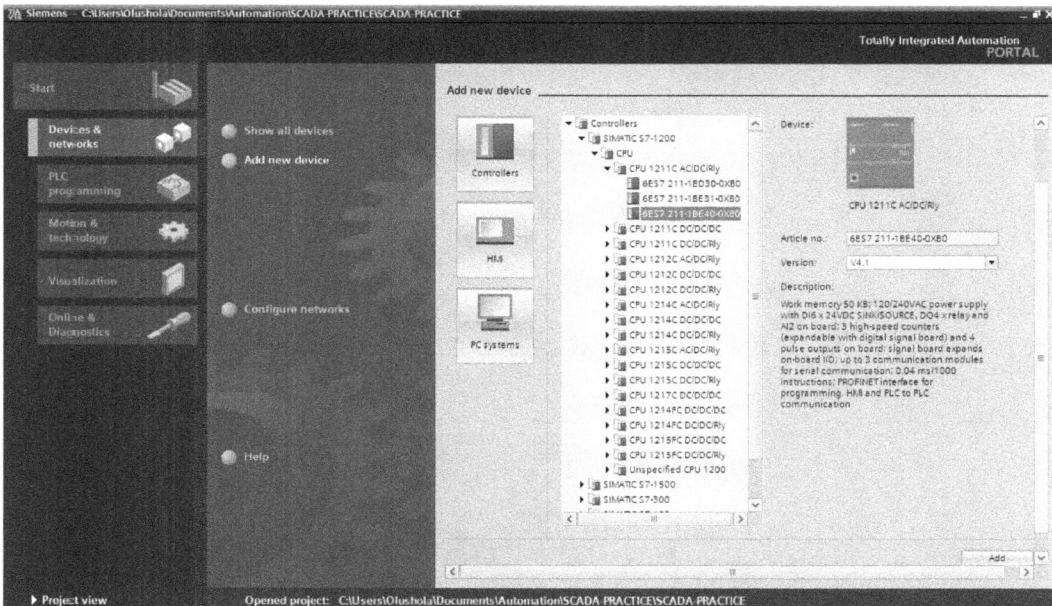

Figure 11.11 – Selecting your controller (PLC)

4. Click on **Add**, and you should see the following screen:

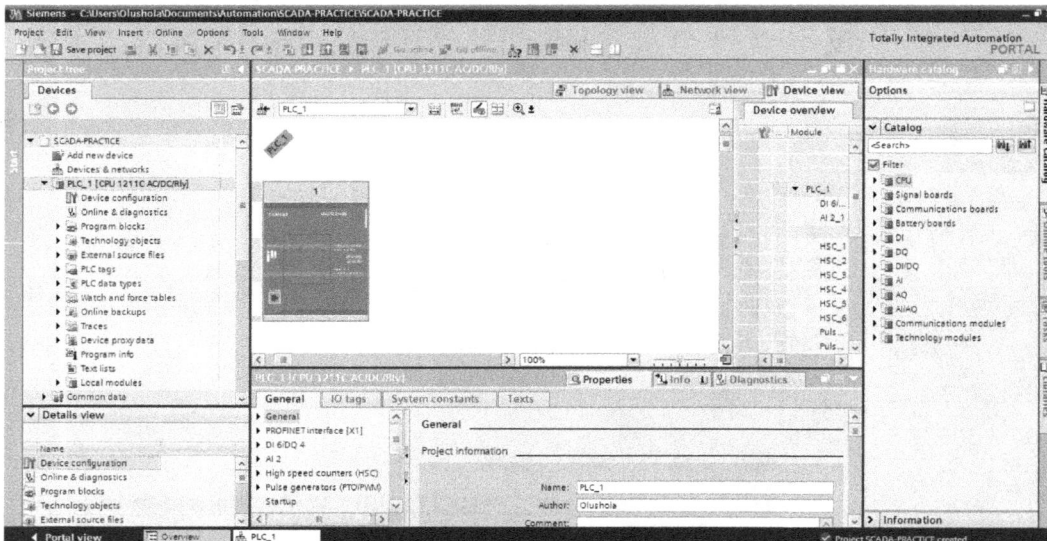

Figure 11.12 – The PLC added to your project

5. Expand **Program blocks** and double-click on **Main [OB1]** to have an environment to write the program, as shown in the following screenshot:

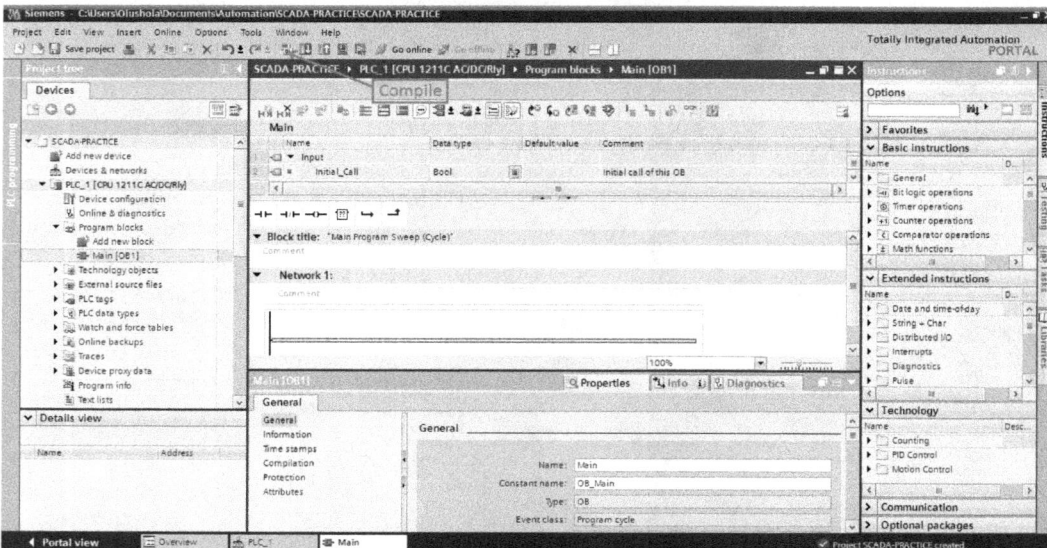

Figure 11.13 – A ladder logic programming environment

6. Write your program and rename all tags, as shown in the following screenshot (*Figure 11.14*). We are using a simple program to understand the operation of SCADA. Click on the **Compile** icon (the icon pointed to in *Figure 11.13*):

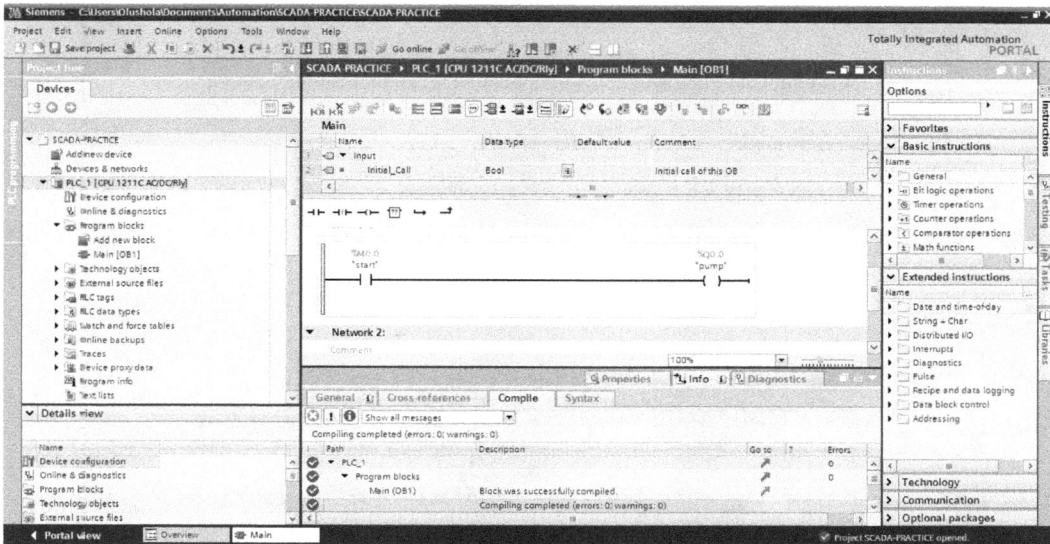

Figure 11.14 – The program written and all tags renamed

7. Right-click on the PLC on the left-hand side (the project tree). Click on **Properties**:

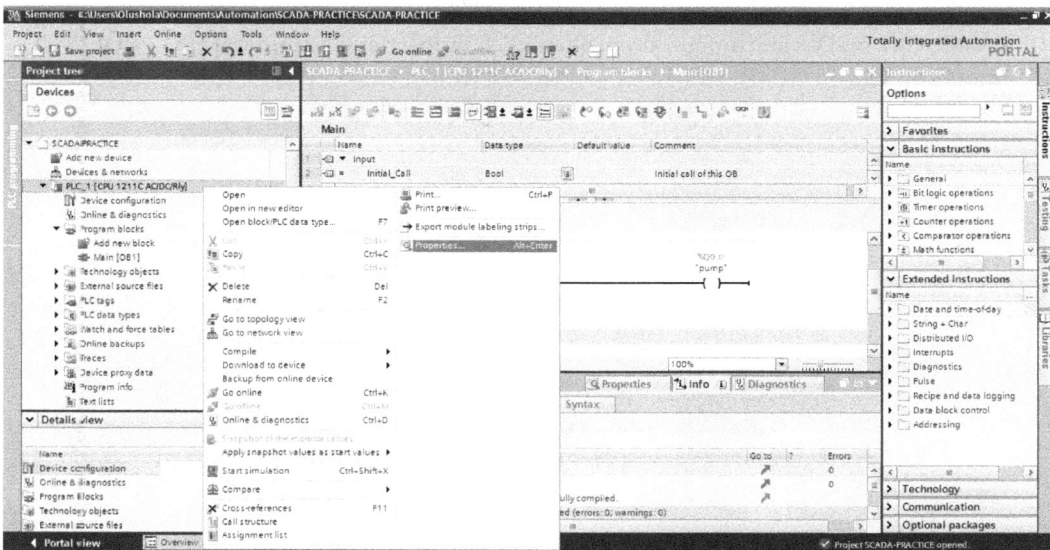

Figure 11.15 – The Properties option

8. Select **Protection**, check the **Permit access with PUT/GET communication from remote partner (PLC, HMI, OPC, …)** box, and click on **OK**:

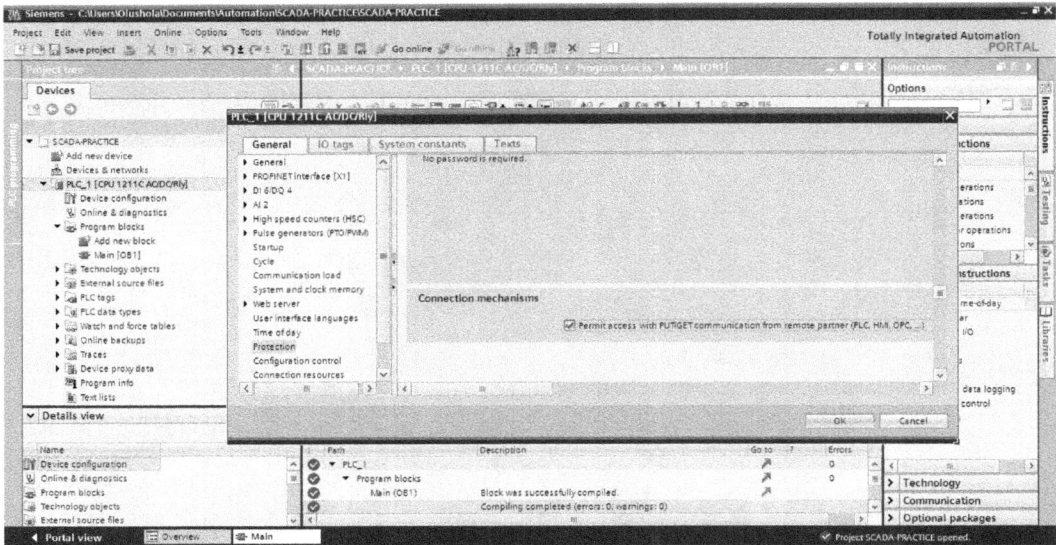

Figure 11.16 – The property dialog box

Let's now proceed to load the program onto a PLC.

9. Connect power to the PLC and use an Ethernet cable to connect the PLC to your PC.

10. Double-click on **Device configuration**. Then, click **Properties** | the **General** tab | **PROFINET interface [X1]**:

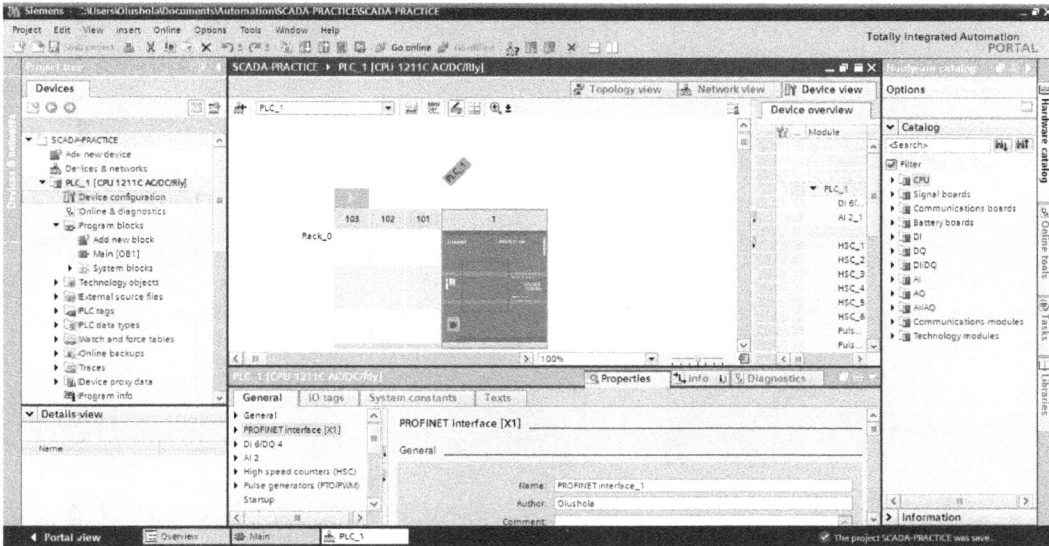

Figure 11.17 – The Device configuration screen

11. Scroll down and type an IP address and subnet mask, and click on **PLC_1** on the image to select the PLC, as shown in the following screenshot:

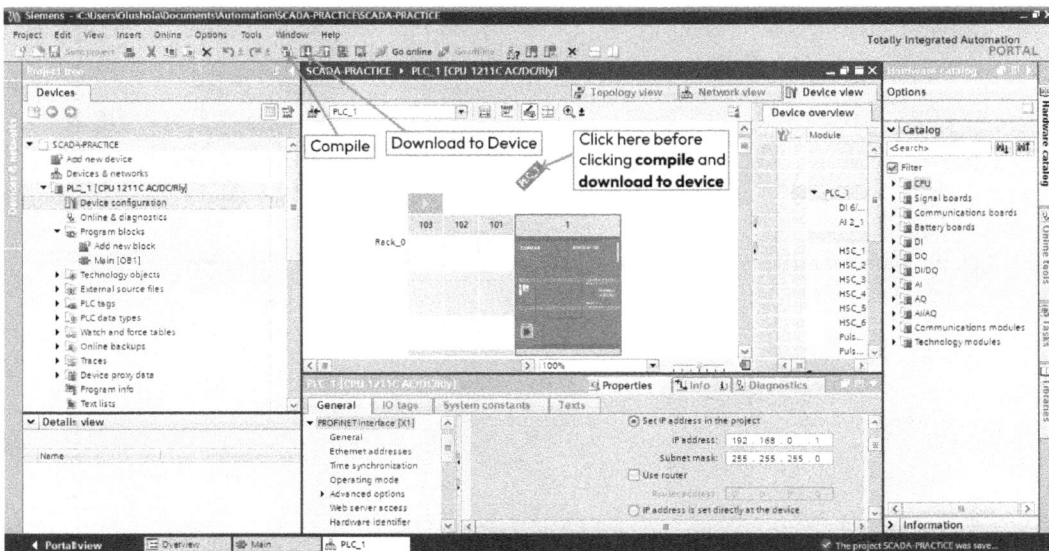

Figure 11.18 – The IP address and subnet mask boxes for PLC_1

12. Click on **Save project**, select **PLC_1**, click on **Compile**, and click on **Download to device**, as shown in the preceding screenshot.

> **Important Note**
> Ensure that PLCSIM is not running before clicking on **Download to device**.

13. In the **Type of the PG/PC interface** drop-down list, select **PN/IE**. In the **PG/PC interface** drop-down list, select your Ethernet card:

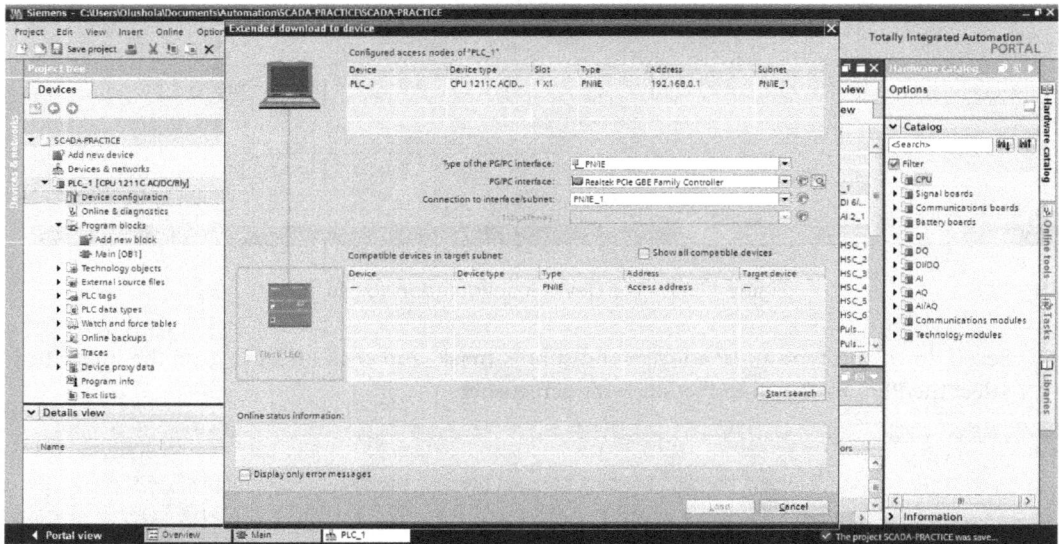

Figure 11.19 – Extended download to the device dialog box

14. Click on **Start search**. It should find the PLC, as shown in the following screenshot. Select the PLC and click **Load**:

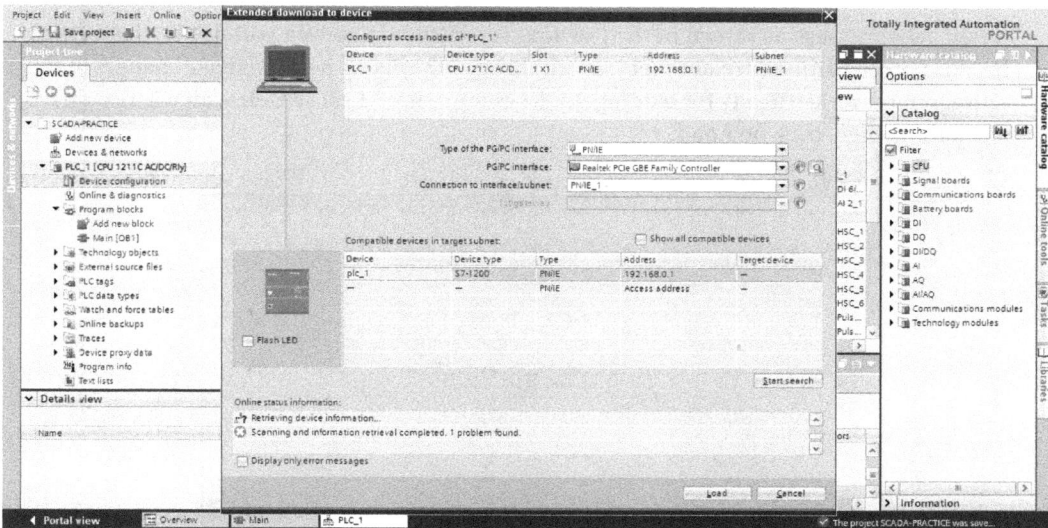

Figure 11.20 – The PLC found after a search

15. Other screens/messages might appear before the following screen; read the messages and respond accordingly. You should be taken to the **Load preview** dialog box if everything is okay:

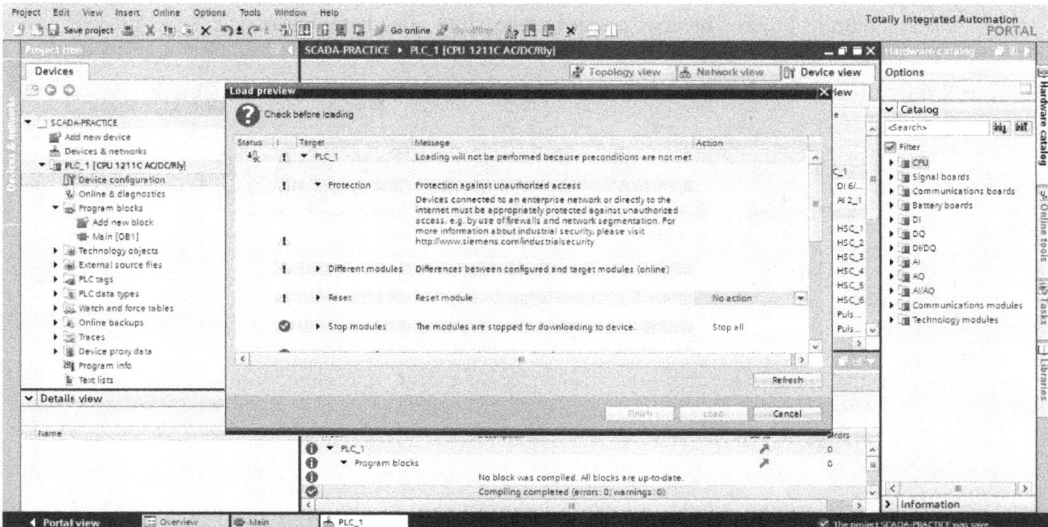

Figure 11.21 – The Load preview dialog box – Loading will not be
performed because preconditions are not met

If you get a message as shown in the previous screenshot (**Loading will not be performed because preconditions are not met**), check through the dialog box and make changes to the settings until you see **Ready for loading**. In the example here, I chose **Delete all** where we have **No action**. Then, click on **Load**:

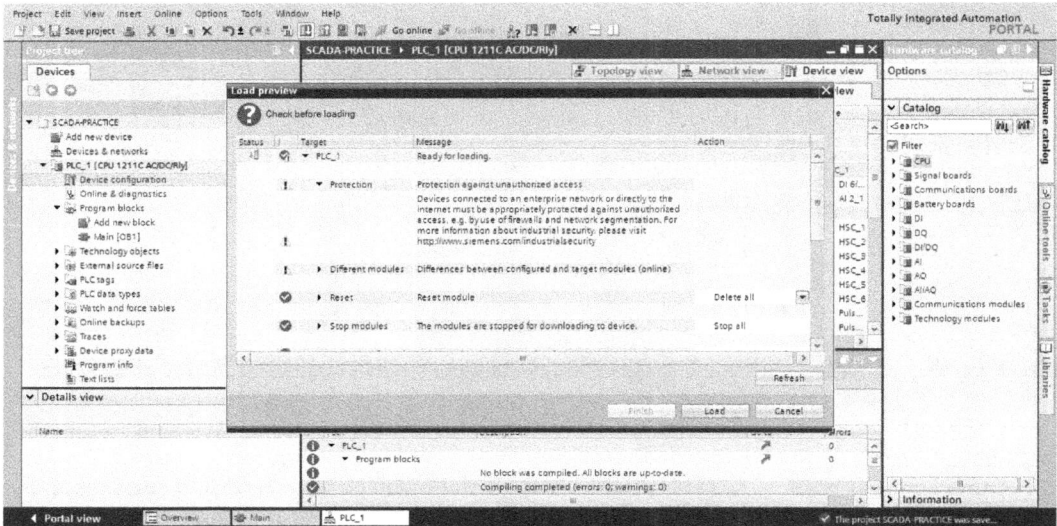

Figure 11.22 – The Load preview dialog box – Ready for loading

16. Check the **Start all** box, as shown in the **Load results** dialog box. Click on **Finish**:

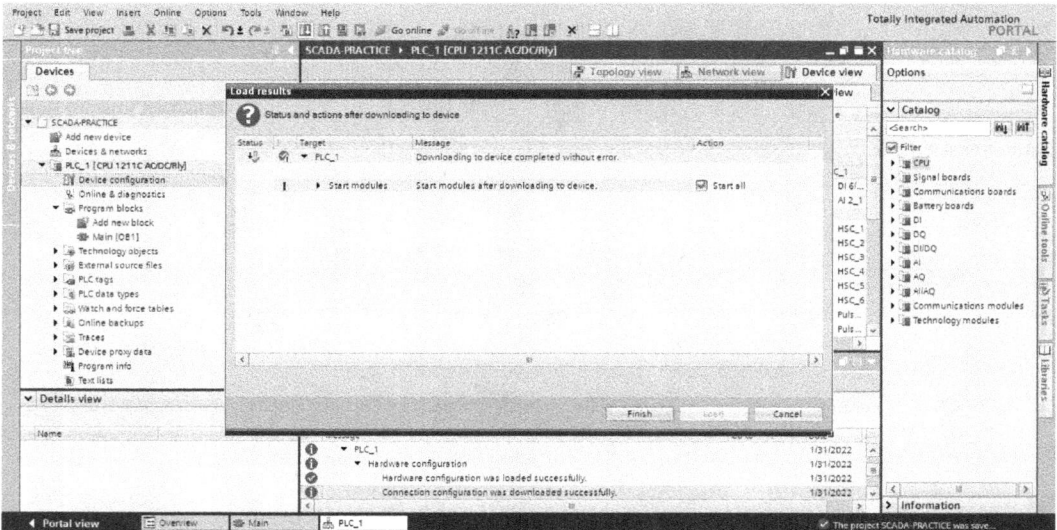

Figure 11.23 – The Load results dialog box

17. If loading is successful, you will see a screen similar to the following, showing **Loading completed** in the **Message** section:

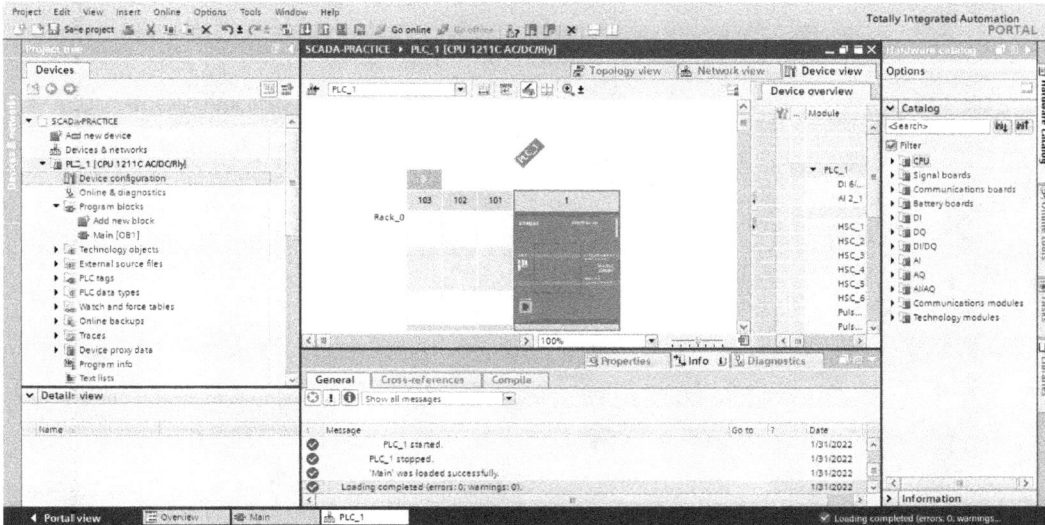

Figure 11.24 – Screen after loading, showing Loading completed in the Message section

Congratulations! You have successfully assigned an IP address (192.168.0.1) to your PLC and loaded your PLC program and configuration to a real PLC (S7 1200, CPU 1211C).

Let's now proceed to assign an IP address to our PC:

1. Right-click on the network icon on the taskbar.

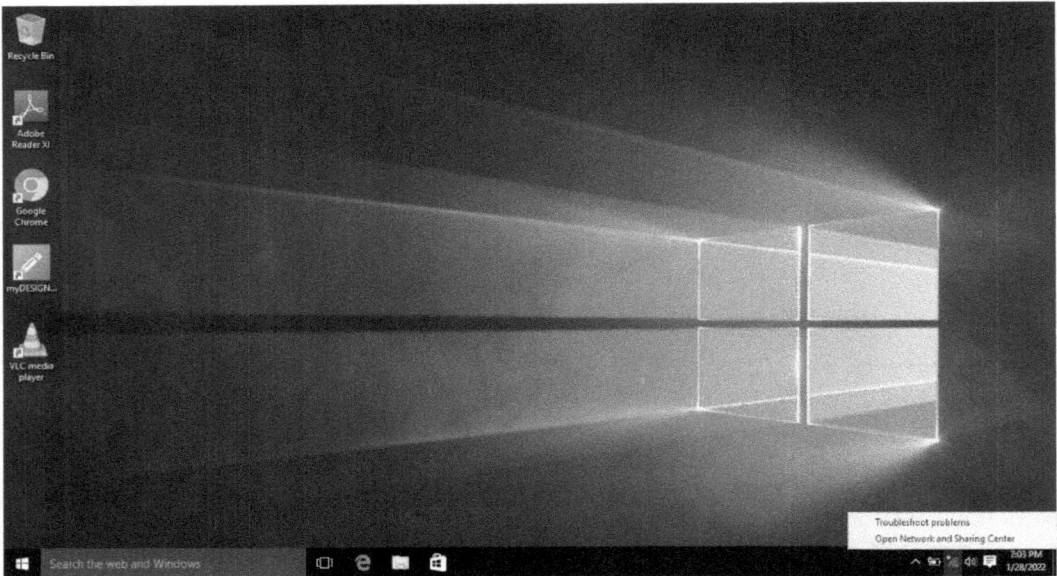

Figure 11.25 – Right-click on the network icon on the taskbar

2. Click on **Open Network and Sharing Center**:

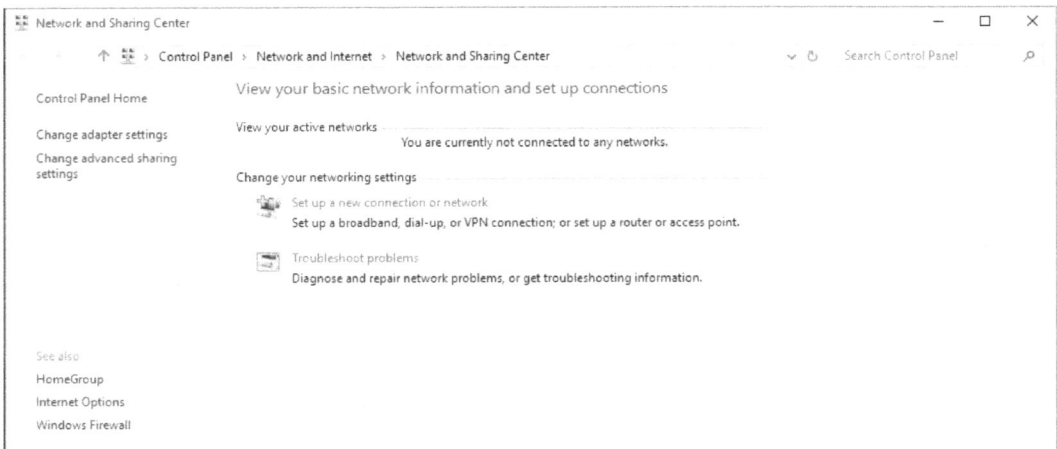

Figure 11.26 – Network and Sharing Center

3.   Click on **Change adapter settings**:

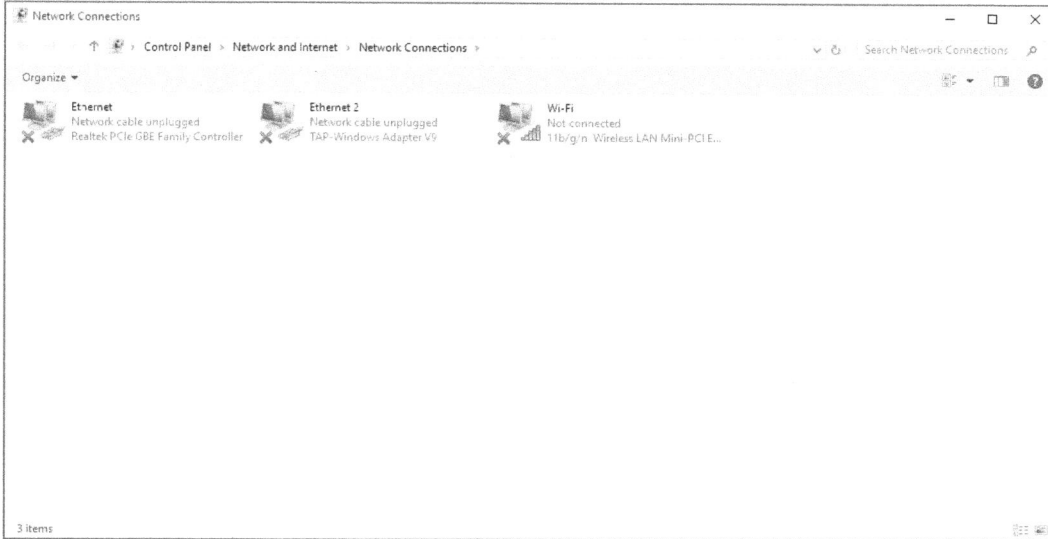

Figure 11.27 – Change adapter settings

4.   Right-click on **Ethernet**:

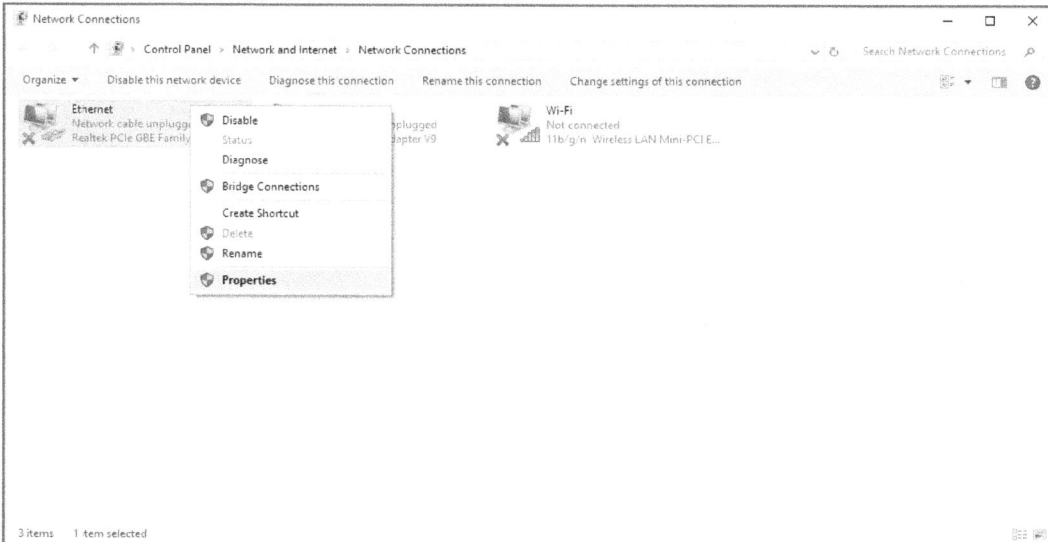

Figure 11.28 – Ethernet – right-click

5.   Click on **Properties:**

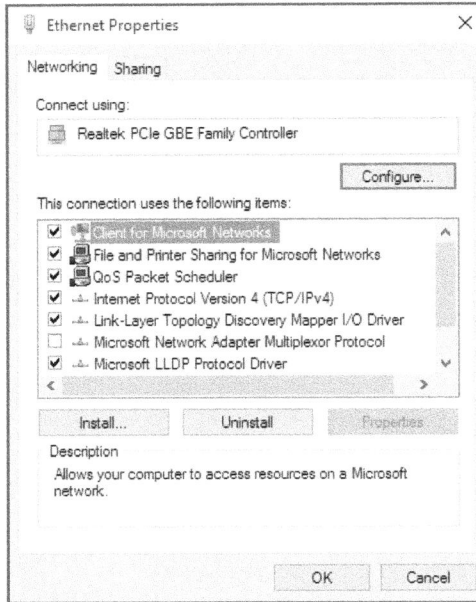

Figure 11.29 – The Ethernet Properties dialog box

6.   Select **Internet Protocol Version 4 (TCP/IPv4)** and click on **Properties**:

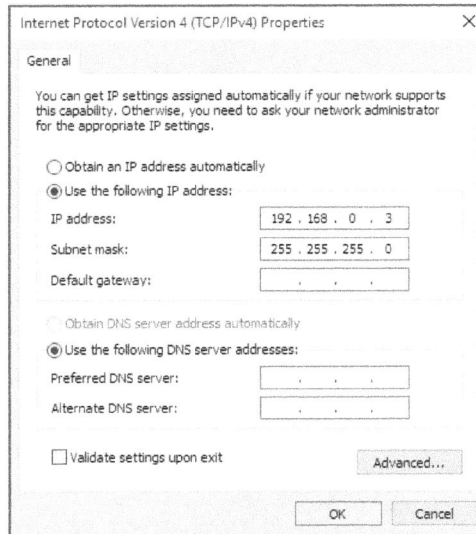

Figure 11.30 – Internet Protocol Version 4 (TCP/IPv4) Properties – Use the following IP address selected

7. Select **Use the following IP address** and type the IP address (for example, 192.168.0.3) and subnet mask (for example, 255.255.255.0). Click on **OK**:

Congratulations! You have successfully assigned an IP address (192.168.0.3) with a subnet mask (255.255.255.0) to the Ethernet card of your PC.

We will now create a new project in myDESIGNER and design our screen to show the button that will act as input and a light to show the status of the pump.

Let's get started:

1. Launch the myDESIGNER application and click on **Empty Project**:

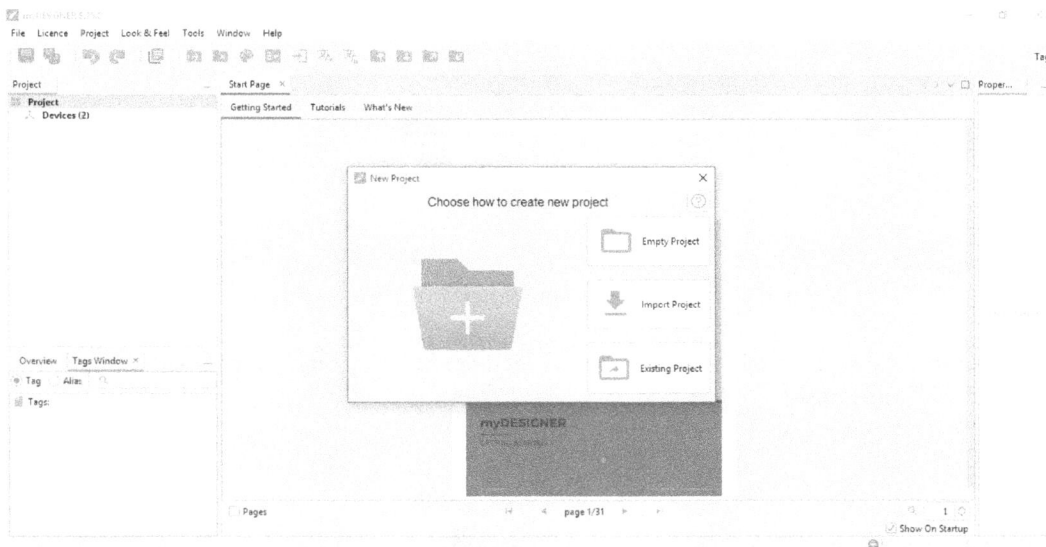

Figure 11.31 – The myDESIGNER application launched

2.   Enter a new name – for example, SCADAsample – and click on **Finish**:

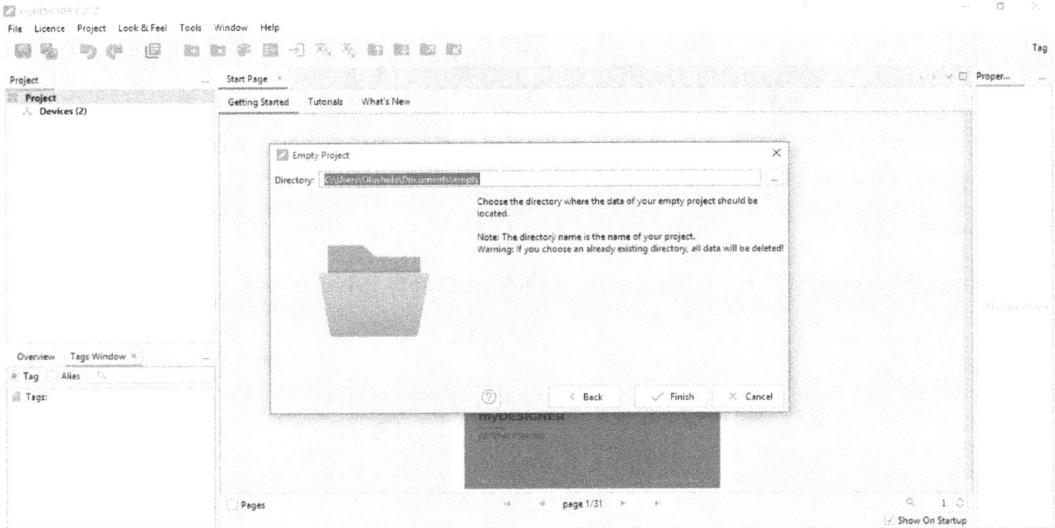

Figure 11.32 – Creating a new project in myDESIGNER

3.   Double-click on the project name on the left and double-click on **Connections**. Then, click on **Add Connection**:

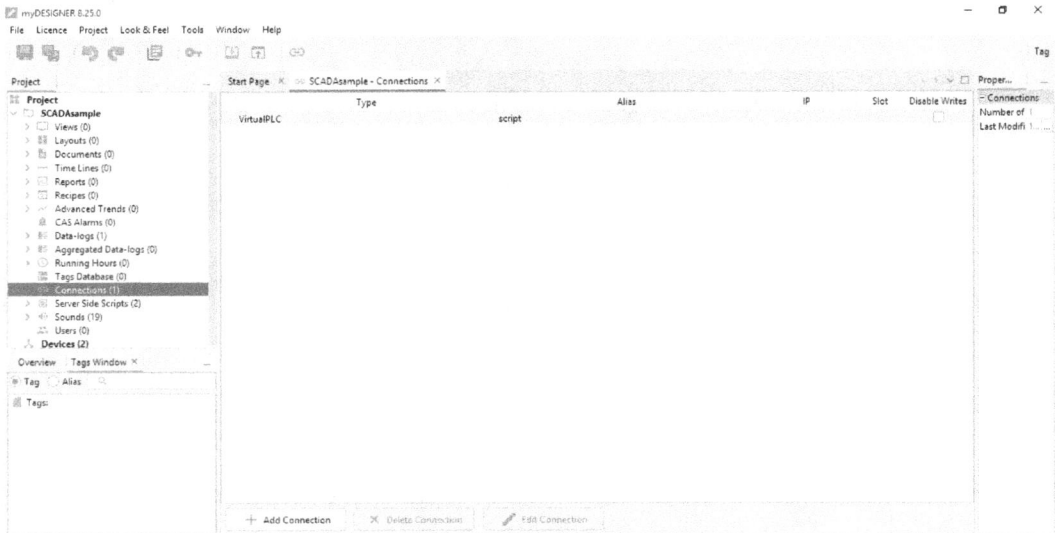

Figure 11.33 – Project (SCADAsample) – Connections

4. Click on **PLC** and then **Add Connection**:

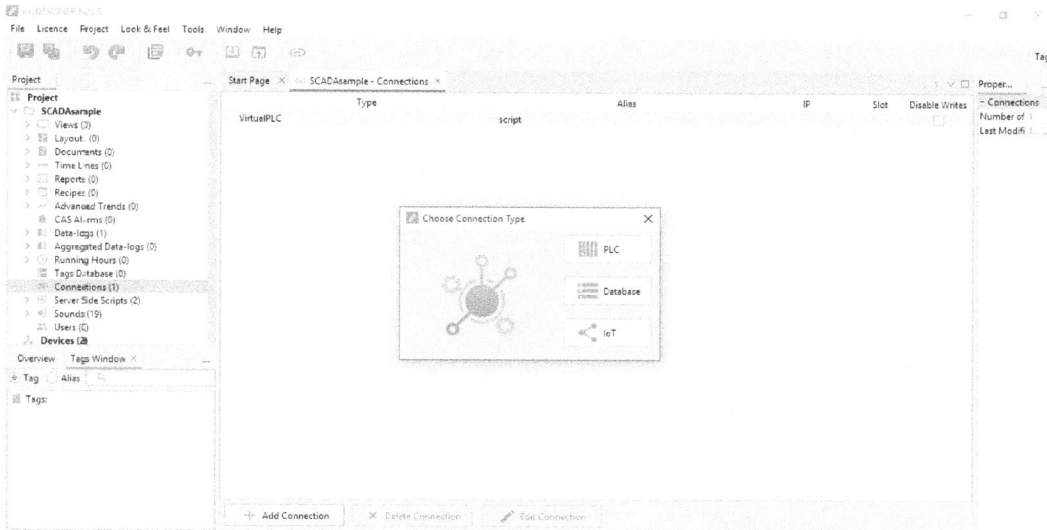

Figure 11.34 – Project (SCADAsample) – Add Connection

5. Make the necessary settings, as shown in the following screenshot, and then click on **Add**:

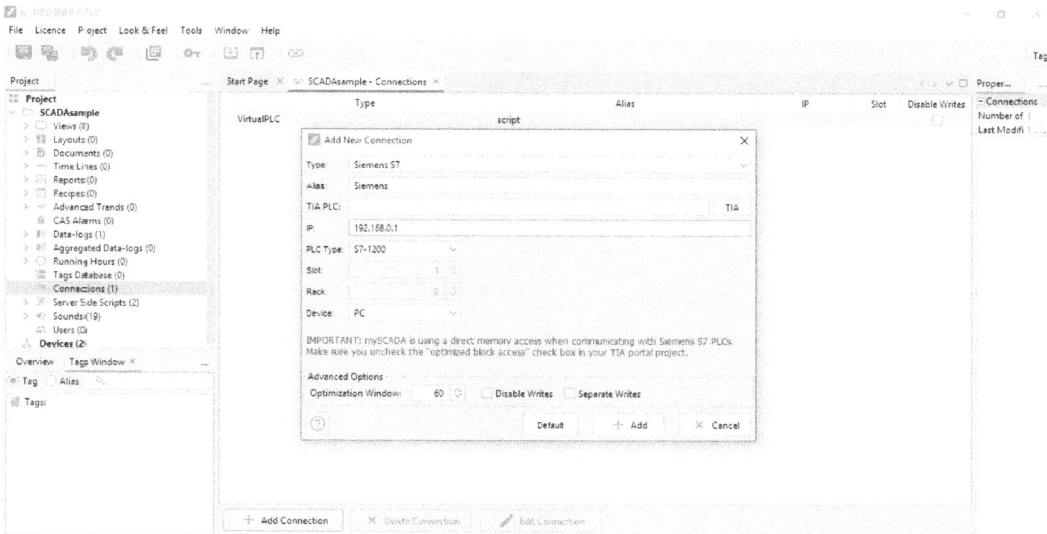

Figure 11.35 – Project (SCADAsample) – the Add New Connection dialog box

You should see the newly added connection:

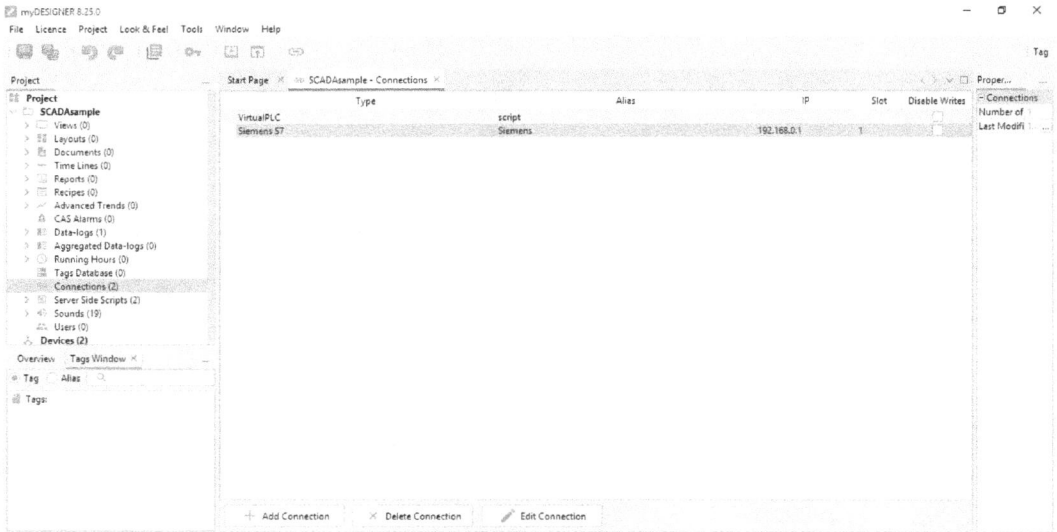

Figure 11.36 – Project (SCADAsample) – a new connection is added

6. Next, double-click on **Devices** and click on **Add Device**. Configure the settings as shown in the following screenshot and then click on **Test Device**:

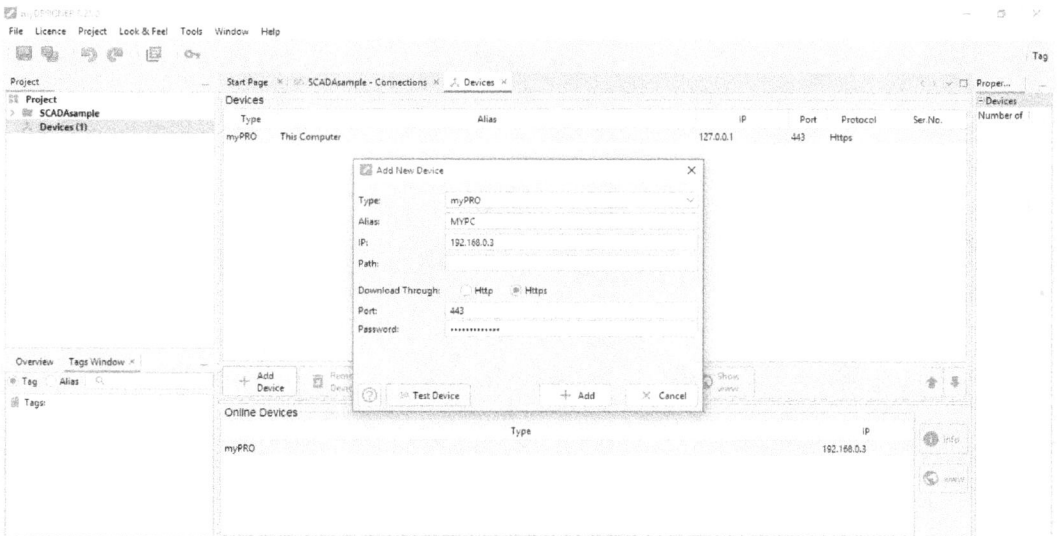

Figure 11.37 – Project (SCADAsample) – adding devices

7. Click on **Cancel** and then **Add**:

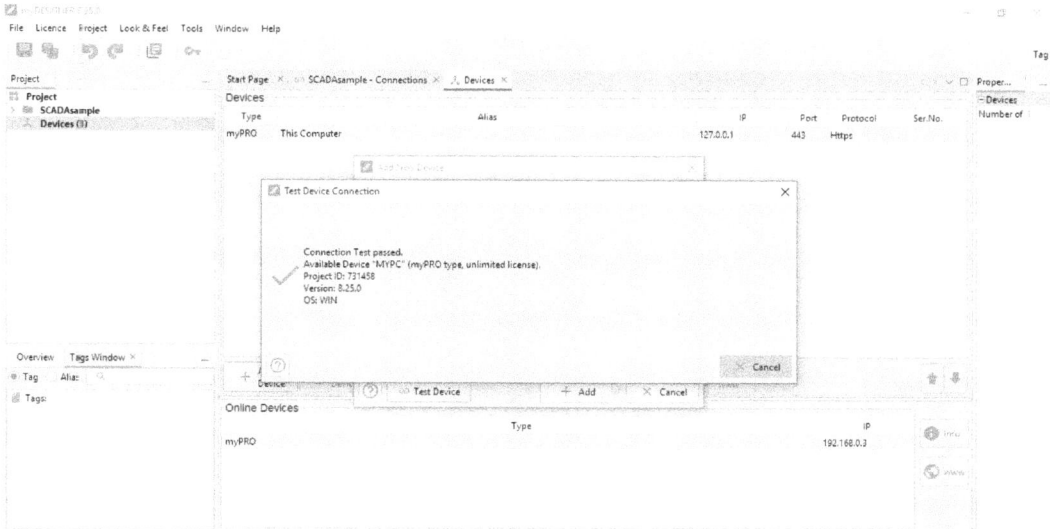

Figure 11.38 – Project (SCADAsample) – the connection between the PLC and the PC is tested and is okay

You should see the newly added device:

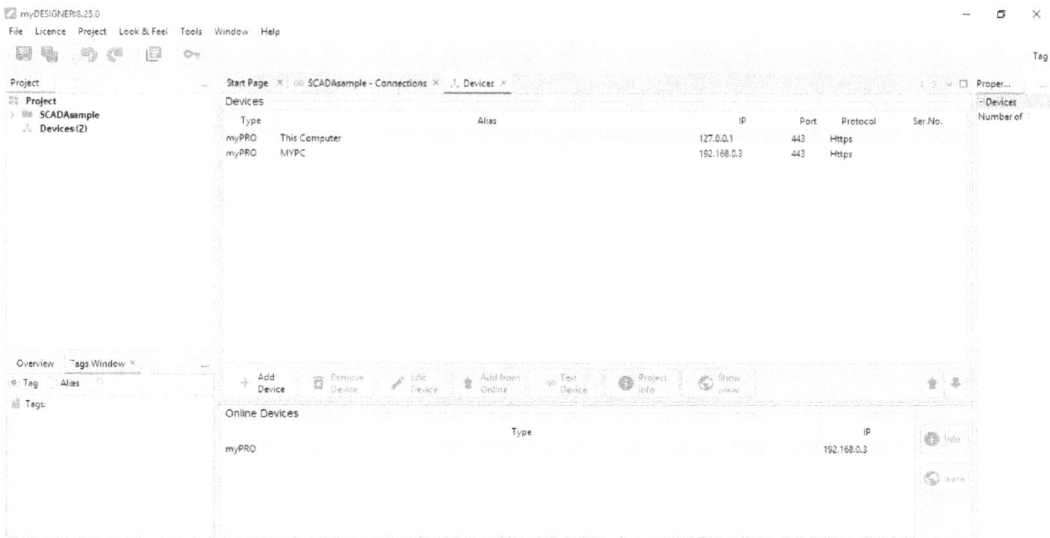

Figure 11.39 – Project (SCADAsample) – devices added

8. Double-click on the project title and double-click on **Tags Database**. Click on **Add**:

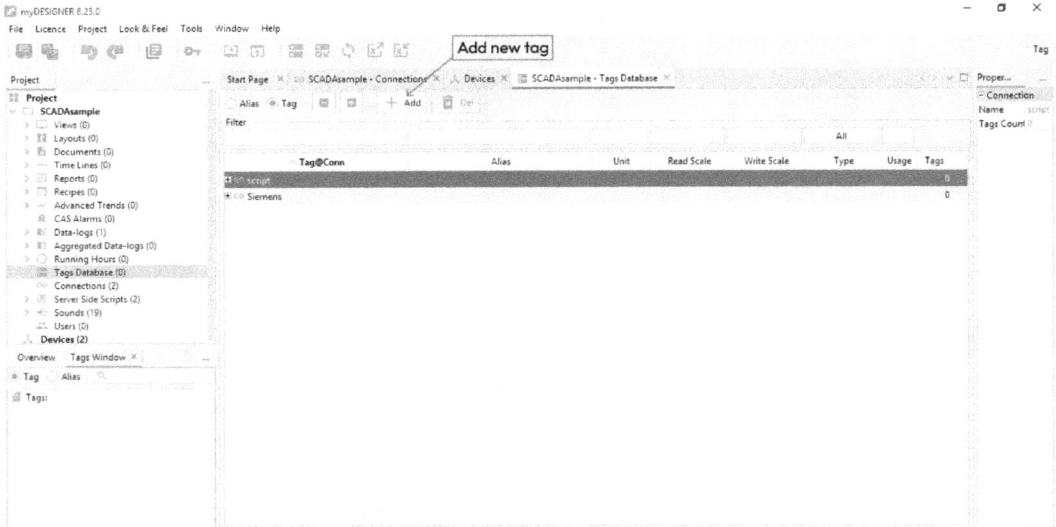

Figure 11.40 – Project (SCADAsample) – Tags Database

9. Select the name of your PLC, as indicated in the following screenshot:

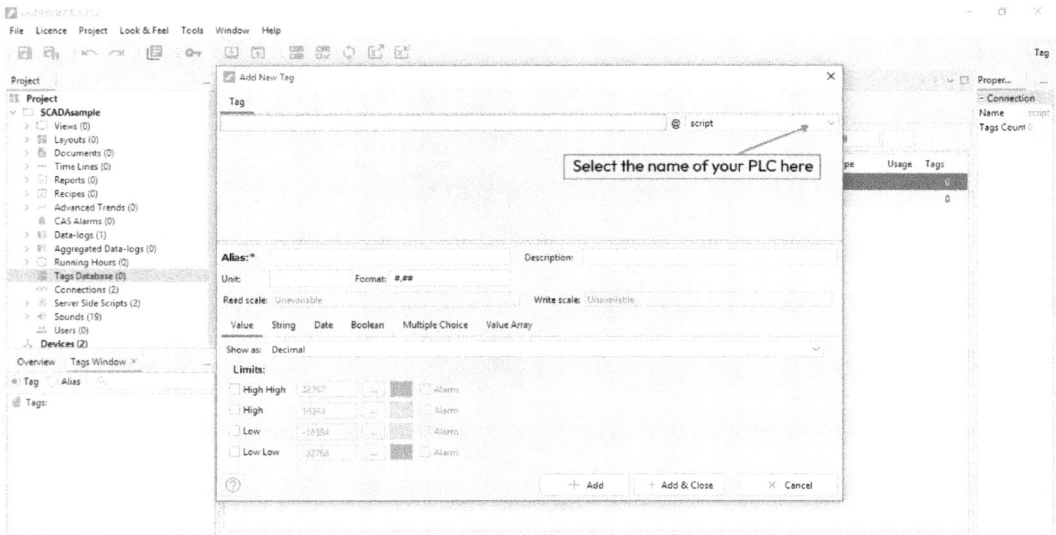

Figure 11.41 – Project (SCADAsample) – the Add New Tag dialog box (1)

10. Select a memory type, datatype, address, and bit so that the tag will be displayed as MX0 . 0, which will be for the M0 . 0 tag used in our PLC program. Also, type the tag name used in your PLC program in the **Alias** box, as shown in the following screenshot:

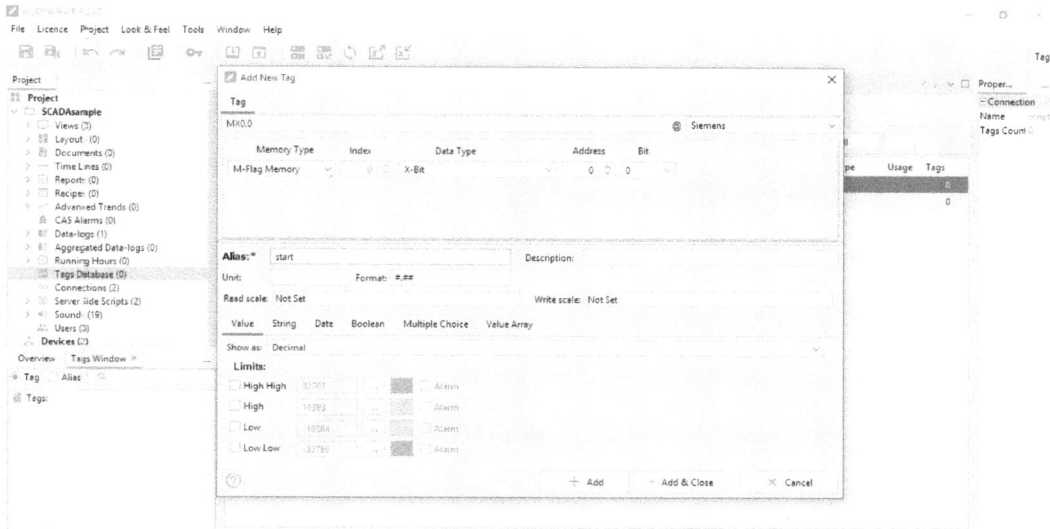

Figure 11.42 – Project (SCADAsample) – the Add New Tag dialog box (2)

11. Click **Add & Close**. You should see the following screen:

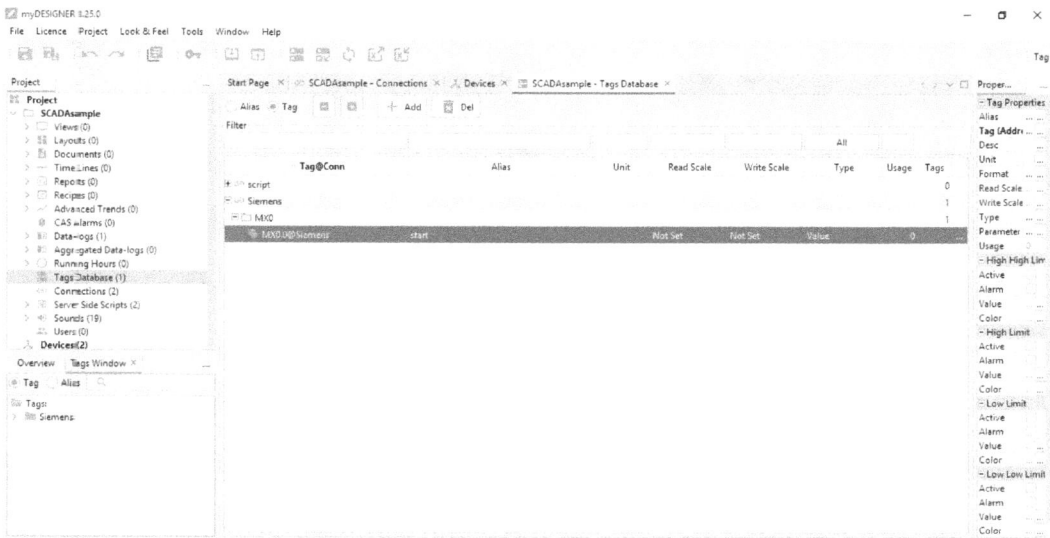

Figure 11.43 – Project (SCADAsample) – New tag added

12. Click **Add** and specify the necessary details for the tag, as shown in the following screenshot:

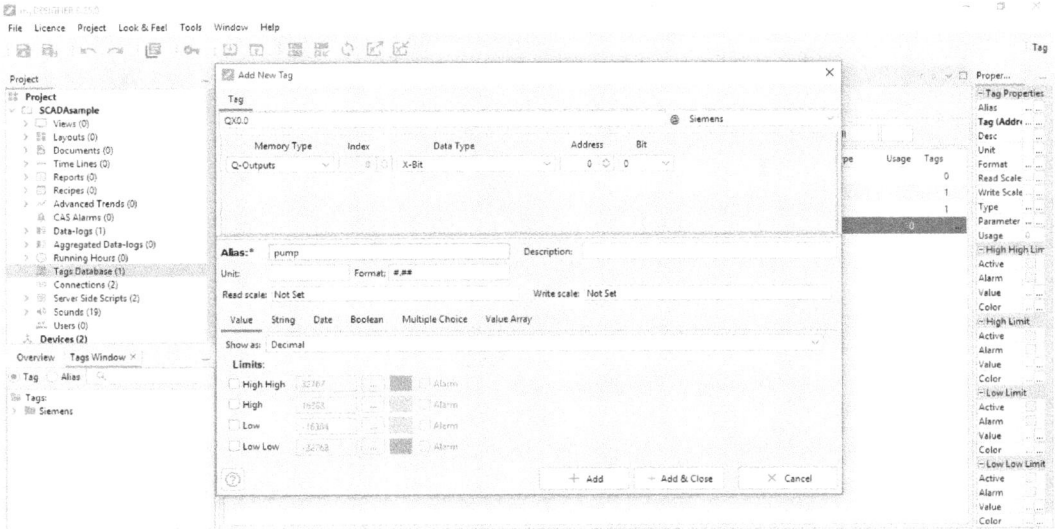

Figure 11.44 – Project (SCADAsample) – the Add New Tag dialog box

13. Click **Add & Close**. You should see the following screen:

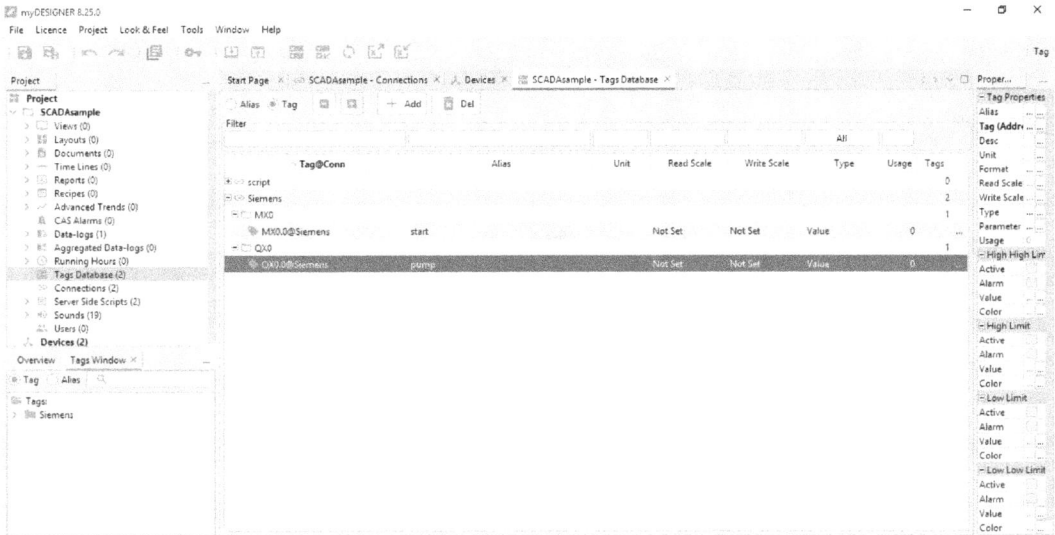

Figure 11.45 – Project (SCADAsample) – new tag added

Congratulations! You have successfully added the necessary tags to the tags database.

Let's now proceed to create a screen using the necessary components and tags:

1. Right-click on **Views** and select **Add New**:

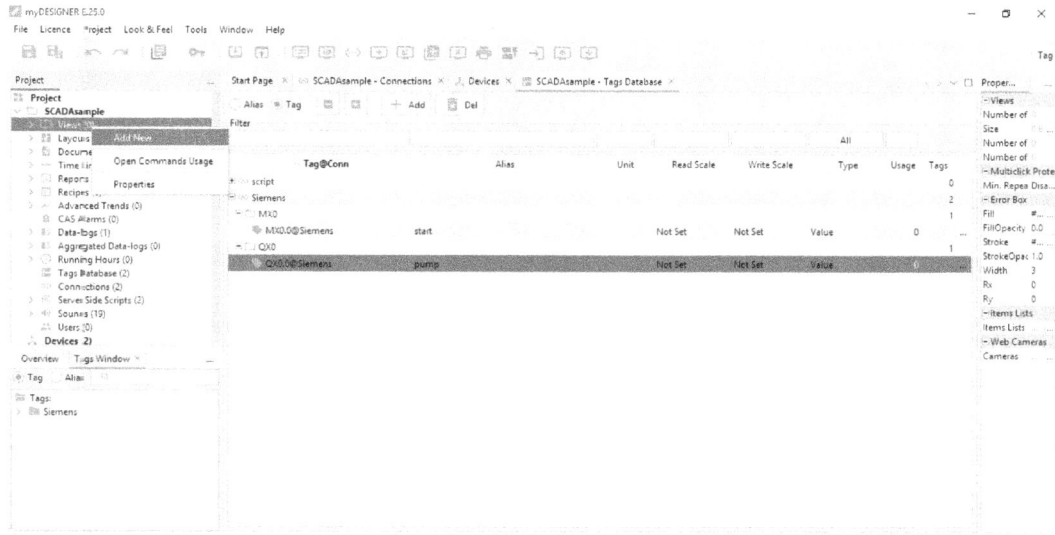

Figure 11.46 – Project (SCADAsample) – right-click on Views

2. Make the necessary settings, as shown in the following screenshot. Once done, click on **Add**:

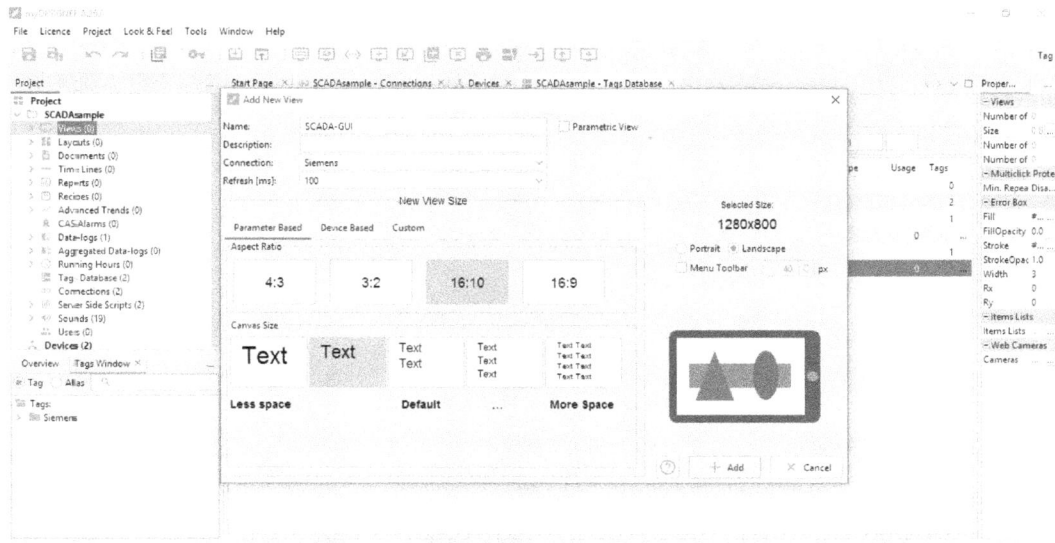

Figure 11.47 – Project (SCADAsample) – Add New View

3. Let's now start to add components. Click on the **Components Edit** icon:

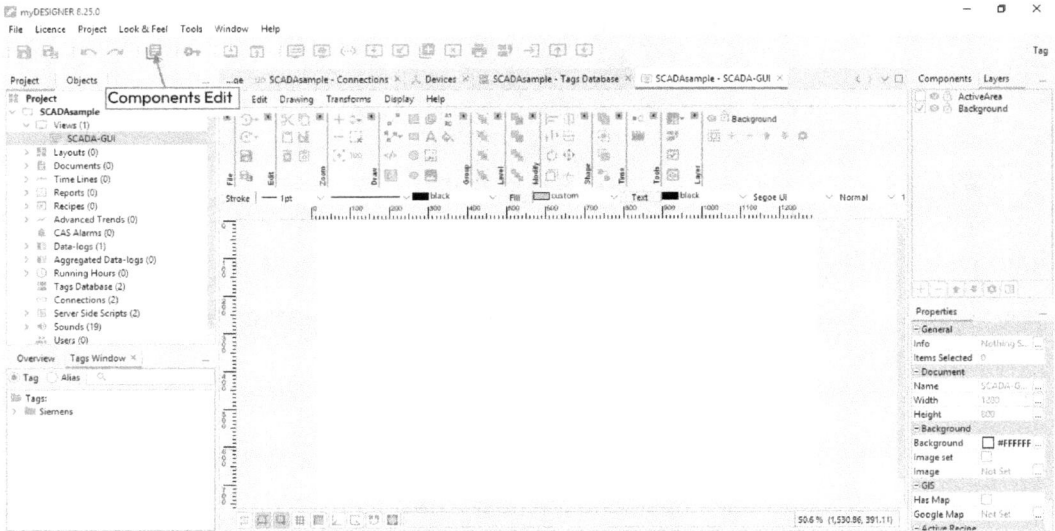

Figure 11.48 – Project (SCADAsample) – a new view (SCADA-GUI) is added

4. Select **Industrial switches** from the categories of components:

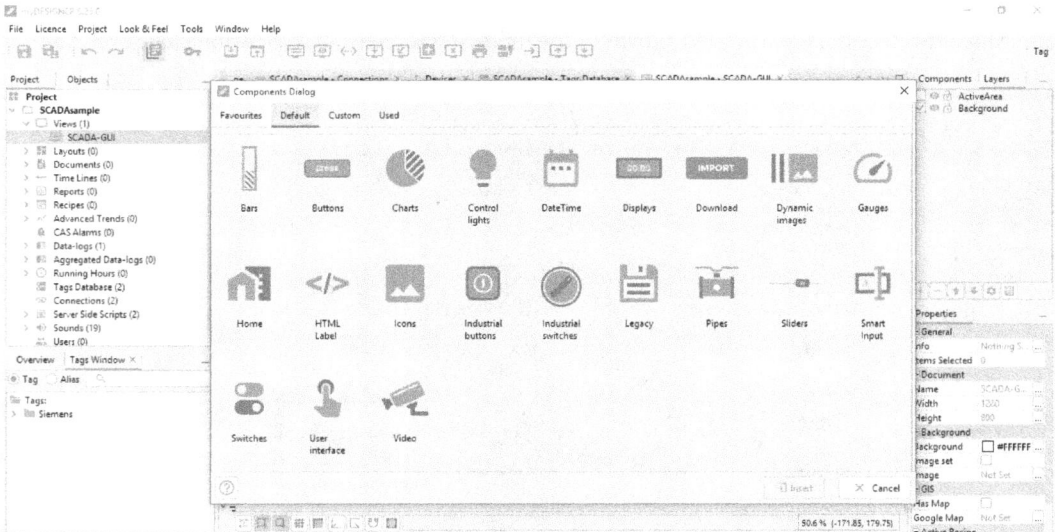

Figure 11.49 – Project (SCADAsample) – the new view (SCADA-GUI) – the Components Dialog box (1)

5. Select **Industrial Switch Button** from the list of switches and click on ... to specify a tag:

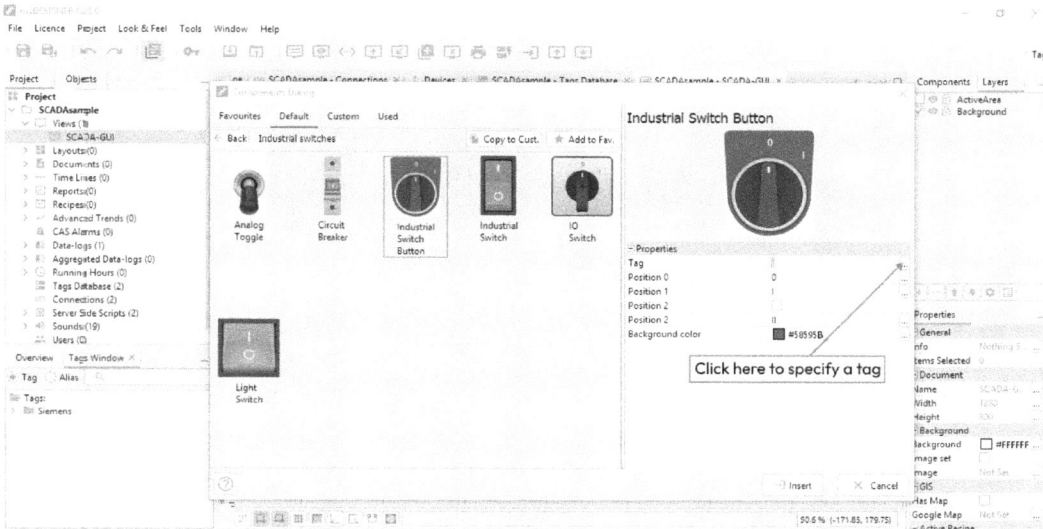

Figure 11.50 – Project (SCADAsample – the new view (SCADA-GUI) – the Components Dialog box (2)

6. Expand Siemens | MX0 and then select the tag for the switch – for example,MX0.0@Siemens [*start]. Then, click on **OK**:

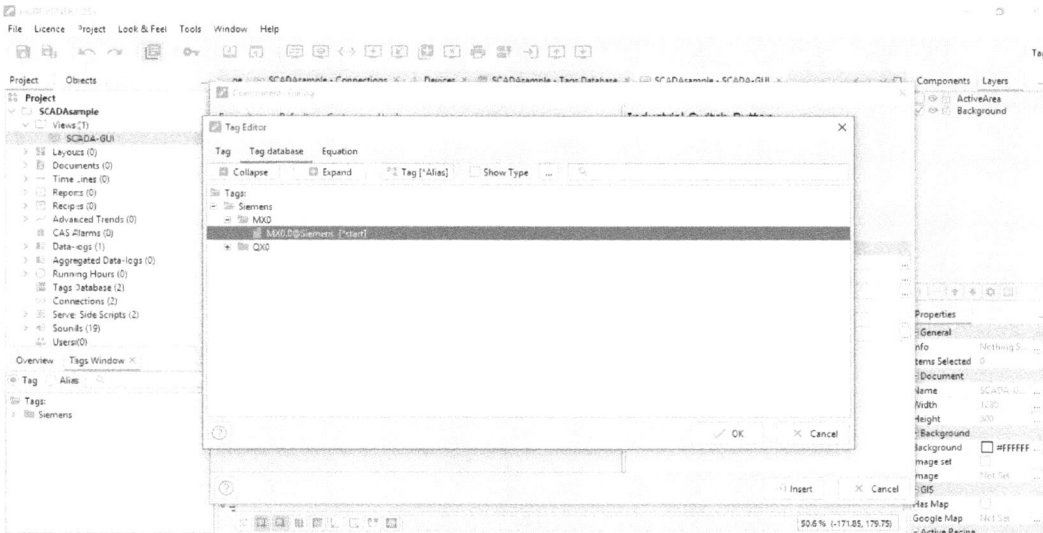

Figure 11.51 – Project (SCADAsample) – the new view (SCADA-GUI) – specifying a tag for the component

7.   Click on **Insert** and click on your screen:

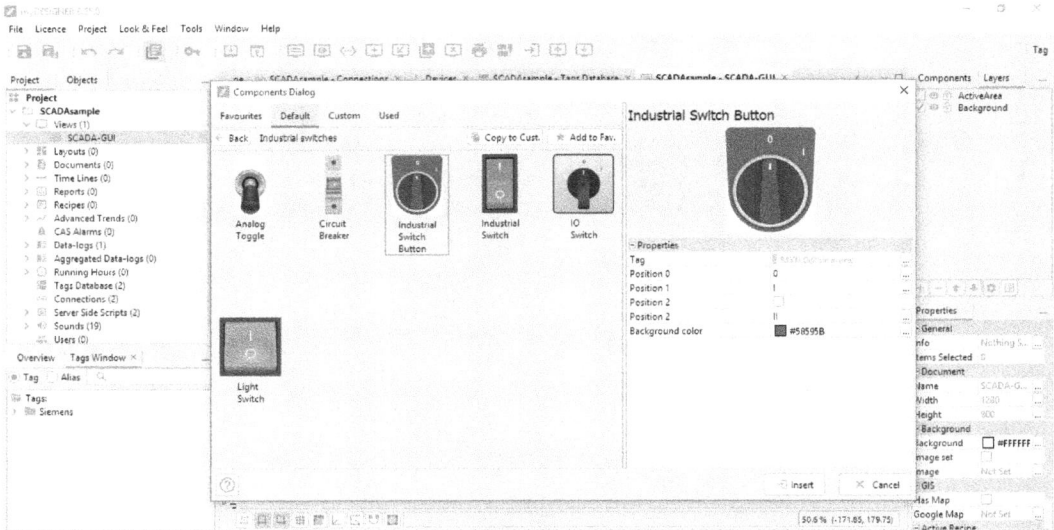

Figure 11.52 – Project (SCADAsample) – the new view (SCADA-GUI) – the Components Dialog box

8.   You can select it to resize it or move it to a position of your choice on the screen:

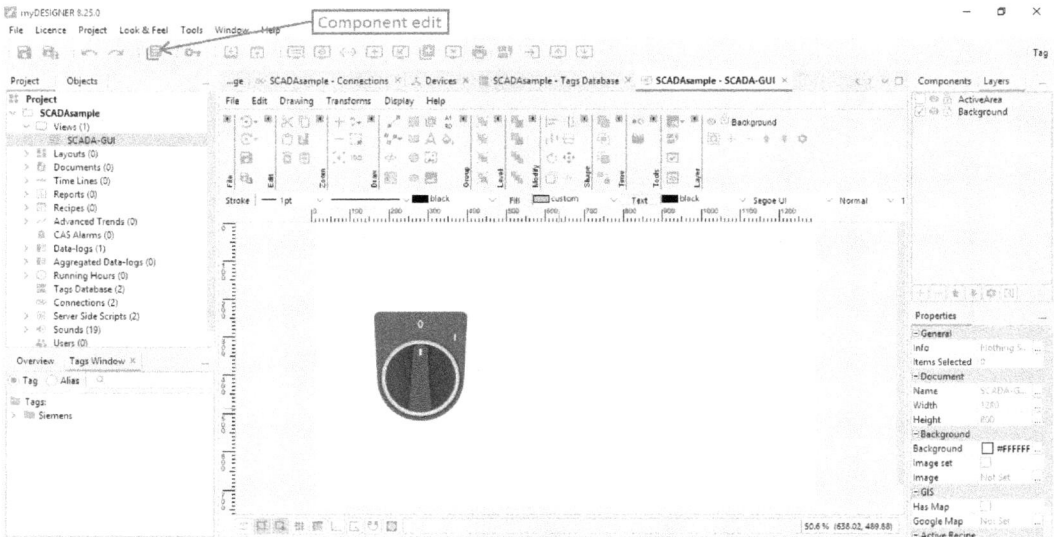

Figure 11.53 – Project (SCADAsample) – the new view (SCADA-GUI) – the component added

9. Click on **Component edit** and then **Control light** in the category:

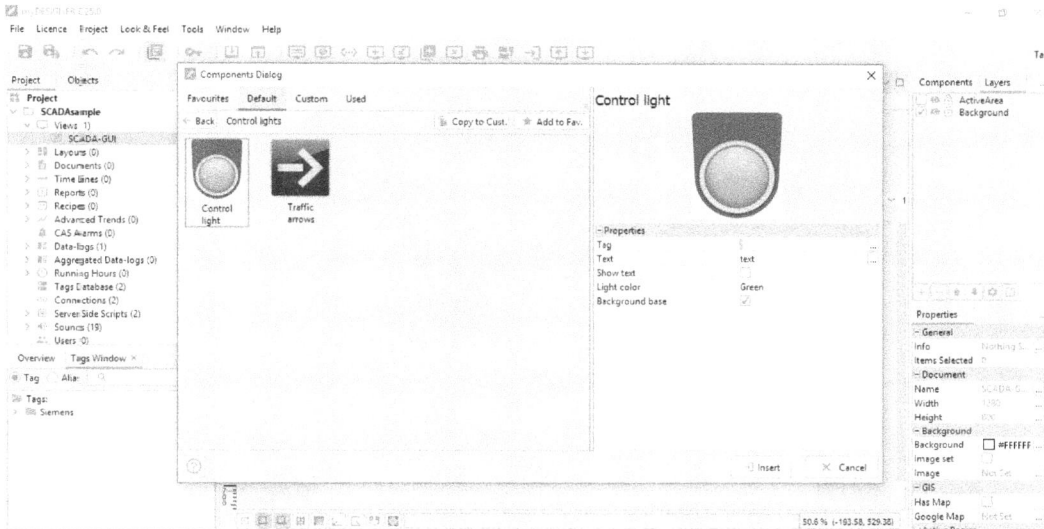

Figure 11.54 – Project (SCADAsample) – the new view (SCADA-GUI) – the Components Dialog box

10. Click on ... to specify a tag. Expand QX0 and select QX0.0@Siemens [*pump], as shown in the following screenshot, and then click **OK**:

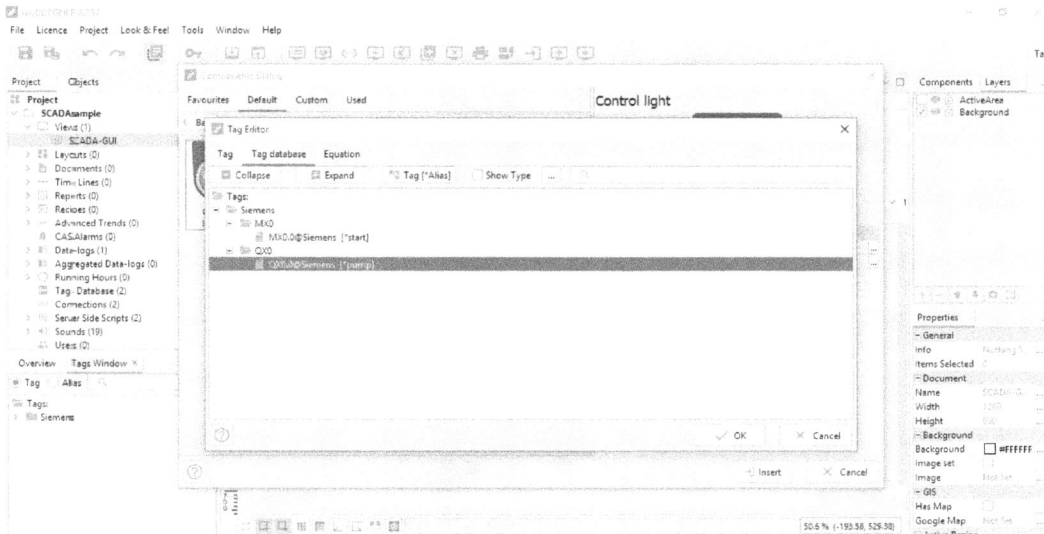

Figure 11.55 – Project (SCADAsample) – the new view (SCADA-GUI) – specifying a tag for the component

11. Click on **Insert** and click on your screen:

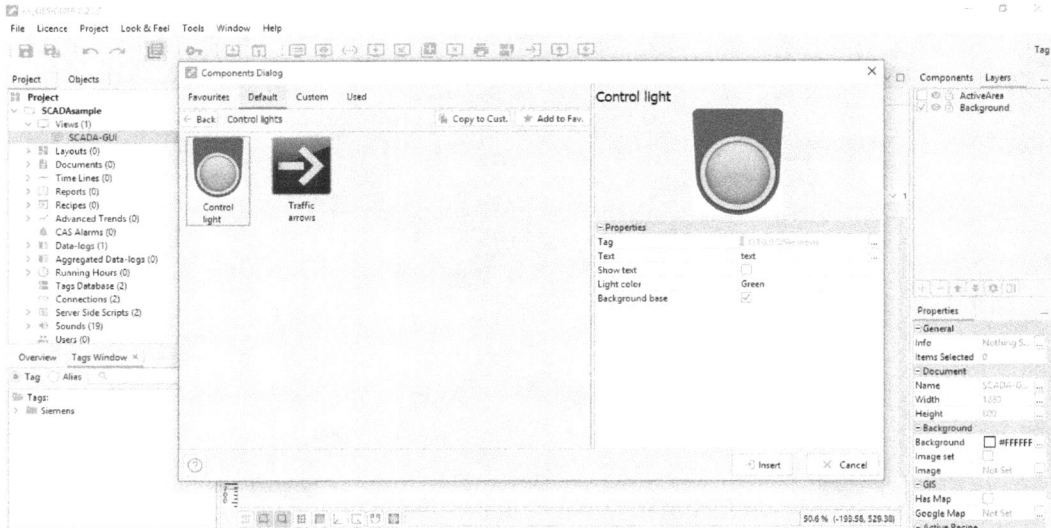

Figure 11.56 – Project (SCADAsample) – the new view (SCADA-GUI) – the Components Dialog box

12. You can now see the control light on the screen. Next, select the switch, as shown in the following screenshot:

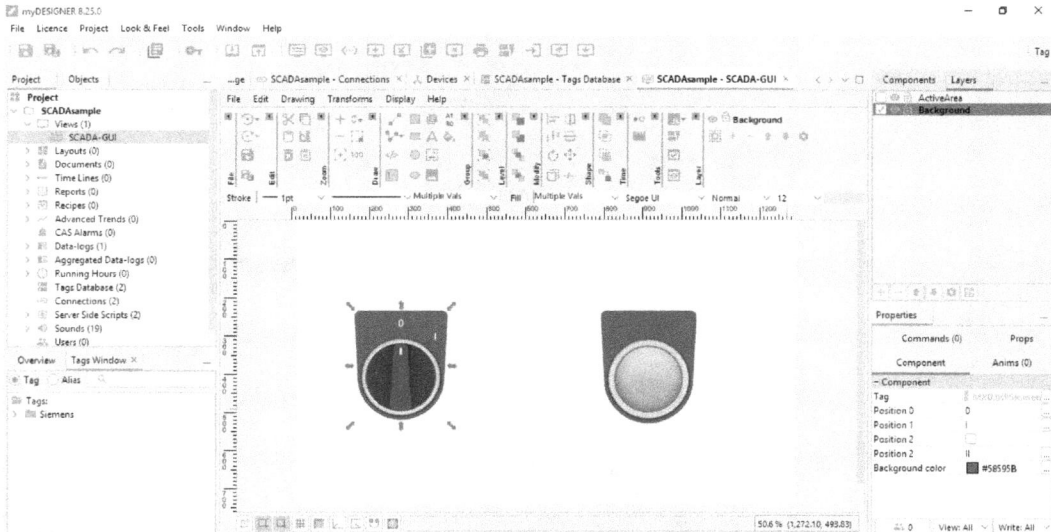

Figure 11.57 – Project (SCADAsample) – the new view (SCADA-GUI) – the switch selected

13. Click on **Commands** and then ...:

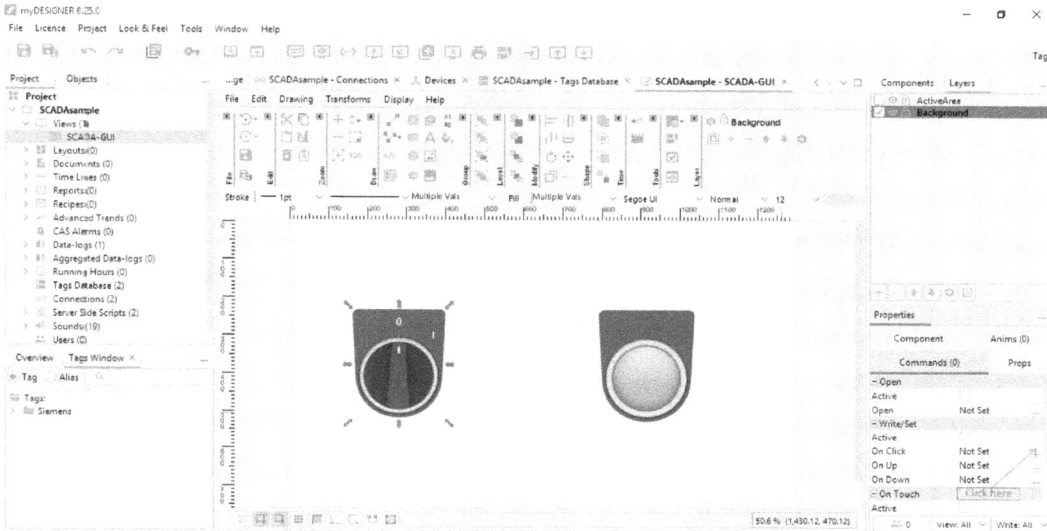

Figure 11.58 – Project (SCADAsample) – the new view (SCADA-GUI) – adding a command to the switch (1)

14. Click **Toggle** and then ...:

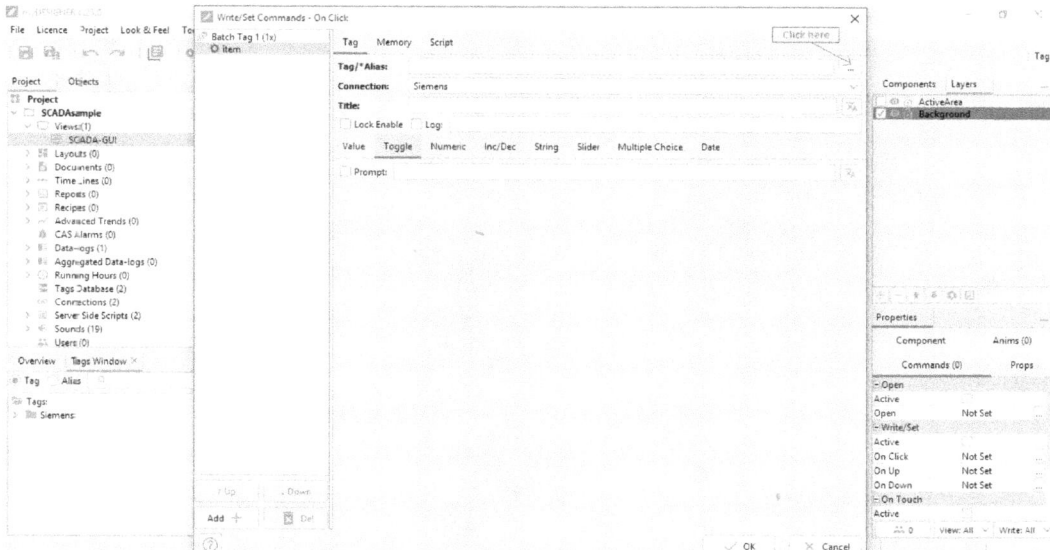

Figure 11.59 – Project (SCADAsample) – the new view (SCADA-GUI) – adding a command to the switch (2)

15. Expand the tags, as shown in the following screenshot:

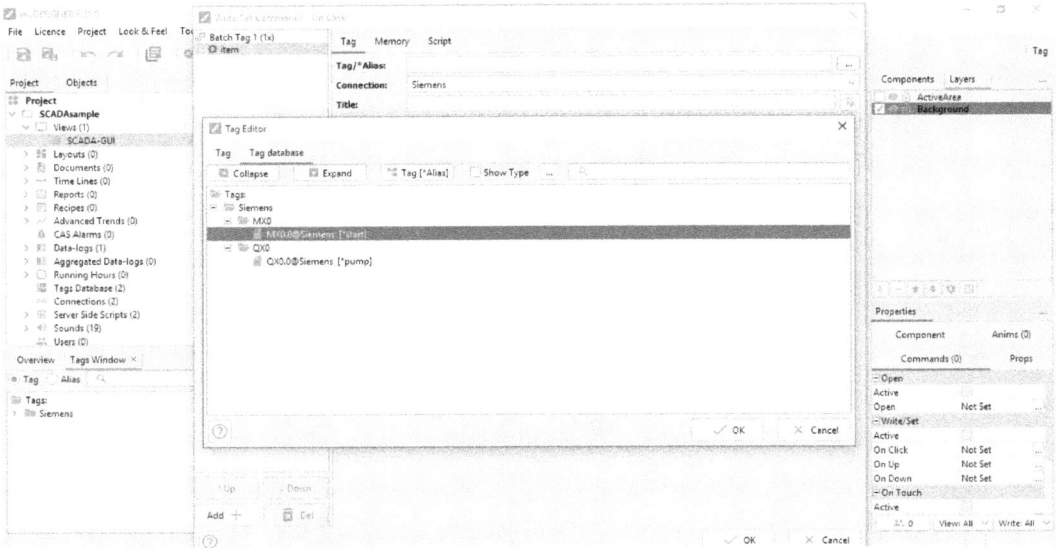

Figure 11.60 – Project (SCADAsample) – the new view (SCADA-GUI) – adding a command to the switch (3)

16. Select the tag MX0.0@Siemens [*start] for the switch.

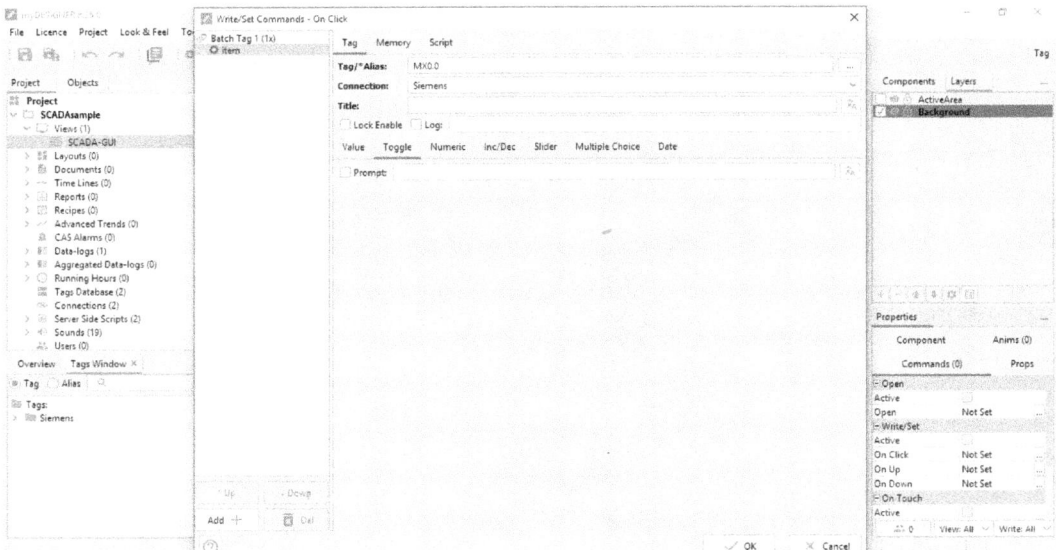

Figure 11.61 – Project (SCADAsample) – the new view (SCADA-GUI) – adding a command to the switch (4)

17. Click on **OK** to return to your screen.

> **Note**
>
> You can use the create text element tool to type text, resize it, and place it in a position of choice on your screen, as shown in the following screenshot:

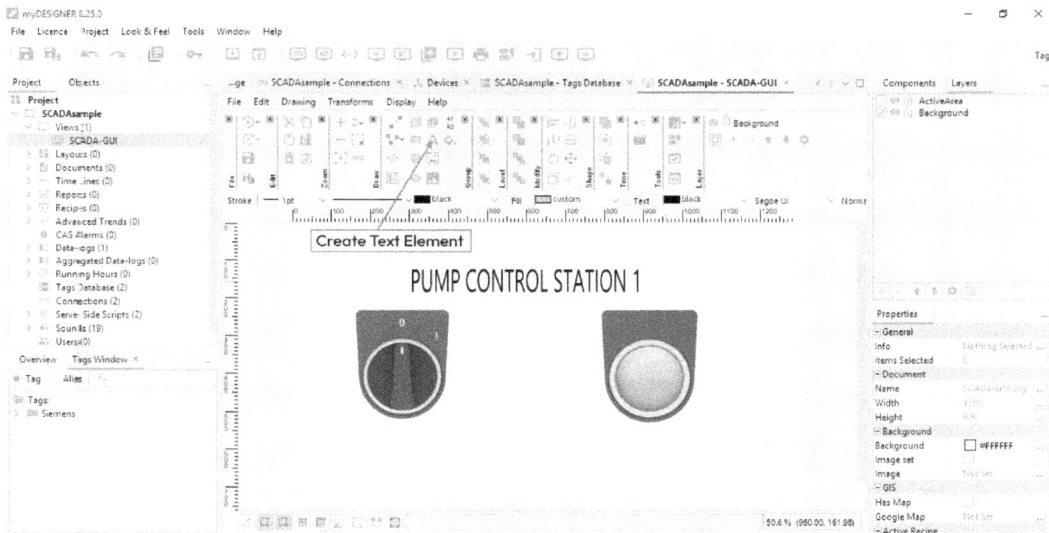

Figure 11.62 – Project (SCADAsample) – the new view (SCADA-
GUI) – the command added and the text element created

Congratulations! You have successfully completed a screen to monitor and control the operation of a pump.

Let's now download the project to our device:

1. Select your project name (SCADAsample) on the left-hand side of the screen (the project tree). Click on **Download to Devices**:

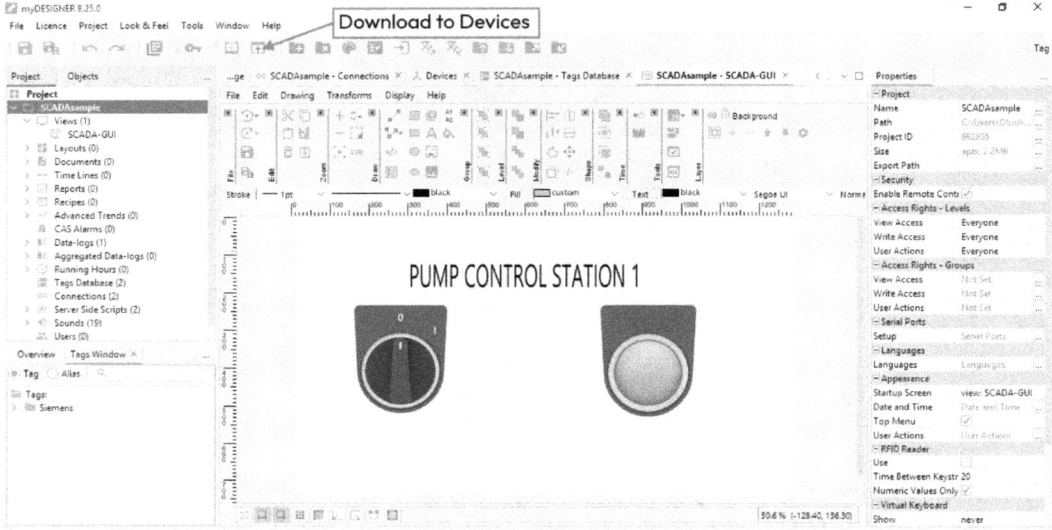

Figure 11.63 – Project (SCADAsample) – the new view (SCADA-GUI) – the project name selected

2. Select the two device options (**This Computer** and **MYPC**):

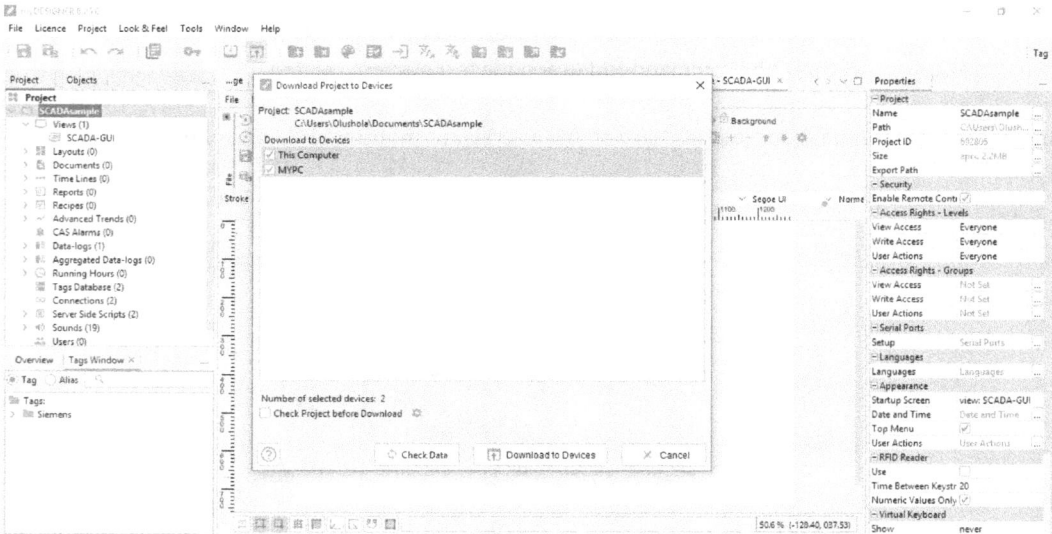

Figure 11.64 – Project (SCADAsample) – the new view (SCADA-GUI) – downloading to devices (1)

3.   Click on **Download to Devices**:

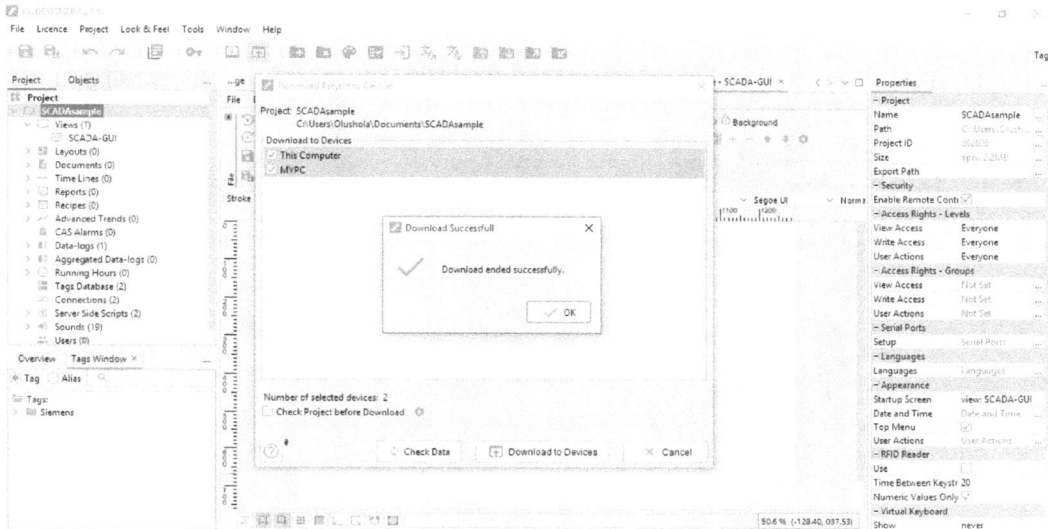

Figure 11.65 – Project (SCADAsample) – the new view (SCADA-GUI) – downloading to devices (2)

Note

If any message comes up, read through it and respond accordingly.

---

4.   Click on **OK** to return to your designed screen.

Note

If your system has been on for up to 2 hours, you will have to restart your system to continue using the software. This is because you are using the free version of mySCADA.

Congratulations! You have successfully completed your SCADA screen and its configuration for monitoring and control.

Let's now view our SCADA for monitoring and control:

1. Start any supported browser for mySCADA – for example, Chrome.

2. Type `localhost` in the address bar and press the *Enter* key:

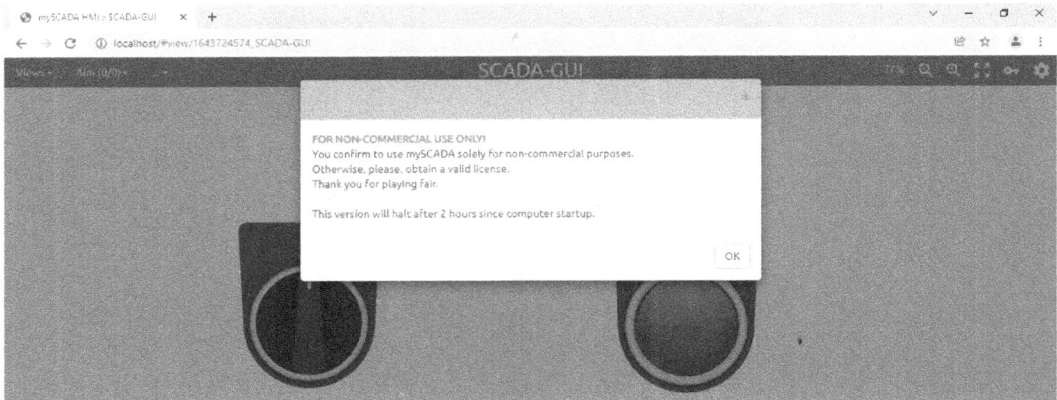

Figure 11.66 – Viewing your SCADA for monitoring and control

> **Note**
> You can also type `127.0.0.1` or the IP address you assigned to the Ethernet card of your PC instead of `localhost`.

3. Click on **OK**. The following figure shows the SCADA screen when the pump is off:

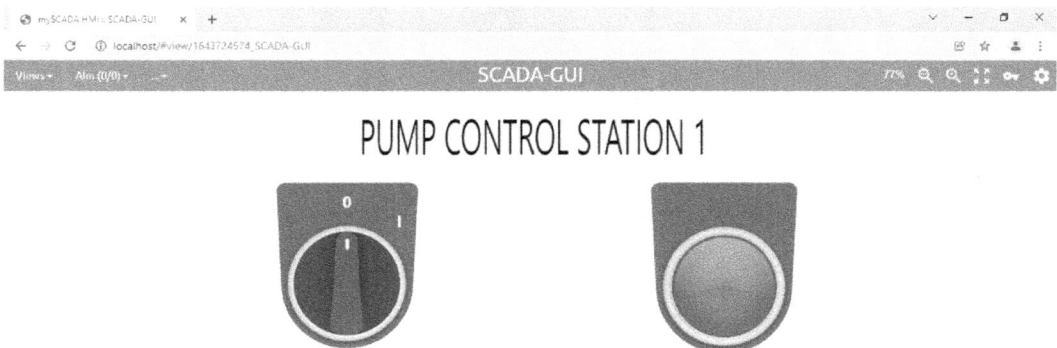

Figure 11.67 – Viewing your SCADA for monitoring and control (pump off)

4. Click on the switch to turn the pump on:

Figure 11.68 – Viewing your SCADA for monitoring and control (pump on)

The following photo shows my setup, which includes a PLC (Siemens S7-1200), a contactor for the pump operation, and a laptop for monitoring and control (when the pump is off):

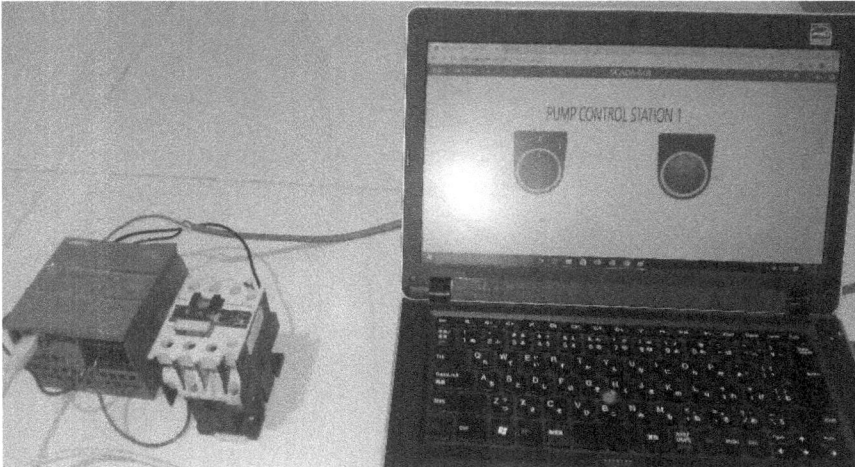

Figure 11.69 – A setup showing a PLC (Siemens S7-1200), a contactor for the pump operation, and a laptop for monitoring and control (when the pump is off)

The following photo shows the connection of the PLC (Siemens S7-1200), the contactor for the pump operation, and my PC for monitoring and control when the pump is on:

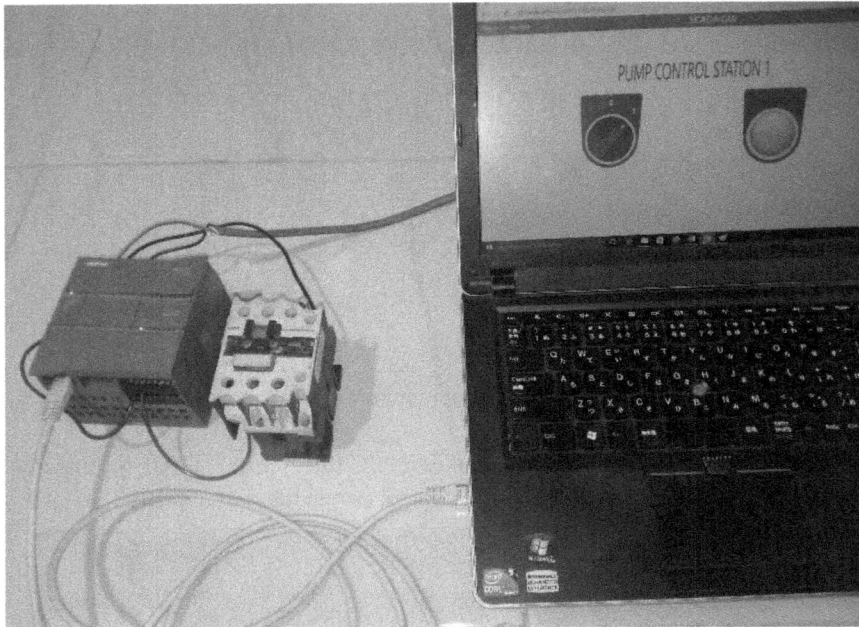

Figure 11.70 – A setup showing a PLC (Siemens S7-1200), a contactor for the pump operation, and a laptop for monitoring and control (when the pump is on)

You have just completed the last section of this book, *Interfacing SCADA with an S7-1200 PLC using mySCADA software* (a practical project), and should be able to demonstrate the SCADA setup on your own if you have the PLC (Siemens S7-1200) and other components and devices used.

Summary

Congratulations! You have successfully completed another chapter of this book. Good job! This chapter gave a good explanation of SCADA in a very simple way. The SCADA hardware (field devices, remote stations, communication networks, and master stations) and software was well explained. You should be able to download and install mySCADA software on your own by following the step-by-step guide. The last section is a very interesting part that shows you how to create a simple SCADA screen. The necessary configuration for monitoring and control was outlined in the step-by-step guide.

The next chapter is another interesting chapter that will give you the fundamentals of process control in various industries.

Questions

The following are questions to test your understanding of this chapter. Ensure you have read and understood the topics in this chapter before attempting the questions:

1. SCADA is an acronym for _____.

2. RTU is an acronym for _____.

3. _____ refers to the sensors, transmitters, and actuators that are directly connected to a machine or plant and generate a digital or analog signal for monitoring.

4. _____ refers to the master computer that runs SCADA software that provides the graphical presentation of a system.

5. _____ is usually installed on a master station to create an interface (GUI) required to monitor and control industrial equipment, and perform all necessary configurations.

Part 3:
Process Control,
Industrial Network,
and Smart Factory

This part provides the fundamental knowledge of process control, industrial network, and the smart factory. It covers topics on instrumentation and control engineering that will be beneficial for engineers in manufacturing, oil and gas, and other industries. You will understand process control including analog signal processing, industrial networks and protocols, and finally, the current trend in manufacturing (the smart factory and its associated technologies, such as IoT, AI, robotics, cloud computing, and so on).

This part has the following chapters:

- *Chapter 12, Process Control – Essentials*
- *Chapter 13, Industrial Network and Communication Protocols Fundamentals*
- *Chapter 14, Exploring Smart Factory (Industry 4.0) with 5G*

12
Process Control – Essentials

Many of the processes carried out in industry require control. You can't have a quality product without control. Manufacturing, oil and gas, agriculture, and other industries require chemical or mechanical operations that need to be controlled to reach the finished product or desired results. Even in our homes, process control takes place in many of the devices or machines that we use. Here are some examples of home appliances that have process control integrated:

- **Refrigerator**: It includes a process control system that makes it cycle on and off to keep your food and drinks cold at a certain temperature.

- **Air conditioner**: This is another device that includes an integrated process control. You set a certain temperature using your remote and it ensures that the temperature of the room is maintained at that setpoint.

- **Pressing iron**: It includes a thermostat or temperature switch that automatically turns off when the temperature has reached a certain level, which can be referred to as a setpoint.

- **Room heater**: It keeps the temperature of a room at a certain temperature using a heater and a temperature switch.

Process control in any industry will include sensors, actuators, and controllers, which we have learned about in previous chapters – this control can also sometimes involve communication networks. They all work together to ensure the required product quality is achieved.

Upon the completion of this chapter, you should understand process control and the various measurement and devices applicable in industrial process control. You will learn hands-on how to control the temperature using a temperature controller. The last section provides practical knowledge about analog input signal processing in a **Programmable Logic Controller** (**PLC**), which is an advanced part of PLC programming. You will understand the wiring and programming of analog input for Siemens S7 1200 PLCs hands-on using the step-by-step guide provided.

In this chapter, you will cover the following topics:

- Introducing process control
- Exploring process control terms
- ISA symbology
- Temperature measurement and transmitters
- Pressure measurement and transmitters
- Level measurement and transmitters
- Flow measurement and transmitters
- Understanding the process control loop
- A practical example of a single loop process control
- Wiring and programming the analog input of a PLC (Siemens S7 1200) for process control (hands-on)

Introducing process control

In a very simple way, we can define process control as the technique of monitoring and adjusting a process to yield the desired result.

In industry today, raw materials undergo a series of procedures (processes) before they become a finished product. These processes need to be carefully monitored and adjusted to ensure quality products are produced efficiently, economically, and safely. The monitoring of processes and the adjustments necessary are usually automated using control systems. If the monitoring and adjustments were to be done manually, there would be a need for more workers and we cannot achieve the best possible results this way due to human errors.

Types of process control systems

Automatic process control can be categorized as one of the following:

- Open-loop
- Closed-loop (feedback)
- Feedforward
- Feedforward-feedback

Let's have a look at each of the types.

Open-loop control

In open-loop process control, the output does not change the control action. It has no sensor or feedback system. An operation that depends on time is a typical example of an open loop – for example, a washing machine has a timer that an operator will set to a desired time. The machine operates and performs the necessary processes to get the clothes cleaned within the specified time. It has no sensor that determines whether the cloth is cleaned or not. It will stop when the set time duration has been reached. Here, the output does not affect the control action. The following figure shows a block diagram for an open-loop control system:

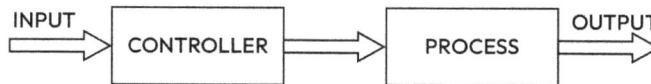

Figure 12.1 – A block diagram for an open-loop process control system

Closed-loop (feedback) control

In closed-loop process control, the output affects the control action. It has a sensor at its output that is fed to its input as a feedback signal. The feedback signal is compared with a setpoint to generate an error signal, which it uses to determine the required action necessary to drive the process to the desired output or result. A closed-loop control system can also be referred to as a feedback control system for this reason. The output is fed back to the input and it uses the measured output (the process variable) and the given setpoint to control the input that keeps the output to the desired value. The following figure shows a block diagram for a feedback process control system:

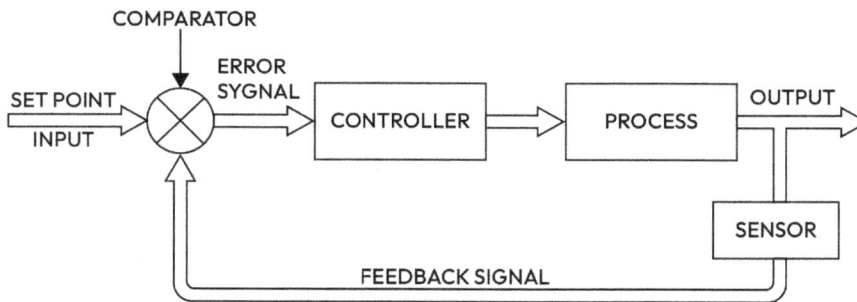

Figure 12.2 – A block diagram for a feedback process control system

Feedback control is a common technology used in various industries such as aerospace (for example, autopilot and rockets), processing (for example, oil and gas, chemical, nuclear reactors, food and beverages, pharmaceutical, and water treatment), and automated manufacturing (for example, robotics and CNC machines).

In the milk manufacturing industry, one of the production processes involves heating the milk to about 70 degrees Celsius within 15 seconds and cooling it immediately to about 3 degrees Celsius to kill germs and increase the shelf life (this process is known as pasteurization). **High Temperature, Short Time (HTST)** pasteurization equipment is usually used.

This equipment uses sanitary plate heat exchangers, utilizing hot water or steam to increase the temperature of the milk. This heating stage is then followed by a rapid cooling stage. HTST pasteurization equipment uses a process control system, usually a feedback control system, to ensure that the milk is pasteurized well.

Feedback control is also common in our homes. Air conditioning is one common example. It has a sensor that reads the temperature of the environment – this reading is fed to a controller that has a setpoint (the desired temperature value you set with your remote – for example, 16 degrees Celsius) and the setpoint is compared with the feedback signal to generate an error signal. The controller uses the generated error signal to decide the control action needed to drive the compressor to ensure that the desired temperature value is maintained. The sensor at its output enables it to determine whether the desired temperature has been reached or not, unlike open-loop.

Feedforward control

With feedforward control, a disruption is measured and corrective action is taken to correct the error before it affects the system. It is a control system that foresees a disturbance and provides a control action before the disruption affects the system. Feedforward control bases its control action on the state of the load variable or input rather than the state of the output, unlike feedback control. It monitors the load variable and takes any necessary action before the process variable (or output) deviates from a setpoint. The output is not fed back, unlike feedback control.

The following figure shows a block diagram for a feedforward process control system:

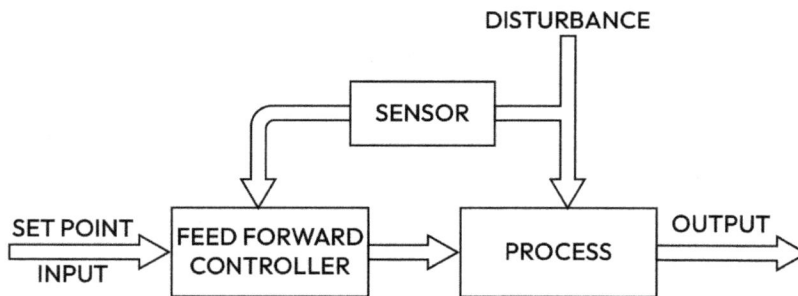

Figure 12.3 – A block diagram for a feedforward process control system

Feedforward-feedback control

Feedforward is not used by itself in industry – it is usually used with feedback control. Hence, feedforward-feedback control can be seen as a type of process control. The feedforward element provides a quick response to any disruption while the feedback element provides the remainder of the response by accurately acting on the error that eventually occurs. It utilizes both the feedforward and feedback controller, as shown in the following figure:

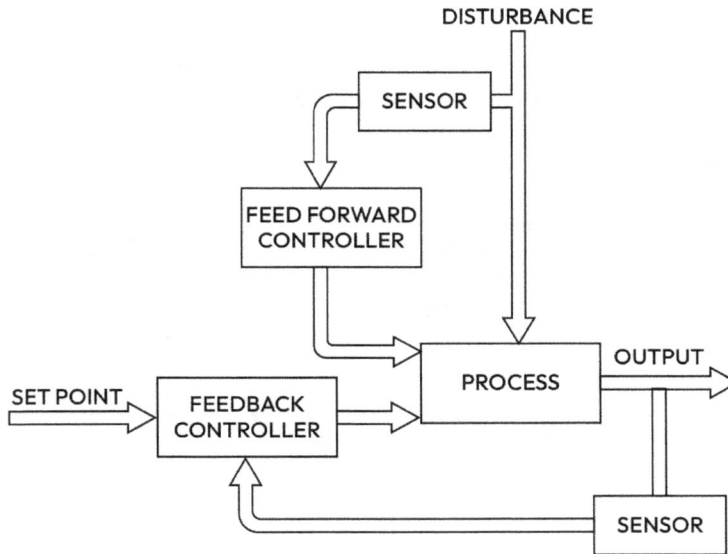

Figure 12.4 – A block diagram for a feedforward-feedback process control system

We have just been introduced to process control and the various types available. Let's proceed to the following section to learn about some process control terms.

Exploring process control terms

Here, we are going to look into some of the common terminology related to process control:

- **Process**: A process is any operation or event (or sequence of operations or events) that causes a physical or chemical change to an input. Raw materials in industry undergo some kind of series of operations, whether being heated, ground, or mixed, for example, before they resemble a finished product.

- **Sensors**: A sensor is a device that senses or detects something – just as we have ears that hear, eyes that see, a nose that smells, and a tongue that tastes. We have sensors that sense or detect various physical properties such as temperature, pressure, level, or flow. Sensors basically convert a physical property into an electrical quantity.

Hence, the four common sensors used in process control are a temperature sensor, a pressure sensor, a level sensor, and a flow sensor – some others include a pH sensor, speed sensor, or a position sensor. In *Chapter 2, Switches and Sensors – Working Principles, Applications, and Wiring*, we discussed sensors but our focus was on sensors that produced digital output, as in **LOW** or **HIGH** (0 or 1). Here in this chapter, we will focus on sensors that produce analog signals.

- **Transmitters**: A transmitter is a device that converts a signal produced by a sensor into a standard signal (4 to 20 mA, 0 to 10 V, or 3 to 15 psi) that is applicable to process control. The output signal from a sensor may be too small and need amplification or conditioning to produce a standard signal. Hence, sensors are usually connected to a transmitter to produce a standard signal output – for example, the following diagram shows how an RTD temperature sensor and an RTD temperature transmitter are wired to produce an output of 4 to 20 mA as the temperature varies:

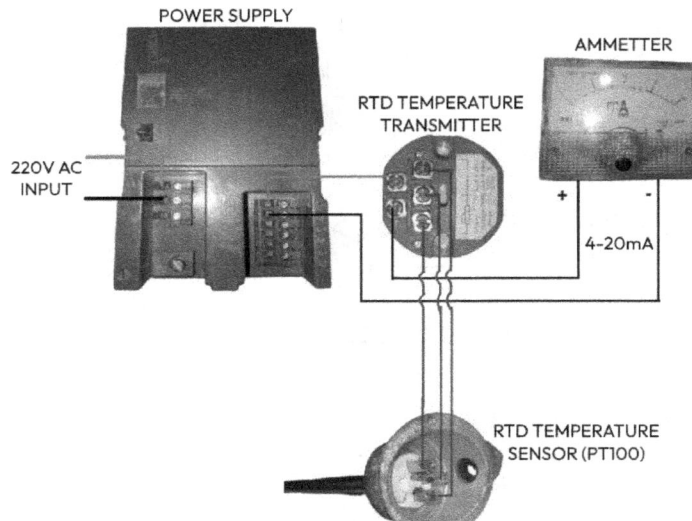

Figure 12.5 – How an RTD temperature sensor and an RTD temperature transmitter are wired to produce an output of 4 to 20 mA as the temperature varies (Credit: Showlight Technologies Ltd.: www.showlight.com.ng)

- **Process variables or process values (PVs)**: This is the actual value or current value of the specific quantity we are measuring. This can be temperature, pressure, level, flow, or another quantity.

- **Setpoints**: This can be referred to as the target value of the process variable. It is the value at which we want the process variable to be maintained. For example, the target temperature of a boiler can be referred to as the setpoint of the boiler.

- **Errors**: This is the difference between the process variable or measured value and the setpoint.

- **Controllers**: This is a device that provides the necessary control action to keep the process variable at the setpoint. It acts on an error signal and provides the necessary control action to the final control element to keep the error at zero so that the setpoint can be maintained.

- **Control elements or final control elements**: These are the devices that the controller operates to keep the process variable at a setpoint. Examples include contactors, relays, solenoid valves, pneumatic control valves, and variable frequency drives.

So, we have just learned about some of the common process control terms. It is important to get familiar with the terms because you will find them in frequent use in any documents or discussions related to process control. Let's now proceed to the following section where we will be learning about ISA symbology.

ISA symbology

The **International Society of Automation** (**ISA**) is a non-profit professional association of engineers, technicians, and managers engaged in industrial automation. ISA is one of the leading standard organizations in the process control trade. They developed a set of symbols for use in the design of process control loops and other engineering diagrams. This set of symbols is referred to as ISA symbology and is detailed in one of their publications (*ISA5.1, Instrumentation Symbols and Identification*), which you can purchase at their website (`https://www.isa.org/`).

ISA symbology is used when creating **Process (or Piping) and Instrumentation Diagrams (P&IDs)**.

A P&ID is a diagram that depicts the details of the piping and instruments used in a processing plant. *Figure 12.14* shows a simple P&ID of a process control loop for a basic level of control. It is usually developed from a **Process Flow Diagram** (**PFD**) at the design stage of an industrial process plant. The PFD shows the basic flow of an industrial process plant. It only shows the major components in the process plant.

The following are some of the ISA symbols extracted from *ISA5.1, Instrumentation Symbols and Identification*:

◯	This is used to represent a device or instrument such as a trans-mitter, sensor, or detector mounted in the field. Examples: (TT) A temperature transmitter mounted in the field. (PT) A pressure transmitter mounted in the field.
⊖	This is used to represent a device or instrument such as a trans-mitter, sensor, or detector mounted in the field. Examples: (LIC / 101) A level indicator and controller. (FIC / 100) A flow indicator and controller.
⊝	This is used to represent a device or instrument at the rear of the main panel (not visible).
⊖	This is used to represent a device or instrument mounted or in-stalled at the front of a secondary or auxiliary panel (visible).

Table 12.1 – ISA symbols for devices and instruments

Note

Individual devices and measuring instruments such as transmitters, sensors, and detectors are represented by a circle.

The following table shows the ISA symbol for PLCs or DCSs:

	This is used to represent a PLC installed in the field.
	This is used to represent a PLC at the front of the main panel (vis-ible at the front of the panel).
	This is used to represent a PLC at the rear of the main panel (not visible).
	This is used to represent a PLC installed at the front of a second-ary or auxiliary panel (visible).

Table 12.2 – ISA symbols for a PLC or DCS

Note

PLCs or DCSs are represented by a square and a diamond inside.

The following table shows the ISA symbols for devices that perform both display and control functions:

	This symbol is used to represent a field-mounted device that both displays readings and performs some kind of control function.
	This is used to represent a device that both displays readings and performs some kind of control function at the front of the main panel (visible at the front of the panel).
	This is used to represent devices that both display readings and perform some kind of control function at the rear of the main panel (not visible).
	This is used to represent a device that both displays readings and performs some kind of control function at the front of a secondary or auxiliary panel (visible).

Table 12.3 – Display and control instruments

> **Note**
> Instruments that both display and perform some kind of control function are represented by a square with a circle inside it.

The following table shows the ISAs symbol for a computer system:

⬡	This is used to represent a computer system installed in the field.
⬡	This is used to represent a computer system at the front of the main panel (visible at the front of the panel).
⬡	This is used to represent a computer system at the rear of the main panel (not visible).
⬡	This is used to represent a computer system installed at the front of a secondary or auxiliary panel (visible).

Table 12.4 – ISA symbols for a computer system

> **Note**
> A computer system that performs some kind of control function is represented by a hexagon.

The ISA symbols in *Table 12.1* to *Table 12.4* can be summarized here for easier understanding:

- A shape with no line running through its center represents an instrument or device located in the field.

- A shape with a single horizontal line running through its center means the instrument or device is located at the front of the main panel (visible at the front of the panel).

- A shape with a dotted line running through its center indicates that the instrument or function is located at the rear of the main panel (not visible).

- A shape with a double line running through its center indicates that the instrument or function is located at the front of a secondary or auxiliary panel (visible).

The following are other ISA symbols extracted from *ISA5.1, Instrumentation symbols and identification*:

– – – – – – – – –	Electronic signals
—//——//—	Pneumatic signals
—⊥——⊥—	Hydraulic signals
▷◁	Venturi tubes
‖	Orifice plates
∿	Coriolis flow meters
⟨	Ultrasonic flow meters (sonic)
▷	Vortex shedding flow meters
⟨⟩	Turbine flow meters
▷◁ / ▷●◁	Gate valves / Globe valves
⊤	Spring diaphragm actuators
⊤	Spring diaphragm actuators with a positioner
Ⓜ	Motor-operated actuators
Ⓢ	Modulating solenoid actuators or solenoid actuators for processing an on-off valve
⊤	Manual actuators or hand actuators

Table 12.5 – Other ISA symbols

More information is available in the *ISA5.1, Instrumentation symbols and identification* document.

We have just learned about the ISA symbology. This knowledge will help us interpret P&ID diagrams. Learn more using the *ISA5.1, Instrumentation symbols and identification* document and other resources available online.

In the following section, you will learn about the measurement of various quantities (temperature, pressure, level, and flow) and their associated transmitters.

Temperature measurement and transmitters

Temperature is the measure of the degree of hotness or coldness of something. Temperature measurement is an important task in the process industry and can be achieved with the aid of a sensor and transmitter. Common sensors are **resistant temperature detectors (RTDs)** and **thermocouples**.

An RTD is a device whose resistance changes as the temperature changes. It has a temperature range of -260 to 850 degrees Celsius. It has better sensitivity than a thermocouple but it is more expensive. The types of RTD sensors, when categorized by the resistant element material, are Pt100, Pt1000, Ni120, and Cu100. The letters represent the material of the element while the number represents the resistance at 0 degrees Celsius. Pt100 is the most common RTD. The resistance element is Platinum and it has a resistance of 100 Ohms at 0 degrees Celsius.

An RTD temperature sensor requires an RTD transmitter to produce a standard signal output (for example, 4 to 20 mA) as the resistance changes with the temperature. *Figure 12.6* shows an RTD temperature transmitter for PT100. The temperature range is 0 to 300 degrees Celsius with an analog signal output of 4 to 20 mA. This means that the transmitter will produce 4 mA at 0 degrees Celsius and 20 mA at 300 degrees Celsius.

Figure 12.6 – An RTD temperature transmitter for PT100 (Credit: Showlight
Technologies Ltd.: www.showlight.com.ng)

Figure 12.5 shows how an RTD temperature sensor and RTD transmitter are wired.

A thermocouple consists of two dissimilar wires joined together at one end to form a sensing joint. A voltage (in mV) is produced and varies as the temperature changes. It has a temperature range of -270 to 1800 degrees Celsius. The thermocouple temperature sensors available include Type K, Type J, type T, Type E, Type N, and Type S. Each type has a temperature range. Type K is the most common thermocouple temperature sensor with a temperature range of -270 to 1,260 degrees Celsius.

A thermocouple temperature sensor requires a thermocouple temperature transmitter to produce a standard signal output (for example, 4 to 20 mA or 0 to 10 V), as the mV produced changes with temperature. *Figure 12.7* shows a thermocouple temperature transmitter for a Type K thermocouple. The temperature range is 0 to 400 degrees Celsius with an analog signal output of 4 to 20 mA. This means that the transmitter will produce 4 mA at 0 degrees Celsius and 20 mA at 400 degrees Celsius.

Figure 12.7 – A thermocouple temperature transmitter for a Type K thermocouple sensor (Credit: Showlight Technologies Ltd.: www.showlight.com.ng)

You have just learned about temperature measurement and transmitters. Let's proceed to the following section to learn about pressure measurement and transmitters.

Pressure measurement and transmitters

Pressure can be defined as the amount of force applied per unit area (*F/A*). The processes in industry require accurate and reliable pressure measurements for safe operations and the production of quality products. The effects of pressure include position movements, changes in resistance, or other physical changes that can be measured. Common pressure sensors employ a Bourdon tube, diaphragm, bellows, force balance or variable, and a capacitance arrangement.

A pressure transmitter is required to produce a standard signal (4 to 20 mA or 0 to 10 V) from whatever pressure effect the sensor is based on. Mostly, a pressure transmitter comes with an inbuilt sensor and is sold as a single package, as shown in the following figure:

Figure 12.8 – A front view of a pressure transmitter (Credit: Showlight
Technologies Ltd.: www.showlight.com.ng)

The following figure shows a simple connection to a pressure transmitter to produce a signal of 4 to
20 mA as the pressure varies:

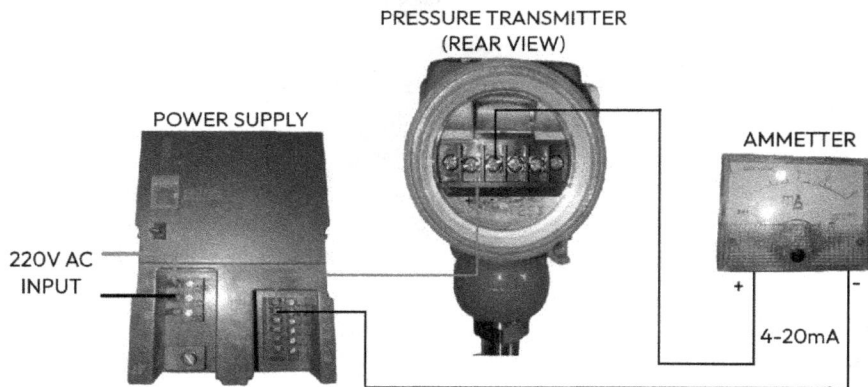

Figure 12.9 – A connection to a pressure transmitter (Credit: Showlight
Technologies Ltd.: www.showlight.com.ng)

The ammeter can be replaced with the analog input of a PLC that supports the current input (4 to 20
mA) if the signal will be fed to a PLC to perform a control function. If the analog input of the PLC
only supports a voltage input (0 to 10 V), there will be a need to convert the current input of 4 to 20
mA to a voltage by connecting a 500-Ohm resistor across the signal. The last section of this chapter
demonstrates this with a practical example.

You've now learned about pressure measurement and transmitters. Let's proceed to learn about level
measurement and transmitters in the following section.

Level measurement and transmitters

The accurate measurement of levels is required in industry to maintain the proper fluid levels for safe operation, quality production, or the smooth running of equipment. Various sensor technologies are available for measuring levels in liquid and also in solids. Capacitance level sensors measure the change in capacitance to determine liquid levels. An ultrasonic level sensor emits a sound wave. The liquid level is directly proportional to the time delay between the wave emission and its reflection. Differential pressure sensor uses the difference in pressure between two points in a tank to determine the level. Level sensors mostly have a transmitter integrated that enables them to produce a standard signal of 4 to 20 mA or 0 to 10 V as a level varies.

There are various level transmitters. The following figure shows a hydrostatic level transmitter that is usually immersed in water or some liquid whose level is to be measured:

Figure 12.10 – A hydrostatic level transmitter
(Credit: Showlight Technologies Ltd.: www.showlight.com.ng)

The following figure shows the wiring of a level transmitter that will produce an output of 4 to 20 mA as the level of water varies:

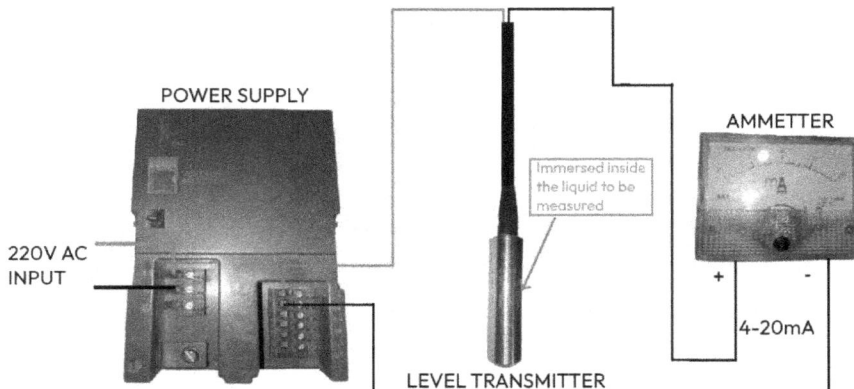

Figure 12.11 – The wiring of a level transmitter that will produce an output of 4 to 20 mA as the level of water varies (Credit: Showlight Technologies Ltd.: www.showlight.com.ng)

The ammeter can be replaced with an analog input of PLC that supports the current input (4 to 20 mA) if the signal will be fed to a PLC to perform a control function. If the analog input of the PLC only supports a voltage input (0 to 10 V), there will be a need to convert the 4 to 20 mA input to a voltage by connecting a 500-Ohm resistor across the signal. The last section of this chapter demonstrates this with a practical example.

You have now learned about level measurement and transmitters. Let's proceed to the next section to learn about flow measurement and transmitters.

Flow measurement and transmitters

Flow measurement is an operation carried out to measure the flow of fluids in a pipe using an instrument. Several flow-measuring instruments exist, which include the following:

- **A differential pressure flow meter**: This uses a pressure difference created by a sensing element (for example, an orifice plate or a Venturi tube) to determine the rate of a flow. A differential pressure transmitter has a high- and low-pressure port to which the high- and low-pressure side of the sensing element will be connected and it produces a standard signal of 4 to 20 mA that corresponds to the flow rate.

The following figure shows a Rosemount differential pressure transmitter:

Figure 12.12 – A Rosemount differential pressure transmitter (this file is licensed under the Creative Commons Attribution-Share Alike 4.0 International license)

- **A velocity flow meter**: This determines a flow rate by measuring the flow velocity. Velocity flow meters include the following:

  - *A turbine flow meter*: This measures the flow rate by measuring the speed of an internal rotating turbine or gears.

  - *A vortex flow meter*: This measures the flow rate by measuring the frequency of vortices created within the flow element. It is suitable for measuring steam, vapors, or clean liquids. The following figure shows a view of a mounted vortex flow meter:

Figure 12.13 – A mounted vortex flow meter (this file is licensed under the
Creative Commons Attribution-Share Alike 4.0 International license)

- **A mass flow meter**: This measures the flow rate by measuring the density of a fluid using a sensing element. Various mass flow meters based on a sensing element include the following:

  - *A Coriolis mass flow meter*: This relates the mass flow to the fluid's inertia sensed when induced by vibration within the meter tubes. It is suitable for measuring liquids but can also measure gases. The following figure shows a Coriolis mass flow meter:

Figure 12.14 – A Coriolis mass flow meter (this file is licensed under the Creative
Commons Attribution-Share Alike 4.0 International license)

- *A thermal mass flow meter*: This relates the mass flow rate to the quantity of energy transferred from a heated probe to the flowing fluid. It is suitable for measuring gases. *Figure 12.15* shows a thermal mass flow meter.

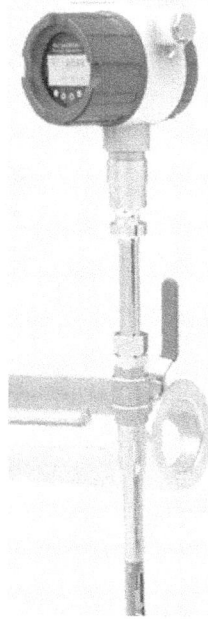

Figure 12.15 – A thermal mass flow meter (this file is licensed under the Creative Commons Attribution-Share Alike 4.0 International license)

We have just learned about flow meters (differential pressure, velocity, and mass flow meters), which should give you a basic idea of flow measurement. Let's proceed to the following section to compound our knowledge about process control.

Understanding the process control loop

The process control loop is a set of devices and tools designed to maintain the desired output from a process variable. Control loop components and instruments measure the variable, respond to it, and control it to maintain a setpoint.

A control loop system can be either an open-loop or closed-loop system, which we explained in an earlier section of this chapter. It usually consists of a sensor or transmitter, a controller, and an actuator. The following figure shows a process control loop for level control:

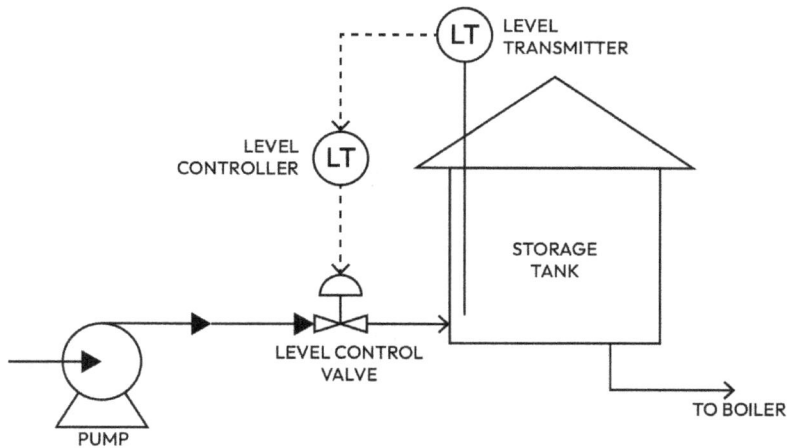

Figure 12.16 – The process control loop for level control

The process control loop in the preceding figure is explained briefly as follows: the pump supplies water to the tank and a control valve (the final control element) controls the flow rate depending on what it learns about the tank level through the signal it receives from the level controller, which has a setpoint. The level transmitter provides a feedback signal to the level controller. The level controller compares the feedback signal from the level transmitter to the setpoint and decides the control signal to give to the control valve to ensure an appropriate flow rate depending on the level of the liquid in the tank.

A practical example of single-loop process control

Here, we will use a temperature sensor, a temperature controller, and a contactor (actuator) to keep the temperature of a heater at the desired level (or setpoint).

Materials/devices

The required materials include the following:

- A thermocouple temperature sensor (Type K)
- A temperature controller (JTC903)
- A contactor
- A heating element

Wiring

Make the connection as shown in the following diagram (*Figure 12.17*):

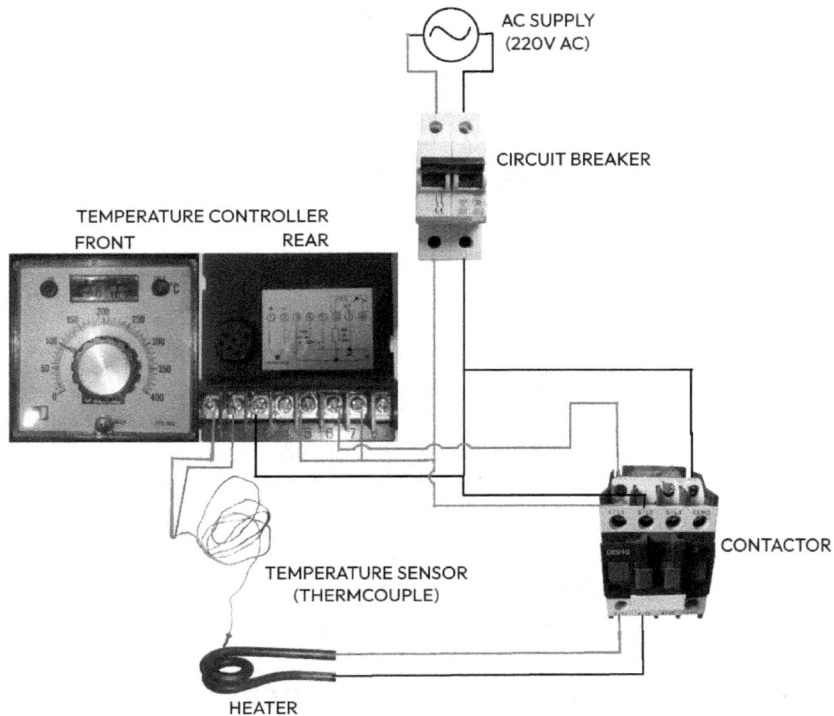

Figure 12.17 – Pictorial diagram of a temperature control system (Credit: Showlight Technologies Ltd.: www.showlight.com.ng)

Operation:

1. Set the dial on the front of the temperature controller to the setpoint of your choice, that is, the temperature you want to maintain – for example, 100 degrees Celsius.

2. Supply power to the circuit by switching on the breaker.

3. When the temperature of the heater is below the setpoint, the common terminal (C) of the temperature controller (that is, terminal 7) will be connected to the normally open contact (that is, terminal 6) – this will energize the contactor because the A2 coil terminal of the contactor is connected to Neutral (N) of the AC supply while the A1 coil terminal that is connected to terminal 6 will now be on Live (L).

4. When the temperature of the heater is equal to the setpoint, the common (C) terminal of the temperature controller (terminal 7) will be disconnected from terminal 6 and connected to terminal 8. This will de-energize the coil of the contactor and power will no longer flow to the heater. Hence, the heater cools gradually and as it falls below 100, the coil is energized again and the process repeats.

This is a simple control that is not very effective, as you cannot maintain the exact setpoint because there will always be an overshoot. A better controller to use is a **Proportional-Integral-Derivative** (**PID**) temperature controller, which can be properly tuned to maintain the desired temperature.

In this section, we have practically demonstrated a simple process control for temperature. You can do this on your own by sourcing the required devices and making the connections as provided in the wiring (*Figure 12.17*). This should give you hands-on experience with a simple process control system.

In the following section, we will learn how to wire and program the analog input of a Siemens PLC to control a process.

Wiring and programming the analog input of a PLC (Siemens S7 1200) for process control (hands-on)

In this section, we are going to dig deeper into our PLC wiring and programming by learning how to wire and program the analog input of PLC to control a process. Analog programming is often seen as the complicated aspect of using a PLC. I have made it simple and easy to understand here. The practical part (hands-on) will give you a better understanding. While every part of this book is important, in particular, *Chapter 7, Understanding PLC Hardware and Wiring, Chapter 8, Understanding PLC Software and Programming with TIA Portal,* and *Chapter 9, Deep Dive into PLC Programming with TIA Portal,* are all required to easily understand this section.

There are specific controllers designed for temperature control – for example, the JTC903 controller used in the previous section for temperature control. We also have controllers designed specifically to measure pressure or level. However, a PLC can be wired and programmed to control any measurable quantity such as temperature, level, or pressure. We just need to use the right sensor or transmitter and actuator, wire it in the right way, and program it to perform the desired control function. Hence, the PLC is a key component in process control.

Let's have a brief look at analog signals for a better understanding of the practical part that follows.

Understanding analog signals

Unlike digital signals, which can only be in two states (0 or 1), analog signals have a continuous range of values from the minimum to the maximum. They can present what's between 0 (minimum, for example, 0 V or 4 mA) and 1 (maximum, for example, 5 V, 10 V or 20 mA). When a voltage varies from 0 V to 10 V, or when a current varies from 4 mA to 20 mA, it is called an analog signal. Examples of sensors and transmitters that produce an analog signal output are temperature transmitters, pressure transmitters, and flow transmitters. Before a PLC can process analog values, it must be converted into digital information first – this is done by the **Analog-to-Digital Converter** (**ADC**) in the analog module of a PLC. One important feature of an analog module is its resolution in bits.

Basically, an 8-bit analog module will have 0 to 254 representing 4 mA to 20 mA or 0 V to 10 V depending on whether the analog signal connected to it is a current or a voltage. A 10-bit analog module will have 0 to 1023 representing 4 mA to 20 mA or 0 V to 10 V. A 12-bit analog module will have 0 to 4095 representing 4 mA to 20 mA or 0 V to 10 V:

Analog values (4 to 20 mA)	4 mA	8 mA	12 mA	16 mA	20 mA
Analog values (0 to 10 V)	0 V	2.5 V	5 V	7.5 V	10 V
Analog values (2 to 10 V)	2 V	4 V	6 V	8 V	10 V
Digitized values	**0**	**256**	**512**	**767**	**1023**

Table 12.6 – The digitized values of analog signals

We are going to use an S7-1200 (CPU 1211C, AC/DC/RLY) PLC in our practical example. It has two inbuilt analog input channels, as shown in the following image. The analog module has a resolution of 10 bits and supports an analog voltage (0 to 10 V). Hence, there is a need to convert a transmitter with an output of 4 mA to 20 mA to a voltage (2 to 10 V).

This can be done with the aid of a 500-Ohm resistor:

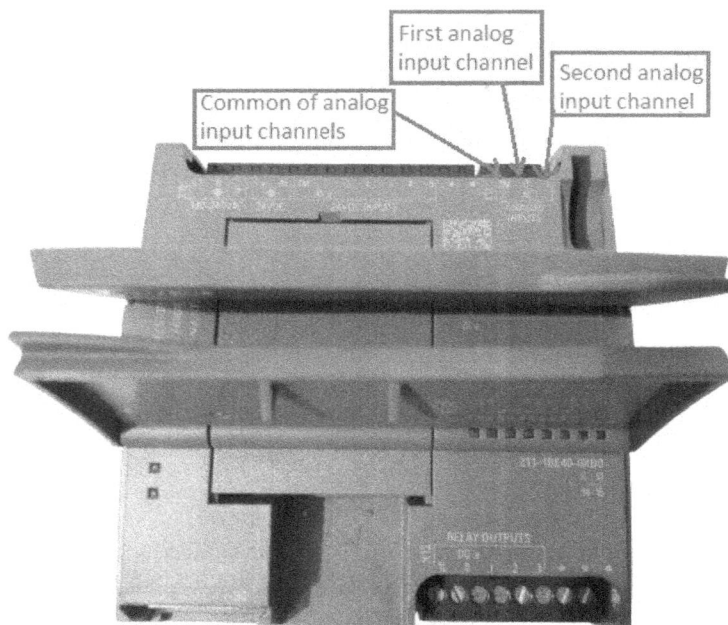

Figure 12.18 – An S7-1200 (CPU 1211C, AC/DC/RLY) PLC showing analog input channels
(Credit: Showlight Technologies Ltd.: www.showlight.com.ng)

There are various data types for the Siemens S7-1200 controller, which include `bool`, `byte`, `word`, `INT`, and `REAL` types. The most common data types used in analog signal processing are `INT` and `REAL` because read-in analog signal values exist as 16-bit integers, which `INT` can hold since `INT` is 16 bits. To avoid rounding errors, `REAL` should be used to get more precise results since `REAL` can hold floating-point numbers or decimals.

Reading analog values

With S7-1200, the address of the first analog input channel is `%IW64`, while the second is `%IW66`. An analog output of 4 to 20 mA, 0 to 10 V, or 2 to 10 V from the transmitter is connected to the first or second channel.

Regardless of resolution, the Siemens PLC always converts the voltage or current into a word value (digitized value) in which the rated maximum value is 27,648 and the data type is `INT`. Hence, 0 to 10 or 4 to 20 mA will be rated 0 to 27,648, as shown in the following table:

Analog values (4 to 20 mA)	4 mA	8 mA	12 mA	16 mA	20 mA
Analog values (0 to 10 V)	0 V	2.5 V	5 V	7.5 V	10 V
Digitized values	**0**	**6,912**	**13,824**	**20,736**	**27,648**

Table 12.7 – The digitized values of five analog signals from minimum
(0 mA or 0 V) to maximum (20 mA or 10 V) for Siemens

0 to 27,648 is a Siemens standard (rated range) for representing the analog signal input of 0 to 10 V or 4 to 20 mA in their analog module. It is not dependent on resolution and cannot be changed. However, the resolution explained earlier affects the accuracy of your reading. The higher the resolution, the more detail you can capture in your reading. 27,648 is not actually the maximum value – it is just the rated maximum. The real maximum value is 32,768 (that is, 15 bits = 2^{15}) and can be used in a calculation to determine the effect of the resolution, as indicated here:

- 8-bit resolution is 2^8 = 256. The smallest change you can have will be 32,768/256 = 128. Hence, you can change in increments of 128 from 0 to 27,648 for 0 V to 10 V, as in, 0, 128, 256 .... 27,648.

- 12-bit resolution is 2^{12} = 4,096. The smallest change you can have will be 32,768/4,096 = 8. Hence, you can change in increments of 8 from 0 to 27,648 for 0 to 10 V, as in, 0, 8, 16 .... 27,648.

- 15-bit resolution is 2^{15} = 32,768. The smallest change you can have will be 32,768/32,768 = 1. Hence, you can change in increments of 1 from 0 to 27,648 for 0 to 10 V, as in, 0, 1, 2, 3, 4 .... 27,648.

The digitized value of 0 to 27,648 will need to be normalized for further processing.

Normalization and scaling

Normalize is an inbuilt function in S7-1200 for converting any given data from a minimum to a maximum to 0 to 1 in float values. It requires you to enter the minimum and maximum values as well as the register containing the data to be converted – for example, `%IW64` if the analog signal is connected to the first analog input channel. The minimum is 0 while the maximum is 27,648. Our practical session will show the way to use **normalize**.

Scale is an inbuilt function in S7-1200 for converting the normalized value from 0 to 1 into the required range. You will learn how to **scale** in the practical session.

Analog input signal processing (a practical demonstration)

Having gone through the basics of analog signal processing, let's get our hands dirty to get a practical understanding of the theories. We will start with a simulation using a virtual PLC to enable those without a real PLC to demonstrate it. We will then use a real PLC that can be practiced by anyone that can afford a PLC and the other materials required.

Simulation (using a virtual PLC)

The following step-by-step guide requires that you have TIA Portal installed on your laptop or PC. Refer to *Chapter 8*, *Understanding PLC Software and Programming with TIA Portal*, to learn how to download and install TIA Portal. You can download and use it for 21 days:

1. Start TIA Portal and create a new project for S7-1200 (CPU 1211C AC/DC/RLY) to generate the following environment:

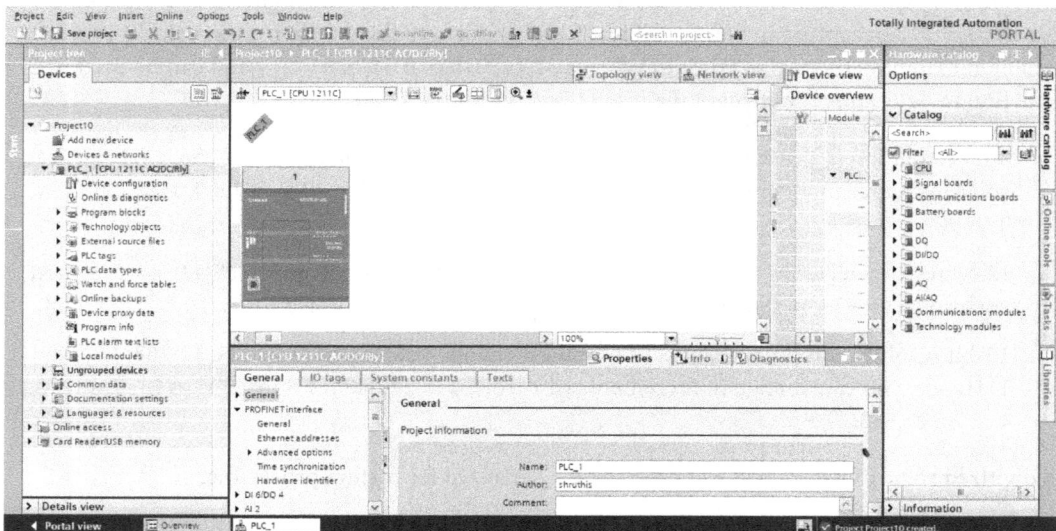

Figure 12.19 – The created project

2. Expand **Program blocks** and double-click **Main [OB1]** to get to the programming environment:

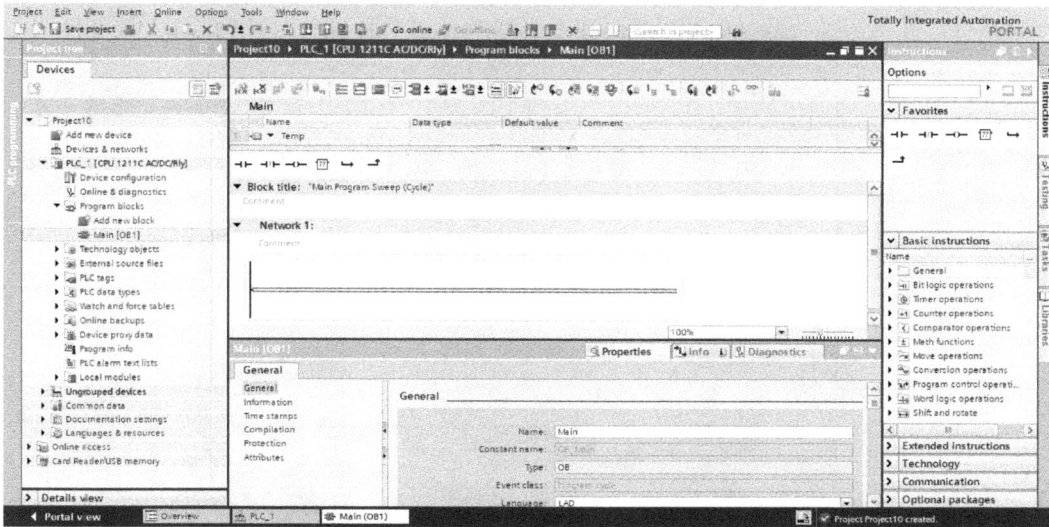

Figure 12.20 – The programming environment

3. Expand conversion operation in the basic instruction list on the right-hand side of the screen and drag the NORM_X command to **Network 1**:

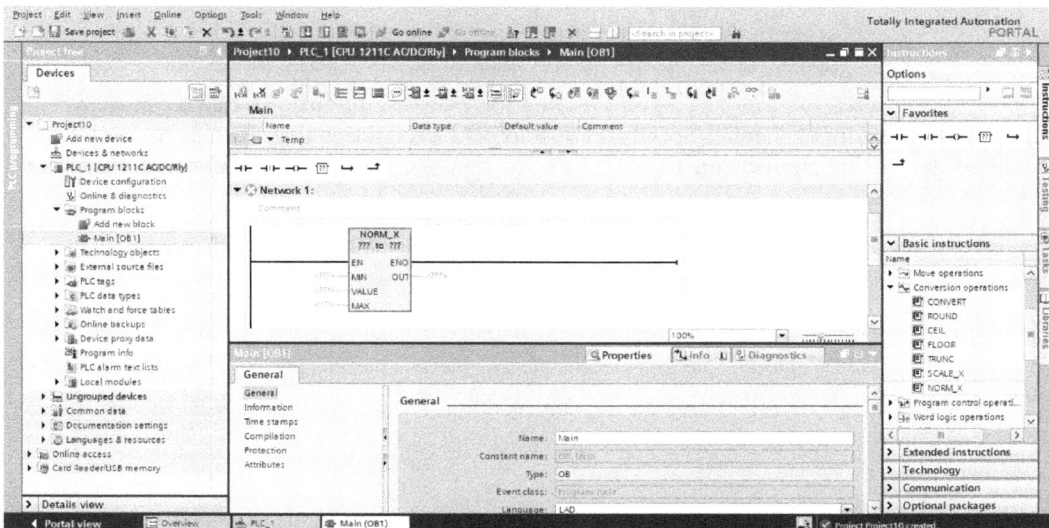

Figure 12.21 – NORM_X instruction 1

4. Set the data types as INT to REAL in the **???** to **???** fields underneath NORM_X in the instruction:

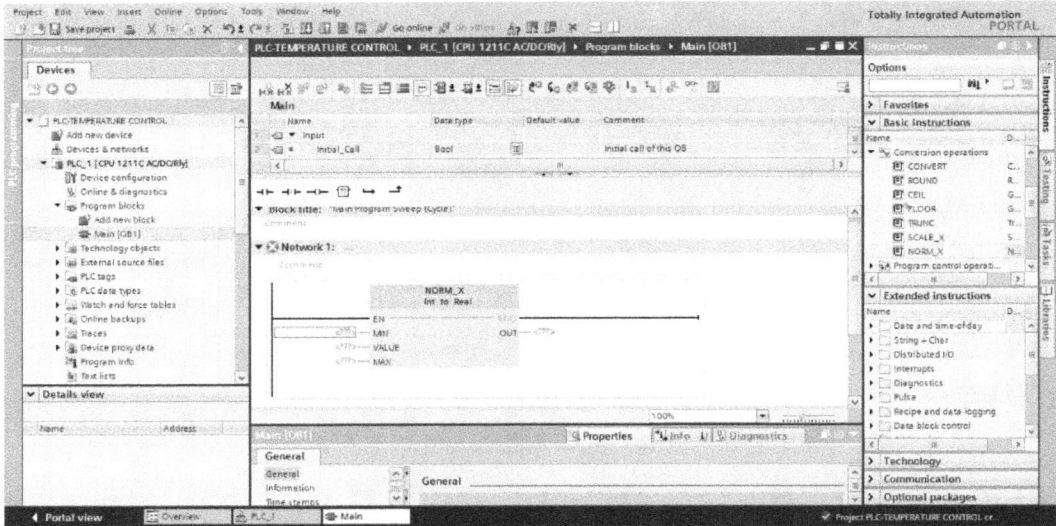

Figure 12.22 – NORM_X instruction 2

5. Input 0 for MIN and 27648 for MAX.

6. Specify the register that will hold the data you want to use to test for VALUE – for example, MD0.

> **Note**
>
> If working with a real S7-1200 PLC, specify %IW64 for VALUE if the analog signal is connected to the first analog input channel. Specify %IW66 for VALUE if the analog signal is connected to the second analog input channel.

7. Specify the register that will hold the normalized data for OUT – for example, MD4:

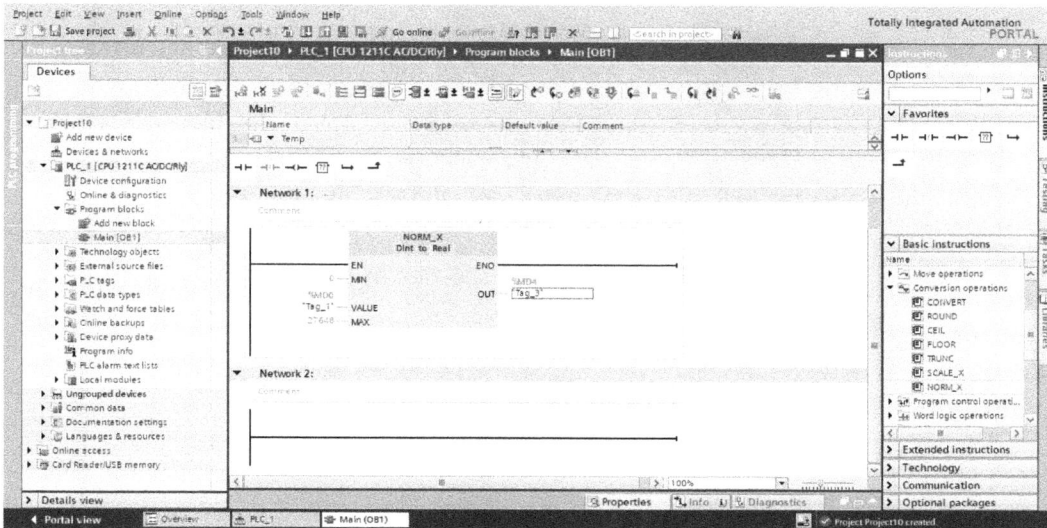

Figure 12.23 – NORM_X instruction 3

> **Note**
>
> MD0 will be replaced with the address of the analog input channel of the PLC when using a real PLC.

8. Expand conversion operation in the basic instruction list on the right-hand side of the screen and drag the SCALE_X command to **Network 1**:

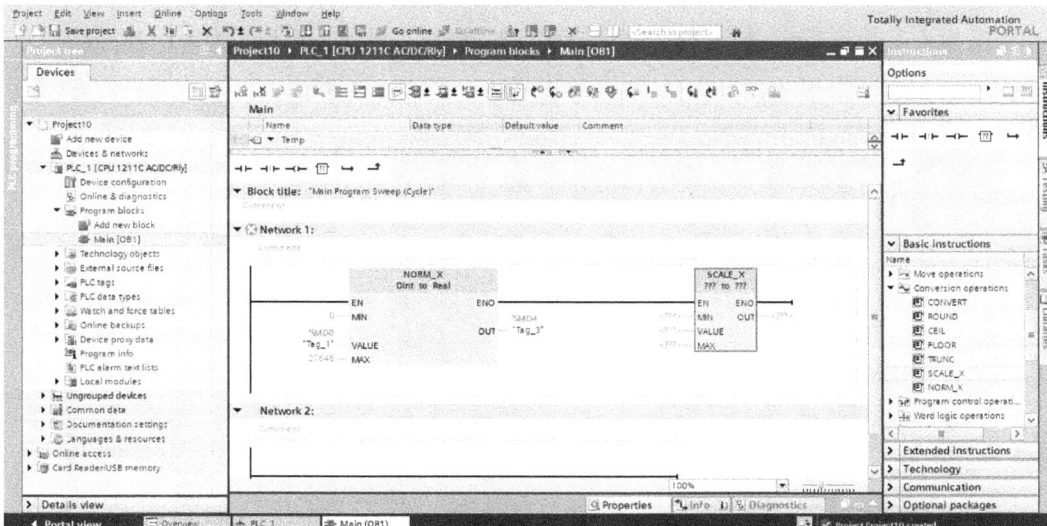

Figure 12.24 – SCALE_X instruction 1

9. Set the data types as REAL to REAL in the **???** to **???** fields underneath SCALE_X in the instruction.

10. Input 0 for MIN and 300 for MAX since the transmitter we will use has an output of 4 to 20 mA with a range of 0 to 300 degrees Celsius.

11. Specify the register of the normalized data (for example, MD4) for VALUE.

12. Specify the register that will contain the scaled result for OUT – for example, MD8.

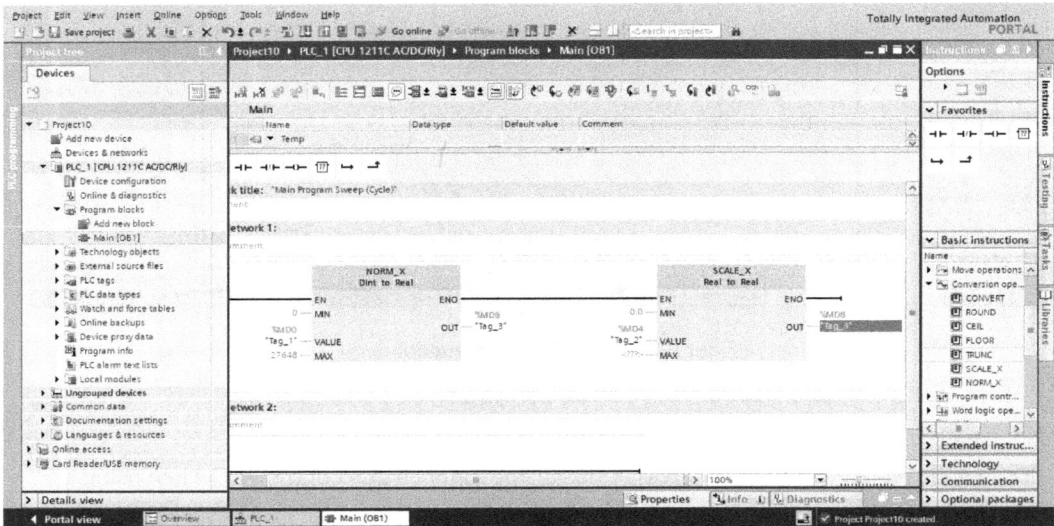

Figure 12.25 – SCALE_X instruction 2

13. Click **Compile** to ensure there are no errors.

Let's now simulate it to see how it works. Refer to the steps in *Chapter 9, Deep Dive into PLC Programming with TIA Portal*, to learn about how to get to the simulation environment:

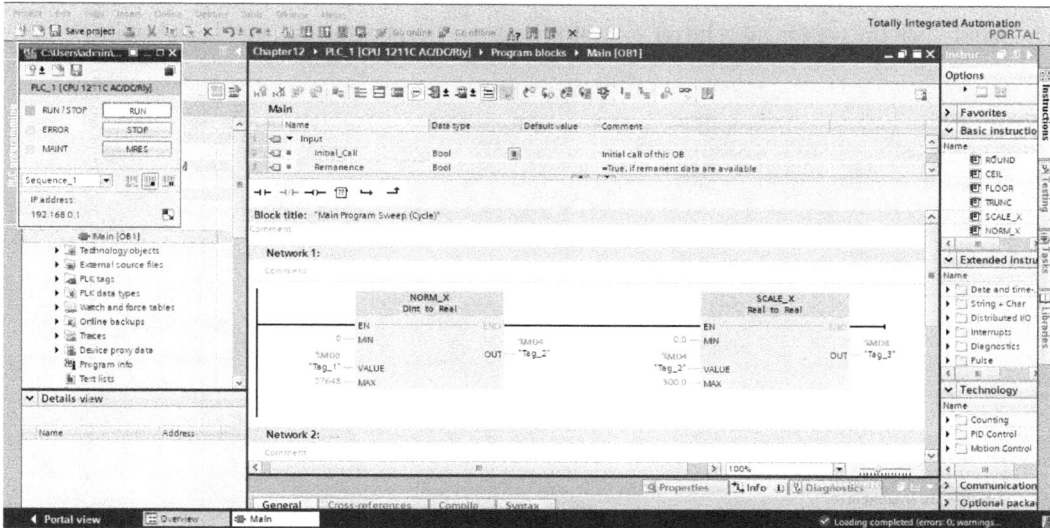

Figure 12.26 – The simulator started

14. Click **Monitoring on/off**:

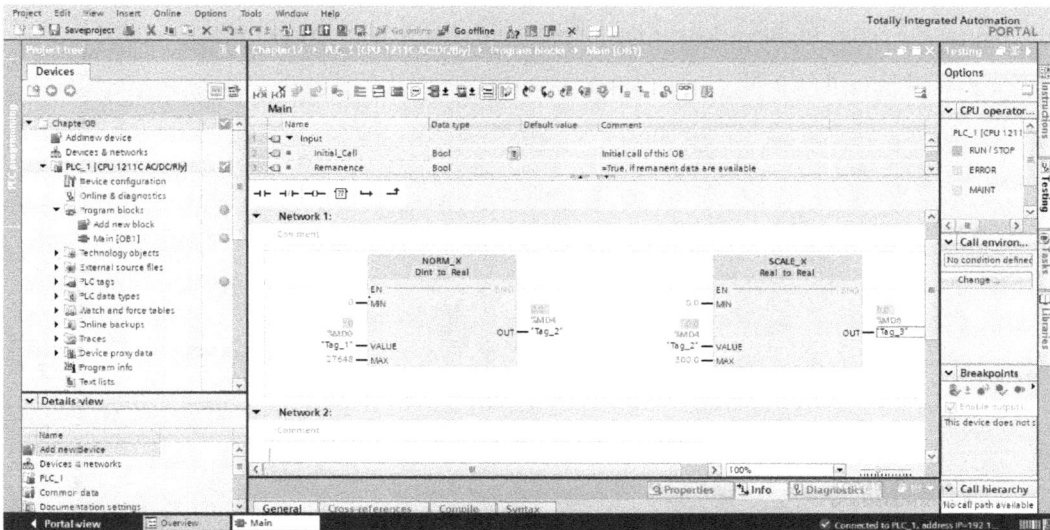

Figure 12.27 – Monitoring on/off

15. Right-click the tag for MD0, hover over **Modify**, then **Display format**, and click **Decimal**:

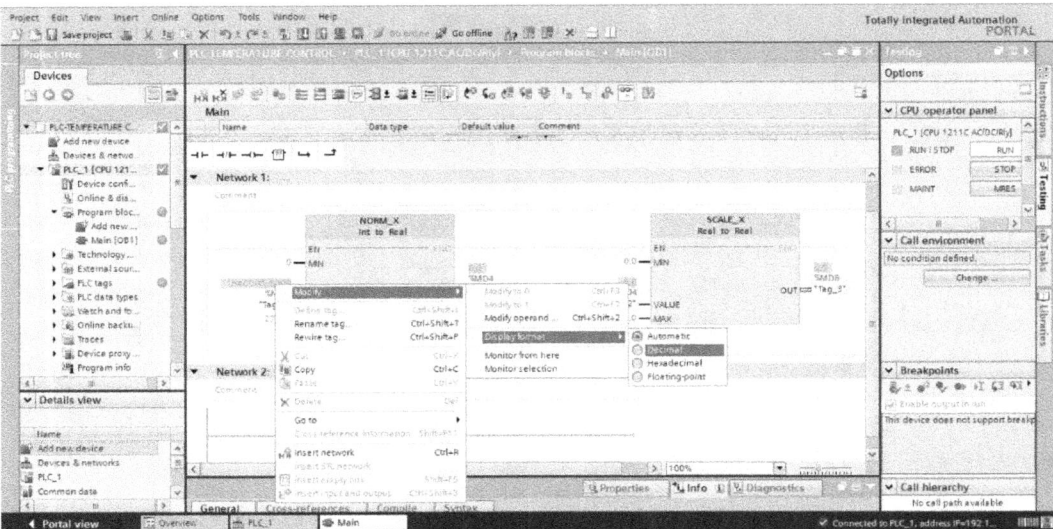

Figure 12.28 – Changing the display format

16. Repeat the preceding steps for the tag for MD8 and choose **Floating point**:

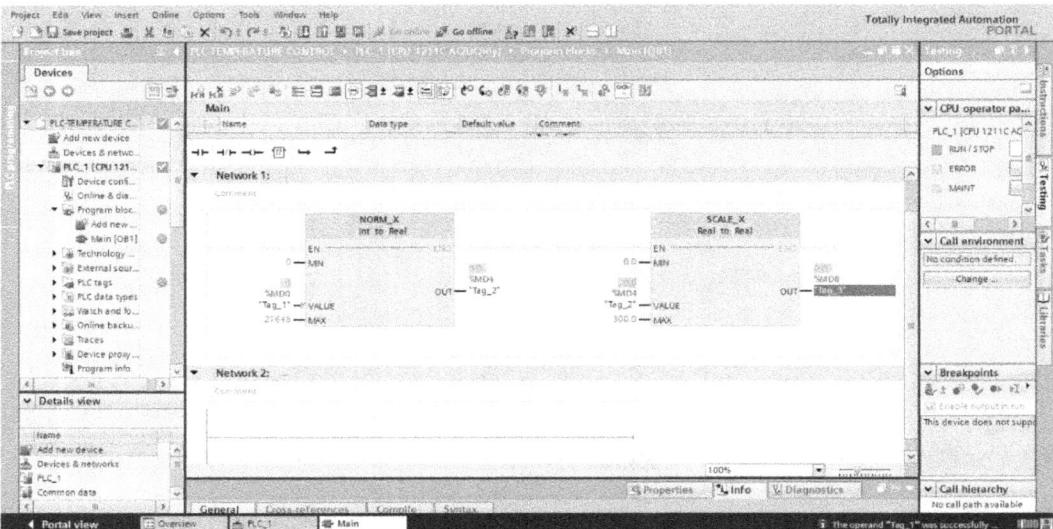

Figure 12.29 – The simulation running

17. Now, right-click the tag for the input value (tag_1), hover over **Modify**, and click **Modify operand**:

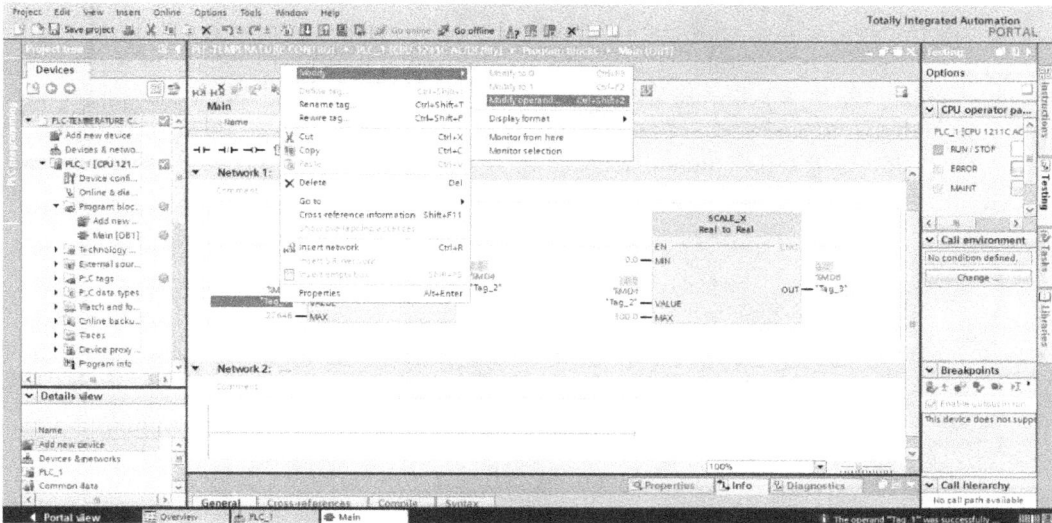

Figure 12.30 – Modifying operand-1

18. Change the format from HEX to DEC (decimal) and then fill in **Modify value** – for example, 13824. Click **OK**:

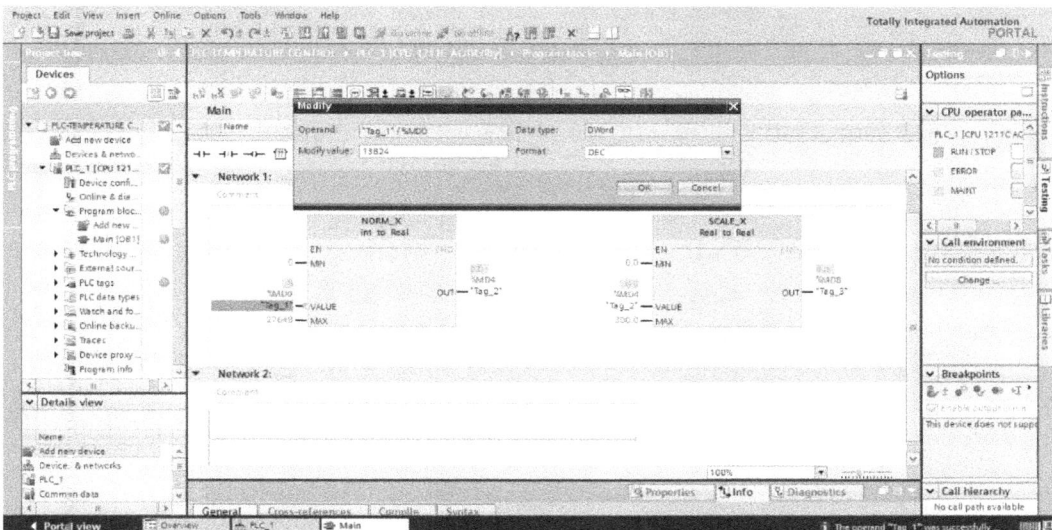

Figure 12.31 – Modifying operand-2

It can be seen from the simulation result in figure 12.32 that when the digitized value (the value at MD0) in NORM_X is half the full value (that is, 13824), the output became 0.5, which means half of 1 since **normalize** (NORM_X) converts the input to a value from 0 to 1. It can also be seen that **scale** (SCALE_X) converted the normalized value of 0 to 1 to 0 to 300, as we can see that the 0.5 (half of 1) gave 150.0 (half of 300) at the output of SCALE_X:

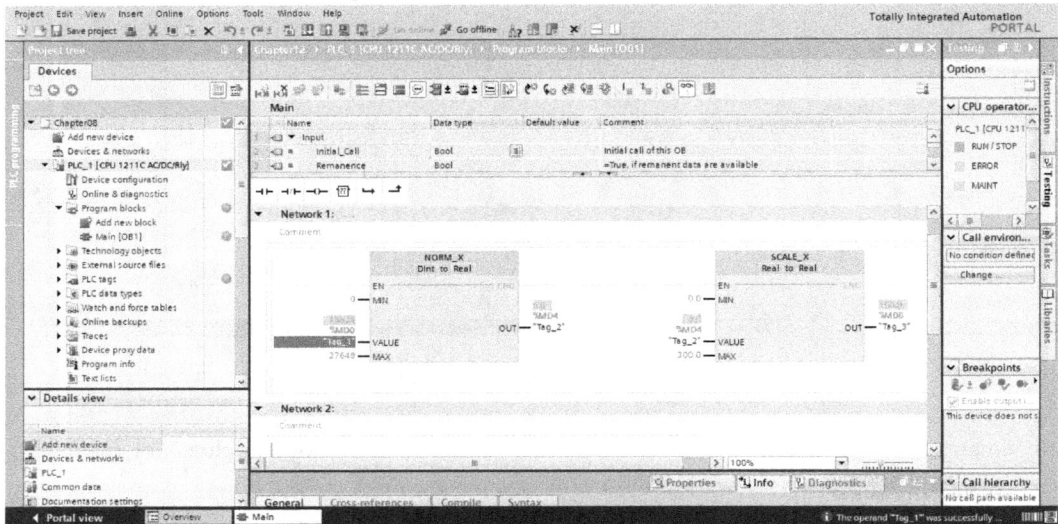

Figure 12.32 – Simulation result 1

You can play with various other values at MD0 to see how **normalize** and **scale** work.

Try 4608 as the digitized value in MD0 – you should get 50.0 at MD8, as shown here:

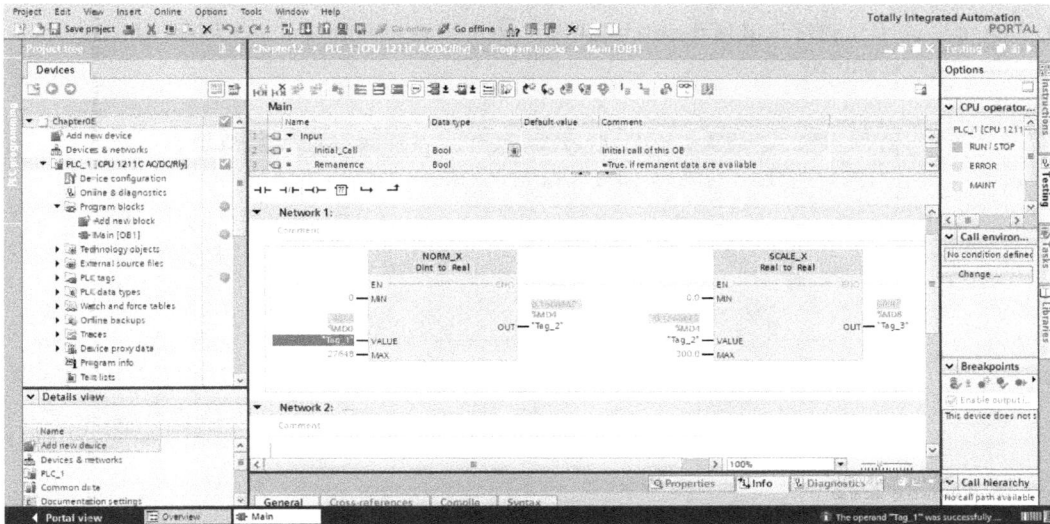

Figure 12.33 – Simulation result 2

Let's go further.

We will now use the analog signal result to energize an Assignment (coil) to get an understanding of using an analog signal input to control a digital output:

1. Stop the simulation by clicking **Monitor on/off** and **Yes** to go offline in the message that appears.

2. Expand **Comparator operations** in the **Basic instructions** list on the right-hand side of the screen and drag the CMP >= instruction to **Network 2**:

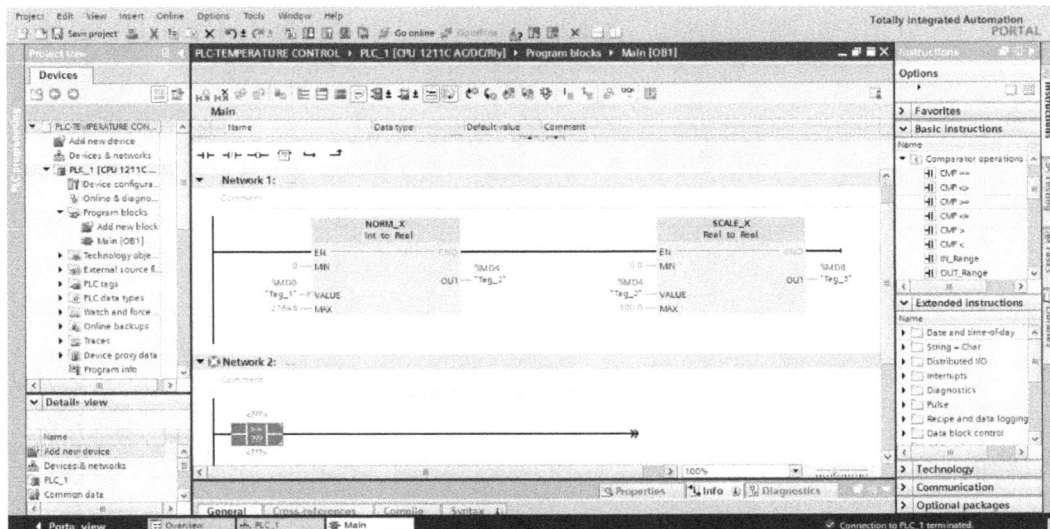

Figure 12.34 – CMP>= instruction 1

3. Double-click the **???** field in the middle of the comparator command and select Real as the data type, as shown:

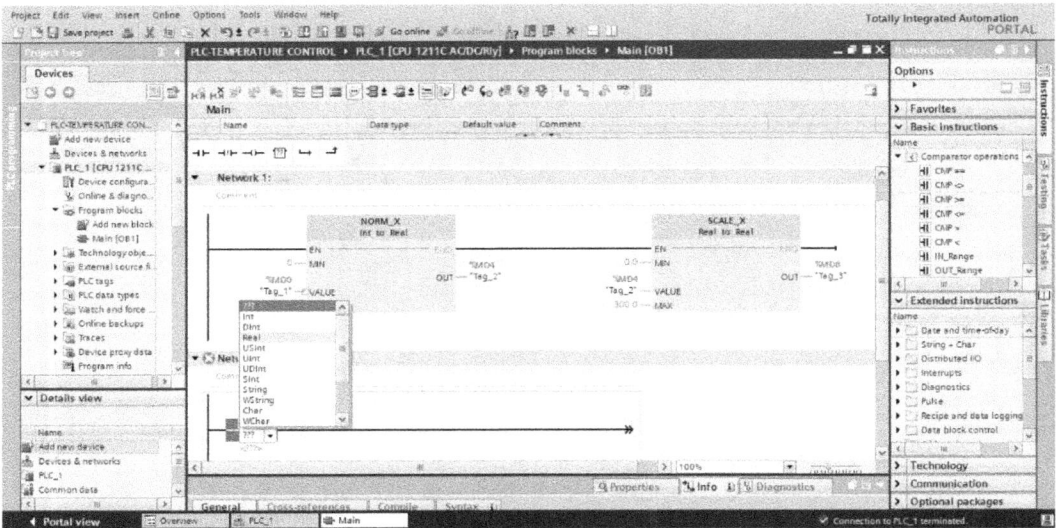

Figure 12.35 – CMP>= instruction 2

4. Input %MD8 in the **???** field at the top of the comparator instruction.

5. Input 50 at the **???** field at the bottom of the comparator instruction:

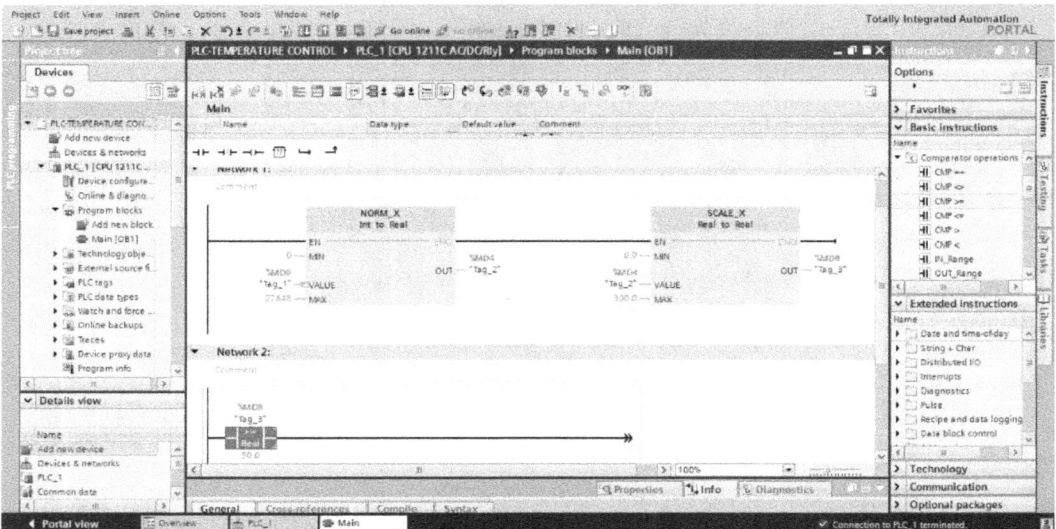

Figure 12.36 – CMP>= instruction 3

6. Expand **Bit logic operations** under the **Basic instructions** list on the right-hand side of the screen and drag the Assignment (coil) instruction to your **Network 2**:

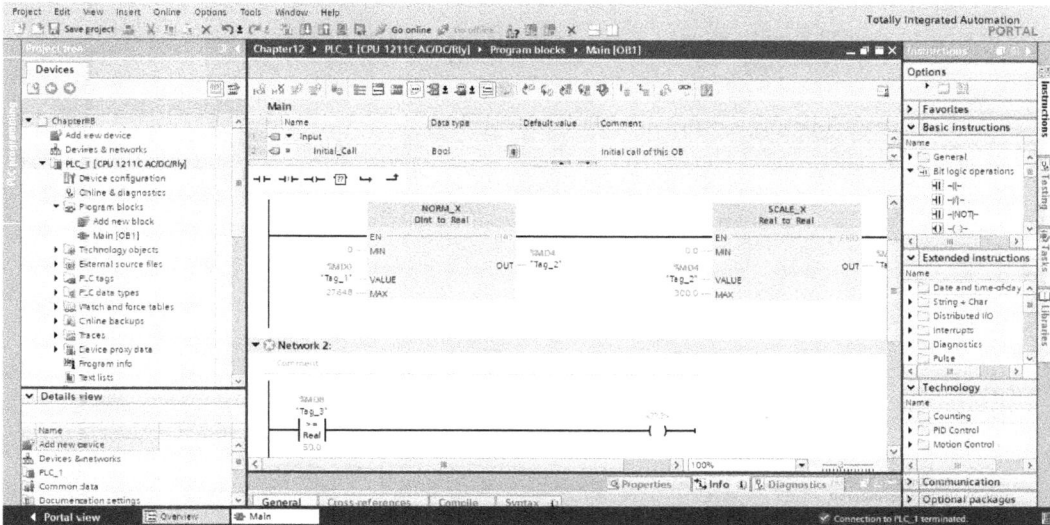

Figure 12.37 –Assignment (coil) instruction 1

7. Input %Q0 . 0 as the output coil address for the coil:

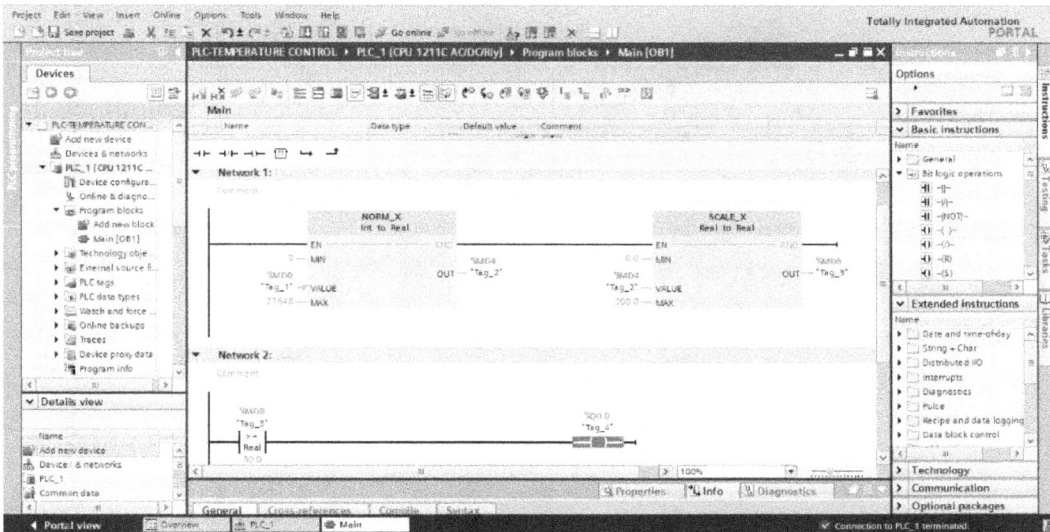

Figure 12.38 – Assignment (coil) instruction 2

Now, let's simulate:

1. Click **Compile** and click **Download** since our simulator has already started:

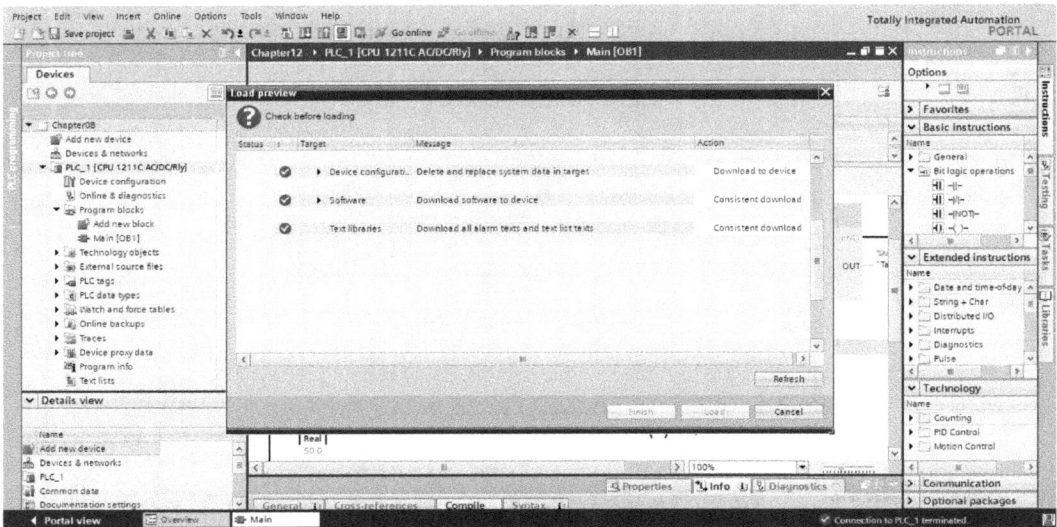

Figure 12.39 – The Load preview pop-up window

2. Click **Load** and click **Monitoring on/off**:

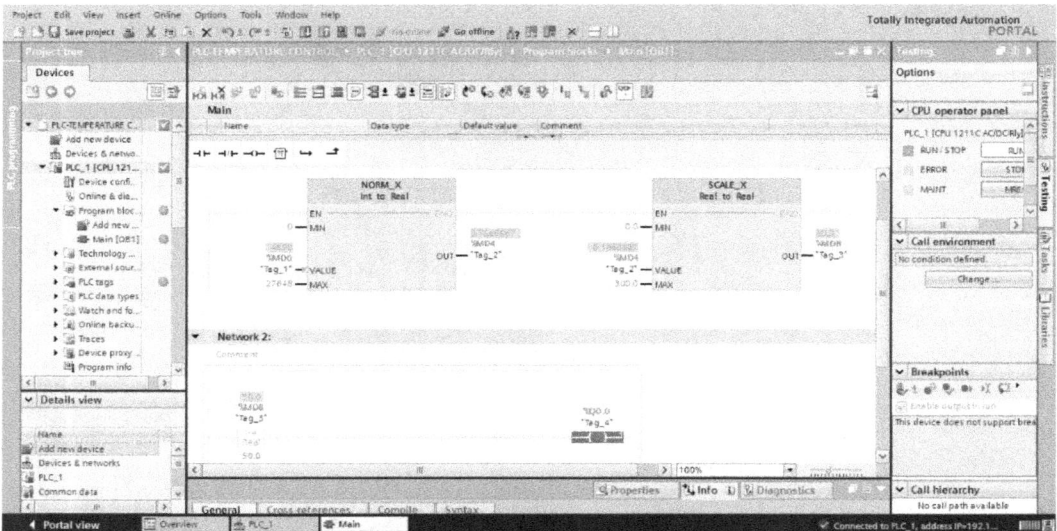

Figure 12.40 – Simulation result 1

The simulation result here shows that the coil (Q0.0) became energized when the value at MD0 (the digitized value) is 4608, which was normalized to 0.1666667 and scaled to 50.0.

If we input 5000 as the digitized value at MD0, it will be normalized to 0.1808449 and scaled to 54.25347. Q0.0 will still be energized because the condition is still met (that is, 54.25347 is greater than 50), as shown here:

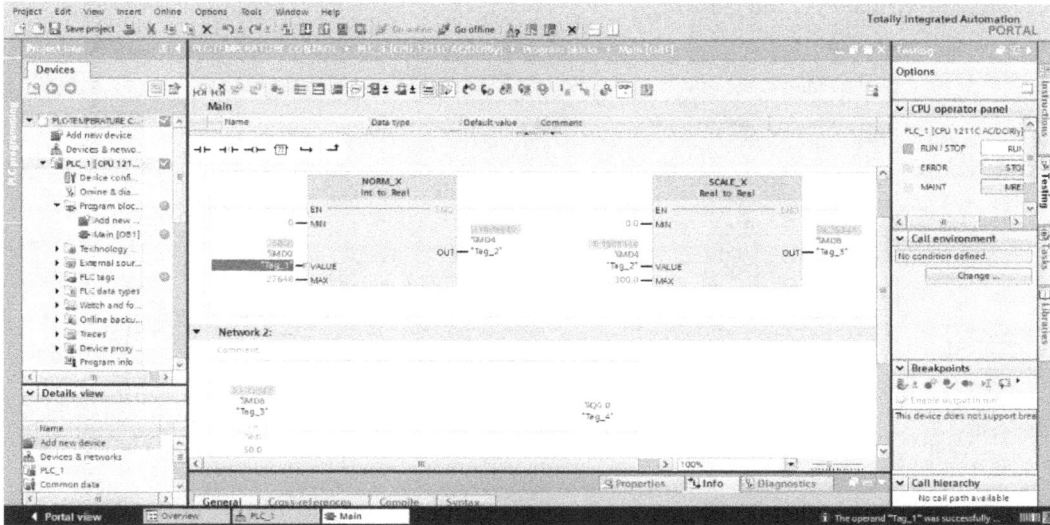

Figure 12.41 – Simulation result 2

Now, input 3000 as the digitized value at MD0 – it will be normalized to 0.1085069 and scaled to 32.55209. Q0.0 will be de-energized because the condition is no longer met (32.55209 is now less than 50), as shown here:

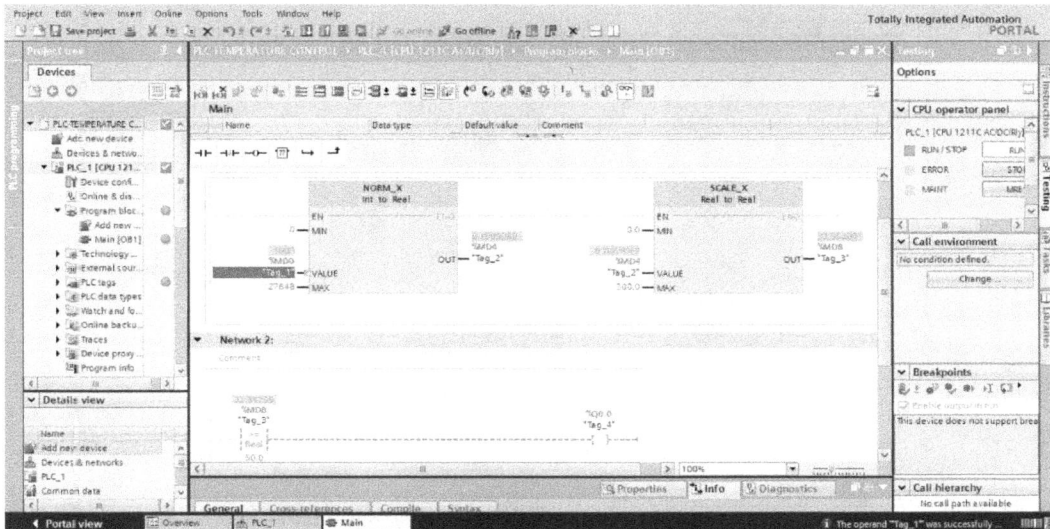

Figure 12.42 – Simulation result 3

3. Stop the simulation by clicking **Monitor on/off** and **Yes** to go offline:

4. Save the changes to your project.

5. Stop the virtual PLC and close it:

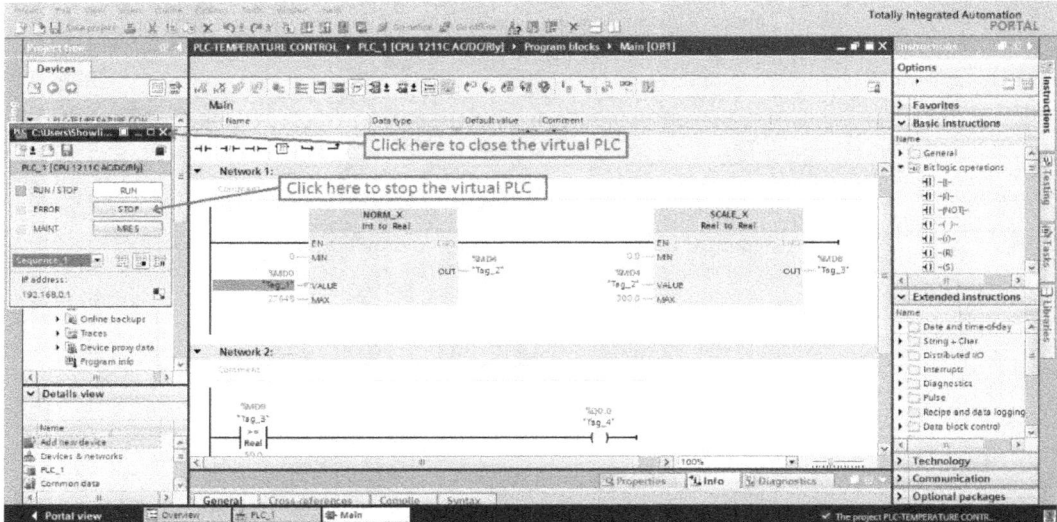

Figure 12.43 – Closing the simulator

6. You will be asked to save changes to the current project – click **Yes**:

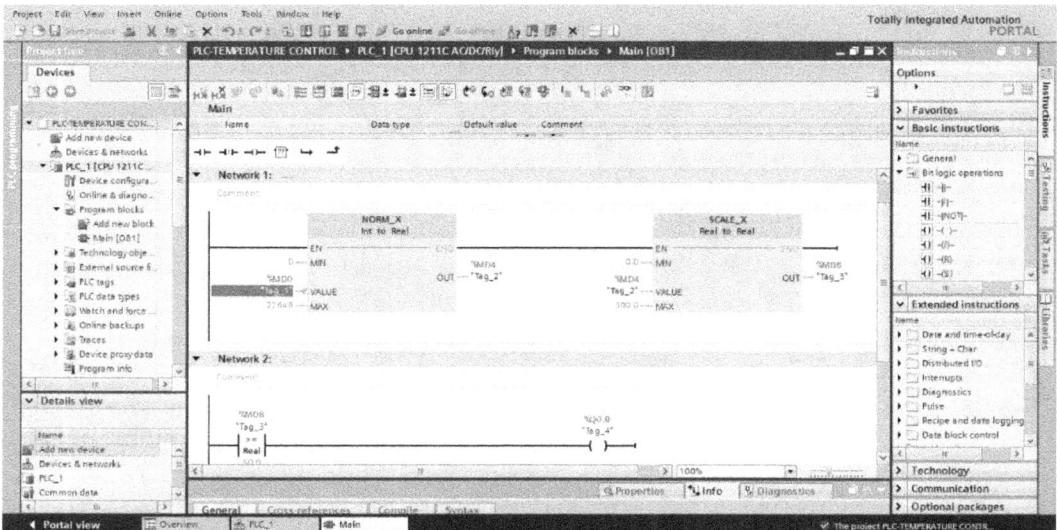

Figure 12.44 – Closing the simulator

You can now close the main project (PLC-TEMPERATURE CONTROL).

> **Note**
>
> This same project will be used to learn about how to use a real PLC to process analog input signals.

Using a real PLC (Siemens S71200 – CPU 1211C – AC/DC/RLY)

In this example, we will wire and program a PLC to keep a heater turned on until the value at MD8 (the scaled value) is greater than or equal to 100. This is a simple control system technique that can be applied to temperature, pressure, or level control.

Requirements

In addition to TIA Portal used in the simulation, the following devices are also required to demonstrate analog input signal processing with a real PLC:

- An RTD temperature sensor (Pt100)
- An RTD temperature transmitter
- A power supply (220 V AC to 24 V DC)
- A PLC (Siemens S7-1200)
- A contactor
- A heating element
- An Ethernet cable
- A PC or laptop with TIA Portal installed

Wiring:

1. Connect the power to the AC input of the power supply using a breaker. Connect the power to the AC input of the PLC through the circuit breaker. Also connect AC power to terminal 1 (L1) and terminal 3 (L2) of the contactor as shown in the diagram.

2. Connect the two terminals of the heater to terminal 2 (T1) and 4 (T2) of the contactor as shown.

3. A three-wire Pt100 RTD comes with three wires – two of the wires have a resistance of 100 Ohms between them. The two wires should be connected to the two terminals with a resistance symbol in between them while the last (third wire) should be connected to the other terminal with no resistance sign as shown.

4. Connect the positive terminal of the 24 V DC to the positive terminal of the transmitter.

5. Connect the negative terminal of the transmitter to the first analog input channel (0).

6. Connect the negative terminal of the power supply unit to the negative terminal of the 24 V DC output of the PLC (M) and also connect it to the common terminal of the analog input channel (2M) as shown.

7. Connect a 500-Ohm resistor between the first analog input (0) and 2M as shown.

8. Connect Q0 . 0 on the PLC to A1 on the contactor and connect 1L on the PLC to live (L) on the breaker. Also, connect A2 on the contactor to Neutral (N) on the breaker as shown:

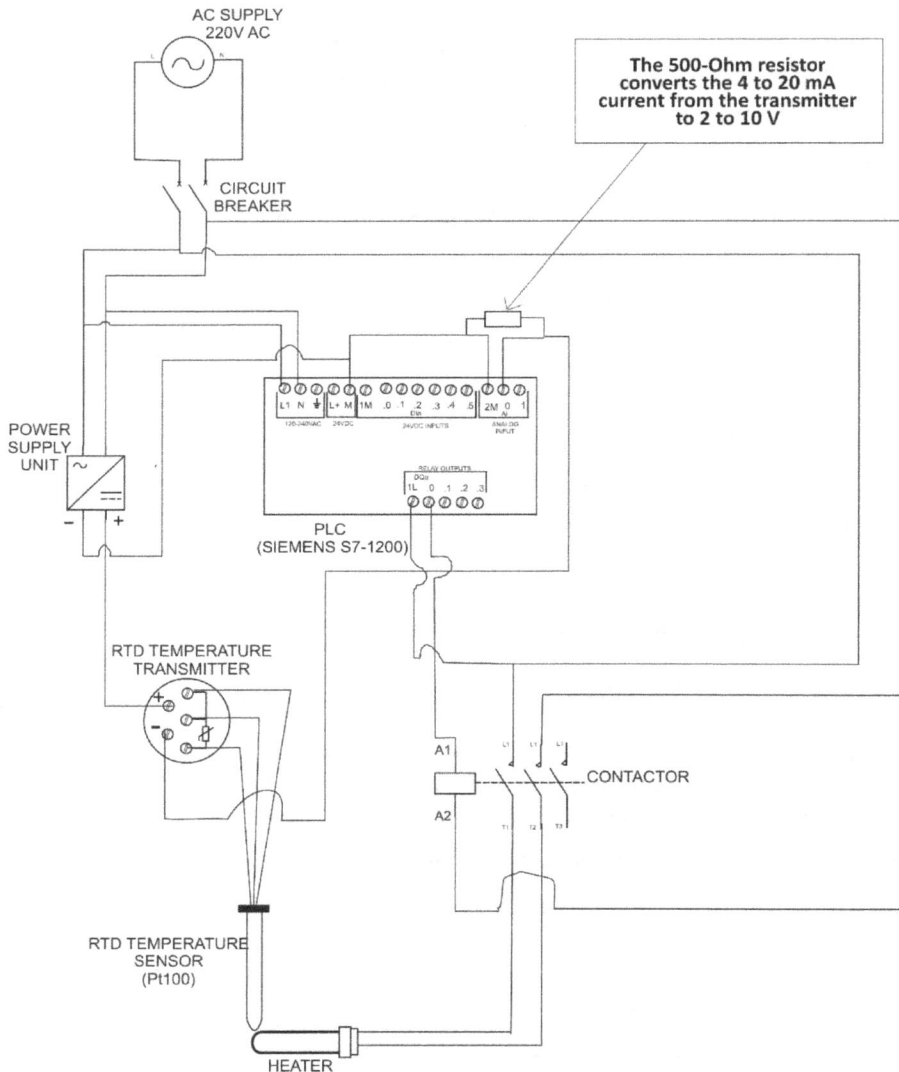

Figure 12.45 – A wiring diagram showing the wiring of a temperature control system using a PLC, a temperature sensor, and a 4 to 20 mA transmitter

Programming

We are going to open the previous program (PLC-TEMPERATURE CONTROL) and modify it to turn on a heater when the value in register MD8 is greater than or equal to 100:

1. Open the saved project (PLC-TEMPERATURE CONTROL). You can refer to the steps for opening a saved project in *Chapter 9, Deep Dive into PLC Programming with TIA Portal*:

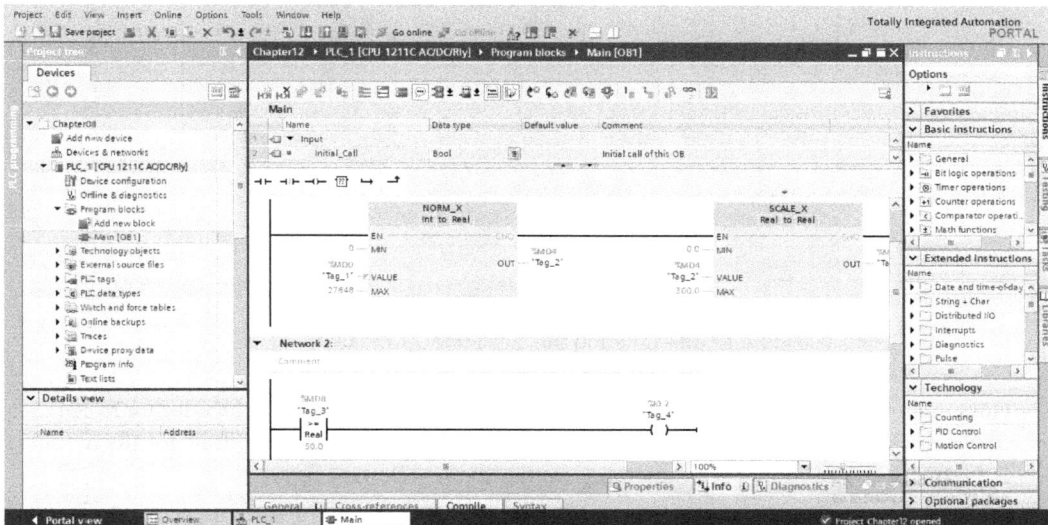

Figure 12.46 – Opening the saved project

2. Select the tag name for MD0 and press **Delete** on the keyboard to delete both the tag name (Tag_1) and register (MD0). Also, select the tag name for Q0.0 and press **Delete** on the keyboard to delete both the tag name (Tag_4) and register **Q0.0**. You should see something similar to the following figure:

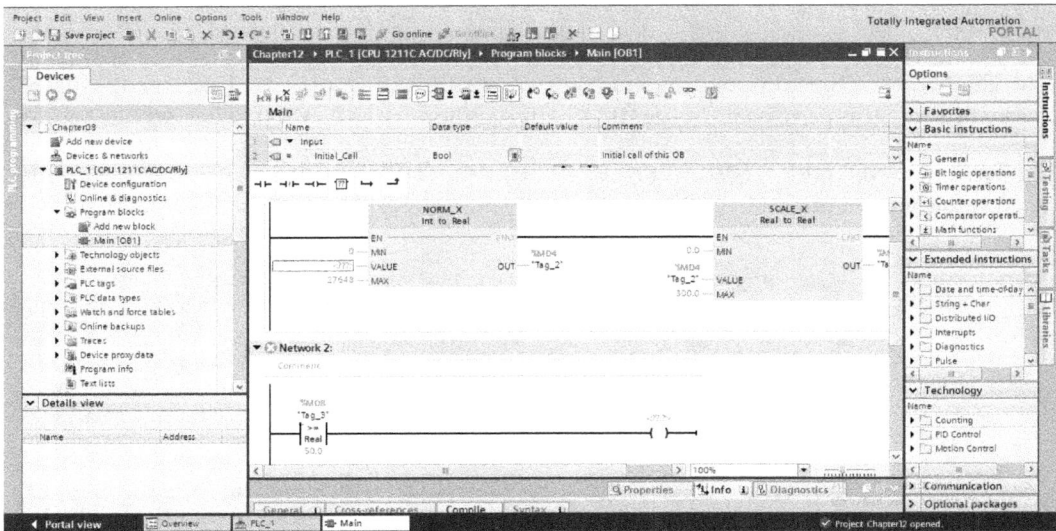

Figure 12.47 – Deleting the tags for MD0 and Q0.0

3.   Now, input the address of the first analog input channel (that is, %IW64) as the value for NORM_X in **Network 1**. Also, input a memory bit address (that is, M0.0) as the address for the output in **Network 2**. You should see something similar to the following figure:

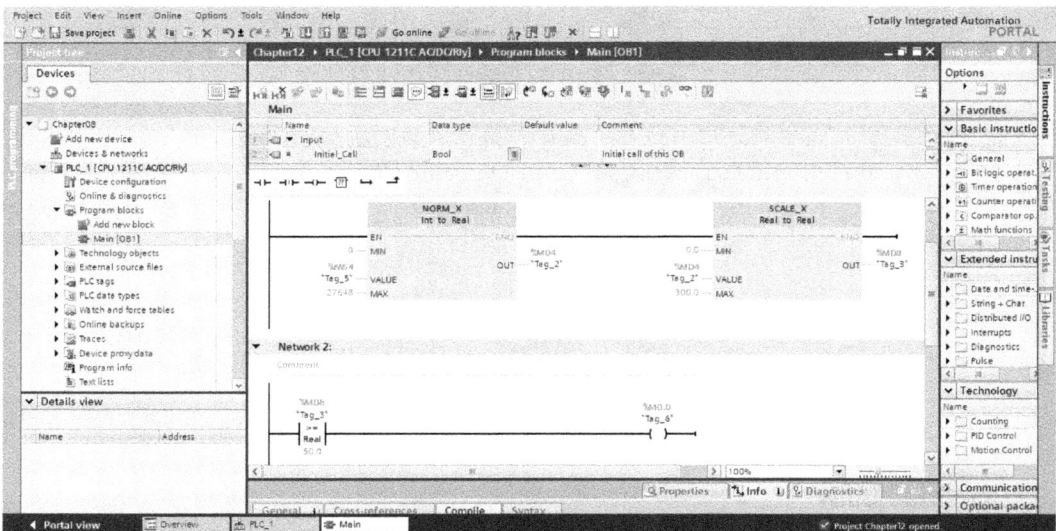

Figure 12.48 – %IW64 now used as the value for NORM_X and M0.0
now used as the register for Assignment (coil)

4. Double-click the 50.0 value at the bottom of the **compare** instruction and input 100 to change the set point to 100:

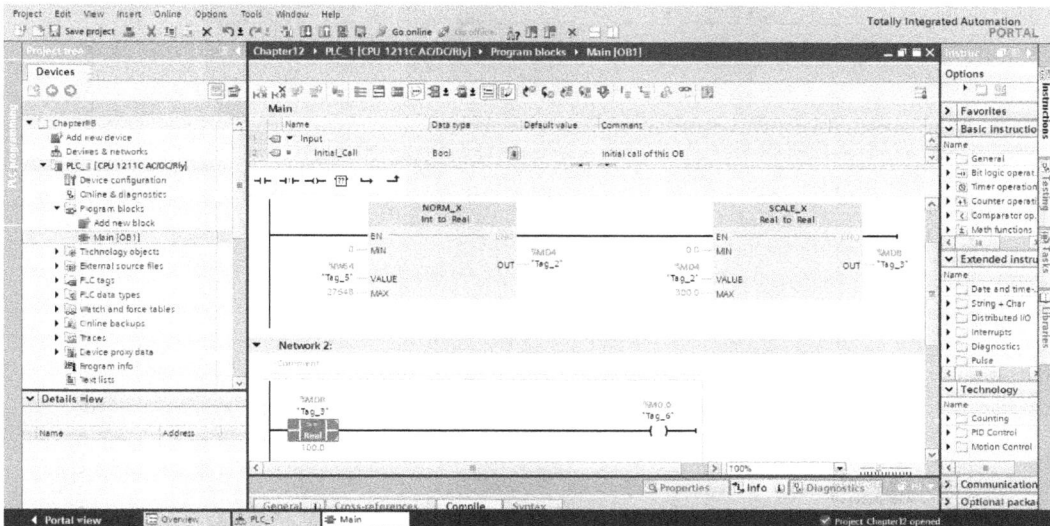

Figure 12.49 – 50 changed to 100

5. Expand **Bit logic operations** under the **Basic instructions list** on the right-hand side of the screen and drag the normally closed instruction to **Network 3**:

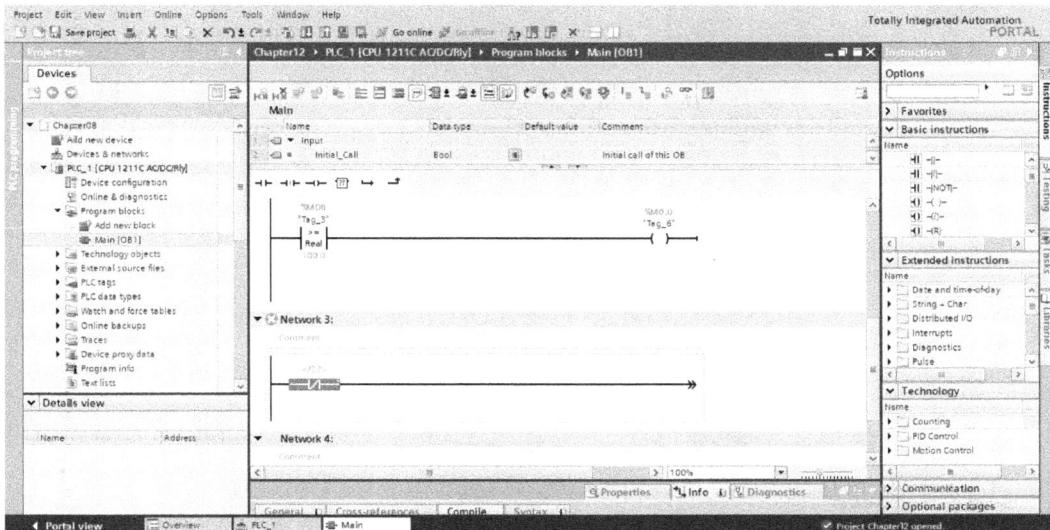

Figure 12.50 – The normally closed contact instruction added to Network 3

6. Now, input %M0 . 0 in the **???** field of the normally closed contact:

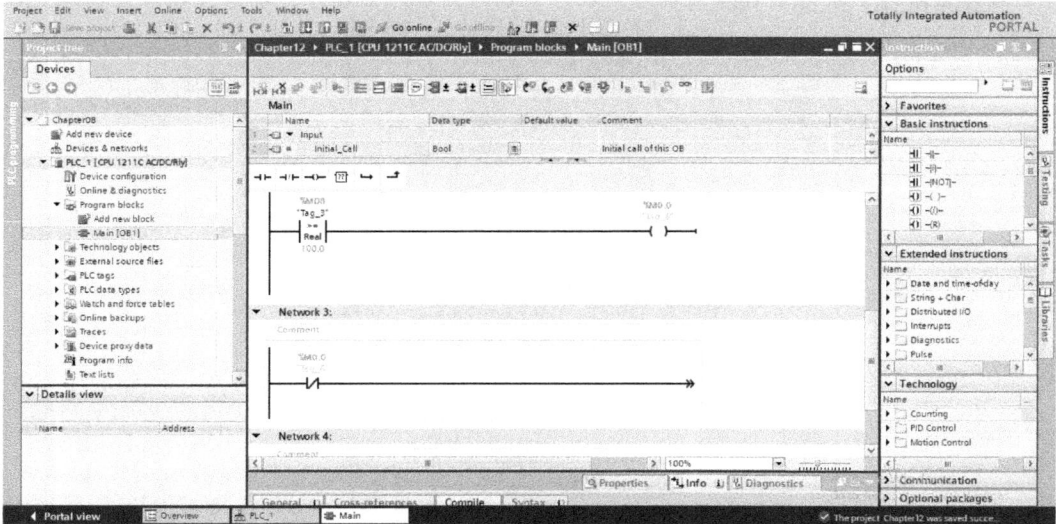

Figure 12.51 – The normally closed contact now has a M0.0 register

7. Expand **Bit logic operations** under the **Basic instructions** list on the right-hand side of the screen and drag the assignment (Coil) instruction to **Network 3**:

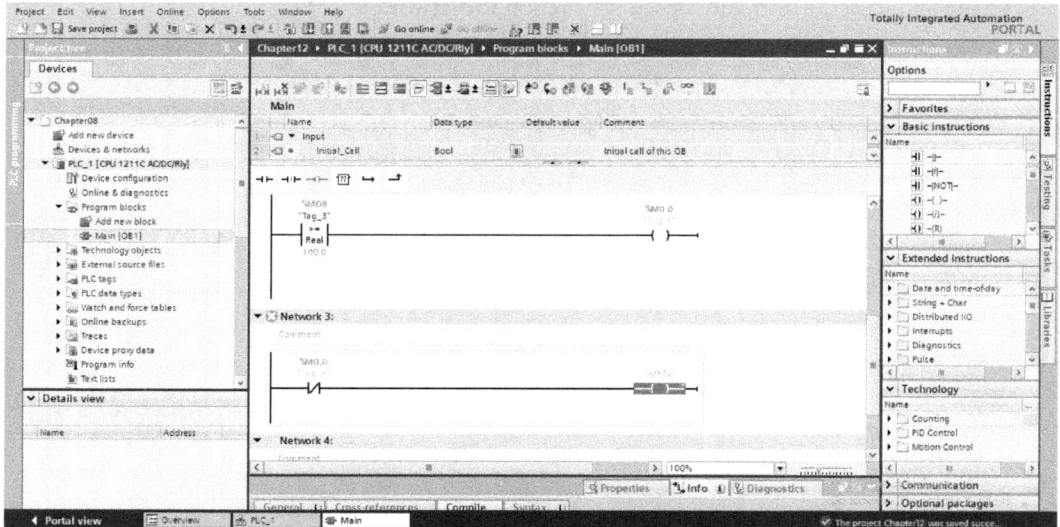

Figure 12.52 – The Assignment (Coil) instruction added to Network 3

8. Now, input %Q0.0 in the **???** field of assignment (coil) to get something similar to the following figure:

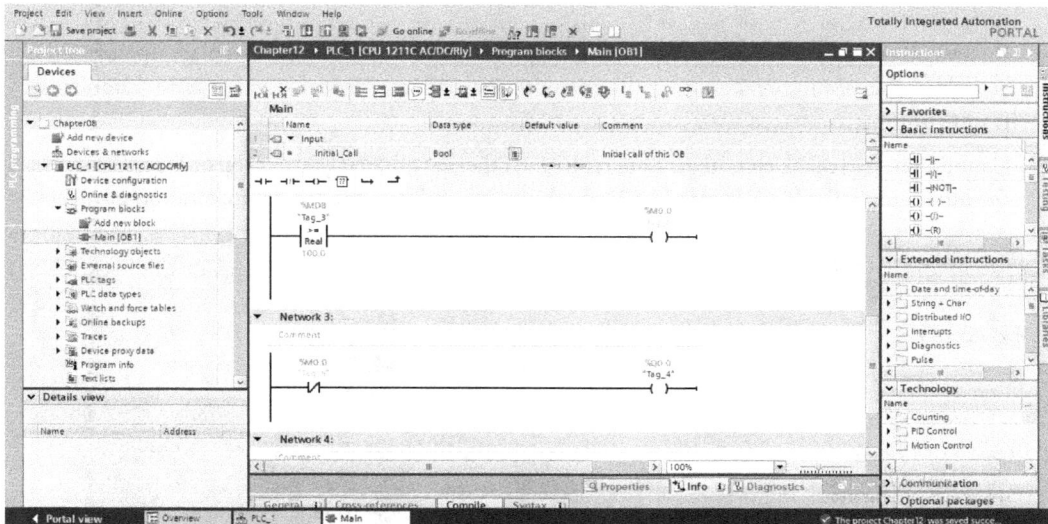

Figure 12.53 – Assignment (Coil) given an address of Q0.0

We have just completed the program. The following screenshot shows the complete program from **Network 1** to **Network 3**:

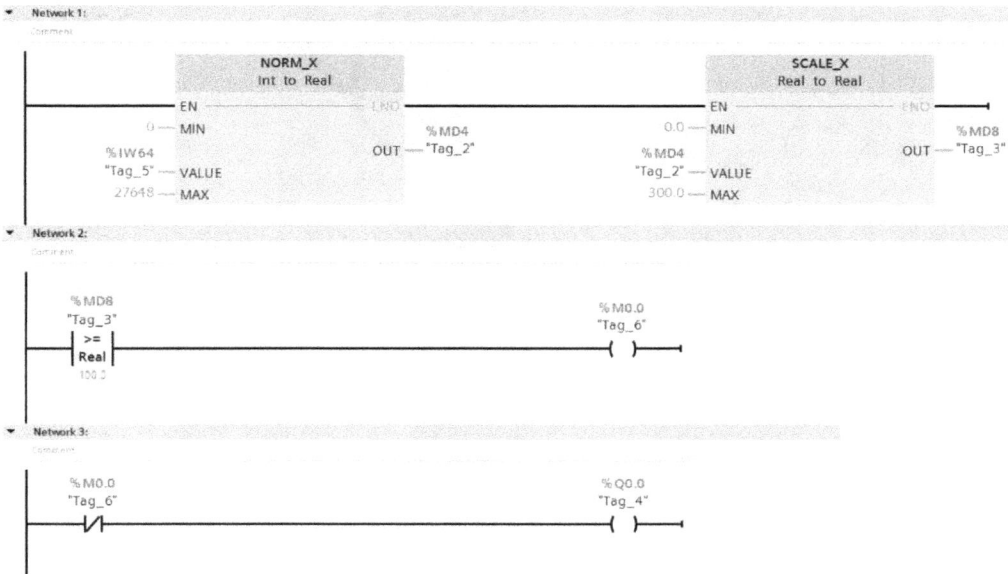

Figure 12.54 – The complete program (an analog input program)

Configuring the IP address and downloading the program to a real PLC

Now, let's proceed to download the program to a real PLC:

1. Ensure all the connections in the wiring diagram (*Figure 12.45*) have been set up correctly. Switch on the circuit breaker to supply power to the devices. Also, ensure PLCSIM (simulator) is not running.

2. Connect one end of the communication cable (the Ethernet cable) to your laptop or PC and connect the other end to the PLC. *Figure 12.55* here shows my simple setup for demonstration purposes:

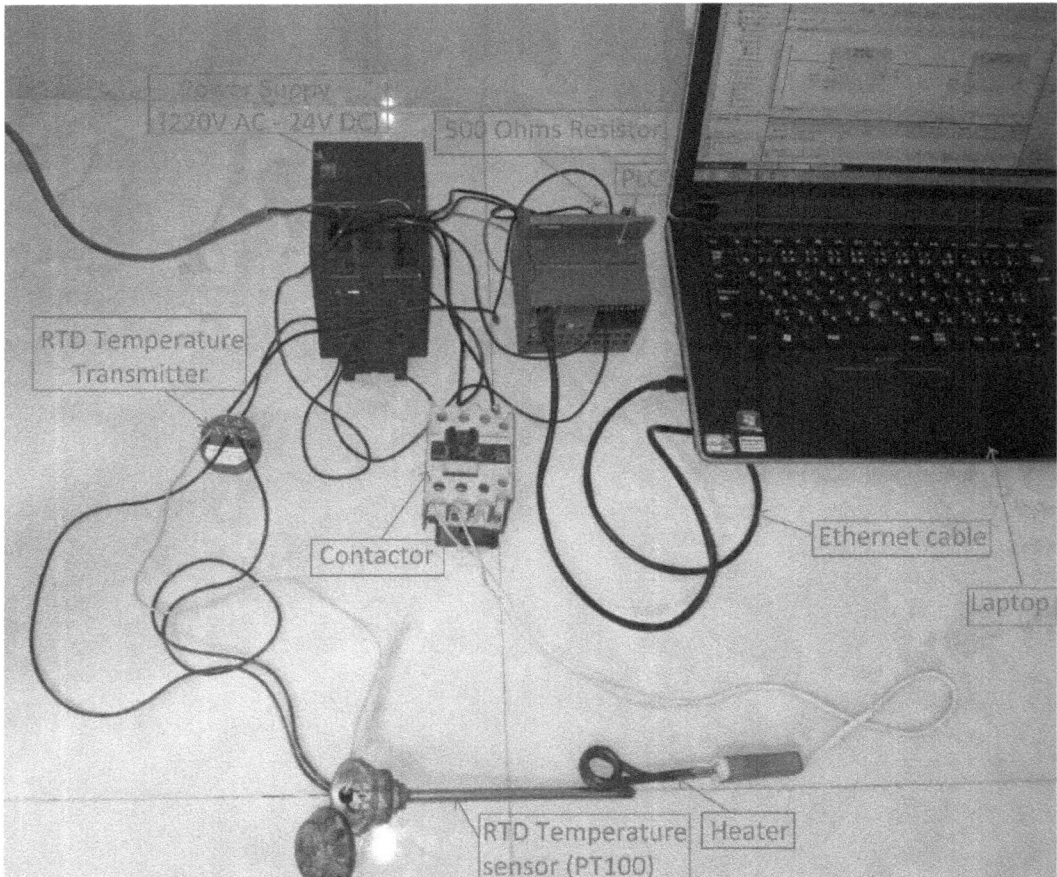

Figure 12.55 – My setup for demonstration purposes

3. Double-click **Device configuration**:

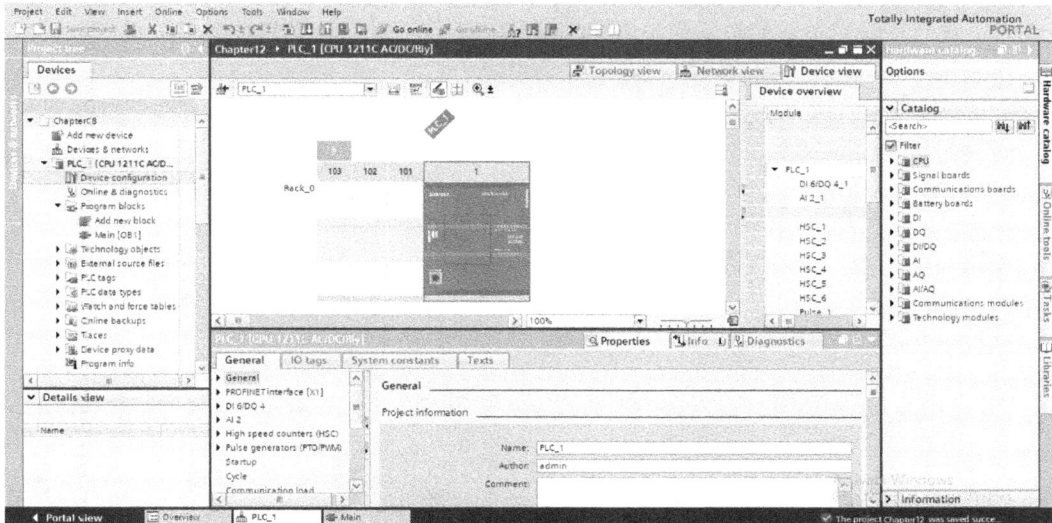

Figure 12.56 – Device configuration

4. Navigate through the **Properties | General** tab and then click **PROFINET Interface [X1]**:

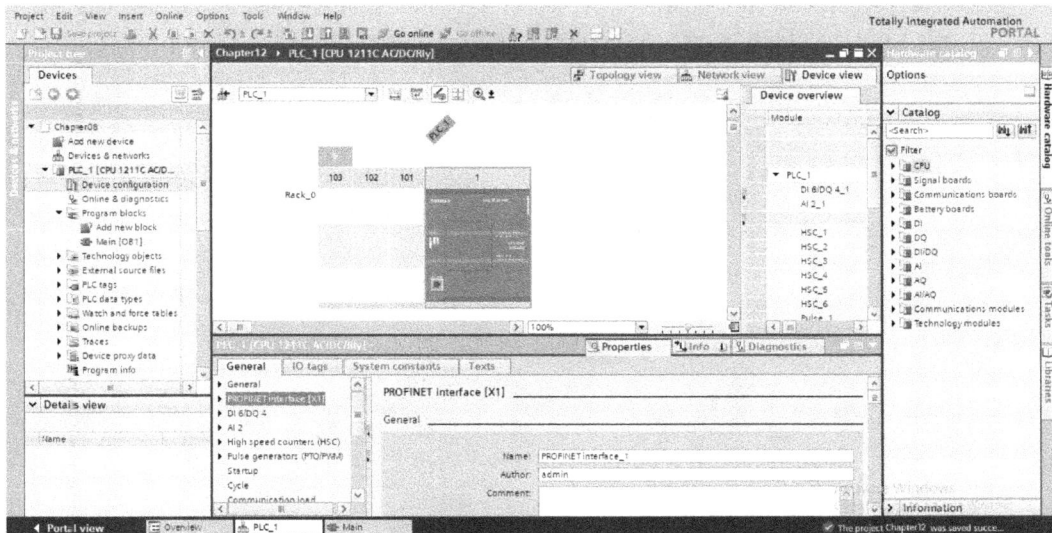

Figure 12.57 – PROFINET Interface [X1]

5. Scroll down to the IP protocol and input an IP address of your choice (for example, `192.168.0.6`) and a subnet mask (for example, `255.255.255.0`) as shown:

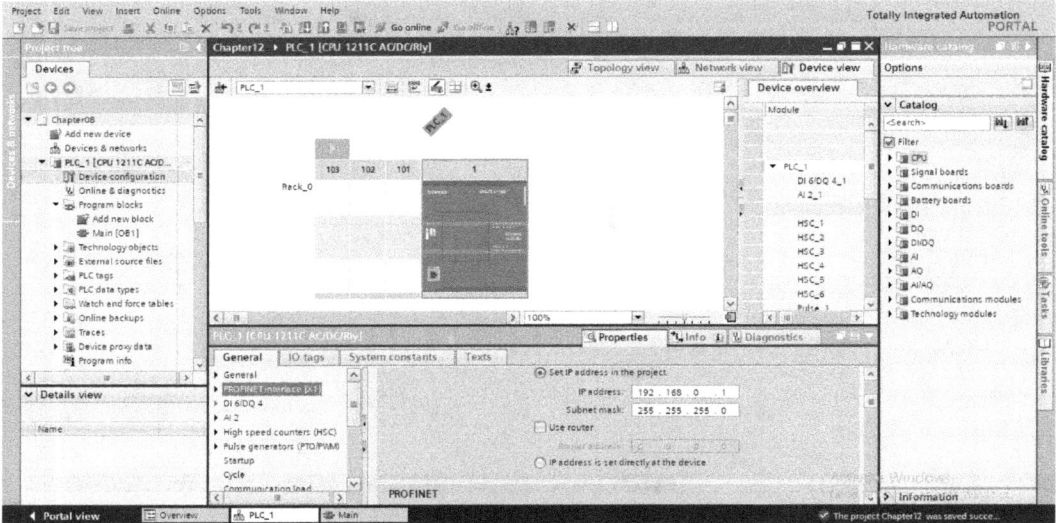

Figure 12.58 – Adding an IP address and a subnet mask

6. Click **Save project** and click on the PLC to select it. You should have the PLC highlighted. The **Compile** and **Download to device** commands will become active as shown here:

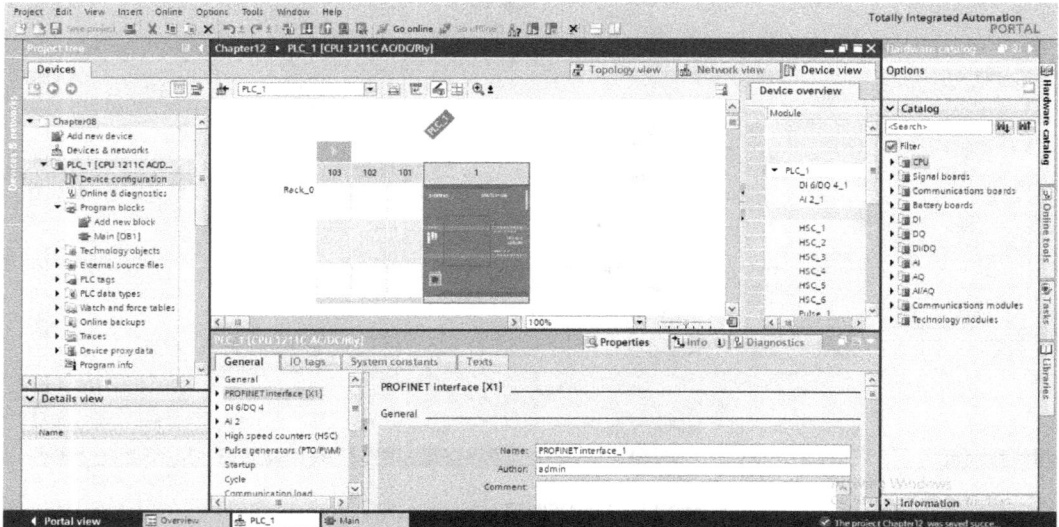

Figure 12.59 – The Compile and Download to device commands

7. Click **Compile** to compile the program and ensure there are no errors.

8. Click **Download to device** and select your PC or laptop's Ethernet controller card from the **PG/ PC interface** list as shown.

> **Note**
>
> I have selected **Realtek PCIe GBE Family Controller** because my laptop's Ethernet controller is Realtek.

9. Click **Start search**:

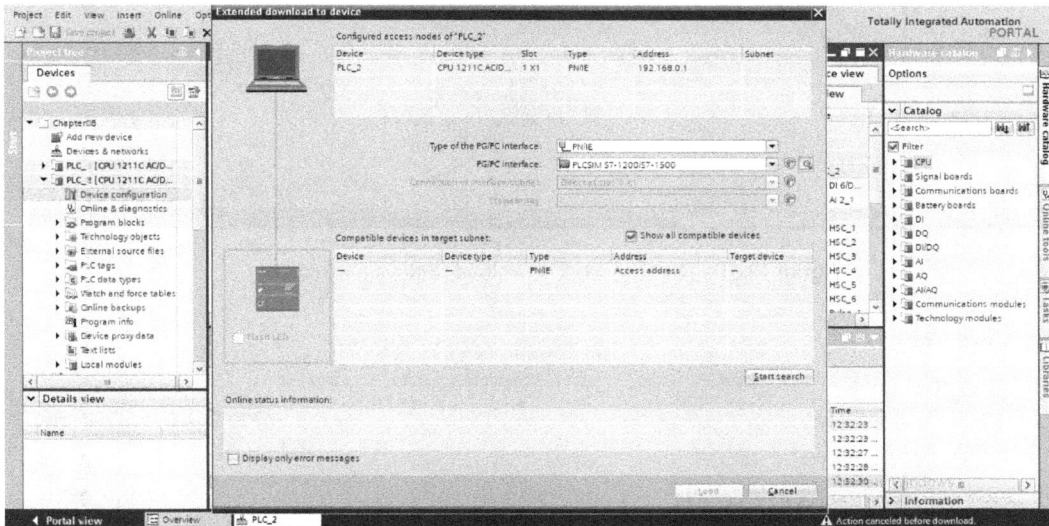

Figure 12.60 – The selected PG/PC interface

10. Click **Load**:

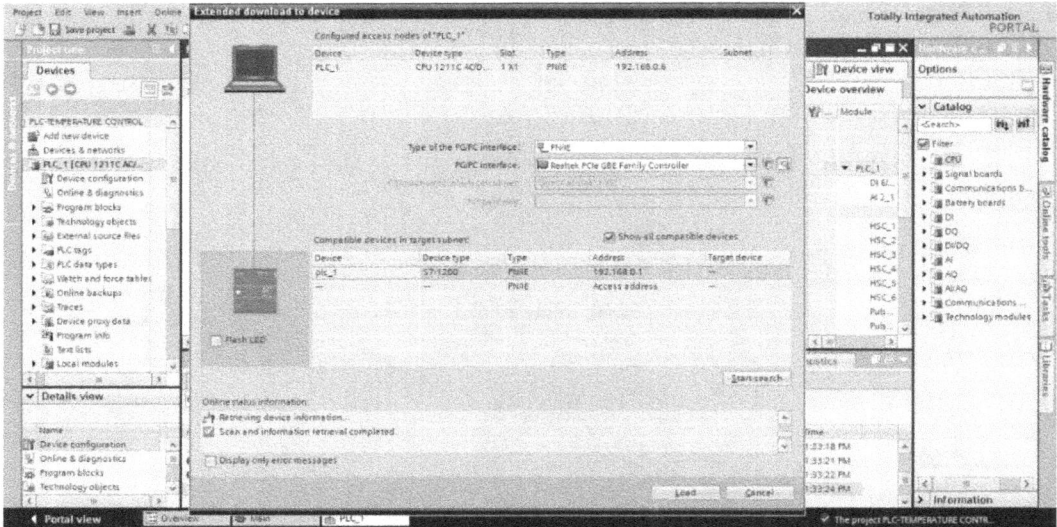

Figure 12.61 – The Extended download to device window

11. Click **Yes**:

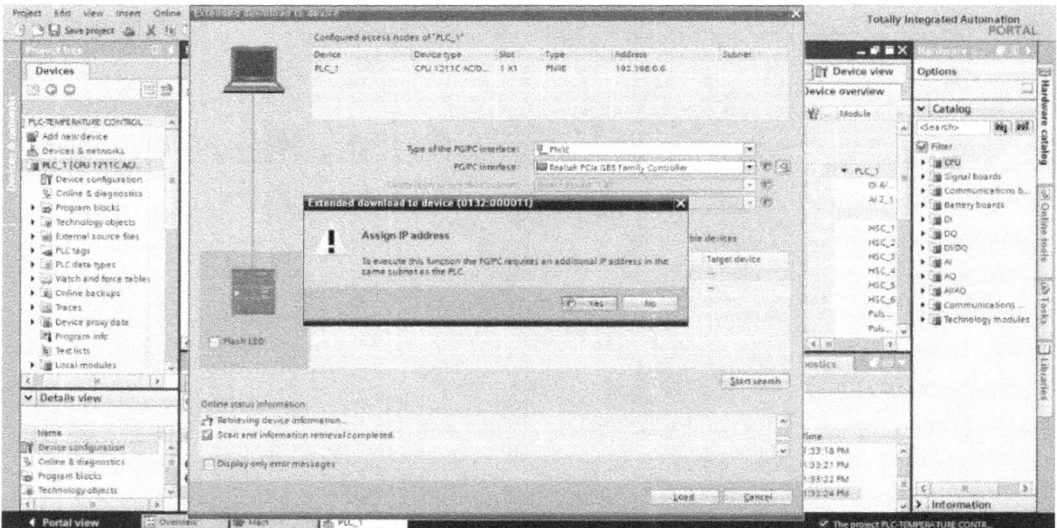

Figure 12.62 – Extended download to device (message 1)

12. Click **OK**:

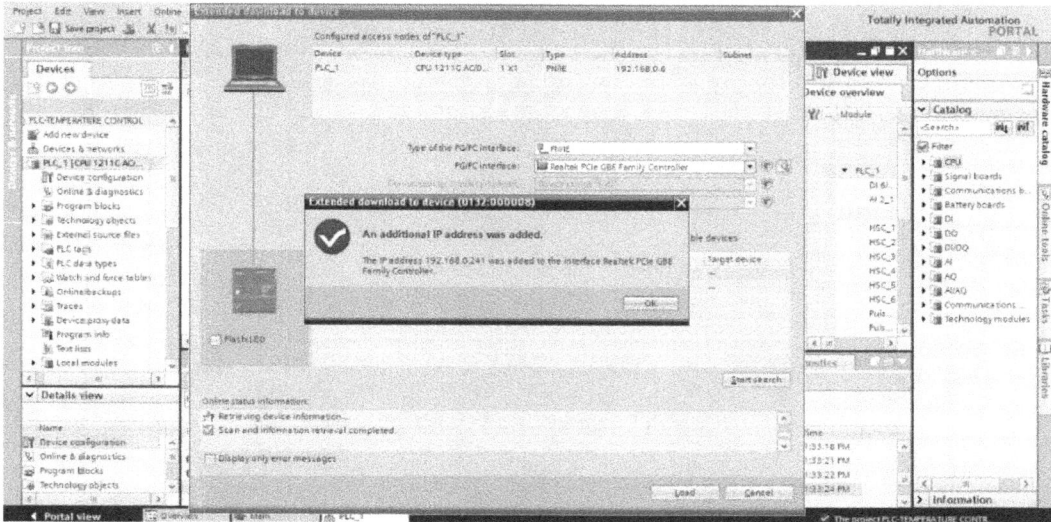

Figure 12.63 – Extended download to device (message 2)

13. Select **Stop all**:

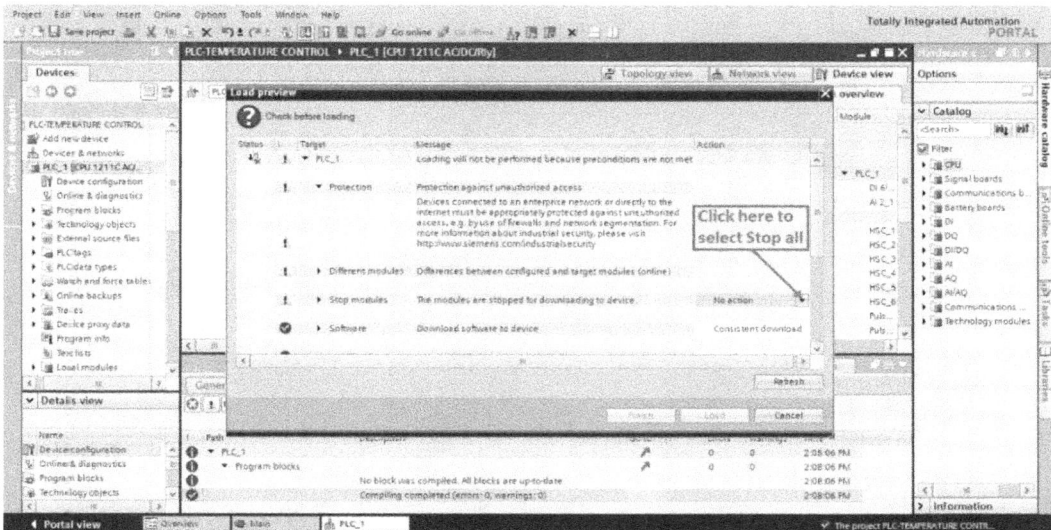

Figure 12.64 – Load preview (1)

14. Click **Load**:

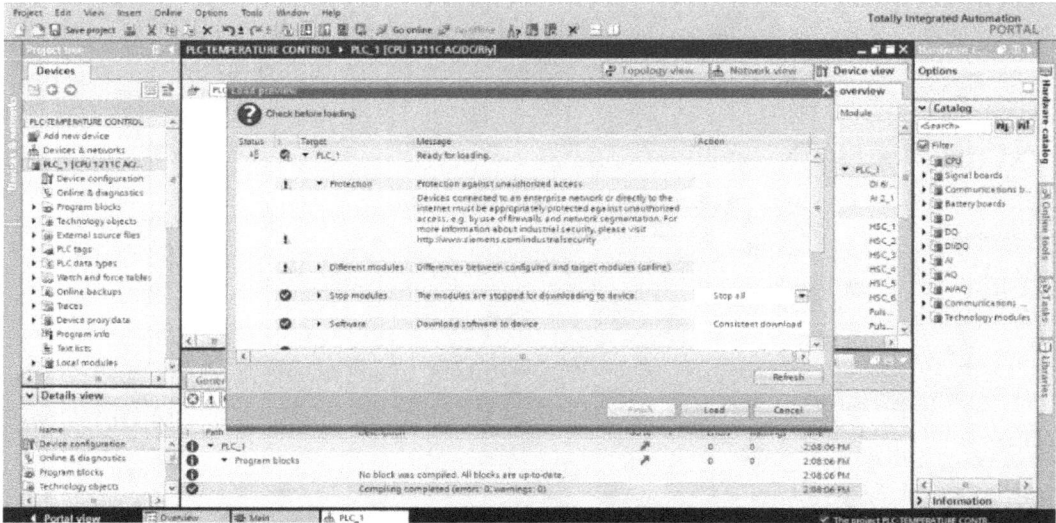

Figure 12.65 – Load preview (2)

15. Check **Start all** as shown here and click **Finish**:

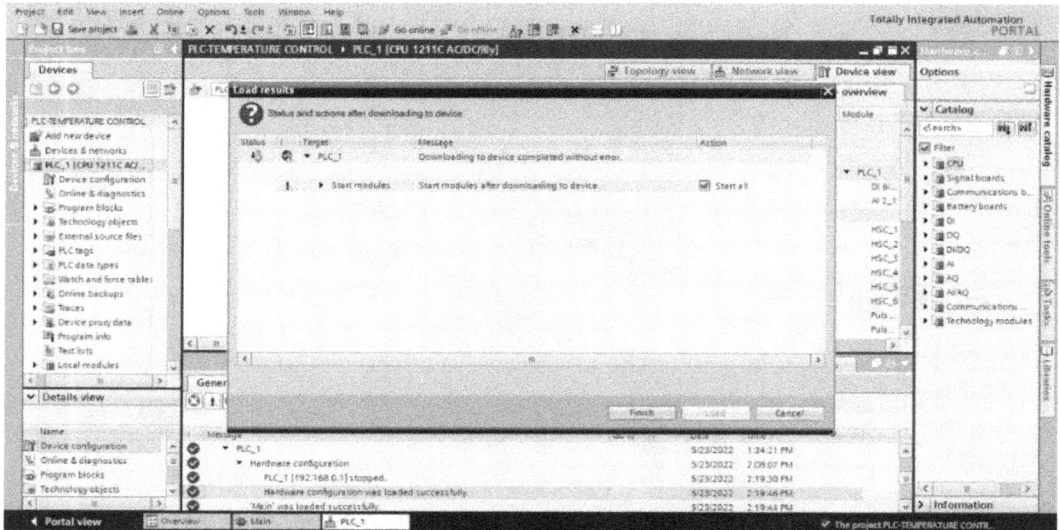

Figure 12.66 – Load results

Congratulations! You have successfully downloaded the program to your PLC and the PLC will begin to function as programmed:

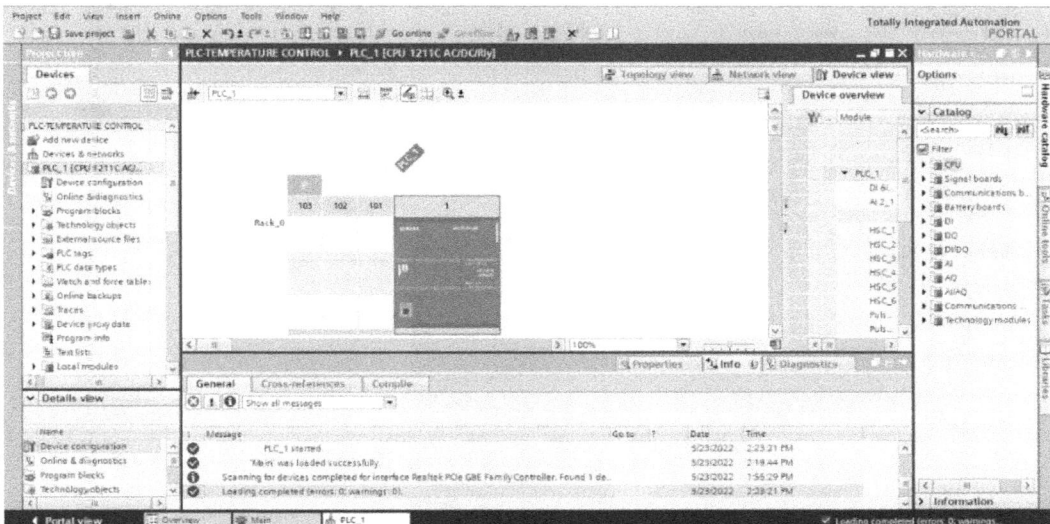

Figure 12.67 – Loading completed with no errors

Let's go online to monitor what happens in the PLC as it processes the analog signal and executes the program. Follow these steps to monitor your PLC program as it executes the instructions:

1. Double-click **Program blocks | Main [OB1]**.

2. Click **Monitoring on/off**.

An observation and explanation of the program

You will observe that the value at %IW64 varies when the heater is on. This is the digitized value of the analog signal connected to the first channel of the analog input module. As the temperature increases, the analog signal increases and the digitized value increases as well. The value at MD4 is the normalized value while the value at MD8 is the scaled value.

In the following figure, **Network 1** shows the digitized value at %IW64 is 9215. The normalized value at MD4 is 0.3332972 while the scaled value at MD8 is 99.98915.

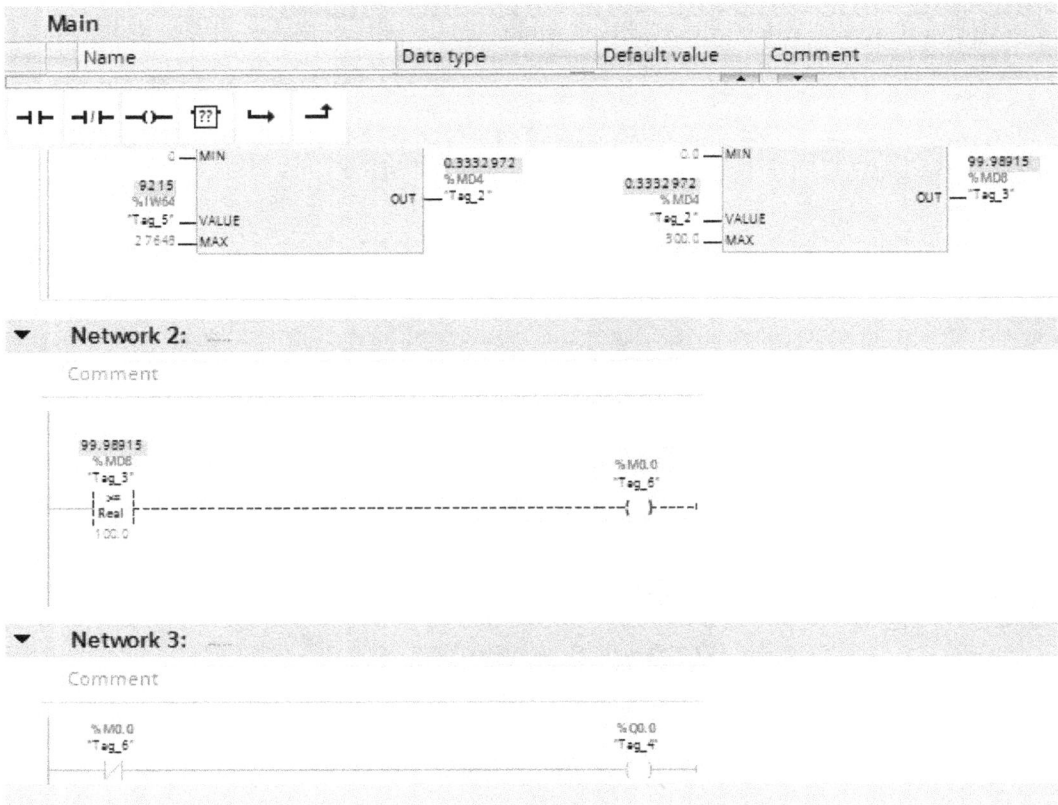

Figure 12.68 – Monitoring the PLC program as it executes instructions (1)

In **Network 2**, there was no output in the comparator to energize or turn on the M0.0 Assignment (coil) because the condition is not met, that is, 99.98915 at MD8 is less than 100.

In **Network 3**, the normally closed contact (M0.0) remains closed because M0.0 is OFF. This allows the signal to flow to Q0.0, which energizes the contactor connected to it and the heater turns ON.

When the value at MD8 is greater than or equal to 100, the comparator will have an output that energizes or turn on the M0.0 Assignment (Coil) and this will de-energize Q0.0 and turn OFF the heater.

In the following figure, **Network 1** shows that the digitized value at %IW64 is 9692. The normalized value at MD4 is 0.3505498 while the scaled value at MD8 is 105.1649.

In **Network 2**, there was an output in the comparator to energize or turn ON the M0.0 Assignment (Coil) because the condition is met, that is, 105.1649 at MD8 is greater than 100.

In **Network 3**, the normally closed contact (M0.0) becomes open because M0.0 is ON. This stops the signal from flowing to Q0.0, which de-energizes the contactor and the heater turns OFF.

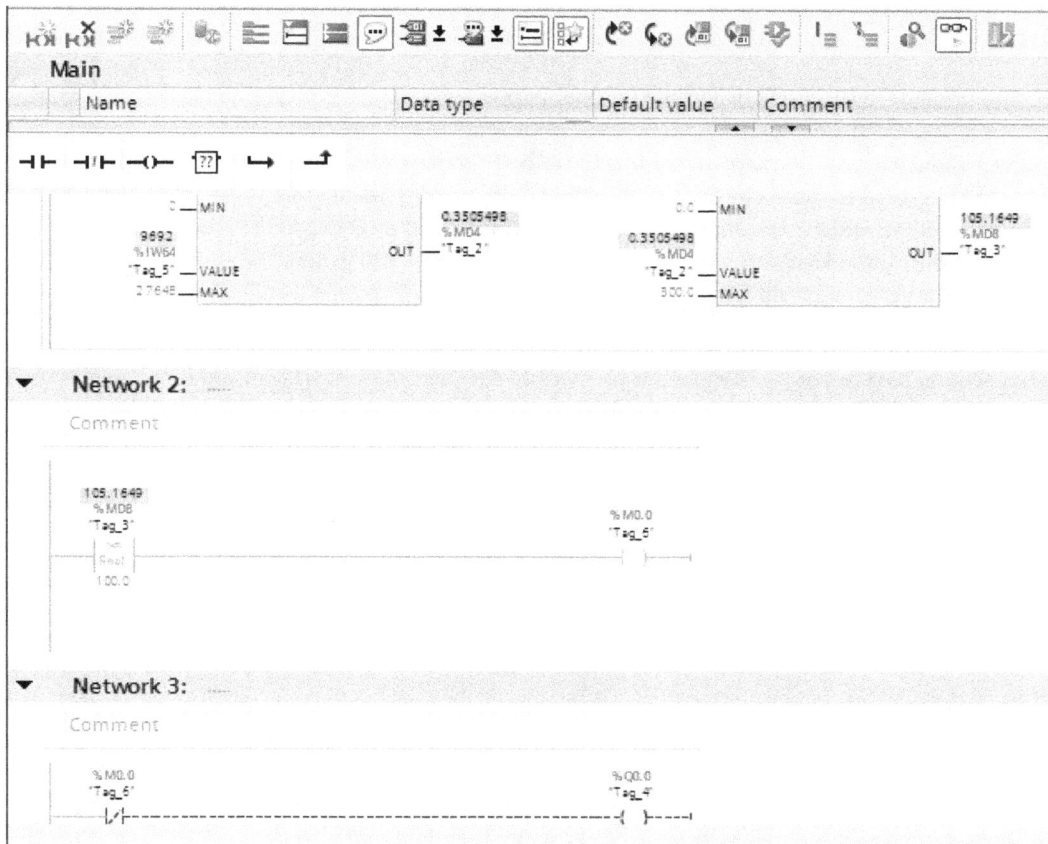

Figure 12.69 – Monitoring the PLC program as it executes instructions (2)

In this section, you have learned how to work with an analog input signal in a PLC. You should now be able to program and simulate a PLC (S7-1200) for analog input signal processing. You should also be able to connect a transmitter with 4 to 20 mA output to an analog input of a real PLC and program it to perform a simple control (for example, stop a heater) at desired setpoint.

This section simplifies analog signal processing which many industrial automation engineers or PLC specialists finds difficult. This book only focuses on analog input signals and we have used it to send out a digital control signal when a condition is met. It is also possible to send out an analog signal (4 to 20 mA) or use a PLC with a PID feature for better control than the example here.

Summary

Well done! You have successfully completed this chapter, which is an important stage of industrial automation. You should now understand process control and be able to explain various process control terms and understand ISA symbols according to *ISA5.1*. You have also learned about the sensors and transmitters required for temperature measurement, level measurement, pressure measurement, and flow measurement, which are common in industry. You should now be able to wire a temperature sensor and transmitter, level transmitter, and pressure transmitter to generate a 4- to 20-mA analog output, which is very important for process control. We have also learned how to keep the temperature of a heater at the desired level by using a temperature controller.

The last section covered more advanced topics within the field of industrial automation, where we discussed the basic theory you need to know about analog input signal processing in a PLC (S7-1200), which includes reading analog signals, normalizing, and scaling. We learned how to program a PLC (S7-1200) to process analog input signals using a step-by-step guide (simulation) and we also used a real PLC to learn how to wire and program the analog input of a real PLC to perform a control action following a step-by-step guide.

The next chapter involves other vital knowledge you will need as an industrial automation engineer. Don't miss out.

Questions

The following are some questions to test your understanding of this chapter. Ensure you have read and understood the topics in this chapter before attempting the questions:

1. In a/an _____ process control, the output does not change the control action.

2. In a/an _____ process control, the output has effect on the control action.

3. In a/an_____ control, disruption is measured and corrective action is taken to correct any errors before they reach the system.

4. A _____ is a device that converts a signal produced by a sensor into a standard signal (4 to 20 mA, 0 to 10 V, or 3 to 15 psi) applicable in process control.

5. _____ can be referred to as the target value of the process variable.

6. _____ is the difference between the process variable or measured value and the set point.

7. The devices that the controller operates to keep the process variable at a setpoint are called _____.

8. ISA is an acronym that stands for _____.

9. _____ is a flow meter that measures the flow rate by measuring the speed of an internal rotating turbine or gears.

10. _____is an inbuilt function of an S7-1200 controller for converting any given data from the minimum to the maximum into 0 to 1 in float values.

11. _____is an inbuilt function of an S7-1200 controllers for converting the normalized value from 0 to 1 into the required range.

12. PID is an acronym for _____.

13. P&ID stands for _____.

13
Industrial Network and Communication Protocols Fundamentals

Industrial networking is important in industrial automation and process control. Industrial automation provides interconnection between devices (sensors, actuators, controllers, and PCs) and also enables communication between them. An industrial network differs from traditional communication, which allows communication between computers and their peripherals such as printers, scanners, and more. An industrial network is a powerful network that connects industrial automation and control devices (PLCs, HMIs, PCs, VFDs, transmitters, and more) and provides real-time monitoring, control, and data integrity in a harsh environment.

By reading this chapter, you should have a better understanding of industrial networks and the various topologies and communication media. You will learn about the basics of wireless networks and understand what 5G is. You will also learn about industrial network protocols, which are important for communication to occur. Finally, we will provide a practical demonstration of a simple industrial network using the PROFINET protocol.

In this chapter, we will cover the following topics:

- Understanding industrial networks
- Understanding network topology
- Network media – wired and wireless (Bluetooth, Wi-Fi, and cellular communication – 1G, 2G, 3G, 4G, and 5G)
- Network connectors and other network components
- Understanding network protocols
- Common industrial network protocol – Foundation Fieldbus

- Common industrial network protocol – PROFIBUS
- Common industrial network protocol – Modbus
- Common industrial network protocol – HART
- Common industrial network protocol – PROFINET

Understanding industrial networks

When two or more devices are connected for data exchange, it is referred to as a network. Each device in the network can be referred to as a node. Industrial networks consist of controllers (PLC, PAC, or DCS), field devices, PCs, and other industrial devices that are connected to share and exchange data. A node in an industrial network can be a controller (PLC, PAC, or DCS), field device (sensor/ transmitter or actuator), HMI, server, or workstation. The nodes in an industrial network utilize a special protocol for communication and to automate processes in the industry.

Understanding network topology

Network topology is the bedrock of a network. It shows how the nodes are arranged and interconnected. It refers to the physical and logical arrangement of a network. Hence, network topology can be grouped into two areas:

- **Logical topology**: This refers to the pattern in which data is being transferred between the nodes in the network. It defines the path that data takes when it is being transmitted. It describes the way data travels in the network.

- **Physical topology**: This refers to the physical layout or arrangement of the nodes in the network. It shows how various nodes in the network are connected.

The following are some common physical topologies:

- **Point-to-point topology**: In a point-to-point topology, a connection is made between two devices with a single wire. There is a direct link between two devices in a point-to-point network, as shown in the following diagram. An example of a point-to-point topology in a home or office is a PC connected to a printer:

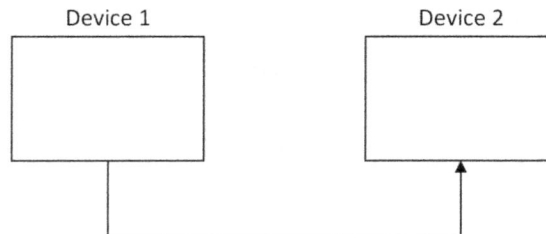

Figure 13.1 – Point-to-point network between device 1 and device 2

In industrial automation, an example of a point-to-point topology is a PLC connected to a PC when downloading a program to the PLC:

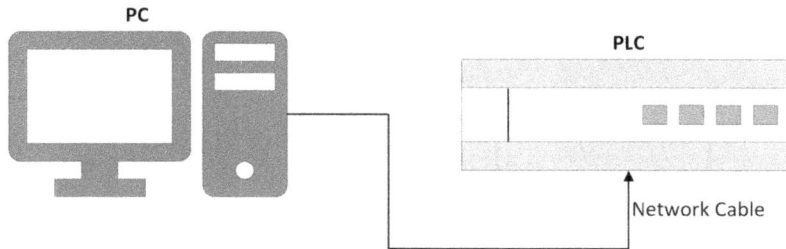

Figure 13.2 – PC connected to a PLC when downloading a program

The following diagram shows a PLC connected to an HMI for monitoring and control purposes:

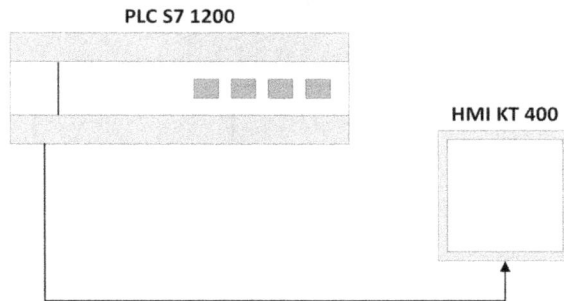

Figure 13.3 – PLC connected to HMI for monitoring and control purposes

- **Bus topology**: In a bus topology, each node is connected to a single cable segment. This is called the trunk, backbone, or bus. Termination resistors are used at both ends to improve signal quality. When data is to be transmitted by a node to another node, the data is sent out to all the nodes in the network. However, this will be ignored by all other nodes and accepted by the destination node – that is, a node will only respond to the data that is directed to it. A common example of the bus topology in industrial automation is a network of PLCs, HMIs, and other devices using the PROFIBUS protocol:

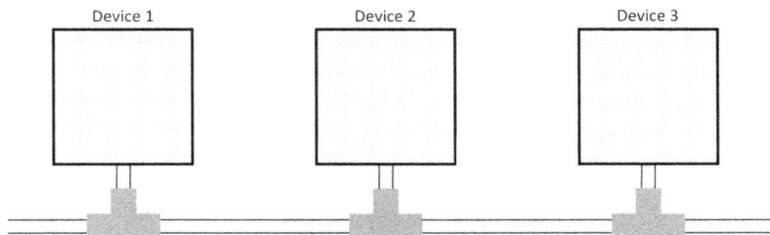

Figure 13.4 – Bus topology

- **Star topology**: In a star topology, each node is connected to a central device. This can be a switch, router, or hub. When a node wants to send data to another node, it sends the data to the central device, which will then pass the data to the destination node. Devices communicate with each other indirectly through the central device:

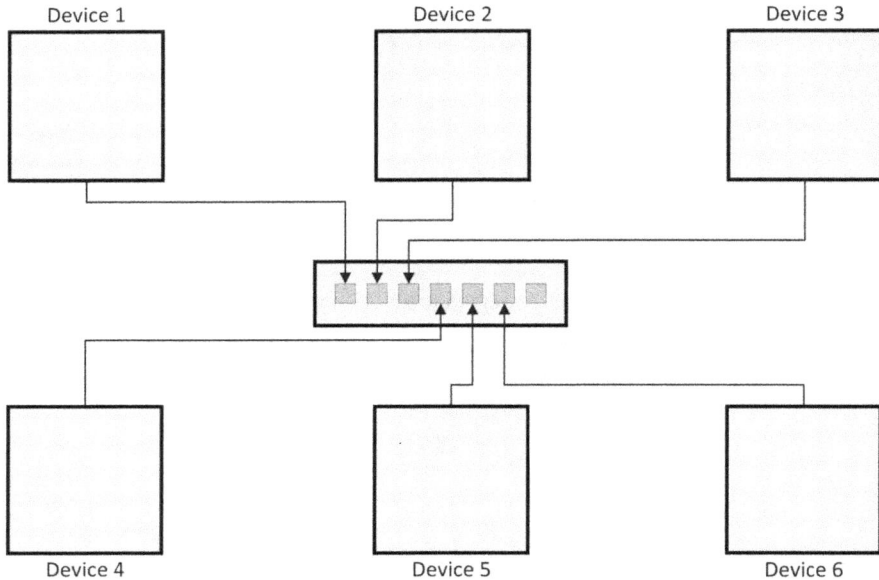

Figure 13.5 – Star topology

- **Ring topology**: As the name suggests, when the nodes in the network are connected, they form a ring. The last node is connected to the first node to form a closed loop, as shown in the following diagram. In a ring topology, data is transmitted from one node to another until it reaches the destination node.

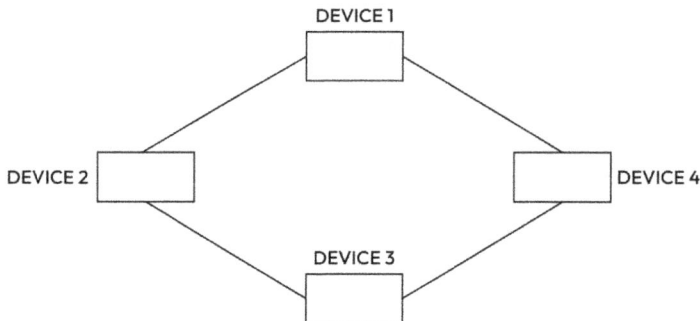

Figure 13.6 – Ring topology

In this section, you learned about network topology. You should now understand the point-to-point, star, bus, and ring topologies. In the next section, we will learn about network media.

Network media – wired and wireless (Bluetooth, Wi-Fi, and cellular communication – 1G, 2G, 3G, 4G, and 5G)

Network media can be defined as the channel of communication used to interconnect nodes on a network. Examples of network media include copper twisted pair cables, copper coaxial cables, and fiber optic cables, which are used in wired networks, and radio waves (Bluetooth, Wi-Fi, cellular communication, and so on), which are used in wireless networks. Hence, a network can be wired or wireless depending on the communication channel being used:

- **Wired network**: A wired network uses cables such as copper twisted pair cables, copper coaxial cables, and fiber optic cables as a communication channel.

  The following figure shows a coaxial cable:

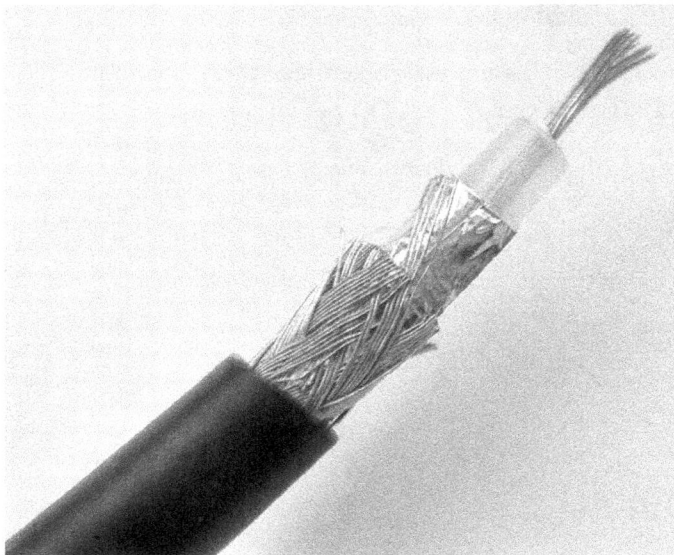

Figure 13.7 – Coaxial cable (RG59) (source: `https://commons.wikimedia.` `org/wiki/File:Coaxial_cable_cut.jpg`)

The following figure shows a PROFIBUS cable:

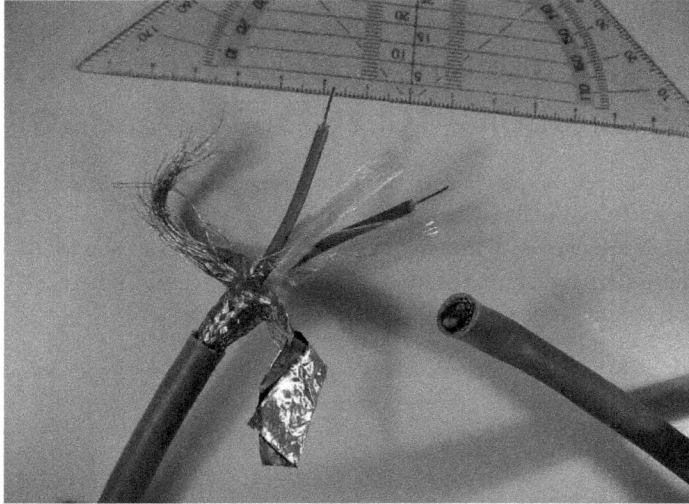

Figure 13.8 – PROFIBUS cable (source: `https://commons.wikimedia.org/wiki/File:0x-pb-kabel.jpg`)

The following figure shows a twisted pair cable:

Figure 13.9 – Twisted pair cable (CAT 6) (source: `https://commons.wikimedia.org/wiki/File:CAT6_twisted_pair.JPG`)

The following figure shows a fiber optic cable (SMF):

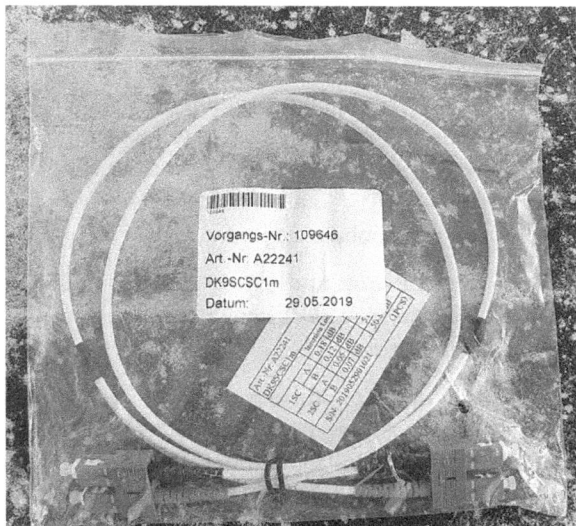

Figure 13.10 – Fiber optic cable (Single Mode Fiber) (source: `https://
commons.wikimedia.org/wiki/File:Optical_fiber_
connectors-optical_patch_cable-01ASD.jpg`)

- **Wireless network**: A wireless network uses radio waves through Bluetooth, Wi-Fi, and cellular communication as a communication channel. A radio wave can be referred to as an electromagnetic wave. It can propagate through air, vacuums, liquids, or even solid objects. It can be used to establish communication between two or more devices without cables:

  - **Bluetooth**: This is a wireless technology standard for short-range. It is used to communicate or exchange data between fixed and mobile devices over short distances using ultra-high frequency radio waves from 2.402 to 2.48 GHz to build a **personal area network** (**PAN**).

  - **Wi-Fi**: This is a wireless communication technology that uses radio waves to establish communication between devices. It can also be referred to as a standard **wireless local area network** (**WLAN**) technology that allows computers and various devices to connect to each other as well as to the internet without cables. Computers and other devices can connect to exchange data via Wi-Fi. Wi-Fi is not an acronym but a brand name.

  - **Cellular Communication** (**Mobile Telephone System**): This is a wireless communication technology that uses radio waves to establish communication between devices. Here, the radio network is distributed over the land through cells, where each cell includes a fixed location transceiver (base station). The cells provide radio coverage over larger areas. Mobile phones or any devices with a radio transmitter and receiver installed in their circuitry can communicate, even if the equipment is moving through cells.

The following are some of the generations of cellular communication/mobile telephone systems:

- **1G**: This is the first generation of cell phone technology. It uses an analog signal. Its maximum speed is 2.4 Kbps.

- **2G**: This is the second generation of cell phone technology. The network is digital, unlike the first generation, which was analog. During the era of 2G, cellular phones were used for data and voice. The maximum speed of 2G with GPRS is 50 Kbps or 1 Mbps with **Enhanced Data Rates for GSM Evolution** (**EDGE**). GPRS is also called 2.5G – that is, the technology between the second and third generations of mobile telephony.

  GPRS is an acronym for **General Packet Radio Service**. It is a packet-switching technology that uses radio waves to enable data transfers through cellular networks. It is used for mobile internet and other data communications.

- **3G**: This is the third generation of cell phone technology. It was introduced in 2001 with the main objective of facilitating greater voice and data capacity, increasing data transmission at low costs, and supporting a wide range of applications. Its speed ranges between 2 Mbps and 21.6.6 Mbps for HSDPA+.

- **4G**: This is the fourth generation of cell phone technology. It was developed to provide high quality, high capacity, and high speed with improved security and lower cost for voice, data, multimedia, and internet over IP (internet protocol). It is suitable for gaming, high-definition mobile TV, video conferencing, cloud computing, and more. Its speed range is between 100 Mbps and 1 Gbps.

- **5G**: This is the fifth generation of cell phone technology. It was designed to provide reliable connectivity to mobile users and connected devices at a speed range of 10 Gbps to 20 Gbps.

  5G provides lots of outstanding features, including the following:

  - Improved mobile phone capabilities and a great capacity to support virtual reality.

  - The ability to control physical infrastructures, cars, and other devices remotely with ultra-reliable, high-availability, and low-latency connections.

  - Seamlessly connects a great number of sensors and devices to the internet (IoT). This is a key feature that is useful for Industry 4.0 (Smart Factory), which will be discussed in the next chapter.

This section discussed network media. You should now understand both wired and wireless networks and the various generation of cellular networks – that is, 1G, 2G, 3G, 4G, and 5G. Now, let's learn about network connectors and other network components.

Network connectors and other network components

Let's have a brief look at network connectors.

Network connectors allow wired network media (coaxial cable, twisted pair, fiber optic, and so on) to connect to a network such as hubs, switches, routers, controllers, PCs, and other devices in a network. They are used to terminate a segment of network cable.

The following are some examples:

- **BNC Connector**: **BNC** is an acronym for **Bayone-Neill-Concelman**. A BNC connector is used for coaxial cables:

Figure 13.11 – BNC connector (source: `https://commons.wikimedia.` `org/wiki/File:BNC_connector_20050720_001.jpg`)

- **PROFIBUS connector**: This is usually used with a PROFIBUS cable to establish communication between devices using the PROFIBUS protocol:

Figure 13.12 – PROFIBUS connector (source: `https://commons.wikimedia.` `org/wiki/File:0x-pb-stecker-verschieden.jpg`)

- **RJ45 connector**: This is one of the most common network connectors in the world. **RJ45** is an acronym for **Registered Jack 45**. It is usually used for twisted pair cables (CAT 5 or CAT 6):

Figure 13.13 – RJ45 connector (source: `https://commons.`
`wikimedia.org/wiki/File:KonektorRJ45.jpg`)

Some other networking components are as follows:

- **Switch**: A switch is a device that connects different devices such as PCs, printers, and more in a network and allows them to share information and resources. In industrial automation, a switch can connect two or more PLCs, HMIs, PCs, and other automation and control devices and allow them to communicate with each other. The following figure shows a switch:

Figure 13.14 – Switch (source: `https://commons.wikimedia.org/wiki/`
`File:5_Port_Gigabit_Netzwerk-Switch_TL-SG1005D_02.jpg`)

There are two types of switch, as follows:

- **Unmanaged switch**: A type of switch that does not require any configuration for it to work. They are used for basic connectivity.

- **Managed switch**: This is a type of switch that requires configuration to make it fit our network or needs. It has many security features that can be configured, enabled or disabled. Some of these features include web authentication, access control, VLAN, and port enabling/disabling. A managed switch allows statistical analysis of packets and provides more special features than an unmanaged switch.

- **Router**: A router is a network device that routes data packets from the source to the destination in a network. Devices that are networked via a switch can be connected to the internet via a router. A router works with IP addresses and can connect two or more different networks or subnetworks, unlike a switch, which works with **Media Access Control** (**MAC**) addresses and can connects devices such as PCs, printers, and more in the same network. Routers receive incoming data packets, analyze them and direct them to another network.

There are two types of routers, as follows:

- **Wireless router**: This does not require any wire or cable to connect devices

- **Wired router**: This requires wire or cable to connect network devices

- **Repeater**: This is a device that amplifies or increases signal strength, so the signal can travel more distance with good quality. In addition, we can adapt different physical media to each other, such as a coaxial cable to an optical fiber, using a repeater.

Now that we have learned about a few network components, let's understand network protocols.

Understanding network protocols

A network protocol can be referred to as a method of data communication between two or more devices on a network. It is a set of rules that should be adhered to for communication to occur between two or more connected devices. Industrial network protocols enable communication between industrial network devices such as PLCs, HMIs, VFDs, PCs, transmitters, actuators, and more. This protocol encompasses the network topology, network media, network connectors, and more.

Many protocols are used around the world for communication. We will focus on industrial network protocols in this book.

Common industrial network protocols can be grouped as wired and wireless networks.

Wired network protocols

Wired network protocols can be grouped into two areas, as follows:

- **Fieldbus**: Fieldbus is a group of industrial network protocols used for real-time distributed control, standardized as IEC 61158. One great advantage of the Fieldbus protocol is that network cable use is reduced – that is, less wiring is required. It supports various network topologies such as bus, ring, and star. It is mostly used in industrial networks due to its simplicity and reliability. The following are popular Fieldbus protocols that are used in industrial automation or process control:

    - Foundation Fieldbus H1

    - PROFIBUS

    - HART

    - DeviceNet

    - CC-Link

    - AS-i

    - Modbus RTU

- **Industrial Ethernet**: This is another group of industrial network protocols that utilize standard Ethernet protocols for process control and automation. They have high performance and can integrate easily into an office network. The following protocols are in this group:

    - PROFINET

    - Foundation Fieldbus HSE

    - Ethernet/IP

    - Ethernet CAT

    - Modbus TCP/IP

Wireless network protocols

Wireless network protocols are a group of protocols that allows communication between devices without the use of wires. They are mostly used by sensors and measuring devices. The following are common wireless network protocols that are used in industrial automation:

- Bluetooth

- Wi-Fi

- Cellular

In the next few sections, we will learn about some common industrial network protocols.

Common industrial network protocol – Foundation Fieldbus

This protocol is developed by the **International Society of Automation (ISA)**. Its application areas include petrochemicals, food and beverages, power, and other heavy-process applications. It is used to connect foundation Fieldbus-enabled devices such as sensors/transmitters to a control system with less wiring. It allows several Fieldbus devices to use a single cable.

Two types of foundation Fieldbus implementation are Foundation Fieldbus H1 and Foundation Fieldbus **High-Speed Ethernet (HSE)**:

- **Foundation Fieldbus H1**: This version provides speeds of up to 31.25 Kbps. Twisted pair cables and fiber optic cables can be used as network media. It provides digital communication and DC power over the same twisted pair. It is implemented using a bus or star topology. Its network can have up to 32 nodes per segment. It is used for communication between several sensors/transmitters and actuators (field devices) and a PLC, DCS, or any other form of controller.

- **Foundation Fieldbus HSE (High-Speed Ethernet)**: This version provides speeds up to 100 Mbps. It is used to connect higher-level devices (for example, PCs, controllers, and remote I/O). Twisted pair cables and fiber optic cables can also be used as network media. It is an industrial Ethernet-based protocol that can be implemented using a star or tree topology. You can easily add or remove devices without affecting the network.

A linking device is usually used to connect a Foundation Fieldbus H1 to a Foundation Fieldbus HSE network, as shown in the following diagram:

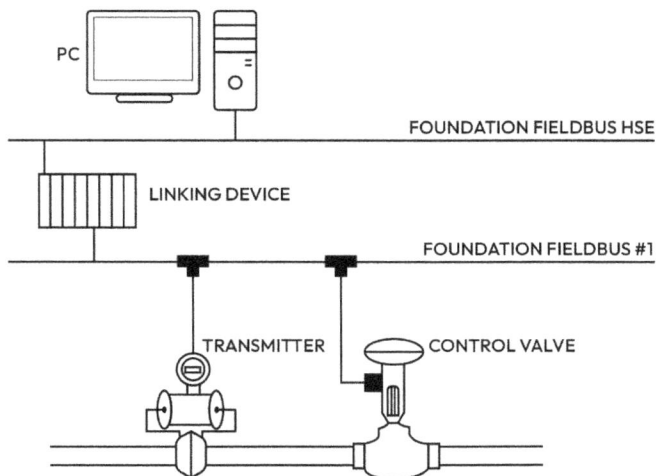

Figure 13.15 – Foundation Fieldbus H1 and HSE

Now, let's learn about the PROFIBUS protocol.

Common industrial network protocol – PROFIBUS

PROFIBUS is an acronym for **Process Fieldbus**. It is a standard protocol that can be used to connect sensors/transmitters, actuators, controllers, HMIs, VFDs, and more. Two-core twisted, shielded cables called PROFIBUS cables (*Figure 13.8*) are used for communication. A PROFIBUS connector (*Figure 13.12*) is used to connect to each device in the network. The PROFIBUS protocol comes in two forms, as follows:

- **PROFIBUS DP (Decentralized Peripherals)**: PROFIBUS DP is among the most commonly used network protocols in the industry. It is an industrial network protocol that reduces the number of wires required to connect sensors and actuators by decentralizing the **input and output (I/O)** modules.

  Before PROFIBUS DP, I/Os were centralized in the control room, as shown in the following diagram, leading to lots of cables being used to wire sensors and actuators (which could be far away from the control room) to the input and output modules in the control room. Instead of connecting individual sensors and actuators in the field directly to the PLC's input and output modules in the control room, we can install an **Interface Module (IM)**; then, the input and output modules far away from the control room and the sensors and actuators can connect to it in the field area. Then, a single cable (PROFIBUS cable) can connect the faraway I/Os in the field (remote I/Os) to an IM in the PLC located in the control room, as shown in *Figure 13.17*. This greatly minimizes the cables required during wiring. We can have up to 32 devices in the PROFIBUS DP network per segment. The data transfer speed varies from 9.6 Mbps to 12 Mbps:

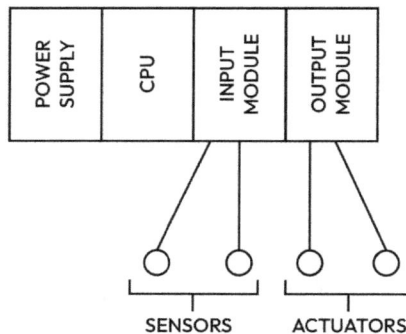

Figure 13.16 – Centralized control system

The following diagram shows a decentralized control system in which a PROFIBUS cable connects remote I/Os in the field to the PLC in the control room through an IM:

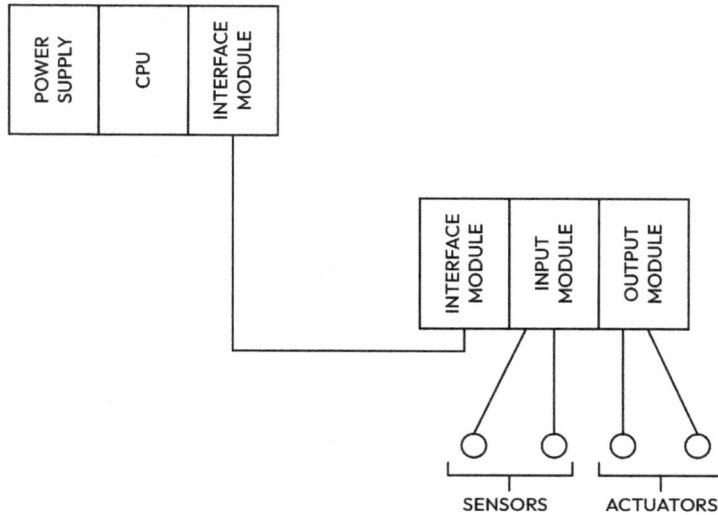

Figure 13.17 – Decentralized control system

- **PROFIBUS PA (Process Automation)**: This is used to monitor process measuring devices through a control system. Here, the I/O module that's used in PROFIBUS DP is no longer required, nor is any wiring between each sensor and I/O module. The sensors are connected in a linear form to a single PROFIBUS PA bus. The network of sensors and actuators are then connected to a PLC via a coupler and a single cable (PROFIBUS cable), as shown in the following diagram. The segment coupler converts the PROFIBUS PA signals into a PROFIBUS DP signal. The segment coupler is required because the PLC does not have a PROFIBUS PA port. It only has a PROFIBUS DP port. PROFIBUS PA has a fixed data transfer speed of 32 Kbps:

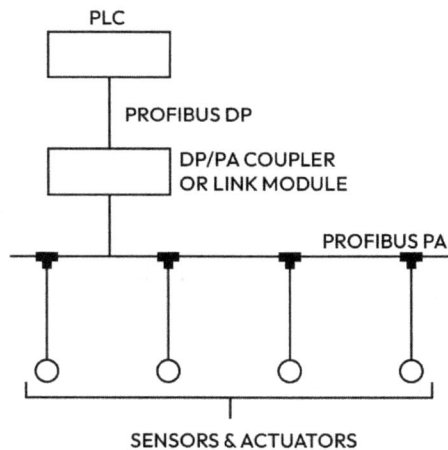

Figure 13.18 – PROFIBUS PA

Now, let's learn about the Modbus protocol.

Common industrial network protocol – Modbus

Modbus is an open protocol that can be used by any manufacturer or vendor without restriction. It was originally designed in the mid-1970s by Modicon as a way to connect intelligent devices with PLCs using a simple master/slave concept. It was initially designed for Modicon PLCs only (that is, a proprietary protocol) but it was later made available for use by any manufacturer without any restriction (that is, an open protocol).

Various vendors now manufacture their PLCs and other industrial devices to support the Modbus protocol. Hence, you will find the Modbus protocol available for use in industrial equipment from various manufacturers, such as Siemens, Allen Bradly, Mitsubishi, and Omron. The Modbus protocol solves the problem of communication between industrial equipment from different manufacturers. With the Modbus protocol, a Siemens PLC can communicate with an Omron PLC, Allen Bradley PLC, Omron VFD, or any other industrial equipment from any manufacturer that supports the Modbus protocol. Modbus can be referred to as the most commonly used industrial protocol. Modbus has various memory areas for storing data, as follows:

- **Discrete input**: Stores digital input values (high or low) from field devices
- **Coil output**: Stores digital output values (high or low)
- **Input registers**: Stores analog input values (continuously varying signals such as temperature, level, flow, and more)
- **Holding register**: Stores analog outputs values (continuously varying signals such as temperature, level, flow, and more)

Now, let's discuss the different types of Modbus.

Modbus Remote Terminal Unit (RTU)

This is the most commonly used type of Modbus protocol. It uses binary coding and CRC error checking, as well as a master and slave concept. A master device is a device that can request information from the slave, while a slave is a device that provides information to the master. A slave will only transmit information when it receives a command from the master to transmit. There can only be one master and up to a maximum of 247 slaves, each of which has a unique address called a unit ID. In the following diagram, the PLC is the master and will communicate with the field devices and VFD (slaves), each of which has a unit ID. We can also have a SCADA/HMI as the master while the PLCs connected to the SCADA/HMI are regarded as the slaves.

The physical communication medium of Modbus RTU is RS232, RS422, and RS485.

RS232 is used for point-to-point topology. It allows communication between two devices (master and slave) at a distance of less than 50 ft.

RS485 and RS422 are used to connect more than two devices on the same line at a distance greater than 50 ft, while RS485 is preferred for multiple devices connected over a long distance of about 4,000 ft. The following diagram shows a master PLC connected to field devices and VFD as slaves:

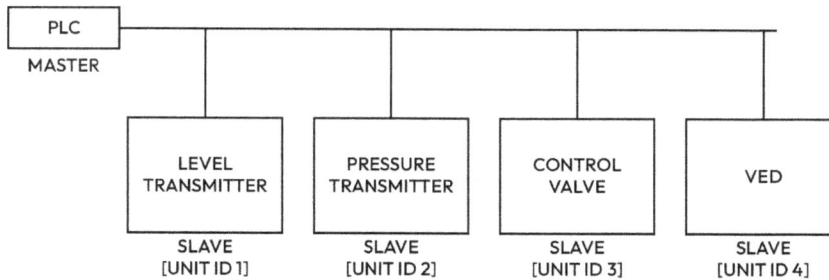

Figure 13.19 – Modbus RTU network with one PLC as master and field devices and VFD as slaves

The following diagram shows a SCADA/HMI as the master and PLCs as the slaves:

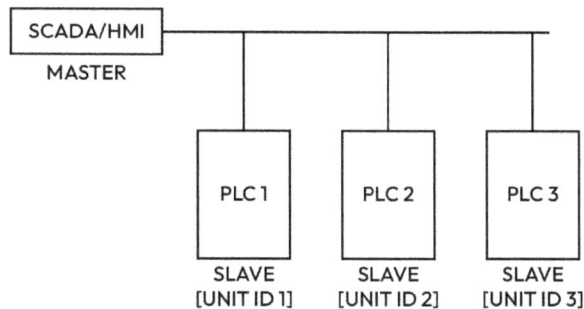

Figure 13.20 – Modbus RTU network with SCADA/HMI as the master and PLCs as the slaves

Modbus Transmission Control Protocol/Internet Protocol (TCP/IP)

This protocol uses the server-client architecture. You can have more than one device acting as the server and more than one device acting as the client, as shown in the following diagram. The clients and server are connected via a regular switch; the physical medium of communication is a regular Ethernet cable. Each device in a Modbus TCP/IP network will have a unique IP address instead of a unit ID:

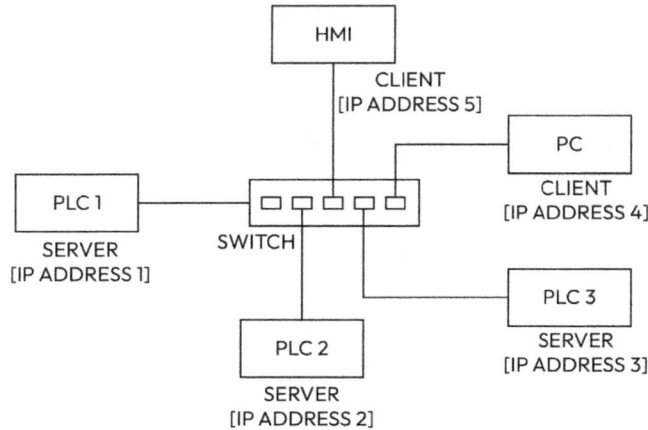

Figure 13.21 – Modbus TCP/IP network with servers and clients

> **Note**
> An open protocol is a protocol that can be used by any company or organization. The manufacturer of an open protocol provides the protocol and allows anyone to use it freely without any charge or conditions of use, while a proprietary protocol is a protocol developed and owned by a specific manufacturer and used by the devices that they manufacture.

Now, let's learn about the HART protocol.

Common industrial network protocol – HART

The **Highway Addressable Remote Transducer** (**HART**) protocol is the world standard protocol and provides data exchange between smart/intelligent field devices and a controller (PLC, DCS, and so on) or handheld communicator. You can control and monitor HART-enabled field devices (sensors, transmitters, and actuators) with the HART protocol. In HART communication, a digital signal is superimposed on the traditional 4-20mA analog signal, allowing the 4-20mA analog signal to co-exist with the digital signal on the same two-wire loop without distortion. Hence, two signals are transmitted simultaneously – that is, the 4-20mA signal and the digital signal. The 4-20mA analog signal represents the measured variable regarding the temperature, pressure, level, or flow while the digital signals carry other device details, such as the device's health, status, diagnostics alerts, and more.

Each sensor/transmitter and actuator (transducer) in the network (highway) has a unique address and can be accessed remotely using that unique address, hence the name HART. We don't have to be at the location of the sensor to configure or access its data. It can be done remotely from the control room or anywhere the HART data is accessible – that is, either from a programming PC or a handheld device (HART communicator) connected anywhere in the network for configuration and other purposes.

It is the largest installed protocol worldwide and its benefits cannot be overemphasized. It provides a reliable and easy way to communicate with field devices. It can be used for various purposes, such as the following:

- Configuration and reconfiguration of HART-enabled SMART field devices
- Commissioning and calibrating HART-enabled SMART field instruments using a handheld device called a SMART communicator
- Troubleshooting control loops
- Getting details about a device's health and status
- Changing device ID or tags online

When a HART-enabled SMART field device or instrument is used with a HART-compatible controller in the control room, more benefits can be obtained from the field instrument.

Apart from the wired HART that has been explained here, WirelessHART also exists. It is a wireless communication protocol that add wireless capability to HART technology. WirelessHART provides a cost-effective way to deliver new measurement values to control systems without the need to run additional wires.

> **Note**
>
> SMART field devices are transmitters or actuators that consist of a processor and other components. They provide more benefits than the traditional analog field devices, which include diagnostics and other features of control that would have been required in the main controller in the control room. Diagnostics in a smart field device can indicate out-of-specification operations, communication link failures, and predict failures that could develop. Most SMART field instruments are HART, Foundation Fieldbus, or PROFIBUS protocol-enabled. **SMART** is an acronym for **Single Modular Auto-ranging Remote Transducer**.

There are two types of HART networks:

- **Point-to-point**: A point-to-point connection can be seen in the following diagram. Here, the digital signal is superimposed on the 4-20mA analog signal. The measured or process variable is transmitted as a 4-20mA analog signal and all additional information is transmitted using a digital signal:

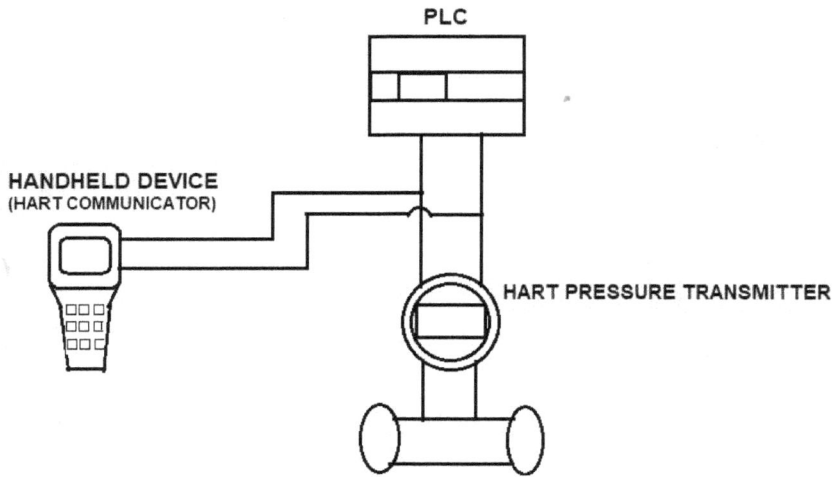

Figure 13.22 – HART transmitter connected point-to-point with the HART communicator

- **Multi-drop**: This allows several devices to be connected on the same pair of wires, as shown in the following diagram. Communication in multi-drop mode is entirely digital:

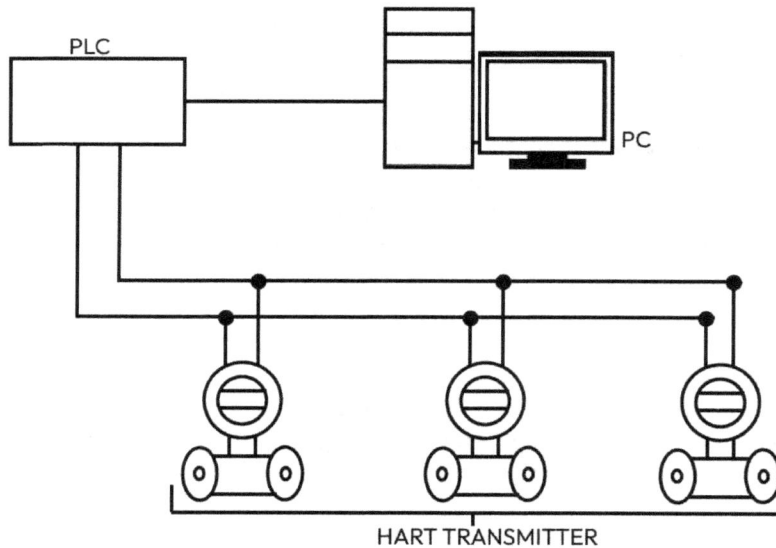

Figure 13.23 – HART transmitter connected in a multi-point mode with a PC for configuration

Now, let's learn about the most common industrial network.

Common industrial network protocol – PROFINET

PROFINET is an industrial Ethernet protocol designed to exchange data between devices and controllers in industrial automation. It is the most well-adopted industrial Ethernet protocol. It can operate at a higher speed than PROFIBUS because it is Ethernet-based. The connector that's used for PROFINET is a standard RJ45 connector (*Figure 13.13*) and the network cable can be a standard Ethernet cable such as those you find in offices, such as twisted pair CAT 6 cables (*Figure 13.9*). A standard Ethernet cable can be damaged easily due to the challenging environmental conditions in the industry. For that reason, an industrial Ethernet cable is suitable for use in PROFINET. It is characterized by shielding protection and ruggedized and proper jacketing. A standard Ethernet switch (*Figure 13.14*) can also be used because it uses the same physical connection standard as Ethernet. PROFINET is now becoming a popular protocol that's implemented in industries due to its high speed of operation (100 Mbps) and flexibility. The PROFINET protocol can be implemented using the point-to-point, bus, or star topology.

Some of the differences between PROFINET and PROFIBUS are as follows:

- A PROFINET cable is a four-core cable, though a standard Ethernet cable can be used. A PROFIBUS cable is usually purple and has two cores.

- PROFINET uses a standard RJ45 connector while PROFIBUS uses a PROFIBUS connector, which looks like a standard DB-9 serial connector.

- PROFINET can have unlimited devices, while each device in a PROFIBUS network has a unique address ranging from 1 to 127. Hence, 127 devices can be connected in a PROFIBUS network.

- The data transfer rate in PROFINET is 100 Mbps, while the maximum data transfer rate in PROFIBUS is 12 Mbps.

The following diagram shows a PLC and HMI connection using the point-to-point topology:

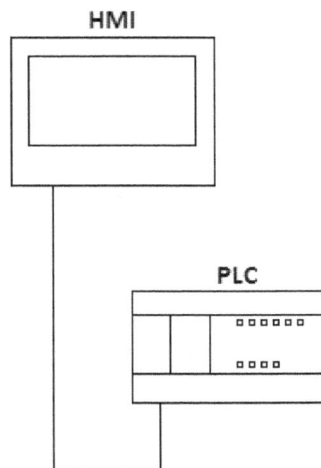

Figure 13.24 – PLC and HMI connections using the point-to-point topology

If more than two nodes are needed in the network, we can implement a star topology, where each node (PLC 1, PLC2, PLC 3, HMI 1, HMI 2, and any other devices) will connect to a central component (switch). PROFINET also supports the bus topology.

Devices in a PROFINET network have three addresses (IP address, MAC address, and device name). We implemented a very simple network using the PROFINET protocol in *Chapter 10, Understanding Human Machine Interfaces (HMIs)*, where we connected a Siemens S7 1200 PLC (CPU 1211C) to a Siemens HMI (KTP 400) for monitoring and control. Here, the PLC and HMI, which are the nodes in the network, had unique device names – that is, PLC_1 and HMI_1 (*Chapter 10, Understanding Human Machine Interfaces (HMIs)*, in *Figure 10.24*). Each device in the network was assigned a different IP address (192.168.0.4 for the PLC and 192.168.0.5 for the HMI).

The following figure shows a PLC connected to an HMI for monitoring and control using PROFINET. Refer to *Chapter 10, Understanding Human Machine Interfaces (HMIs)*, to learn more about this configuration:

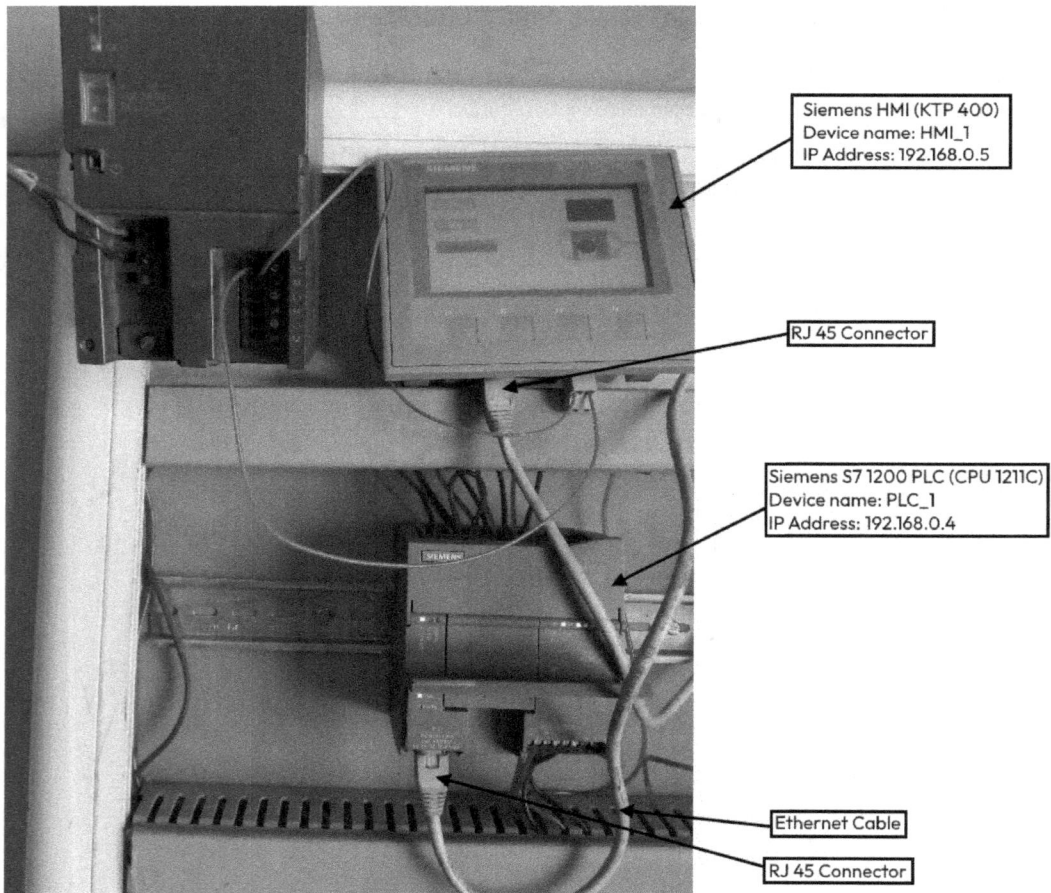

Figure 13.25 – PLC connected to an HMI (point-to-point using the PROFINET protocol) (credit for this image goes to Showlight Technologies LTD: www.showlight.com.ng)

The S7 1200 PLC and KTP400 HMI already have inbuilt PROFINET ports, making it easy to establish communication between them and other devices using the PROFINET protocol. Configuration can be done easily with the TIA portal. Refer to *Chapter 10, Understanding Human Machine Interfaces (HMIs)*, for more information.

The following diagram shows more than two devices connected via a standard Ethernet switch using the PROFINET protocol. Each device will have a unique device name and unique IP address:

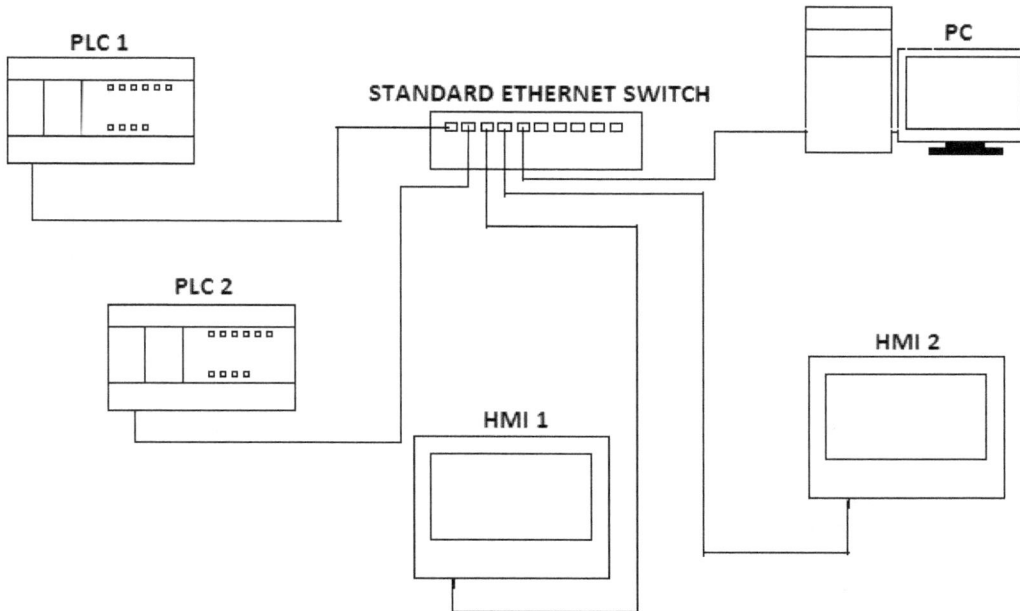

Figure 13.26 – Devices on a network using the PROFINET protocol with a standard Ethernet switch

The following diagram shows devices on a network using the PROFINET protocol with an industrial Ethernet switch from Siemens (SCALANCE). SCALANCE XB005 is an unmanaged industrial Ethernet switch from Siemens that is simple to use. It has five RJ45 ports for up to five different devices. It operates at 10/100 Mbps:

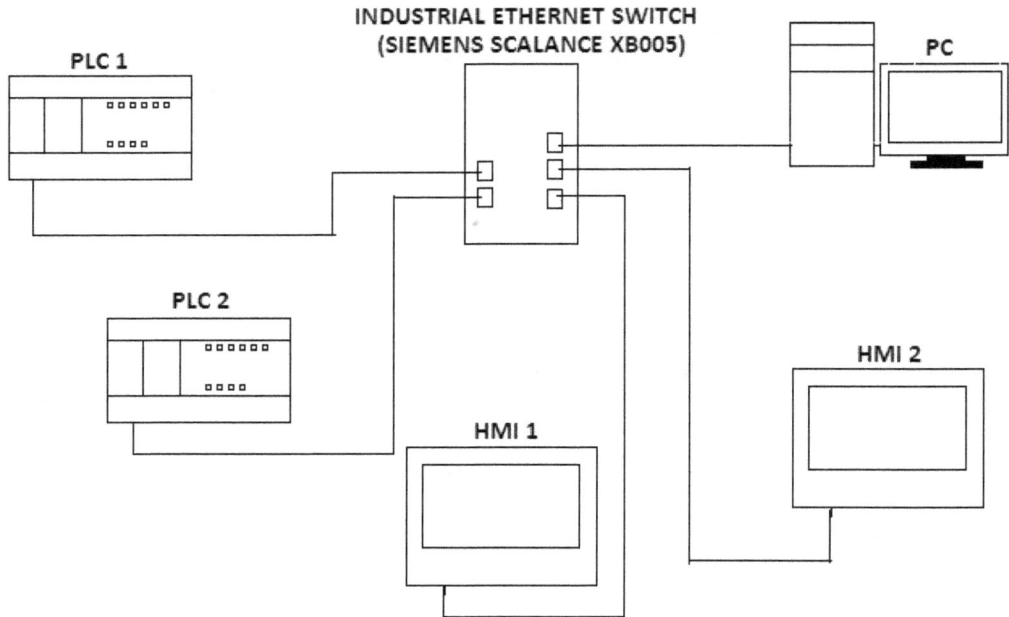

Figure 13.27 – Devices on a network using the PROFINET protocol with an industrial Ethernet switch

With that, we have covered the PROFINET protocol. You should now know about the basic components and devices required to set up a network using the PROFINET protocol.

Summary

Congratulations! You have completed this chapter. You should now understand what is meant by an industrial network and be able to explain the basic physical topologies available (point-to-point, bus, star, and ring). You also learned about network media and should now understand wired and wireless network media. This chapter also explained various cellular communication generations (1G, 2G, 3G, 4G, and 5G). You should also have an understanding of common network connectors and other components used in industrial networks. Most importantly, we covered the network protocol and discussed some of the commonly used protocols in the industry (Foundation Fieldbus, PROFIBUS, Modbus, HART, and PROFINET).

In the next chapter, we will cover another important topic you cannot afford to miss, smart factories (Industry 4.0).

Questions

The following questions will help test your understanding of this chapter. Ensure you have read and understood the topics in this chapter before attempting these questions:

1. The term _____ describes the connection between two or more industrial automation/control devices such as controllers, sensors/transmitters, actuators, PCs, and more to exchange data for real-time monitoring and process control purposes in harsh environments.

2. Each device in a network can be referred to as a _____

3. _____ refers to the physical and logical arrangement of a network.

4. _____ refers to the pattern in which data is being transferred between the nodes in the network.

5. In the _____ topology, each node is connected to a central device, which can be a switch, router, or hub.

6. _____ can be defined as the channel of communication that's used to interconnect nodes on a network.

7. A _____ network uses radio waves such as Bluetooth, Wi-Fi, and cellular communication as a communication channel.

8. _____ allow wired network media (coaxial cable, twisted pair, fiber optic, and so on) to connect to a network via a hub, switch, router, controller, PC, or other devices in a network.

9. RJ45 is an acronym for _____.

10. A _____ is a type of switch that needs to be configured to make it fit our network or need.

11. A _____ can be referred to as a method of data communication between two or more devices on a network.

12. PROFIBUS is an acronym for _____.

13. HART is an acronym for _____.

14

Exploring Smart Factory (Industry 4.0) with 5G

A smart factory is an intelligent factory that is characterized by connectivity and digitalization in the production or manufacturing structure that depends on the utilization of machinery connected to the internet. Data is stored in the cloud and used to monitor and control production or manufacturing processes. The stored data is also used to make decisions and resolve issues that may arise.

Industry 4.0 is the current trend in manufacturing that combines computation, network, and physical process integrations (cyber-physical systems), automation, and the **Internet of Things** (**IoT**) to create a smart factory. In a smart factory, machines are connected to the cloud via smart sensors and other smart devices. Smart factories merge the virtual world with the physical world.

Industry 4.0 is a new phase in the industrial revolution that increases production and efficiency with reduced cost. Many industries are now looking toward upgrading or switching to a smart factory due to its many benefits, which we will look at in this chapter. A smart factory can be seen as an advancement in industrial automation that allows automated machines or equipment in a factory to be controlled and monitored from anywhere in the world via the internet. Industry 4.0 is powered by data and other key technologies, which we will also look at in this chapter.

Upon completing this chapter, you should be able to understand what smart factories are, their benefits, and the steps for building one. You will also understand how 5G, which we learned about in *Chapter 13, Industrial Network and Communication Protocols Fundamentals*, will be of benefit in a smart factory.

In this chapter, we will cover the following topics:

- Understanding Industry 4.0
- Exploring the key technologies in Industry 4.0
- The benefits of Industry 4.0
- Steps for building a smart factory

- The benefits of a 5G-enabled smart factory
- Connecting your PLC or machine data to the cloud

Understanding Industry 4.0

To have a better understanding of Industry 4.0, we need to understand the industrial revolution.

The industrial revolution refers to the transition and changes in industrial or manufacturing processes with new innovative technologies.

Let's take a look at a brief history of the industrial revolution and its transitions:

- The **first industrial revolution (Industry 1.0)** started around 1784. Here, machines were used in industries and powered by steam.

- The **second industrial revolution (Industry 2.0)** started around 1870. Here, electricity was introduced in powering industrial machines. The mass production process also came into play during the second industrial revolution. Assembly lines operated by humans were used for mass production.

- The **third industrial revolution (Industry 3.0)** started around 1969. In this revolution, automation was introduced into industrial processes. It was during this revolution that industrial processes started using sensors, actuators, controllers, and robots, which made production or industrial processes require little or no human effort. This book focuses on the use and implementation of automation in the industry.

- The **fourth industrial revolution (Industry 4.0)** was introduced around 2011. This is the new phase of the industrial revolution. Here, industrial machines are connected to the internet. These machines can think, make decisions, and communicate with each other over the internet with little or no human intervention. You can access any machine from anywhere in the world via the internet.

  Industry 4.0 involves the intelligent networking of machines and industrial processes with the help of information and communication technology. Machines, computers, and other devices are connected and communicate with each other to make decisions regarding industrial or production processes with little or no human intervention. It incorporates IoT, cloud computing, big data, and other technologies.

Industry 4.0 enhances what was started in Industry 3.0 (the use of automated machines and computers) with smart and autonomous systems powered by IoT, artificial intelligence, cloud computing, and other technologies associated with Industry 4.0. These will be explained in the next section.

Exploring the key technologies in Industry 4.0

Industry 4.0 implements various existing technologies in a more advanced way. The key technologies in Industry 4.0 are as follows:

- **Internet of Things (IoT)**: This refers to a network of physical objects or devices that collect data with the help of sensors and communicate with each other via the internet. In the manufacturing or other production industry, IoT is a technology that allows field devices (sensors and actuators) to communicate with each other with little or no human intervention. This is called **Industrial Internet of Things (IIoT)**. IIoT is a component of Industry 4.0. It enables manufacturers to monitor the status, performance, and health of devices, schedule predictive maintenance, or perform remote troubleshooting and improve operation safety and security.

- **Robotics**: Robots are programmed machines that can carry out a series of tasks with little or no human intervention. They can make decisions based on input and carry out specific tasks on their own. Robots are important components in Industry 4.0.

- **Artificial intelligence (AI)**: This is a technology or computer program that enables a computer or robots to perform tasks that are usually carried out by humans. With AI, machines can now learn, think, and act like humans. Note that there is still no program that matches human flexibility and capacity to operate in various areas. However, some are available that make machines work and operate like human experts in some specific areas.

  Machines can now learn from past experiences, and reason and plan just like humans, which enables them to perform complex tasks. AI is concerned with the development of smart machines that would have ordinarily required human intelligence. AI requires machine learning, data analytics, and more to function. Machine learning is a method used to train a computer or machine to learn from its inputs without explicitly programming it for every situation.

  Data analytics is the science of analyzing raw data to obtain meaningful information or draw out a conclusion. Data analytics helps make sense of data. Information obtained from data analytics can be used to optimize processes and increase efficiency. In the manufacturing or production industries, the runtime, downtime, and job queues for various machines can be analyzed and used to plan the workload for the machines to operate effectively.

  When robotics is combined with AI, we have what's called an intelligent robot – that is, a programmed machine that can behave like a human or mimic human behavior, such as learning, reasoning, planning, problem-solving, and more.

- **Cloud computing**: This involves storing data on the internet and having access to the stored data when required. Data from various devices is stored in a remote server so that it can be accessed from anywhere in the world via the internet when required.

- **Cyber security**: This is a technology or practice that involves defending data, servers, computers, and other devices in a network from malicious attacks.

- **Augmented reality and virtual reality**: **Augmented reality (AR)** is a technology that provides an interactive 3D experience by combining a real-world view with computer-generated elements in real time. It can come in the form of goggles (that is, AR goggles) or a tablet or smartphone that uses the camera of the respective device to see the real world in front of the viewer and manipulate it by superimposing a computer-generated element on it. AR can interpret, manipulate, and enhance real-world views in real time. It allows you to interact with the real world in a meaningful way. It can instantly provide useful information to maintenance engineers wearing the AR headset. **Virtual reality (VR)**, on the other hand, is a technology that places a user inside a 3D computer-generated environment/element. The user can move around and interact with these computer-generated 3D elements. VR creates a complete virtual world or computer environment within the headset. Engineers, designers, or clients using VR headsets can move around a facility or product, viewing it under different lighting conditions without the need to produce a prototype.

We'll look at the benefits of Industry 4.0 in the next section.

The benefits of Industry 4.0

Industry 4.0 comes with various benefits:

- **It increases productivity**: Industry 4.0 increases productivity because more can be produced faster. Industry 4.0 reduces downtime, leading to more production.

- **It increases efficiency**: Since Industry 4.0 reduces downtime, more products are produced faster.

- **Flexibility**: Industry 4.0 makes your industry or manufacturing process flexible as you can easily scale production up or down. It is also easy to introduce new products to a production line.

- **Cost reduction**: Industry 4.0 reduces production or manufacturing costs due to automation, IoT, and other technologies that have been integrated into the manufacturing processes. However, you will need to invest some amount of money at the setup stage.

- **Increased revenue**: Industry 4.0 increases revenue due to automation and other industry 4.0 technologies that have been implemented. Less staff will be required to carry out tasks in many cases.

- **Worker safety**: Industry 4.0 leads to worker safety as the machines being used on the production line require little or no human intervention and can be operated remotely via the internet.

Now that you know the benefits of Industry 4.0, let's have a look at the steps for building a smart factory.

Basic steps for building a smart factory

The following points will help in developing a smart factory:

- **Know your needs and define your goals**: Just like any other investment, you must first identify the reasons why you want to build a smart factory, define the goals, and prioritize them. A smart factory is usually created in phases. Understanding your needs and goals and prioritizing them will help you invest in the right equipment or tools at the right time in each phase.

- **Start with the areas that will be most beneficial**: You don't have to start with all areas of your facility at once. You can make your smart factory implementation a gradual process.

- **Train your personnel on the new technology**: A smart factory involves new technologies that your existing staff might not be conversant with. It is important to get your staff equipped with the required skills to work with the new equipment and devices required in a smart factory.

- **Hire a skilled workforce**: Besides training your existing staff on the new technologies in a smart factory, you may also need to bring in a skilled workforce. For example, a data analyst will help turn your factory data into usable data that can help in planning and making meaningful decisions. Other skilled workforces that are required in a smart factory include IT experts with networking and network security skills, robotics experts with a good understanding of the basic operation and programming of robots, PLC programmers, HMI programmers, and more.

- **Guard against cyber attacks**: A smart factory connects equipment and devices to the cloud. Your factory is prone to cyber attacks from cyber criminals, which are on the increase. Take necessary precautions to protect your data and equipment from cyber attacks.

- **Invest in new equipment and tools**: New equipment will be required to facilitate a smart factory. For instance, smart sensors and other smart devices will be required to help in getting machine data to the cloud. Computers and software will also be required to store, analyze, and manage data effectively.

- **Keep updating your smart factory implementation**: You are not likely to fully implement a smart factory in all areas of your facility at once. You will need to keep updating your smart factory implementation by expanding to other areas of your facility after one area is completed.

- **Develop a good wireless connection**: Wireless connectivity is key in building a smart factory. A good wireless network, such as 5G, will enable fast and reliable data exchange that will make a smart factory that relies on internet connectivity a reality.

In the next section, we will discuss the benefits of a 5G-enabled smart factory in more detail.

The benefits of a 5G-enabled smart factory

The following are the benefits of using 5G in a smart factory:

- With 5G, more smart sensors and smart devices can be connected to the internet, boosting IoT capacity. This leads to a massively connected factory.

- 5G enables staff using AR/VR headsets to move freely on the factory floor and stay connected to the private 5G network. A 5G network can provide the high-speed requirement for AR/VR to function effectively.

- 5G enables staff to monitor machine performance, analyze production processes, and share that information between various smart factory locations for proper decision-making and planning. A 5G network enables large amounts of data to be analyzed at high speed.

- Robots are used extensively in industries to move products, and some are also designed to work alongside humans. Most of these industrial robots are powered by a wired network connection because the current wireless speed is not good enough to make them perform effectively. With the emergence of 5G, robots can now be connected wirelessly without the limitation of cables, allowing them to be well utilized and maximized in a smart factory. 5G helps collaborative robots to perform better.

- 5G enables the use of AI in a smart factory to offer more tangible features that lower-speed networks cannot provide.

In summary, 5G connectivity enables automation and other key technologies of Industry 4.0 such as IoT, AI, robotics, cloud computing, AR/VR, and more to function effectively, leading to an increase in efficiency, an increase in the productivity and quality of the factory, and an improvement in the working conditions of the staff in a smart factory.

Let's proceed to the next section, where we will briefly look at connecting PLCs to the cloud.

Connecting your PLC or machine data to the cloud

It's important to discuss how our PLC data can get to the cloud since it's key in a smart factory. In this section, we will briefly talk about the IIoT gateway, which is a key component required to get PLC or machine data to the cloud.

An IIoT gateway can be a piece of hardware or software that can integrate various devices such as sensors, actuators, PLC, and lots more, collect data from them, and transmit it to the cloud or local servers. It allows connected devices and the outside world (the cloud) to communicate. It can be referred to as a bridge that sits between connected devices in an industrial environment and the cloud. With the IIoT gateway, you can monitor and control machines or devices in the factory from anywhere in the world since the device's data is available in the cloud. You can monitor the data via a dashboard on a PC, tablet, or smartphone connected to the internet; you can also send commands from your

PC, tablet, or smartphone to control your equipment or machine. In each case, data passes through the IIoT gateway. It is one of the required pieces of equipment when setting up a smart factory. The IIoT gateway understands different protocols used by IoT devices, such as Bluetooth, Zigbee, MQTT, Wi-Fi, cellular, and others. The IIoT gateway processes critical data close to the source of data, which leads to quick decision-making and less load on the central platform. This is known as edge computing.

There are various models of the IIoT gateway from different manufacturers. Examples of IIoT gateways from Siemens include SIMATIC IOT2020, SIMATIC IOT2040, and SIMATIC2050. The Siemens SIMATIC IOT2050 is an intelligent IoT gateway that connects internal IT, manufacturing, cloud department, and combines data from various sources. It is built with powerful Texas instrument ARM processor (AM6528). It has 1 GB of DDR4 RAM. SIMATIC IOT2050 comes with multiple interfaces such as mPCIe slots for wireless card, Gbit LAN, USB port, Serial port, Arduino port to connect to Arduino UNO Rev3 and DP Port to connect directly to a monitor.

The following are some other IIoT gateway manufacturers and their models:

- **Advantech**: UNO-125IG
- **Welotech**: EG602W, EG503W, EG602L
- **Softing**: UAGATE SI, UAGATE MB, and edgeGate
- **WLINK**: WL-D80 and G510-LF4 (compatible with 5G)

Each IIoT gateway has manuals and guides that will help with installing and configuring them. Some will also have a forum where you can learn from those that have used the device.

The following diagram shows how an IIoT gateway link devices or things (sensors, actuators, PLCs, and so on) to the cloud:

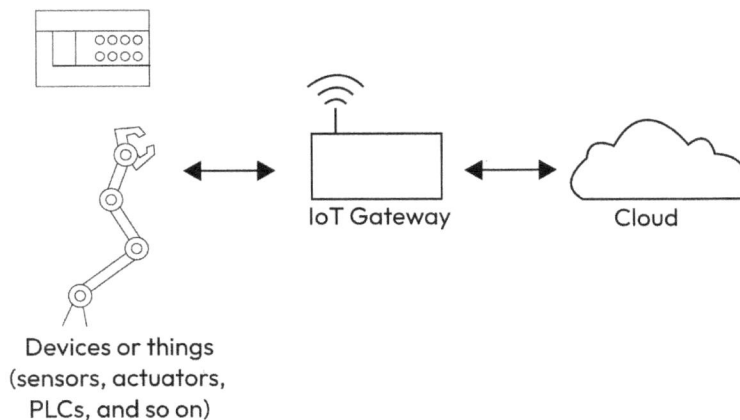

Figure 14.1 – IoT gateway (the bridge between devices and the cloud)

We should now have a basic understanding of IIoT gateways and how our PLC data can get to the cloud. This is a good start for industrial automation engineers that want to go into the world of Industry 4.0 or smart factories.

Summary

Congratulations – you have completed the last leg of this book's journey. Well done! You should now understand Industry 4.0. You should also be able to explain the key technologies of Industry 4.0, which include (IoT), AI, robotics, cloud computing, AR, and VR. You have also learned the benefits of Industry 4.0, the steps for building a smart factory, and the benefits of a 5G-enabled smart factory. The last section, *Connecting your PLC or machine data to the cloud*, explained an important component (IIoT gateway) that's required to connect PLC data, sensors, and actuators to the cloud.

Thank you for taking the time to read this book. I hope that the information provided has helped you gain new skills in the field of industrial automation or improved your industrial automation skills. Effort has been made to make this book hands-on and practical enough. I advise you to download the necessary software as instructed, install it, and follow the steps as you read for better understanding. You can take advantage of the simulation demonstrations in some of the chapters and if you can afford the equipment/devices used, I recommend that you purchase them, connect them as instructed, and practice to gain hands-on experience with the actual component or device.

By going through this book and continuing to practice, you can further improve your skills in industrial automation to become an expert. You can do this by learning more online, practicing, or enrolling at a training center that specializes in providing industrial automation, robotics, IIoT, and other related training in a practical way using real-life industrial devices that you will be able to practice with while learning. I aim to see experts emerge after reading this book. Industrial automation is such an interesting field with a good market. Most machines in industry are now automated and there will always be a need for experts with good knowledge of developing, maintaining, and troubleshooting such machines. You have started well by reading this book. I look forward to seeing you at the top of the field of industrial automation and process control.

Questions

The following questions have been provided to test your understanding of this chapter. Ensure you have read and understood the topics in this chapter before attempting these questions:

1. The _____ industrial revolution involves mass production through the use of assembly lines consisting of machines that are powered by electricity.

2. In the _____ industrial revolution, industrial machines can think, make decisions, and communicate with each other over the internet with little or no human intervention.

3. The _____ refers to a network of physical objects or devices that collect data with the help of sensors and communicate with each other via the internet.

4. With _____, machines can now learn, think, and act like humans.

5. _____ is a method used to train a computer or machine to learn from its inputs without explicit programming it for every situation.

6. _____ involves storing data on the internet and having access to the stored data when required.

7. _____ is the practice of defending data, servers, computers, and other devices in a network from malicious attacks.

8. _____ is a technology that provides an interactive 3D experience by combining a real-world view with computer-generated elements in real time.

9. _____ is a technology that places a user inside 3D computer generated environments/elements.

10. IIoT is an acronym for _____.

Assessments

Chapter 1

1. Industrial Automation
2. Flexible
3. Control
4. Industrial Automation
5. A production line or assembly line
6. Fixed automation systems, programmable automation systems, and flexible automation systems
7. Batch production
8. Automated guided vehicles (AGVs)
9. Field
10. Actuators
11. Sensors
12. Computer Numerical Control
13. Programmable Logic Controller
14. Supervisory Control And Data Acquisition
15. Human Machine Interface

Chapter 2

1. Switches and sensors
2. Sensors
3. Sensors
4. Manually operated switches
5. Mechanically operated
6. Single Pole Double Throw (SPDT)
7. A level switch
8. Inductive proximity sensor

Assessments

9. Temperature switch

10. Limit switch

11. Double Pole Double Throw (DPDT) switch

12. Single Pole Double Throw (SPDT) float switch

Chapter 3

1. Electric actuator

2. Stepper motor

3. Pneumatic actuator

4. Stator and rotor

5. Hydraulic actuator

6. Double Pole Double Throw (DPDT) relay

7. Single-acting cylinder with spring return

8. Contactor

Chapter 4

1. AC motor and DC motor

2. Synchronous AC motor

3. Electromagnetic induction

4. Slip

5. Series wound

6. Stepper motor

7. Servo

8. Direct On Line (DOL)

9. Star

10. Delta

Chapter 5

1. The frequency of the voltage source and the number of poles
2. A variable frequency drive, an adjustable speed drive, or an inverter
3. A converter or rectifier, a DC filter, and an inverter
4. A DC filter
5. Harmonics
6. The RPM
7. A VFD parameter

Chapter 6

1. Computer aided design
2. Ladder diagram
3. Pictorial diagram
4. Survey window
5. Program bar

Chapter 7

1. PLC
2. Programmable Logic Controller
3. Power supply module, CPU module, Input module, Output module
4. Central Processing Unit (CPU)
5. Input module
6. Output module
7. Scan cycle
8. Scan time
9. Sinking and sourcing

Chapter 8

1. Software or program
2. A PLC user program
3. Ladder Diagram (LD), Function Block Diagram (FBD), Sequential Function Chart (SFC), Instruction List (IL), Structured Text (ST)
4. Ladder Diagram (LD)
5. International Electrotechnical Commission
6. Function Block Diagram (FBD)
7. Instruction List (IL)
8. Handheld devices and Personal Computers (PCs)
9. Rungs

Chapter 9

1. Latching
2. Timer
3. Counter
4. Compare

Chapter 10

1. Hardware, software
2. Man-machine interface
3. Operator interface terminal
4. User interface

Chapter 11

1. Supervisory Control and Data Acquisition
2. Remote Terminal Unit or Remote Telemetry Unit
3. Field devices
4. Master station
5. SCADA software

Chapter 12

1. Open-loop
2. Closed-loop
3. Feedforward
4. Transmitter
5. Setpoint
6. Error
7. Control element or final control element
8. International Society of Automation
9. Turbine flow meter
10. Normalize
11. Scale
12. Proportional-Integral-Derivative
13. Piping and Instrumentation Diagram or Process and Instrumentation Diagram

Chapter 13

1. Industrial network
2. Node
3. Network topology
4. Logical topology
5. Star
6. Network media
7. Wireless
8. Network connectors
9. Registered Jack 45
10. Managed switch
11. Network protocol
12. Process Fieldbus
13. Highway Addressable Remote Transducer

Chapter 14

1. second
2. fourth
3. Internet of Things
4. Artificial intelligence
5. Machine learning
6. Cloud computing
7. Cyber security
8. Augmented reality
9. Virtual reality
10. Industrial Internet of Things

Index

‹packt›

Packtpub.com

Subscribe to our online digital library for full access to over 7,000 books and videos, as well as industry leading tools to help you plan your personal development and advance your career. For more information, please visit our website.

Why subscribe?

- Spend less time learning and more time coding with practical eBooks and Videos from over 4,000 industry professionals
- Improve your learning with Skill Plans built especially for you
- Get a free eBook or video every month
- Fully searchable for easy access to vital information
- Copy and paste, print, and bookmark content

Did you know that Packt offers eBook versions of every book published, with PDF and ePub files available? You can upgrade to the eBook version at packtpub.com and as a print book customer, you are entitled to a discount on the eBook copy. Get in touch with us at customercare@packtpub.com for more details.

At www.packtpub.com, you can also read a collection of free technical articles, sign up for a range of free newsletters, and receive exclusive discounts and offers on Packt books and eBooks.

Other Books You May Enjoy

If you enjoyed this book, you may be interested in these other books by Packt:

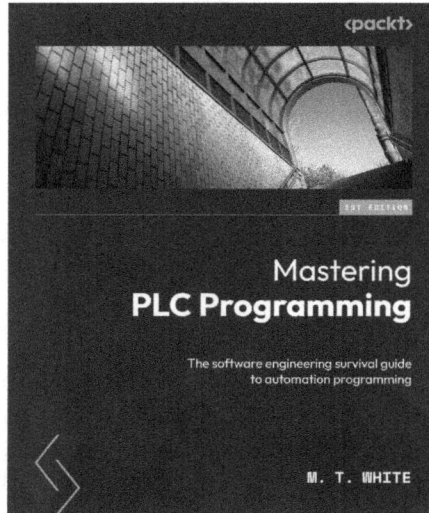

Mastering PLC Programming

M. T. White

ISBN: 978-1-80461-288-0

- Find out how to write PLC programs using advanced programming techniques
- Explore OOP concepts for PLC programming
- Delve into software engineering topics such as libraries and SOLID programming
- Explore HMIs, HMI controls, HMI layouts, and alarms
- Create an HMI project and attach it to a PLC in CODESYS
- Gain hands-on experience by building simulated PLC and HMI projects

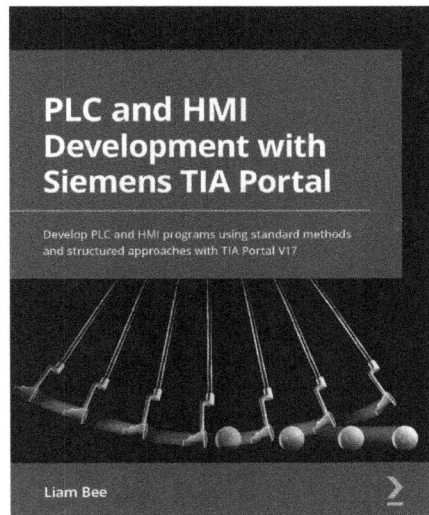

PLC and HMI Development with Siemens TIA Portal

Liam Bee

ISBN: 978-1-80181-722-6

- Set up a Siemens Environment with TIA Portal
- Find out how to structure a project
- Carry out the simulation of a project, enhancing this further with structure
- Develop HMI screens that interact with PLC data
- Make the best use of all available languages
- Leverage TIA Portal's tools to manage the deployment and modification of projects

Packt is searching for authors like you

If you're interested in becoming an author for Packt, please visit `authors.packtpub.com` and apply today. We have worked with thousands of developers and tech professionals, just like you, to help them share their insight with the global tech community. You can make a general application, apply for a specific hot topic that we are recruiting an author for, or submit your own idea.

Share your thoughts

Now you've finished *Industrial Automation from Scratch*, we'd love to hear your thoughts! Scan the QR code below to go straight to the Amazon review page for this book and share your feedback or leave a review on the site that you purchased it from.

`https://packt.link/r/1800569386`

Your review is important to us and the tech community and will help us make sure we're delivering excellent quality content.

Download a free PDF copy of this book

Thanks for purchasing this book!

Do you like to read on the go but are unable to carry your print books everywhere? Is your eBook purchase not compatible with the device of your choice?

Don't worry, now with every Packt book you get a DRM-free PDF version of that book at no cost.

Read anywhere, any place, on any device. Search, copy, and paste code from your favorite technical books directly into your application.

The perks don't stop there, you can get exclusive access to discounts, newsletters, and great free content in your inbox daily.

Follow these simple steps to get the benefits:

1. Scan the QR code or visit the link below

https://packt.link/free-ebook/9781800569386

2. Submit your proof of purchase

3. That's it! We'll send your free PDF and other benefits to your email directly

www.ingramcontent.com/pod-product-compliance
Lightning Source LLC
Chambersburg PA
CBHW080124220326
41598CB00032B/4948